U0181972

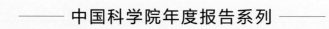

——— 中国科学院年度报告系列 ———

2021
科学发展报告

Science Development Report

中国科学院

科学出版社

北京

内 容 简 介

本报告是中国科学院发布的年度系列报告《科学发展报告》的第 24 部，旨在全面综述和分析 2020 年度国际科学研究前沿进展动态，研判和展望国际重要科学领域研究发展趋势，揭示和洞察科技创新突破及快速应用的重大经济社会影响，观察和综述国际主要科技领域科学研究进展及科技战略规划与研究布局，评述和介绍国内外主要科学奖项的获奖工作，报道我国科学家具有代表性的重要科学研究成果，概括我国科学研究整体发展状况，并向国家决策部门提出有关中国科学的发展战略和科技政策咨询建议，为国家促进科学发展的宏观决策提供重要依据。

本报告对国家各级科技决策部门、科研管理部门等具有连续的重要学术参考价值，可供国家各级科技决策和科研管理人员、科研院所科技研究人员、大专院校师生以及社会公众阅读和参考。

图书在版编目(CIP)数据

2021 科学发展报告/中国科学院编 . —北京：科学出版社，2022.6
（中国科学院年度报告系列）
ISBN 978-7-03-072124-2

Ⅰ.①2… Ⅱ.①中… Ⅲ.①科学技术-发展战略-研究报告-中国- 2021
Ⅳ.①N12②G322

中国版本图书馆 CIP 数据核字（2022）第 065385 号

责任编辑：侯俊琳　牛　玲　朱萍萍 / 责任校对：杨　然
责任印制：师艳茹 / 封面设计：有道文化

科 学 出 版 社　出版
北京东黄城根北街 16 号
邮政编码：100717
http://www.sciencep.com

北京九天鸿程印刷有限责任公司 印刷
科学出版社发行　各地新华书店经销
*
2022 年 6 月第 一 版　开本：787×1092　1/16
2022 年 6 月第一次印刷　印张：30 3/4　插页：2
字数：530 000
定价：198.00 元
（如有印装质量问题，我社负责调换）

专家委员会

总体策划

课题组

审稿专家

自觉履行高水平科技自立自强的使命担当

侯建国

（代序）

习近平总书记在中国科学院第二十次院士大会、中国工程院第十五次院士大会和中国科协第十次全国代表大会上发表重要讲话，向全国科技工作者发出"加快建设科技强国，实现高水平科技自立自强"的伟大号召，并对强化国家战略科技力量作出部署要求。国家战略科技力量作为国家科技实力的主要载体和集中体现，要以高度的思想自觉和行动自觉，切实履行好高水平科技自立自强的使命担当。

一、深刻认识高水平科技自立自强的重大意义和丰富内涵

习近平总书记重要讲话，站在全面建设社会主义现代化国家的战略高度，紧扣新时代发展要求，准确把握国际发展大势和科技经济社会发展规律，全面部署了五个方面重大战略任务，深刻阐明了实现高水平科技自立自强"干什么""谁来干""怎么干"的总体要求，为推动我国科技创新工作提供了根本遵循和行动指南。

高水平科技自立自强具有深刻的历史逻辑。独立自主是我们党长期坚持的基本方针。新中国成立之初，党中央提出自力更生、艰苦奋斗的方针，发出向科学进军的号召。随着我国改革开放不断深入和经济全球化发展，党中央与时俱进提出科学技术是第一生产力，坚持走中国特色自主创新道路。党的十八大以来，以习近平同志为核心的党中央坚持创新是引领发展的第一动力，部署实施创新驱动发展战略，推动我国科技事业取得历史性成就、实现历史性变革，进入创新型国家行列。加快实现高水平科技自立

自强，既体现了与自力更生、自主创新、创新驱动一脉相承的精神实质，也体现了充分发挥我国科技发展已有良好基础和独特优势，在新的历史起点上向更高水平迈进的必然趋势和内在要求。

高水平科技自立自强具有鲜明的时代特征。当前，世界百年未有之大变局加速演进，世纪疫情影响深远，世界各国都把强化科技创新作为实现经济复苏、塑造竞争优势的重要战略选择，积极抢占未来科技制高点，科技创新成为大国博弈的主要战场。新一轮科技革命和产业变革正处于蓄势跃迁、快速迭代的关键阶段，加快重塑产业形态和全球经济格局，与中华民族伟大复兴进程形成历史性交汇，既为我国加快实现赶超提供了难得历史机遇，也提出了新课题新挑战。我们必须树立底线思维，把高水平科技自立自强作为应对各种风险挑战的"定海神针"，作为赢得可持续发展竞争优势的"制胜法宝"。

高水平科技自立自强具有明确的实践导向。立足新发展阶段、贯彻新发展理念、构建新发展格局、推动高质量发展，实现高水平科技自立自强是必然要求。当前，我国科技创新依然面临一些突出问题，如集成电路、高端制造、关键材料、基础软件等领域，一些底层技术和关键核心技术依然受制于人。加快实现高水平科技自立自强，构建自主、完备、高效、开放的新时代科技创新体系，形成基础牢、能级高、韧性强、可持续的科技创新能力，才能为增强国家发展力、引领力、生存力提供强大科技支撑。

高水平科技自立自强具有丰富的内涵要求。实现高水平科技自立自强，要求必须做到原创和引领力强，关键核心技术自主和安全性强，对经济社会发展的支撑和带动作用强，应急应变和应对重大风险挑战的能力强，既要能够解决"心腹之患"，也要能够解决"燃眉之急"。高水平科技自立自强是主动的战略选择，而不是被动的应对之策；是创新体系和创新生态的系统比拼，而不是单项的、点上的竞争；是更加开放的而不是封闭的，必须充分利用全球创新资源，深度融入全球创新网络。

二、国家战略科技力量在高水平科技自立自强中担当重要使命

习近平总书记深刻指出，"世界科技强国竞争，比拼的是国家战略科技

力量。国家实验室、国家科研机构、高水平研究型大学、科技领军企业都是国家战略科技力量的重要组成部分，要自觉履行高水平科技自立自强的使命担当"，明确了四类国家战略科技力量的战略定位和任务要求。加快建设一支引领力强、战斗力强、组织力强的国家战略科技力量，是实现高水平科技自立自强的关键支撑。

各类国家战略科技力量要充分发挥优势、加强协同配合。党的十八大以来，以习近平同志为核心的党中央高度重视国家战略科技力量建设。近年来，国家实验室建设加快推进，中科院"率先行动"计划深入实施，高水平研究型大学"双一流"建设持续推进，企业创新主体地位不断强化，初步形成具有中国特色国家战略科技力量发展格局。当前，科技创新的复杂度越来越高，重大创新突破往往需要多学科领域深度交叉融合、多主体开放协同。各类国家战略科技力量由于各自的定位和特点，任务类型上各有分工，在创新链环节上各有侧重，要在明确分工的基础上强化协同、优势互补，加强从基本原理、原型、产品到规模市场化的有机衔接和紧密配合，共同履行好解决国家重大战略需求的使命任务。

国家战略科技力量要筑牢高水平科技自立自强的人才根基。国家战略科技力量承担国家重大科技任务、建设运行重大科技基础设施和创新平台，是培养集聚高水平科技人才的最好实践载体。近年来，我国通过载人航天、探月工程、北斗导航、深海探测等一批国家重大科技任务，迅速培养出一大批科技领军人才。实现高水平科技自立自强关键靠人才，国家战略科技力量要紧紧围绕新时代人才强国战略部署和要求，在创新实践中培养战略科学家、科技领军人才和创新团队，特别要大胆选拔和放手使用优秀青年人才，支持青年人才挑大梁、当主角，培养壮大卓越工程师队伍，加快打造培养高水平科技人才的战略支点。

国家战略科技力量要在创新体系建设中发挥骨干引领作用。国家战略科技力量和其它各类创新主体一起，共同构成了国家创新体系，各类创新主体间的联系互动水平，特别是国家战略科技力量的带动作用，决定着创新体系的整体效能。国家战略科技力量要做基础研究的引领者，产出更多重大原创性成果，加快建设原始创新策源地；要做关键核心技术攻关和产学研合作的领头雁，产出重大战略产品和系统解决方案，带动我国创新链产业链向中高端攀升；要打造集约高效、开放共享的创新平台，加强公共科研平台、中试转化平台、重大科技基础设施等建设，促进科学数据共享共用；做参与国际科技合作的主要代表，牵头发起大科学计划和大科学工程，参与国际科技组织和重点领域与新兴技术标准、规则制定，深度融入全球创新网络，加快提升我国国际科技治理话语权。

三、国家科研机构要努力打造国家战略科技力量主力军

习近平总书记2013年到中国科学院考察工作时，要求中国科学院要牢记责任，率先实现科学技术跨越发展、率先建成国家创新人才高地、率先建成国家高水平科技智库、率先建设国际一流科研机构；在2019年致中国科学院建院70周年的贺信中，要求中国科学院"加快打造原始创新策源地，加快突破关键核心技术，努力抢占科技制高点"。中国科学院深入贯彻落实习近平总书记重要指示要求，明确"定位"、准确"定标"、科学"定事"，加快打造国家战略科技力量主力军。

明确"定位"，以国家战略需求为导向强化使命担当。始终坚持"国家科学院""人民科学院"的定位，时刻牢记作为"国家队""国家人"，必须心系"国家事"、肩扛"国家责"，始终胸怀"国之大者"，秉持国家利益和人民利益至上，树立创新科技、服务国家、造福人民的思想，坚决与习近平总书记对科技创新的重要指示和要求、与党中央国务院的重大决策部署对标对表，聚焦主责主业，把精锐力量整合集结到原始创新和关键技术攻关上来。

准确"定标"，以更高标准衡量创新贡献。将充分体现国家意志、有效满足国家需求、代表国家最高水平，作为做好"国家事"、担好"国家责"的衡量标准，打造专业领域科技"国家队"。充分发挥体系化、建制化优势，在国家最紧急最紧迫的关键核心技术攻关上，产出用得上、有影响的重大技术和战略性产品；在基础交叉前沿研究上，努力取得一批世界领先水平的重大原创突破，始终走在科学的前沿，体现独特优势和核心竞争力，在国际科技竞争赛场上与强者同台竞技、奋勇夺金。

科学"定事"，聚焦主责主业狠抓落实。明确将"强基础、抓攻关、聚

人才、促改革"作为重点任务。在基础研究方面，将定向性、体系化基础研究作为主攻方向，改革选题机制、组织模式和管理方式，加快产出"从0到1"的重大原创成果。在关键核心技术攻关方面，加强多学科交叉和大兵团协同作战，积极承担和高质量完成国家重大科技任务。在人才队伍建设方面，重点抓好战略科学家和青年科技人才，带动人才队伍整体能力提升和全面协调发展。在深化改革方面，推动体系化重构、系统化重塑、整体性变革，加快提升科研院所治理体系和治理能力现代化水平。

四、努力在高水平科技自立自强中做好 "四个表率"

习近平总书记要求两院院士做"四个表率"，这也是对广大科技人员的共同要求，更是履行好高水平科技自立自强使命担当的精神力量。我们要不断增强"四个意识"，坚定"四个自信"，做到"两个维护"，在思想上政治上行动上始终同以习近平同志为核心的党中央保持高度一致，以实际行动践行"两个确立"。

做胸怀祖国、服务人民的表率。我国科技事业70多年取得辉煌成就，最根本、最重要的一条经验，就是广大科研人员在党的领导下，始终把国家和人民的需要作为科技创新的根本出发点和落脚点。实现高水平科技自立自强，必须传承好老一辈科学家心系人民、爱国奉献的优良传统，大力弘扬科学家精神，以国家民族命运为己任，始终保持深厚的家国情怀和强烈的社会责任感，把个人的科技追求融入到国家发展和民族振兴的伟业之中，努力在新时代建功立业。

做追求真理、勇攀高峰的表率。实现高水平科技自立自强，需要有敢为天下先的创新自信和勇气，树立攻坚克难、勇攀高峰的雄心壮志，凝聚敢为人先、追求卓越的"精气神"，在破解重大科技难题上担当作为，在解决国家重大需求中要敢于啃硬骨头、打硬仗，摆脱跟踪惯性和跟风惰性，走出"舒适区"，勇闯"无人区"，做出更多重大战略性和原创性贡献。

做坚守学术道德、严谨治学的表率。良好的科研道德和作风学风是科技创新的生命线。要积极倡导重实干、重实绩的评价导向，坚决破除"四唯"，让想创新、能创新的人才有舞台，使科研队伍中的老实人不吃亏、投机者难得利。引导广大科研人员坚守科研伦理，发扬学术民主，践行学术规范，对学术不端"零容忍"，让学术道德和科学精神真正内化于心、外化于行，努力营造风清气正的科研环境。

做甘为人梯、奖掖后学的表率。青年科技人才决定着科技创新的活力和未来。要紧紧围绕2035年建成世界重要人才中心和创新高地的战略目标，把政策重心放在青年科技人才上，为他们提供更多成长机会、搭建更好的创新平台，着力解决好工作生活方面的实际困难，让青年科技人才全身心干事创业，加快脱颖而出，为高水平科技自立自强打下坚实的人才基础。

（本文刊发于2022年5月1日出版的《求是》杂志，收入本书时略作修改）

前　言

当前，新一轮科技革命和产业变革深入发展，全球科技创新进入空前密集活跃期，交叉融合、协同联合、包容聚合的特征愈发明显，为经济社会发展注入新动能，不断催生出新技术、新模式和新业态，正在深刻改变着当今世界的发展格局。同时，世界正经历百年未有之大变局，国际环境日趋复杂，不稳定性不确定性明显增加，新冠肺炎疫情影响广泛深远，经济全球化遭遇逆流，世界进入动荡变革期。

2020年10月召开的党的十九届五中全会提出"坚持创新在我国现代化建设全局中的核心地位，把科技自立自强作为国家发展的战略支撑"，会议通过的《中共中央关于制定国民经济和社会发展第十四个五年规划和二〇三五年远景目标的建议》明确指出，将瞄准人工智能、量子信息、集成电路、生命健康、脑科学、生物育种、空天科技、深地深海等前沿领域，实施一批具有前瞻性、战略性的国家重大科技项目，这为我国未来一段时期的科技创新指明了方向。

2020年，我国科学家在一系列前沿科学技术研究中取得重要成果。我国首次火星探测任务"天问一号"成功发射，"嫦娥五号"首次实现月面自动采样返回，"奋斗者"号创造我国载人深潜新纪录，科学家们还在积极应对新冠肺炎疫情方面取得突出进展，实现超越经典算力的光量子计算，找到小麦"癌症"克星，攻克20余年悬而未决的几何难题，等等。这些创新和突破为建设世界科技强国打下了坚实基础。

把握世界科技发展大势，研判科技发展的战略方向，敏锐抓住科技创新的突破口和新的生长点，寻找路径、把握方向、前瞻布局，对于我国在全球科技发展中取得竞争优势具有极其重要的意义。中国科学院作为国家战略科技力量的重要组成部分，作为我国科学技术方面的最高学术机构和国家高端科技智库，把握世界科学技术的整体竞争发展态势和趋势，对科学技术与经济社会的未来发展进行前瞻性思考和布局，促进和提高国家发展决策的科学化水平，是我们义不容辞的使命与责任。

1997 年 9 月，中国科学院决定发布年度系列报告《科学发展报告》，按年度连续全景式综述分析国际科学研究进展与发展趋势，向国家最高决策层和社会全面系统地报告世界和中国科学的发展情况，评述科学前沿动态与重大科学问题，报道介绍我国科学家取得的代表性突破性科研成果。随着国家全面建设创新型国家和推进科技强国建设，《科学发展报告》将进一步系统、全面地观察和揭示国际重要科学领域的研究进展、发展战略和研究布局，服务国家促进科学发展的宏观决策。

《2021 科学发展报告》是该系列报告的第 24 部，主要包括科学展望、科学前沿、2020 年中国科研代表性成果、科技领域与科技战略发展观察、重要科技奖项巡礼、中国科学发展概况和中国科学发展建议等七大部分。受篇幅所限，报告所呈现的内容不一定能体现科学发展的全貌，重点是从当年受关注度最高的科学前沿领域和中外科学家所取得的重大成果中，择要进行介绍与评述。

本报告的撰写与出版是在中国科学院侯建国院长的关心和指导及众多院士专家的参与下完成的，得到了中国科学院发展规划局、中国科学院学部工作局的直接指导和支持。中国科学院科技战略咨询研究院承担本报告的组织、研究与撰写工作。丁仲礼、杨福愉、解思深、陈凯先、姚建年、郭雷、曹效业、汪克强、潘教峰、夏建白、李永舫、付小兵、王友德、聂玉昕、沈电洪、吴学兵、赵刚、习复、叶成、吴善超、龚旭、张利华、叶小梁、吴家睿、黄大昉、章静波、张树庸、杨茂君、张正斌、吕厚远、吴乃琴、顾兆炎、刘文彬等专家参与了本年度报告的咨询与审稿工作，在此一并致以衷心感谢。

中国科学院《科学发展报告》课题组

2021 年 12 月 8 日

目　录

CONTENTS

第一章

科学展望

An Outlook on Science

1.1　碳中和背景下的海洋碳循环及海洋增汇潜力展望

戴民汉　孟菲菲　曹知勉

（厦门大学近海海洋环境科学国家重点实验室）

实现碳中和（carbon neutrality）以应对全球气候变化，其本质是人类对现代地球系统碳循环模式的人为调整和适应性管理。碳循环涉及碳元素在大气、陆地、海洋及生物圈之间的蓄积、储存和交换过程，以及其与气候系统在不同时间尺度和跨尺度的互馈机制，极其复杂，尚未真正获得清楚认识，因而是当今国际前沿科学命题。自工业革命以来，人类活动排放的 CO_2 导致大气 CO_2 浓度急剧增加，目前已突破 410 ppm[①][1]，达到过去 200 万年内的最高值。已有足够的证据表明，由大气 CO_2 浓度增加主导的温室效应是驱动目前全球变暖的主要因子。

2015 年底，第 21 届联合国气候变化大会达成了《巴黎协定》，目前已有 196 个国家和地区加入，《巴黎协定》为 2020 年后全球应对气候变化所实施的行动做出了安排，其主要目标是将 21 世纪全球平均气温上升幅度控制在 2℃ 以内，并力争控制在 1.5℃ 以内。为此，需要全球迅速且显著地减少 CO_2 排放量，而且要从 2020 年开始，有效地从大气中移除 CO_2，到 2050 年前后达到净零碳排放（net-zero carbon emissions）。2020 年 9 月 22 日，习近平主席在第七十五届联合国大会一般性辩论上向全世界郑重承诺：“中国将提高国家自主贡献力度，采取更加有力的政策和措施，力争中国 CO_2 排放于 2030 年前达到峰值，努力争取 2060 年前实现碳中和。”[②]

从全球尺度视之，碳中和的内涵其实十分简单，即人为 CO_2 的排放与去除达到平衡[2]；在区域或国家尺度，由于存在部分自然系统碳汇计量的困难，碳中和可解读为：在一定时期内把由人类活动排放的 CO_2，通过节能减排、植树造林或碳捕获等形式移除掉，从而实现净零碳排放。需要指出的是，CO_2 并非唯一的人为排放的温室气

---

① ppm 指百万分比浓度。

② 新华社．习近平在第七十五届联合国大会一般性辩论上发表重要讲话．http://www.gov.cn/xinwen/2020-09/22/content_5546168.htm? gov.

体。《巴黎协定》第四条提出的目标则对应净零排放（net-zero emissions），即人类活动造成的全球温室气体排放与清除之间达到平衡。2018年联合国政府间气候变化专门委员会（Intergovernmental Panel on Climate Change，IPCC）发布的《全球1.5℃升温特别报告》指出，1.5℃温控目标要求全球CO_2排放在2050年前后达到净零水平，而非CO_2温室气体排放实现净零可以略晚，但在2050年之前需要实施深度减排[2]。因此，净零排放目标在温室气体覆盖范围上广于碳中和目标，在达成时间上允许前者晚于后者。目前，已有超过130个国家立法确认或承诺在2040～2060年达到碳中和。

海洋依托于其巨大的热容吸收了气候系统中约93%的新增热量[3]；同时海洋作为地球系统中最大的活跃碳库，吸收了超过25%人类活动排放（1850～2020年的排放总量）的CO_2[4]，对缓解气候变化起到了至关重要的作用。但海洋的持续吸热储碳也显著地改变着海洋理化环境与生态系统。自工业革命以来，海洋已由碳源（向大气释放CO_2）转化为碳汇（吸收大气CO_2）[5]，并且碳汇效应随大气CO_2浓度的升高逐渐增强[6]。但是，在未来净零排放或负排放情景下，大气CO_2浓度将逐渐降低，海洋碳汇能力减弱，甚至可能将碳重新释放回到大气圈，这将进一步增加碳减排压力[7]。目前，国内外学者已在全球和区域海洋开展了大量海-气界面碳通量观测研究，对其CO_2源汇分布有了基本的认识。但是，如何准确评估海洋碳收支、碳库容量及其动态变化，仍存在极大的不确定性和科学技术方面的挑战；而以碳中和为目标的减排和增汇行动必将给地球气候系统带来一次巨大且前所未有的快速改变，如何预测这些人为干预行动对海洋碳循环和碳收支的扰动及对气候的影响，将是全球变化科学面临的重大新挑战。

一、海洋碳汇过程

海洋是地表系统最大的碳储库，碳储量约是大气的50倍，其中98%以溶解无机碳（dissolved inorganic carbon，DIC）的形式存在[4]。溶解无机碳系统，也称作碳酸盐系统，包括游离二氧化碳（CO_2）/未电离的碳酸（H_2CO_3）、碳酸氢根（HCO_3^-）和碳酸根（CO_3^{2-}）三种形态，这三种形态处于动态平衡[8]。大气CO_2通过海-气交换进入海洋后会发生以下一系列化学反应：

$$CO_2(g) \longleftrightarrow CO_2^*(aq)$$
$$CO_2^*(aq) + H_2O \longleftrightarrow H_2CO_3$$
$$H_2CO_3 \longleftrightarrow H^+ + HCO_3^-$$
$$HCO_3^- \longleftrightarrow H^+ + CO_3^{2-}$$

打破海水碳酸盐系统原有的平衡，导致这三种形态的物质的量浓度发生相应的变化（化学反应方程式中g表示气态，aq表示溶解态）。这样的动态平衡构建了海水碳酸盐缓冲系统，即部分进入海洋的 CO_2 会转换成其他形态，从而使海水中 CO_2 浓度升高的程度低于非缓冲系统[8]。

总体而言，随着深度的增加，海水的"年龄"也随之增大，停留时间亦然，通风减弱，即需要很长的时间才能重新与大气接触，发生交换。表层海水停留时间最短，1000 米以下则可达数百年甚至千年。如果将海洋表层的 DIC 输送至深层，便可长时间与大气隔绝，实现碳封存，同时促进上层海水进一步吸收大气中的 CO_2。海洋碳泵，通常包括溶解度泵、生物泵和碳酸盐泵等，承担了输运、"封存"储碳的任务（图1）。

溶解度泵通常指高纬海区在冷空气和强风的作用下，表层海水快速降温，海水中 CO_2 溶解度增大，海洋通过海-气交换从大气中吸收大量 CO_2；同时随着深层水的形成，高密海水携带吸收的 CO_2 下沉进入大洋热液环流，脱离海-气交换层，从而实现对大气 CO_2 的封存。生物泵始于海洋真光层，浮游植物吸收海水中的游离 CO_2 和 HCO_3^- 进行光合作用，将无机碳转化为有机碳，其中部分以颗粒有机碳（particulate organic carbon，POC）的形式通过沉降等过程输送至深海，部分以溶解有机碳（dissolved organic carbon，DOC）的形式向下扩散至深海，进而实现碳封存。需要指出的是，在输送过程中部分有机碳会被再矿化成 DIC 释放到周围水体中。碳酸盐泵是控制海洋碳循环的另一重要过程。海水碳酸盐系统具有一定的缓冲作用：一方面，CO_2 进入海水使 HCO_3^- 和 CO_3^{2-} 之间的比例发生变化而减缓 pH 的降低；另一方面，海水中碳酸盐沉淀的形成会放出 CO_2（如图1碳酸盐泵中红色箭头所示），而碳酸盐溶解会从大气吸收 CO_2（如图1碳酸盐泵中蓝色箭头所示），这就是碳酸盐泵。在海底沉积物中存在大量的碳酸盐，因此，深海碳酸盐的沉积与溶解在长时间尺度上可调节大气 CO_2 浓度。

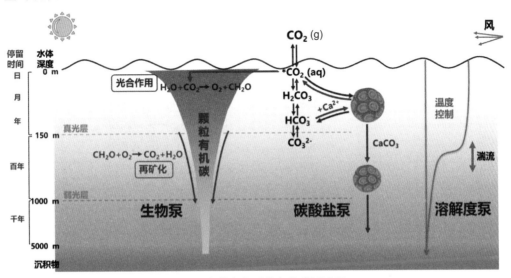

图1 海洋中主要的自然碳汇过程

二、海洋碳通量研究

海洋与大气的交界面是地球系统中物理、化学过程最为活跃的界面之一，在此界面上时刻进行着物质（水蒸气、CO_2 等）和能量（动量、热量等）的交换。海-气 CO_2 交换通量由复杂的物理和生物地球化学过程所驱动，测量方法包括放射性碳同位素 ^{14}C 示踪法、稳定碳同位素比值法、海-气 CO_2 分压差法、涡动相关 CO_2 通量直接观测法等。目前最常用的是海-气 CO_2 分压差法，基于现场观测数据，结合海-气界面气体交换速率计算海-气 CO_2 通量（F），公式如下：

$$F = k \times s \times \Delta p_{CO_2}$$

$$\Delta p_{CO_2} = p_{CO_2 (SW)} - p_{CO_2 (air)}$$

式中 k 是气体交换系数，受海面上风、流、浪等因素的影响；s 是 CO_2 在海水中的溶解度，主要受温度控制，温度越高，CO_2 溶解度越小；$p_{CO_2 (SW)}$ 和 $p_{CO_2 (air)}$ 则分别代表表层海水和大气中 CO_2 的分压，两者间的分压差（即 Δp_{CO_2}）决定了海-气 CO_2 通量的方向。如果 Δp_{CO_2} 大于零，即表层海水 CO_2 分压高于大气，海洋向大气释放 CO_2，即为碳源；反之海洋则为碳汇。总体而言，目前大气 CO_2 分压比海洋高且增长速度快，海洋是大气 CO_2 的汇。2011～2020 年，全球海洋的海-气净通量的年平均值为 2.2 Gt[①] 碳，是大气 CO_2 的汇。海洋每年从大气中吸收的量约为 2.8 Gt 碳，相当于人为 CO_2 总排放量的 26%[4]。

（一）国际研究进展

2009 年联合国环境规划署（UNEP）、联合国粮农组织（FAO）和联合国教科文组织政府间海洋学委员会（IOC-UNESCO，简称 IOC）发布了《蓝碳：健康海洋固碳作用的评估报告》，指出海洋固碳占全球固碳总量的 55%，有着巨大的碳汇容量和增汇潜力。保护国际基金会（Conservation International，CI），IOC 和世界自然保护联盟（International Union for Conservation of Nature，IUCN）发起了"蓝碳倡议"（Blue Carbon Initiative）计划，并成立海洋蓝碳政策工作组和科学工作组，发布了《蓝碳政策框架》（*Blue Carbon Policy Framework*）、《国家蓝碳政策评估框架》（*National Blue Carbon Policy Assessment Framework*）等一系列海洋碳汇报告。美国国家海洋和大气管理局（NOAA）从市场机会、认可和能力建设、科学发展等方面提出了海洋碳汇工作部署。海洋碳通量监测与碳收支估算被国际众多研究计划纳入其关注的核心科学问题，包括世界海洋环流实验（The World Ocean Circulation Experi-

---

① 1Gt＝10^{15}g 。

ment，WOCE）、全球联合海洋通量研究（Joint Global Ocean Flux Study，JGOFS）等。由 IOC 召集，联合国际海洋碳循环协调计划（International Ocean Carbon Coordination Project，IOCCP）、上层海洋-低层大气计划（Surface Ocean-Lower Atmosphere Study，SOLAS）、国际海洋生物地球化学与生态系统整合研究计划（Integrated Marine Biogeochemistry and Ecosystem Research，IMBER）、气候与海洋变化与预测计划（Climate and Ocean：Variability，Predictability and Change，CLIVAR）、全球碳计划（The Global Carbon Project，GCP）及相关领域专家成立"海洋碳综合研究"（Integrated Ocean Carbon Research，IOCR）工作组，旨在规划未来十年海洋碳循环综合研究，以支撑联合国海洋科学促进可持续发展十年规划。2021 年，IOC 发布了《海洋碳综合研究：海洋碳知识摘要及未来十年海洋碳研究和观测协调展望》（下文简称《海洋碳综合研究》）报告，强调了海洋通过吸收人为排放 CO_2 在调节气候变化方面发挥着重要作用，但同时预警海洋碳汇在未来可能被逆转，海洋不但不再吸收 CO_2，还可能释放 CO_2，加剧温室效应。因此，开展海洋碳汇演变研究十分重要与迫切[7]。

1. 大洋

表层大洋二氧化碳地图（Surface Ocean CO_2 Atlas，SOCAT）数据集收集来自科考船、商船船载海水 p_{CO_2} 实测数据以及浮标、自动驾驶船等 CO_2 传感器数据；由超过 100 位成员组成的国际海洋碳研究组对实测数据进行质量控制后每年公开发布更新数据集[9]；数据量由 2011 年第一次发布时的 630 万条增长到 2021 年的 3000 万条。最新数据集收录了 1957～2020 年全球表层海水 p_{CO_2} 的观测数据，其分布如图 2 所示。以 SOCAT 数据集为基础，通过大数据技术构建均匀的大洋 p_{CO_2} 格点数据，以及相关的数据同化、模式研究得到了快速发展。基于此，GCP 最新发布的《全球碳收支》（Global Carbon Budget）报告综合集成了 8 个模式和 8 个大数据产品的研究结果，对海洋碳收支及其误差进行了科学的评估[3]。

2. 近海

近海连接陆地和大洋，面积仅占全球海洋总面积的 7%～10%，有机碳生产却占 30%～50%，有机碳埋藏率更是高达 80%[10]，从而在全球碳循环中扮演着重要的角色。近海受人为活动和全球变化的双重影响，其中的物理与生物地球化学过程远比大洋复杂，除了海-气界面，还受陆-海、近海-大洋界面过程的影响，碳通量及其调控过程均具有高度的时空变异性，碳源汇研究难度更大[5,11-16]。从全球集成研究结果来看，近海整体是大气 CO_2 的汇，年碳汇量为 0.19～0.45 Gt 碳 [17-21]，但这一结果仍存在较大的不确定性。相关研究指出，大部分近海的碳源汇格局可在不同的空间和时间尺

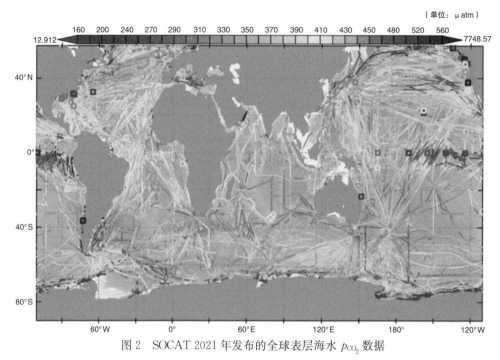

（单位：μatm）

图2　SOCAT 2021年发布的全球表层海水 p_{CO_2} 数据

资料来源：SOCAT网站，https://www.socat.info/

度上进行切换[22]，因此，季节和年际变化对近海碳通量估算很重要，可引入较大误差，目前还没有足够的数据进行准确评估[6]。此外，定量甄别近海的自然与人为碳源/汇贡献也是一个颇具挑战性的科学命题。

3. 海洋碳通量及碳库储量的演变

（1）工业革命前［图3（a）］，大气、陆地、海洋之间的碳收支基本达到平衡。陆地每年通过初级生产和岩石风化从大气净吸收 0.6 Gt 碳；而海洋，包括近海和大洋，则通过海-气交换每年向大气释放 0.6 Gt 碳；因此，大气碳储量维持不变。

（2）18世纪瓦特改良蒸汽机后，工业革命风起云涌，彻底改变了人类的生产和生活方式，人类历史自此进入快速发展阶段。这样的发展是以化石燃料的燃烧作为动力，大量石油、煤炭被开采并投入使用转化成 CO_2，加速了这些本应在地层深处存留数千万年的碳进入大气的过程，大自然的碳收支平衡也被打破［图3（b）］。现今，化石燃料碳储量减少，并转化为 CO_2 在大气、陆地和海洋间进行了再分配。其中，大气和近海碳储量增加50%，变化最大。在大气高 CO_2 浓度的驱动下，陆地碳汇通量增加了4倍之多，海洋则从工业革命前的碳源转变为碳汇。

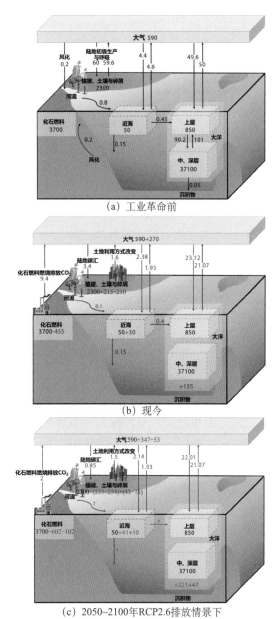

（a）工业革命前

（b）现今

（c）2050~2100年RCP2.6排放情景下

图 3　全球不同储库碳储量（箱内数字，单位：Gt 碳）及
不同储库间碳通量（箭头上数字，单位：Gt 碳/a）的变化

红色数字表示碳储量及碳通量较工业革命前的变化量；紫色和蓝色数字分别表示模式预测到 2050 年大气 CO_2
浓度达到最高值时其后开始下降的第一个 50 年内碳储量及碳通量较工业革命前的变化量

资料来源：此图重绘自文献［23］

（3）未来，在代表性浓度路径（representative concentration pathway，RCP）2.6排放情景下，大气CO_2浓度到2050年达到最高值，之后稳步下降。据模型估算结果显示［图（3c）］，到2100年，化石燃料碳储量减少近1/5，大气碳储量先增大后小幅下降，海洋碳储量则持续增加，近海碳储量较工业革命前增幅超过100%。值得关注的是，2100年大气CO_2浓度在RCP2.6情景下与现今相当，但即使在大气高浓度CO_2的情景下，陆地与海洋由于自身碳储量的持续增加，碳汇能力显著降低，海洋碳汇通量是现今的50%，而陆地仅为现今的1/4。如何保持或增强自然系统碳汇功能，是未来碳中和路径中科技界面临的重大挑战。

4. 基于海洋的碳增汇途径

目前CO_2的排放水平已大大超出自然过程所能清除的量，单靠减排可能不足以应对气候变化。海洋是巨大的碳储库，具有增汇移除、封存CO_2的巨大潜力。海洋CO_2的去除途径主要包括生物途径和化学途径两类（图4）。

1）生物途径

（1）营养施肥。向表层海洋添加氮、磷、铁等营养物质，刺激浮游植物的初级生产，通过增强上层海洋的生物泵过程增加海洋固碳。

（2）人工上升流和下降流。人工上升流是指通过人为干预将营养盐丰富的深层水输送到上层海洋，刺激浮游植物生长，从而吸收更多的大气CO_2；人工下降流则是通过将上层海水加速泵入到海洋深层从而达到碳转移的目的。

（3）大型海藻养殖。通过大规模的大型海藻养殖吸收大气CO_2，再将大型海藻用于制作长寿命产品或转化为生物能以达到碳封存或利用。

（4）滨海生态系统修复（滨海蓝碳）。通过保护和修复沿海生态系统，包括湿地、海草床、红树林等，提高其固碳量，实现碳封存。

2）化学途径

（1）海洋碱化。通过化学方法增加海洋碱度，恢复海洋碳酸盐系统缓冲能力的同时增强溶解度泵的效率，从而增加其对大气CO_2的吸收。

（2）海水碳萃取。通过电化学反应，直接萃取海水中的CO_2并将其转化为长寿命产品；或增加海水的碱度，提高其吸收大气CO_2的能力。

相对于陆地，目前对海洋碳移除的途径、效率、工程应用、生态效应等方面的认知较为匮乏，亟须开展深入研究。美国国家科学院2021年发布的《海洋二氧化碳移除与封存策略》对以上六种方法进行了全面评估，包括其成效、可持续性、可推广性、潜在的环境风险、社会风险以及目前对每种方法的了解程度等，特别提及了每种方法的研究重点，并预估了未来5~10年的成本投入[24]。

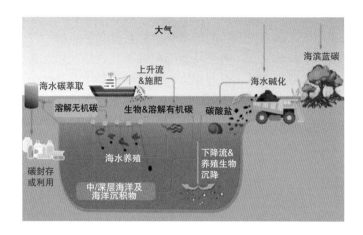

图 4　基于海洋的碳增汇途径

数据来源：重绘自《未来能源倡议》评估报告[25]

（二）我国海洋碳汇研究基础

过去 20 余年来，我国科技界及相关行业部门主持开展了大量碳循环过程研究和通量观测，初步评估了中国海洋生态系统的碳储量及其变化，并且在自然生态系统碳收支及其控制机制研究方面取得显著进展。我国海洋国土面积达 300 万 km²，其中东海是重要的碳汇区，南海受西太平洋深层水输入过量溶解无机碳影响表现为弱源，但具有增汇的潜力，黄海和渤海碳收支基本平衡[26]。中国海整体上则是大气 CO_2 的汇，每年从大气吸收约 1000 万 t 碳。南海是西北太平洋最大的边缘海，也是国际上海-气碳通量研究最为系统的边缘海之一。通过 20 多年的观测积累，我国建立了全球最大的南海区域海-气 CO_2 通量数据库，描绘了南海 CO_2 源汇格局，揭示了其时空变化规律。南海总体上是大气 CO_2 的源，主体海域（不包括泰国湾和北部湾）年释放量为 1330 万 t 碳，平均海-气 CO_2 通量低于其他西边界流的边缘海系统[27]。以此为基础，研究揭示边缘海 CO_2 源汇格局与"边缘海-大洋"和"陆地-边缘海"两个界面的物质交换息息相关，我国科学家构架了物理-生物地球化学耦合、无机碳-营养盐耦合定量解析新方法，建立了大洋主控型边缘海和河流主控型陆架海碳循环理论框架，并成功拓展应用至全球典型陆架边缘海 CO_2 源汇格局及其控制过程的解析[16,18,28]。

三、展　　望

地球气候系统是地球系统各组分跨圈层、跨界面、多尺度耦合作用下的动态平衡系统，海洋和大气则更是处于高度耦合的状态。实现碳中和目标涉及海洋-陆地-大气-人类等多圈层交互作用，而人类活动如何影响地球系统关乎自然、社会、经济的可持续发展，必须充分基于科学认知，解决其中的关键基础科学问题。与此同时，由于海洋碳汇能力可能在碳达峰后转弱，因此探索海洋脱碳的技术途径十分必要。但在实施海洋脱碳和增汇工程之前，亟须在提升自然和人为碳源汇准确监测的基础上，加强各类增汇途径相关科技的基础研究，从而为制定碳中和最优技术路径和方案搭建理论框架。

1. 海洋碳汇的重大基础科学问题

《海洋碳综合研究》报告提出了下列重大科学问题：①海洋中自然发生的非生物与生物过程在碳汇中的作用；②海洋与陆地间的碳交换过程；③人类对海洋碳循环的改变及其反馈[6]。因此，受海洋增暖、酸化及脱氧等全球变化过程与人类干扰的双重影响，海洋生态系统碳循环的定量、模拟与预测依然面临重大科学挑战。其中的重大基础科学问题包括：①准确核算海洋碳库、碳汇清单及其演化；②海洋碳泵的控制机理以及不同碳泵的耦合与分异；③碳中和情景下的海洋碳汇演化趋势。

陆架边缘海位于陆地-大洋交界带，是地表系统中海-陆-气相互作用最剧烈的区域，具有独特的地理、物理、化学和生物特征，其中的碳通量及其调控过程均具有高度的时空变异性，是全球碳循环的关键环节。一方面，陆架边缘海与邻近大洋的交换在一定程度上决定了边缘海的物质收支，而以热力学和生物泵为核心的生物地球化学过程则是驱动碳循环的内因，两者共同调控边缘海碳源汇格局[29]；另一方面，河流携带大量来自陆地的有机碳、营养盐和污染物汇入海洋，不仅导致近海富营养化及生态环境恶化，且陆源碳无法有效保存，进而引发一系列生态环境问题和陆-海-气界面交互系统的碳源汇效应。因此，海洋碳汇与生物多样性等关键生态服务功能的协同提升至关重要，涉及的重大基础科学问题包括：①解析陆-海、海-气、海-洋多界面、全链条碳足迹与碳循环机制；②甄别自然和人为碳源汇格局及其对区域碳汇清单核算的影响；③提出海洋生态系统保护、生物多样性保护与固碳增汇协同增效的最优路径。

2. 海洋自然碳源汇动态监测体系

开展系统的海-气融合联网观测，重视网络观测的标准化、规范化和自动化，提升海洋各类碳源汇观测的时空分辨率；同时，推进关键数据集成和数据平台的建立；通过机器学习、人工神经网络等大数据技术开发海洋 p_{CO_2} 网格化数据集，进一步拓展碳源汇分布的时空范围，降低碳清单核算的不确定性。这里，遥感技术及算法的建立至关重要，可以实现海洋碳源、碳汇动态变化的实时监测与定量估算。同时，观测与数值模式的融合也不可或缺。

3. 基于海洋的人为增汇预研

建立基于温室气体的可监测、可验证、可支撑（monitoring/verification/support，MVS）的新方法体系，以及基于碳盘点的人为碳排放反演方法体系；开展海洋 CO_2 移除和增汇手段研究，包括已知的和未知的（颠覆性技术理论）生物与非生物过程，实现去碳、封存的科学原理和增汇功效，分析技术瓶颈、评估生态和社会风险；建立全面的海洋增汇技术可行性评估体系和法律框架，打通基础研究、前沿技术和工程应用链条。

参考文献

[1] IPCC. 2021：Summary for Policymakers. In：Climate Change 2021：The Physical Science Basis. https://www. ipcc. ch/report/sixth-assessment-report-working-group-i/[2022-02-10].

[2] IPCC. Global Warming of 1.5℃. https://www. ipcc. ch/sr15/[2022-01-20].

[3] von Schuckmann K,Cheng L,Palmer M D,et al. Heat stored in the Earth system：Where does the energy go? Earth System Science Data,2020,12(3):2013-2041.

[4] Friedlingstein P,Jones M W,O'Sullivan M,et al. Global carbon budget 2021. Earth System Science Data. https://doi. org/10. 5194/essd-2021-386[2022-01-20].

[5] Bauer J E,Cai W J,Raymond P A,et al. The changing carbon cycle of the coastal ocean. Nature,2013,504(7478):61-70.

[6] Laruelle G G,Cai W J,Hu X,et al. Continental shelves as a variable but increasing global sink for atmospheric carbon dioxide. Nature Communications,2018,9(1):454.

[7] Aricò S,Arrieta J M,Bakker D C E,et al. ,Integrated ocean carbon research：A Summary of ocean carbon research, and vision of coordinated ocean carbon research and observations for the next decade. https://www. sciencedirect. com/bookseries/elsevier-oceanography-series/vol/65/suppl/C [2022-01-20].

[8] Zeebe R E, Wolf-Gladrow D. CO_2 in Seawater：Equilibrium, Kinetics, Isotopes. Oxford：Gulf Professional Publishing,2001.

[9] 钟国荣,李学刚,曲宝晓,等. 基于广义回归神经网络的全球表层海水 1°×1°二氧化碳分压数据推演. 海洋学报,2020,42(10):70-79.

[10] Gattuso J P, Frankignoulle M, Wollast R. Carbon and carbonate metabolism in coastal aquatic ecosystems. Annual Review of Ecology and Systematics,1998. 29(1):405-434.

[11] Cai W J. Estuarine and coastal ocean carbon paradox：CO_2 sinks or sites of terrestrial carbon incineration? Annual Review Marine Science,2011,3:123-145.

[12] Chen C T A, Borges A V. Reconciling opposing views on carbon cycling in the coastal ocean：Continental shelves as sinks and near-shore ecosystems as sources of atmospheric CO_2. Deep Sea

Research Part II: Topical Studies in Oceanography,2009,56(8):578-590.

[13] Hofmann E E,Cahill B,Fennel K,et al. Modeling the dynamics of continental shelf carbon. Annual Review of Marine Science,2011,3:93-122.

[14] Lacroix F, Ilyina T, Laruelle G G, et al. Reconstructing the preindustrial coastal carbon cycle through a global ocean circulation model: Was the global continental shelf already both autotrophic and a CO_2 sink? Global Biogeochemical Cycles,2021,35(2):e2020GB006603.

[15] Dai M H. What are the exchanges of carbon between the land-ocean-ice continuum? In: Wanninkhof R, Sabine C, Aricò S. Integrated Ocean Carbon Research: A Summary of Ocean Carbon Research, and Vision of Coordinated Ocean Carbon Research and Observations for the Next Decade. Paris: the Intergovernmental Oceanographic Commission of UNESCO(UNESCO-IOC),2021:20.

[16] Dai M H. 2021. Boundary regions: Land-ocean continuum and air-sea interface. In: Wanninkhof R, Sabine C,Aricò S. Integrated Ocean Carbon Research: A Summary of Ocean Carbon Research,and Vision of Coordinated Ocean Carbon Research and Observations for the Next Decade. Paris: the Intergovernmental Oceanographic Commission of UNESCO(UNESCO-IOC),2021:14-15.

[17] Cai W J,Dai M H,Wang Y. Air-sea exchange of carbon dioxide in ocean margins: A province-based synthesis. https://doi. org/10. 1029/2006GL026219[2021-01-10].

[18] Dai M H,Cao Z M,Guo X H,et al. Why are some marginal seas sources of atmospheric CO_2? Geophysical Research Letters,2013. 40(10):2154-2158.

[19] Borges A V,Delille B,Frankignoulle M. Budgeting sinks and sources of CO_2 in the coastal ocean: Diversity of ecosystems counts. https://doi. org/10. 1029/2005GL023053[2021-01-10].

[20] Chen C T A, Huang T H,Chen Y C,et al. Air-sea exchanges of CO_2 in the world's coastal seas. Biogeosciences,2013,10(10):6509-6544.

[21] Laruelle G G,Lauerwald R,Pfeil B,et al. Regionalized global budget of the CO_2 exchange at the air-water interface in continental shelf seas. Global Biogeochemical Cycles,2014,28(11):1199-1214.

[22] Cai W J,Xu Y Y,Feely R A,et al. Controls on surface water carbonate chemistry along North American ocean margins. Nature Communications,2020,11(1):2691.

[23] Dai M H,Su J Z,Zhao Y Y,et al. Carbon fluxes in the coastal ocean: synthesis,boundary processes and future trends. Annual Review of Earth and Planetary Sciences, https://doi. org/10. 1146/ annurev -earth-032320-090746[2022-01-10].

[24] National Academies of Sciences,Engineering,and Medicine. A Research Strategy for Ocean-based Carbon Dioxide Removal and Sequestration. Washington DC: The National Academies Press. 2021.

[25] Energy Futures Initiative. Uncharted waters: Expanding the options for carbon dioxide removal in coastal and ocean environments. https://static1. squarespace. com/static/58ec123cb3db2bd94e057628/ t/5fda3ec0e28fdf61aebf3159/1608138446173/Uncharted＋Waters＋Final＋12. 16. 20. pdf[2022-02-21].

[26] 刘茜,郭香会,尹志强,等. 中国邻近边缘海碳通量研究现状与展望. 中国科学:地球科学,2018, 48:1422-1443.

[27] Li Q,Guo X H,Zhai W D,et al. Partial pressure of CO_2 and air-sea CO_2 fluxes in the South China Sea:Synthesis of an 18-year dataset. Progress in Oceanography,2020,182:102272.

[28] Cao Z M,Yang W,Zhao Y Y,et al. Diagnosis of CO_2 dynamics and fluxes in global coastal oceans. National Science Review,2020,7(4):786-797.

[29] 戴民汉,孟菲菲. 南海碳循环:通量、调控机理及其全球意义. 科技导报,2020,38(18):30-34.

Ocean Carbon Cycle and Ocean-Based Carbon Solutions in the Context of Carbon Neutrality

Dai Minhan,Meng Feifei,Cao Zhimian

The essence of carbon neutrality is the multi-scaleinteractions between the Earth's climate system and carbon cycle. The oceanic carbon cycle is a key component of the global carbon cycle and plays an important role in the Earth's climate system. Increasing ocean carbon sinks is one of the effective ways to achieve carbon neutrality. In this context,this article firstly introduces the basic information and controlling processes of the oceanic carbon sink,followed by a review of the status and progress of both international and domestic ocean carbon research. After pointing out six methods of marine CO_2 removal,we propose suggestions on future marine solutions for carbon neutrality,including the major scientific questions to be answered and the potential technical ways to be implemented.

1.2 感染性疾病的免疫治疗

徐若男　施　明　王福生

（中国人民解放军总医院第五医学中心感染病医学部）

感染性疾病指由病原体感染而导致的疾病，是临床最常见的疾病之一，一般包括传染性和非传染性两类。抗菌药物的涌现和更新，挽救了千百万名感染性疾病患者的生命。但是艾滋病、病毒性肝炎、结核病、各类新突发传染病等仍旧是影响人类健康的重要疾病。在加速传统药物研发的基础上，围绕上述疾病病原体的特点和致病机制，针对机体免疫系统开展的免疫治疗方案极大丰富了现有临床治疗手段，在多种重大传染病、新突发传染病及常见传染病的临床救治中发挥了独特作用。

免疫治疗是指针对机体低下或亢进的免疫状态，人为地增强或抑制机体的免疫功能以达到治疗疾病目的的治疗方法。广义上讲，免疫治疗是基于人类的免疫系统，通过诱导、增强或抑制免疫应答来治疗疾病，通过免疫分子、免疫细胞而开发的治疗方案等都属于免疫治疗的范畴。本文以获得性免疫缺陷综合征（acquired immunodeficiency syndrome，AIDS，又称艾滋病）、慢性乙型肝炎及肝癌、结核病、新型冠状病毒肺炎、巨细胞病毒感染、EB 病毒感染等疾病为例，论述免疫治疗的研究进展。

一、艾滋病的免疫治疗

自 1981 年首次发现人类免疫缺陷病毒（human immunodeficiency virus，HIV，又称艾滋病病毒）以来，截至 2021 年，世界卫生组织（World Health Organization，WHO）官方统计全世界 HIV 感染者已超过 4000 万名，HIV 感染引发的艾滋病会导致人体免疫机能缺陷，从而易发生感染和肿瘤，严重威胁人类健康[1]。在过去的 30 多年里，人类对抗艾滋病的战争先后经历了三次革命性进展。1996 年，联合抗逆转录病毒治疗（combination antiretroviral therapy，cART）的问世，无疑是第一次革命性进展，使艾滋病由致死性疾病转变为某种程度上可控的慢性疾病[2]。cART 能有效降低患者血液中的病毒载量，增加 CD4$^+$T 细胞数量，同时避免了单一用药导致的耐药性。cART 虽然能高效抑制患者体内 HIV 复制，但无法彻底恢复机体的免疫功能，停药后病毒复制快速反弹。1997 年，长期存活的病毒储存库的发现成为 HIV 感染研究历程上的第二次革命性进展，其很好地解释了 cART 无法彻底治愈 HIV 感染的原因，对其后寻求有效策略清除 HIV 具有重要意义[3]。2009 年"柏林病人"的出现为人类实现艾滋病的功能性治愈乃至根除性治愈带来了希望和曙光，人类迎来了抗艾战役的第三次革命性进展[4]。

高效抗病毒治疗是当前治疗 HIV 慢性感染和艾滋病最主要的临床手段，可持久抑制病毒复制，可在一定程度上改善患者的免疫功能，部分实现免疫重建，明显改善 HIV 慢性感染和艾滋病患者的疾病进展和预后，使得艾滋病功能性治愈成为可能。但是，因 HIV 感染者接受 cART 治疗时的 $CD4^+T$ 细胞基线数量不同，导致长期 cART 治疗后 $CD4^+T$ 细胞数量的升高幅度和绝对数量存在差异。同时，cART 介导的免疫重建具有局限性，表现在：①部分患者 $CD4^+T$ 细胞数量恢复不理想，长期 cART 治疗后 $CD4^+T$ 细胞数量仍不能提升至正常水平（>500 个/μl）。即使 $CD4^+T$ 细胞数量提升至>500 个/μl，更高的 $CD4^+T$ 细胞数量仍旧可以让患者受益，而单纯 cART 往往不能达到预期。②即使 $CD4^+T$ 细胞得到有效提升，机体依旧存在持续性免疫细胞过度活化和高水平的炎症反应，发生艾滋病相关事件、非相关事件甚至死亡的风险依旧很高。③病毒储存库的持续存在，使得期望在个体生存期内仅靠 cART 将 HIV 彻底清除的目标不可能实现，甚至实现停药后长期病毒不反弹的目标可能性也较低。④机体肠道黏膜的损伤、淋巴结和胸腺组织的纤维化不能被 cART 有效逆转。⑤部分患者进入艾滋病晚期骨髓造血功能受损，单纯 cART 不能修复，影响免疫重建的效率和效果。

HIV 感染不仅是病毒感染性疾病，同时也是免疫系统疾病，在有效抑制病毒复制的同时，围绕免疫系统过度活化、免疫紊乱开展各类免疫干预治疗，对于有效抑制病毒复制、提高和维持 $CD4^+T$ 细胞的数量和功能、降低艾滋病相关和非相关事件、降低病死率具有重要意义。目前常用的免疫治疗策略包括免疫细胞治疗、干细胞治疗、细胞因子治疗、HIV 疫苗、中和抗体、免疫检查点抑制剂等。

1. 免疫细胞治疗

嵌合抗原受体 T（chimeric antigen receptor T，CAR-T）细胞免疫疗法通过基因编辑使 T 细胞表达嵌合抗原以加强 T 细胞对感染 HIV 细胞的靶向性和杀伤能力。与天然的效应 T 细胞相比，CAR-T 细胞直接识别抗原而不依赖于主要组织相容性复合体（major histocompatibility complex，MHC）分子的呈递，有利于阻止或限制病毒的免疫逃逸。CAR-T 细胞在体外生成后，可以在体外存活并扩增几个数量级，从而提供大量的抗原特异性细胞。CAR-T 细胞中的记忆 T 细胞（memory T cell）数量较多，对再度活跃的 HIV 直接攻击效应强。理想情况下，CAR-T 细胞可以稳定并持久地监视储藏的 HIV。近年来，双共刺激分子（CD28 和 4-1BB）的双 CAR-T 细胞疗法具有更强的体内扩增能力和更长的体内存活时间[5]。在双 CAR-T 结构外加上一层保护膜，使其表面表达 C34-CXCR4 结构，可进一步防止 HIV 对 CAR-T 细胞的感染。虽然 CAR-T 细胞疗法为艾滋病的治疗提供了新的策略，但是如何获得足够的 CAR-T 细胞数量、提升扩增细胞中功能性记忆 T 细胞的数量、避免新生 CAR-T 细胞被感染等问题还需要更多的深入研究。

自体 CCR 基因修饰的 $CD4^+$ T 细胞、病毒特异性细胞毒性 T 细胞（cytotoxic T lymphocyte，CTL 细胞）、抗原刺激树突状细胞（dendritic cell，DC）等治疗方案，在增强机体免疫应答、提高 $CD4^+$ T 细胞数量等方面具有一定的效果。锌指核酸酶（zinc finger nucleases，ZFNs）敲除 *CCR5* 基因的自体细胞疗法，能够在患者体内产生记忆 $CD4^+$ T 细胞，单次给药后可达到持续的 HIV 缓解[6]。该疗法产生新型干细胞样记忆 $CD4^+$ T 细胞亚群的扩增，并促进 T 细胞稳态恢复，使 HIV 感染者长期获益。在 SHIV（具有 HIV 包膜蛋白的嵌合猴病毒）感染的恒河猴模型中，将 SHIV 疫苗免疫与抗病毒治疗相结合，并在 ART 中断后用特异性的 CTL 免疫疗法可有效抑制病毒反弹[7]。

2. 干细胞治疗和细胞因子治疗

间充质干细胞（mesenchymal stem cell，MSC）是一群起源于中胚层的多能干细胞，具有自我更新、多向分化和免疫调控潜能。自 1966 年 MSC 被发现后，其因多潜能和可扩增性对再生医学产生了不可估量的影响。MSC 作为一种新型的免疫治疗方案，具有良好的免疫调节作用和强大的抗炎特性，可通过感知炎症信号（细胞因子/趋化因子受体和整合素的表达）迁移到炎症部位，通过直接接触和旁分泌效应发挥作用。MSC 可以有效降低多种组织器官（如肝、肺、胸腺等）的纤维化，促进组织再生；同时 MSC 可以维护骨髓微环境的稳定，促进造血。人脐带间充质干细胞在治疗 HIV 免疫重建失败的临床研究中同样显示出提升 $CD4^+$ T 细胞数量、降低免疫系统超活化的效应[8]。除此之外，多种细胞因子（如 IL-2、IL-7 等）对于促进 $CD4^+$ T 细胞数量提升也有一定效果[9]。

3. HIV 疫苗

HIV 疫苗的开发经历了漫长的过程。这种疫苗不仅能激发机体产生对 HIV 的免疫反应，还有助于 HIV 抗体的产生。DNA 疫苗是继完整病原体疫苗和基因工程重组蛋白疫苗之后的第三代疫苗，HIV DNA 疫苗多以病毒的结构蛋白基因和调节基因为靶点，将编码这些基因的 DNA 直接注入体内，在机体内产生类似 HIV 活疫苗的效果。该类疫苗一般不良反应小，在动物实验水平上已显示出良好效果。而 HIV 活疫苗由去掉致病基因的 HIV 活病毒制成，具有较高的滴度，但其安全性需慎重考虑。除此之外，应用 HIV 复制所必需的蛋白（如 Gag、Tat、Rev、Nef 等）设计的蛋白疫苗也在进一步的研究中。总而言之，尽管 HIV 疫苗前景乐观，但其安全性和临床效果还有待于深入研究。

4. 中和抗体

近年来，HIV 广谱中和抗体（bnAbs）被证实几乎可以中和所有 HIV 毒株的不同亚型，有望为 HIV 的预防、治疗和疫苗开发带来新的突破。bnAb 不仅可以直接中和病毒，阻断病毒感染靶细胞，还可激发体内机体抗病毒免疫应答反应。bnAb 可有效改善 HIV 感染动物模型和 HIV 感染者的病毒血症，延迟病毒反弹。目前，多种 bnAb 已经进入临床试验阶段，不同 bnAb 联用或 bnAb 与抗病毒药物联用，可以更大程度地减少耐药和抗体逃逸株的产生[10]。

5. 免疫检查点抑制剂

免疫检查点是在免疫细胞上表达、能调节免疫激活程度的一系列分子，抑制性免疫检查点如程序性细胞死亡蛋白 1（programmed death-1，PD-1）和细胞毒性 T 淋巴细胞相关蛋白 4（cytotoxic T lymphocyte-associated antigen-4，CTLA-4）可负性调节免疫细胞功能。在 HIV 感染期间，抑制性免疫检查点分子的上调会导致 T 细胞耗竭，从而丧失效应功能。PD-1 单克隆抗体的应用在提升病毒特异性 CD8$^+$ T 细胞功能、减少病毒库方面也显示出一定优势。PD-1 和 CTLA-4 单克隆抗体联用具有更强的 T 细胞活化作用，并且可以有效缩小 HIV 病毒储存库[11]。伴随着生物技术的进步，相信在不久的将来，免疫治疗手段与其他策略的联合应用能够使更多的 HIV 感染者实现疾病的长期缓解及功能性治愈。

二、慢性乙型肝炎及肝癌的免疫治疗

慢性乙型肝炎病毒（hepatitis B virus，HBV）感染是全球范围内导致肝脏疾病的常见原因，根据 WHO 估计，全球有大约 2.6 亿名 HBV 感染者，慢性乙型肝炎也是我国常见的慢性传染病之一。HBV 感染过程中，机体免疫状态、HBV 以及肝脏组织微环境三大因素相互影响，共同影响着疾病的进展和转归，其中机体的免疫应答反应不仅参与控制病毒复制和清除病毒，而且还参与慢性乙型肝炎的致病过程，是决定慢性乙型肝炎临床转归和治疗效果的重要因素。现有的干扰素和核苷类似物的单药、联合或续贯抗病毒治疗方案仅仅可以帮助部分患者实现乙肝病毒表面抗原（hepatitis B surface antigen，HBsAg）转阴。通过免疫调节治疗打破机体免疫耐受，有效激活免疫系统，加速 HBsAg 转阴的速度和效率必将成为慢性乙型肝炎的重要治疗策略。《慢性乙型肝炎防治指南》中明确指出"免疫调节治疗是慢性乙型肝炎治疗的重要手段之一"[12]。基于一些临床前的实验研究结果，治疗性疫苗、toll 样受体激动剂、凋亡诱导剂、维甲酸诱导基因蛋白 I 激动剂（retinoic acid inducible gene 1 protein，RIG-I）、HBsAg 抗体、PD-1/PD-L1 抗体等免疫治疗正在被探索作为促进 HBsAg 转阴的新手段。从现有的一些结果来看，上述免疫治疗手段在 HBV 感染患者的临床治疗中具有一定的作用。例如，在接受 RIG-I 激动剂和核苷类似物治疗的初治的非肝硬化慢性乙型肝炎（chronical hepatitis B，CHB）患者中，在 12 周或 24 周时，可以观察到 22% 的患者出现 HBsAg 下降超过 0.5 lg IU/ml[13]。接受 PD-1 抗体单药或与治疗性疫苗联合应用的部分患者在 24 周时可以观察到 HBsAg 表面抗原的清除[14]。但是，目前对于免疫治疗在 CHB 患者中的应用还属于早期探索阶段，需要有更多的研究支持。

肝细胞癌（hepatocellular carcinoma，HCC）是原发性肝癌中最常见的一种类型，其主要病因是 HBV 或者丙型肝炎病毒（hepatitis C virus，HCV）感染引起。慢性乙型肝炎及肝细胞癌的免疫治疗主要包括治疗性疫苗、免疫检查点抑制剂以及过继性免疫细胞治疗等。

1. 慢性乙型肝炎的治疗性疫苗

慢性乙型肝炎的治疗性疫苗旨在打破乙肝患者和病毒携带者的免疫耐受状态，启动机体自体内源性清除病毒的过程。蛋白疫苗是目前研究最为深入的乙肝疫苗，主要原理是通过将 HBV 表面蛋白（如 preS1、preS2、S）或核心抗原（HBcAg）进行一定的修饰后诱发机体产生较强的特异性免疫反应。为了进一步提高免疫原性，将 HBsAg 和乙型肝炎免疫球蛋白相结合开发出的免疫复合物疫苗，通过改变对 HBsAg 的呈递方式以诱发免疫反应，消除免疫耐受性。抗体能对抗原的免疫原性起到增强的作用，在于增加抗原被抗原提呈细胞（antigen presenting cell，APC）捕获并加工处理，进而激活抗原特异性淋巴细胞，诱发产生免疫应答[15]。DC 细胞疫苗中，荷载 HBV 相关抗原的 DC 细胞可有效打破免疫耐受，恢复细胞免疫应答以清除 HBV。DNA 疫苗是将荷载表面抗原（HBsAg）的 DNA 在患者体内接种，反复刺激免疫细胞，以产生与减毒活疫苗同样效果的体液免疫和细胞免疫。

2. 肝细胞癌的治疗性疫苗

肝细胞癌的治疗性疫苗包括肽类疫苗、DC 细胞疫苗、全细胞疫苗、溶瘤病毒疫苗等，用以增加机体对肿瘤抗原的免疫反应。目前，一些多肽，如甲胎蛋白（AFP）、多药耐药相关蛋白 3（MRP3）和糖基蛋白 3（GPC3），已经被证明具有良好的耐受性和安全性。表达 AFP 的复制缺陷腺病毒疫苗在肝癌患者体内可引发不同程度的 CD8[+] 和 CD4[+] T 细胞反应以及腺病毒中和抗体反应，且表现出良好的耐受性和安全性[13]。GPC3 衍生的肽疫苗可诱导肝癌患者 GPC3 特异性的 CTL 细胞增加，有效降低术后肿瘤的复发率[14]。MRP3 是 HCC 免疫治疗中具有强免疫原性的肿瘤抗原的潜在候选抗原。MRP3 肽疫苗表现出较好的安全性和较强的免疫原性，对延长患者生存期具有一定效果[15]。使用装载肿瘤细胞裂解物的 DC 细胞疫苗具有抗肿瘤作用，可有效缩小肿瘤体积或者降低血清肿瘤标志物，并且在有些患者体内产生抗原特异性免疫反应[16]。溶瘤病毒以其独特的靶向性和较高的杀伤作用展示出巨大的临床应用潜能，取得了令人瞩目的成果。溶瘤病毒可直接裂解肿瘤细胞，释放可溶性肿瘤抗原，从而诱导抗肿瘤抗原特异性 CTL 反应。一种溶瘤性和免疫治疗性牛痘病毒可在特定的癌细胞中复制，在临床试验中通过静脉注射和瘤内给药显示出良好的耐受性，并能延长患者的生存期，对肝细胞癌治疗具有一定疗效[17]。然而，溶瘤病毒易被机体快速清除，与免疫检查点（CTLA-4、PD-1/PD-L1）抑制剂联用，有望实现溶瘤病毒和免疫细胞的双重抗癌功效。

3. 免疫检查点抑制剂

免疫检查点抑制分子（CTLA-4、PD-1/PD-L1 等）的单克隆抗体在肿瘤治疗领域获得了巨大成功。这些单抗通过抑制免疫检查点受体和配体的相互作用而起到免疫增强的作用。HBV 感染的肝细胞中 PD-L1 表达上调，HBV 抗原特异性 T 细胞 PD-1 表达上调，二者结合激活细胞耗竭通路，使 T 细胞失去效应功能。阻断免疫检查点信号通路可能激活免疫反应，达到抗病毒效应。现有的单克隆抗体类免疫检查点（CTLA-4、PD-1/PD-L1 等）抑制剂等有助于逆转 T 细胞功能耗竭，改善患者体内免疫微环境，抑制肿瘤细胞生长，延长患者的生存时间[18]。但治疗效果存在异质性，与患者自身 PD-1/PD-L1 表达水平、免疫微环境等密切相关，且存在诱发严重不良反应的风险。因此，如何优化免疫治疗手段，使得 PD-1/PD-L1 低表达，且抗体治疗无应答的肝癌患者获得更大的临床受益，需要更加深入的研究。

4. 过继性免疫细胞治疗

过继性免疫细胞包括自然杀伤细胞（natural killer，NK）、肿瘤浸润淋巴细胞（tumor infiltrating lymphocyte，TIL）、细胞因子诱导的杀伤细胞（cytokine-induced killer，CIK）和 CAR-T 细胞等，也显示出对 HCC 的抗肿瘤效应。在肝癌患者中应用免疫治疗方案在安全性和有效性方面都取得了令人鼓舞的结果，不同的免疫疗法或免疫疗法与其他疗法的联合应用必将大大改善肝癌患者的长期预后。

三、结核的免疫治疗

结核病是由结核杆菌感染引起的慢性传染病。截至 2020 年，据 WHO 估算，全球结核潜伏感染人数约有 17 亿，约占全世界人口的 1/4，自 2007 年以来一直位居单一传染性疾病死因之首[19]。通过各项防治措施的有效实施，中国的结核病疫情在几十年间显著下降。2019 年中国的结核病发病率为 50/10 万，远远低于全球平均水平的 130/10 万；发病率每年以 3.2% 的幅度递降，明显高于全球 1.7% 的年递降水平；中国的结核病病死率已降到 2.2/10 万，远低于全球 16/10 万的平均水平[20]。

在 1945 年链霉素问世之前，结核病是不治之症。随着利福平等抗结核药物的问世，结核病的危害已大大降低。但是伴随耐药疫情逐渐严重，单一的化药模式将难以治愈耐多种药的结核病，消灭持留菌或休眠菌，挡住结核病的复燃是治愈结核病的关键。据报道结核病人 50% 以上合并免疫系统损坏，为此，WHO 结核病研究与发展战略规划中提出了化学疗法与免疫疗法相结合的治疗方案[21]。关于结核病免疫治疗研究主要包括细胞因子治疗、疫苗治疗、免疫检查点抑制剂治疗、干细胞治疗以及增强非特异性免疫力的治疗等。

1. 细胞因子治疗

结核分枝杆菌引起的免疫反应主要由 T 细胞介导，Th1 型和 Th2 型细胞因子的动态平衡在结核病的免疫应答中发挥重要作用。Th1 型细胞因子（如 IL-2、IL-12、IFN-γ、TNF-α 等）可促进细胞免疫反应，而 Th2 型细胞因子（如 IL-4、IL-5、IL-10 等）会拮抗 Th1 介导的细胞免疫反应。细胞因子治疗策略旨在增强 Th1 型细胞免疫应答，同时抑制 Th2 型细胞免疫应答。IFN-γ 可激活单核巨噬细胞，对减少分枝杆菌菌落数，减轻肺部病变具有一定的保护作用。IL-12 可诱导 Th1 型细胞免疫应答，与 DNA 疫苗联用治疗可产生良好的免疫保护。TNF-α 参与巨噬细胞的活化与肉芽肿的形成，通过促进巨噬细胞的吞噬阻断病菌播散；但 TNF-α 也可诱发细胞坏死并加剧炎症，在临床治疗中需要充分考虑[22]。IL-7 和 IL-2 的联用可促进 IFN-γ 分泌，抑制 IL-4 合成，进而调节 Th1/Th2 的免疫平衡[23]。

2. 疫苗治疗

目前处于临床试验阶段的候选结核病疫苗主要包括分枝杆菌全细胞衍生疫苗和亚

单位疫苗。全细胞疫苗来源于结核分枝杆菌、卡介苗或非结核分枝杆菌菌株。此类疫苗可进一步细分为基因改造的减毒的活疫苗，或灭活分枝杆菌疫苗。全细胞衍生疫苗包含许多不同的抗原成分，并且由于存在非蛋白抗原（例如在结核分枝杆菌外层普遍存在脂质和糖脂、微生物代谢物和磷酸抗原等），有助于诱导多样化的免疫反应。亚单位疫苗可以进一步细分为佐剂蛋白亚单位疫苗和重组病毒载体疫苗。亚单位疫苗针对少数特定抗原，虽然蛋白质亚单位疫苗 M72/AS01E 的 2b 期临床试验取得了令人满意的保护性免疫效果，但是如何提升这类疫苗的抗原性仍存在诸多挑战[24]。

3. 免疫检查点抑制剂

免疫检查点抑制剂通过抑制免疫检查点的信号转导影响 T 细胞和 B 细胞功能，常见种类包括 PD-1 单克隆抗体、CTLA-4 单克隆抗体等。这些免疫检查点介导的 T 细胞免疫耗竭与结核病的发生发展密切相关。例如：抑制 CTLA-4 虽然不能改善细菌清除率，但是能增强小鼠抗结核病的免疫反应。抑制 PD-1/PD-L1 途径可增强人外周血单核细胞中结核分枝杆菌特异性反应效率。与此同时，免疫检查点抑制剂诱发的过度炎症和局灶性坏死存在加速结核病进展的风险[22]。因此，未来需要在优势动物模型中对免疫检查点抑制剂应用的时机、剂量和疗程等进行进一步优化和评估。

4. 干细胞治疗

基于免疫重建目的的干细胞治疗是一个全新的领域，极大地丰富了结核病的治疗手段。通过干细胞移植技术将可塑性免疫原始细胞输入体内，可实现补充免疫细胞、恢复或增强患者细胞免疫功能的目的。间充质干细胞（mesenchymal stem cell，MSC）是临床常用的干细胞类型，具有免疫调节和抗菌特性，可改善结核病人的外周血免疫功能、促进肺病理组织修复。自体间充质基质细胞治疗耐多药结核病的安全性和有效性已得到证实，细胞输注后有助于受损肺组织修复和免疫应答反应增强[25]。MSC 治疗还可改善耐多药结核病患者的治疗效果，为耐多药结核病的治疗提供了新的方向。

5. 非特异性免疫治疗

维生素 D_3 缺乏是结核病发展的危险因素，将其作为辅助治疗手段可促进临床和影像学改善、宿主免疫激活和痰转化时间加快[26]。黄芩苷、虎耳草糖苷 D、大黄提取物、异甘草素、黄芪多糖等诸多中药成分或中药提取物可通过诱导巨噬细胞自噬、抑制或破坏结核分枝杆菌的生成等不同途径起到辅助抗结核治疗的作用[27]。相信伴随结核病致病机制研究的逐步深入和完善，结核病免疫治疗研究必然取得更大进步。

四、新型冠状病毒肺炎的免疫治疗

新型冠状病毒肺炎（coronavirus disease 2019，COVID-19）是由新型冠状病毒（severe acute respiratory syndrome coronavirus 2，SARS-CoV-2）感染导致的肺炎。2020 年 1 月，WHO 将 COVID-19 列为"国际关注的突发公共卫生事件"，并将其传播和影响的风险评估提高到最高等级。截至 2022 年 1 月，COVID-19 全球患病人数超过 4 亿，死亡人数接近 579 万[28]。我国 COVID-19 的临床诊治主要依据《新型冠状病毒肺炎诊疗方案》①，主要治疗手段包括对症治疗、抗病毒治疗、呼吸支持治疗、免疫治疗以及中医药治疗等[29]。由于 SARS-CoV-2 系新发病毒，目前尚无特效的抗病毒药物。

免疫治疗对于遏制炎症细胞的聚集性浸润、降低炎症因子水平和恢复免疫系统平衡发挥了重要作用。当前，免疫治疗方法主要包括针对炎症因子效应的抗炎症因子治疗、康复者恢复期血浆治疗、中和抗体治疗以及干细胞治疗等。

1. 抗炎症因子治疗

新冠肺炎患者体内的炎症风暴是导致病情转危、呼吸衰竭、多脏器功能障碍甚至死亡的重要原因。IL-6 是引发炎症风暴的重要炎症因子。针对 IL-6 的单克隆抗体可抑制这种免疫系统的过度反应，阻断炎症风暴，阻止患者向重症和危重症发展，从而降低病死率[30]。目前 IL-1 家族阻断剂、抗 IL-6 抗体等正处于临床试验阶段。基于临床试验效果，IL-6 阻断剂托珠单抗已经被我国国家卫生健康委员会写入了新版的COVID-19 诊疗方案中，用于治疗重症或危重症新冠肺炎患者。针对机体系统性免疫紊乱与耗竭，也有利用抗 PD-1 单克隆抗体等免疫检查点抑制剂来逆转 COVID-19 免疫耗竭的临床研究。

2. 康复者恢复期血浆和中和抗体治疗

康复者恢复期的血浆中含有大量特异性中和抗体，在缺乏疫苗和特效治疗药物的情况下，恢复期血浆是治疗新冠病毒感染最有效的方法。目前已开展了多项临床试

---

① 为做好新冠肺炎医疗救治工作，切实提高规范化、同质化诊疗水平，国家卫生健康委员会持续对该诊疗方案进行更新。截至 2022 年 3 月 15 日，已更新至试行第九版。

验，以评估恢复期血浆治疗新冠肺炎的有效性和安全性。初步数据显示，恢复期血浆对治疗新冠肺炎有潜在益处，可大幅降低危重症患者病死率，具有临床应用的潜力[31]。然而血浆成分复杂，是否会引起其他问题还存在不确定性；且其安全性存在隐患，在采集、筛查处理和使用的每一个过程都需要做好严格的质控。这些都限制了其大规模应用。

抗病毒中和抗体通过与 SARS-CoV-2 表面蛋白相结合，可有效阻止 SARS-CoV-2 感染细胞，已成为有效控制病毒复制、短期内迅速改善症状的重要手段。针对新冠病毒中和抗体药物研发的靶点集中在 S 蛋白的受体结合域，候选中和抗体大多数是从康复者血液中筛选出来，然后在体外进行大规模生产，再回输到病人体内，起到治疗的效果。多种中和抗体在临床试验阶段展示出较好的治疗效果[32]。与康复期血清相比，中和抗体通过基因改造的细胞表达，中和滴度高，成分单一，特异性和安全性良好，甚至对变异株也显示一定的中和效果。

3. 干细胞治疗

间充质干细胞不仅可以抑制炎症因子风暴，也可以在免疫力低下时增强人体免疫力，促进新冠病毒感染后肺部内源性修复和再生机制。间充质干细胞能够通过抑制淋巴细胞产生炎性细胞因子来调节免疫系统，并诱导产生抗炎细胞因子。由此可见，干细胞治疗可能为 COVID-19 患者提供一种独特的治疗选择。COVID-19 疫情暴发以来，国内外已有多个团队完成了干细胞治疗新冠肺炎及其后遗症的 I 期和 II 期临床试验，III 期临床试验正在进行中，其细胞来源包括脐带间充质干细胞、牙髓间充质干细胞及一些自体脂肪来源间充质干细胞。间充质干细胞凭借着免疫调节、抗炎作用、修复受损组织等特性，在 COVID-19 重症患者的救治中起到很好的效果，为提高治愈率、降低病死率带来了希望。现有结果表明，间充质干细胞治疗 COVID-19 安全性良好，在缩短病程、减轻肺部损伤、降低炎症因子水平等方面显示出较好的临床疗效[33,34]。除了间充质干细胞移植，间充质干细胞外泌体在治疗新冠肺炎的临床研究中也取得了良好的效果，在治疗重症新冠肺炎患者中成效显著，康复率高达 71%[35]。外泌体临床应用的安全性良好，有助于恢复氧合、降低炎性因子水平，并促进免疫系统重建，是一种很有前途的新冠肺炎治疗候选药物。相信在不久的将来，伴随药物研发、疫苗接种、健康防护等措施，新冠肺炎疫情一定能得到有效控制。

五、巨细胞病毒感染的免疫治疗

巨细胞病毒（cytomegalovirus，CMV）是一种常见的疱疹病毒组 DNA 病毒，在人群中感染率非常高，中国成人感染率达95%以上，通常呈隐性感染，多数感染者无临床症状。在机体的免疫功能极度低下的情况下，针对病毒的特异性细胞免疫也降低，机体无法清除感染或体内激活的病毒，导致病毒活动性感染和相关疾病的发生。病毒可侵入肺、肝、肾、唾液腺、乳腺及其他腺体，以及多核白细胞和淋巴细胞，可长期或间隙地自唾液、乳汁、汗液、血液、尿液、精液、子宫分泌物等多处排出病毒，是引起新生儿出生缺陷、异基因造血干细胞移植（allo-HSCT）后严重并发症的常见原因之一。

CMV特异性CTL细胞是控制CMV感染的关键免疫细胞，它们通过释放颗粒酶和穿孔素杀伤靶细胞，从而达到清除感染和杀伤转化细胞的作用。CTL细胞过继疗法是利用自身静脉血的淋巴细胞，在体外通过靶细胞抗原和淋巴因子的诱导，分化扩增成具有强大杀伤力的CTL细胞，再经静脉回输体内，从而有效地发挥免疫效应，达到清除病毒的目的。过继回输CMV特异性CTL细胞成为治疗造血干细胞移植后CMV感染的新策略，短暂的过继输注可促进体内持久抗CMV感染免疫重建[36]。另外，巨细胞病毒免疫球蛋白是一种CMV特异性多克隆免疫球蛋白制剂，可与CMV表面抗原结合，从而中和CMV，阻止其进入细胞，并具有呈递CMV颗粒的能力[37]。巨细胞病毒免疫球蛋白还可调节免疫细胞功能并与之相互作用，从而发挥积极的免疫平衡，抑制CMV复制。

六、EBV感染的免疫治疗

EB病毒（Epstein-Barr virus，EBV）是疱疹病毒科嗜淋巴细胞病毒属的成员。EBV的持续感染导致B细胞、内皮细胞和上皮细胞破坏，机体无法借助自身免疫功能或者抗病毒药物实现EBV的有效清除。EBV也是异体干细胞移植后感染的重要病原体之一。

细胞免疫治疗作为一种新型治疗方式，能够帮助机体建立抗病毒免疫反应，效率高且副作用少。体外扩增的病毒特异性的CTL细胞通过细胞表面的T细胞受体，识别那些被EBV感染或转化的异常细胞，通过释放颗粒酶和穿孔素杀伤靶细胞，从而达到清除感染和杀伤转化细胞的作用。EBV特异性CTL细胞已被证实为有效的治疗或预防EBV感染的方案[38]。产生EBV特异性CTL的方法先是采用EB病毒转染B细胞，生成EBV转化的B淋巴母细胞样细胞（EBV-lymphoblastoid cell line，EBV-LCL），以此作为抗原提呈细胞，然后再加以射线照射或丝裂霉素灭活，反复刺激外周血中的淋巴细胞，诱导扩增针对EBV特异性CTL。输注体外扩增的EBV特异性CTL被证明是安全和有效的治疗和预防方式。这些发现对同种异体干细胞移植受体的

EBV 免疫治疗具有潜在的意义。针对 EBV 的预防性和治疗性疫苗研究也正在如火如荼地进行中，希望能够有助于减少 EBV 感染和 EBV 相关性肿瘤的发生。

七、展　望

　　免疫治疗分为主动、被动和免疫调控三大方向。2018 年的诺贝尔生理学或医学奖授予了美国免疫学家詹姆斯·艾利森（James Allison）和日本免疫学家本庶佑，以表彰他们在发现负性免疫调节治疗癌症疗法方面的巨大贡献。研究显示针对 CTLA-4 和 PD-1 分子的"免疫检查点疗法"在一些特定肿瘤患者体内产生了良好的效果，根本性地改变了肿瘤患者的长期预后，证实免疫应答的方向可以借助外力改变。各类细胞工程性抗体、治疗性疫苗、体外扩增的自体或者异体的免疫细胞分别以免疫补充、免疫恢复的途径改变着免疫应答结局。免疫治疗具有多元化、多功能的特点，代表未来新的治疗技术和发展方向，拓宽了感染性疾病的治疗手段，也必将对现有传染病的诊疗模式产生重大影响。

参考文献

[1] WHO Media Centre. HIV/AIDS. http://www.who.in/media centre/factsheets/fs360/en/[2015-08-25].

[2] 中华医学会感染病学分会艾滋病学组．艾滋病诊疗指南（2011 版）．中华传染病杂志，2011，10：629-640.

[3] Finzi D，Hermankova M，Pierson T，et al. Identification of a reservoir for HIV-1 in patients on highly active antiretroviral therapy. Science，1997，278(5341)：1295-1300.

[4] Hutter G，Nowak D，Mossner M，et al. Long-term control of HIV by CCR5 Delta32/Delta32 stem-cell transplantation. The New England Journal of Medicine，2009，360(7)：692-698.

[5] Maldini C R，Claiborne D T，Okawa K，et al. Dual CD4-based CAR T cells with distinct costimulatory domains mitigate HIV pathogenesis *in vivo*. Nature medicine，2020，26(11)：1776-1787.

[6] Zeidan J，Sharma A A，Lee G，et al. Infusion of CCR5 gene-edited T cells allows immune reconstitution，HIV reservoir decay，and long-term virological control. bioRxiv，2021，2021.02.28.433290.

[7] Fan J，Liang H，Ji X，et al. CTL-mediated immunotherapy can suppress SHIV rebound in ART-free macaques. Nature Communications，2019，10(1)：2257.

[8] Wang L，Zhang Z，Xu R，et al. Human umbilical cord mesenchymal stem cell transfusion in immune non-responders with AIDS：A multicenter randomized controlled trial. Signal Transduction And Targeted Therapy，2021，6(1)：217.

[9] Group I-E S，Committee S S，Abrams D，et al. Interleukin-2 therapy in patients with HIV infection. The New England Journal of Medicine，2009，361(16)：1548-1559.

[10] Abela I A, Kadelka C, Trkola A. Correlates of broadly neutralizing antibody development. Current Opinion in HIV and AIDS, 2019, 14(4): 279-285.

[11] Henderson L J, Reoma L B, Kovacs J A, et al. Advances toward Curing HIV-1 Infection in Tissue Reservoirs. Journal of Virology, 2020, 94(3). https://doi.org/10.1128/JVI.00375-19.

[12] 王贵强, 王福生, 庄辉, 等. 慢性乙型肝炎防治指南(2019年版). 中华临床感染病杂志, 2019, 12(06): 401-428.

[13] Butterfield L H, Economou J S, Gamblin T C, et al. Alpha fetoprotein DNA prime and adenovirus boost immunization of two hepatocellular cancer patients. Journal Of Translational Medicine, 2014, 12: 86.

[14] Shimizu Y, Suzuki T, Yoshikawa T, et al. Cancer immunotherapy-targeted glypican-3 or neoantigens. Cancer science, 2018, 109(3): 531-541.

[15] Mizukoshi E, Nakagawa H, Kitahara M, et al. Phase I trial of multidrug resistance-associated protein 3-derived peptide in patients with hepatocellular carcinoma. Cancer Letters, 2015, 369(1): 242-249.

[16] El Ansary M, Mogawer S, Elhamid S A, et al. Immunotherapy by autologous dendritic cell vaccine in patients with advanced HCC. Journal Of Cancer Research And Clinical Oncology, 2013, 139(1): 39-48.

[17] Heo J, Reid T, Ruo L, et al. Randomized dose-finding clinical trial of oncolytic immunotherapeutic vaccinia JX-594 in liver cancer. Nature Medicine, 2013, 19(3): 329-336.

[18] Xu J, Zhang Y, Jia R, et al. Anti-PD-1 antibody SHR-1210 combined with Apatinib for advanced hepatocellular carcinoma, gastric, or esophagogastric junction cancer: An open-label, dose escalation and expansion study. Clinical Cancer Research, 2019, 25(2): 515-523.

[19] Schito M, Migliori G B, Fletcher H A, et al. Perspectives on advances in tuberculosis diagnostics, drugs, and vaccines. Clinical Infectious Diseases, 2015, 61(Suppl 3): S102-S118.

[20] 杨绪军. 中国结核病发病率为50/10万, 远低于全球平均水平. http://ie.bjd.com.cn/5b5fb98da0109f010fce6047/contentApp/5b5fb9d0e4b08630d8aef954/AP61728bebe4b08aed9d9c5af7.html [2021-01-20].

[21] Etemadi A, Farid R, Stanford J L. Immunotherapy for drug-resistant tuberculosis. Lancet, 1992, 340(8831): 1360-1361.

[22] Tiberi S, Du Plessis N, Walzl G, et al. Tuberculosis: Progress and advances in development of new drugs, treatment regimens, and host-directed therapies. The Lancet Infectious Diseases, 2018, 18(7): e183-e198.

[23] 杨晓敏, 董德琼, 杨渝浩, 等. IL-7和IL-2协同对肺结核患者Th1/Th2平衡调节作用的研究. 贵州医药, 2006, 30(04): 315-317.

[24] Schrager L K, Vekemens J, Drager N, et al. The status of tuberculosis vaccine development. The Lancet Infectious Diseases, 2020, 20(3): e28-e37.

[25] Skrahin A, Jenkins H E, Hurevich H, et al. Effectiveness of a novel cellular therapy to treat multi-

drug-resistant tuberculosis. International Journal of Mycobacteriology,2016,4:21-27.

[26] Salahuddin N,Ali F,Hasan Z,et al. Vitamin D accelerates clinical recovery from tuberculosis: results of the SUCCINCT Study [Supplementary Cholecalciferol in recovery from tuberculosis]: A randomized,placebo-controlled,clinical trial of vitamin D supplementation in patients with pulmonary tuberculosis. BMC infectious diseases,2013,13:22.

[27] 夏露,卢水华,刘平. 中药对结核病免疫调控机制的研究进展. 中国防痨杂志,2020,42(02): 168-172.

[28] Johns Hopkins Coronavirus Resource Center. COVID-19 Dashboard by the Center for Systems Science and Engineering(CSSE) at Johns Hopkins University(JHU). https://coronavirus. jhu. edu/map. html.

[29] 国家感染性疾病临床医学研究中心,中华医学会感染病学分会,中国研究型医院学会生物治疗学专业委员会. 间充质干细胞治疗新型冠状病毒肺炎专家共识(2021 年,北京). 传染病信息, 2021,34(02):99-106.

[30] Zhang C,Wu Z,Li J W,et al. Cytokine release syndrome in severe COVID-19:Interleukin-6 receptor antagonist tocilizumab may be the key to reduce mortality. International Journal of Antimicrobial Agents,2020,55(5):105954.

[31] Shen C,Wang Z,Zhao F,et al. Treatment of 5 critically ill patients with COVID-19 with Convalescent Plasma. The Journal of the American Medical Association,2020,323(16):1582-1589.

[32] Shi R,Shan C,Duan X,et al. A human neutralizing antibody targets the receptor-binding site of SARS-CoV-2. Nature,2020,584(7819):120-124.

[33] Meng F,Xu R,Wang S,et al. Human umbilical cord-derived mesenchymal stem cell therapy in patients with COVID-19:A phase 1 clinical trial. Signal Transduction And Targeted Therapy, 2020,5(1):172.

[34] Shi L,Huang H,Lu X,et al. Effect of human umbilical cord-derived mesenchymal stem cells on lung damage in severe COVID-19 patients:a randomized,double-blind,placebo-controlled phase 2 trial. Signal Transduction And Targeted Therapy,2021,6(1):58.

[35] Sengupta V,Sengupta S,Lazo A,et al. Exosomes derived from bone marrow mesenchymal stem cells as treatment for severe COVID-19. Stem Cells And Development,2020,29(12):747-754.

[36] Mous R,Savage P,Remmerswaal E B,et al. Redirection of CMV-specific CTL towards B-CLL via CD20-targeted HLA/CMV complexes. Leukemia,2006,20(6):1096-1102.

[37] Germer M,Herbener P,Schuttrumpf J. Functional properties of human cytomegalovirus hyperimmunoglobulin and standard immunoglobulin preparations. Annals of Transplantation, 2016, 21: 558-564.

[38] Prockop S,Doubrovina E,Suser S,et al. Off-the-shelf EBV-specific T cell immunotherapy for rituximab-refractory EBV-associated lymphoma following transplantation. The Journal of Clinical Investigation,2020,130(2):733-747.

Immune Treatment of Infectious Diseases

Xu Ruonan, Shi Ming, Wang Fusheng

Infectious diseases are caused by different pathogen infection and become one of the most common clinical diseases up to now. Accompanying the development of traditional drugs, the outcomes of infectious diseases have been improved, while the HBV, HCV, HIV infection are still the major medical burden to the public health. Immunotherapy aiming at regulation of the immune system has provided available options for the treatment of a variety of major infectious diseases, new outbreak of infectious diseases and common infectious diseases. Immunotherapy has the characteristics of diversification and multi-function, representing a new direction for new technology, which will broaden the choice for treatment, and will certainly have a significant impact on the diagnosis and treatment mode of infectious diseases.

第二章

科学前沿

Frontiers in Sciences

2.1 火星探测的科学研究进展

潘永信[1] 张荣桥[2] 王 赤[3] 李春来[4]

（1. 中国科学院地质与地球物理研究所；2. 国家航天局探月与航天工程中心；
3. 中国科学院国家空间科学中心；4. 中国科学院国家天文台）

2020 年 7 月，阿联酋的"希望号"、中国的"天问一号"、美国的"毅力号"火星探测器先后成功发射。2021 年，这些探测器相继成功抵达火星，火星探测迎来了新一轮热潮。"天问一号"任务在世界航天史上创下纪录，首次实现通过一次任务完成火星环绕、着陆和巡视探测三大目标。这次任务迈出了我国火星探测的第一步，也标志着我国深空探测实现了从地-月系统到行星际的跨越。

火星研究主要涉及以下三方面的重大科学问题：第一，类地行星的形成和演化。火星是离太阳第四近的行星，它记录了太阳系早期演化的历史。第二，行星宜居性。火星是与地球相邻的类地行星，通过火星和地球的比较研究，可更好地理解地球的过去、现今和未来演变趋势。第三，地外生命。过去和现在火星是否存在生命，一直是人们最为关注的问题。

火星探测始于 20 世纪 60 年代初，美苏之间的太空竞争引发了第一轮火星探测热潮，美国和苏联先后成功对火星实施了飞掠、环绕和着陆探测。90 年代后，随着苏联的解体，美国独自开启了第二轮火星探测的序幕，多个国家也相继加入了本轮火星探测热潮。迄今，人类已实施了 47 次火星探测任务。目前在火星周围和火星上运行的探测器达 12 个，包括 8 个环绕探测器、3 个火星车和 1 个着陆器，数量之多居地外行星之冠。

本文简要回顾人类探测火星历程，梳理火星探测的重要科学问题、取得的科学发现，以及最新研究进展。

一、国际火星探测的科学研究进展

1. 20 世纪 60～80 年代：火星表面基本特征

火星表面具有怎样的特征和组成？有没有类似地球的海洋、河流和湖泊？是否存在生命？在 20 世纪 60 年代以前，人类只能依赖地基天文望远镜的远程观测，无法获得火星表面的高分辨率图像。人造探测器近距离观测一举改变了这一状况，也开启了火星科学研究的序幕（图 1）。

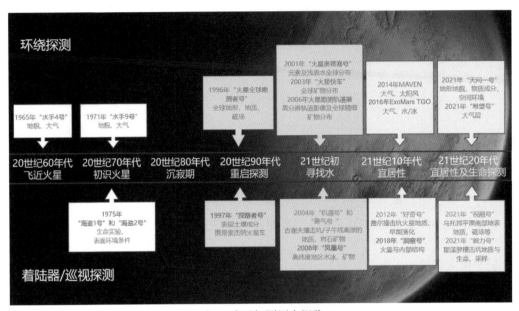

图 1　火星探测历史概览

图中蓝色、褐色和橙色字分别表示环绕器、着陆器和巡视车

1965 年，美国发射的"水手 4 号"（Mariner 4）探测器成功飞掠火星，发现火星大气十分稀薄，并向地球传回了首张火星的近距离照片，展示出火星表面遍布大大小小的撞击坑[1]。1971 年，美国"水手 9 号"（Mariner 9）探测器成功进入环绕火星轨道，成为第一颗人造火星卫星。"水手 9 号"探测器对火星表面 85% 以上的区域进行了遥感观测，发现了大量火山地貌、沙丘、古流水痕迹等，并测量了火星大气压和温度等。基于遥感观测结果，行星科学家绘制出第一张火星地质图[2]。1976 年，美国"海盗 1 号"（Viking 1）和"海盗 2 号"（Viking 2）探测器分别在火星北部克利斯平原（Chryse Planitia）和乌托邦平原（Utopia Planitia）成功着陆，首次在火星表面开展实地探测，获得了着陆区的地表特征、气象、物质组成等科学数据，发现着陆区荒芜、布满砾石[3]；着陆地点的原位生物学试验并没有发现复杂的有机物，没有获得生命存在的确切证据[4]。值得一提的是，包括苏联发射的福波斯（Phobos）在内的多个探测任务都曾尝试探测火星磁场，但均没有测量出准确的磁场特征[5]。

2. 1990～2010 年：火星上的水

水是地球生命之源，在地球上，生命活动离不开水。在火星上寻找水也成为寻找火星生命的一个重要目标。此外，水在维持地球环境宜居性中也发挥重要作用，除了在参与地表形貌改造、气候变化和生命过程之外，还参与深部岩浆活动、板块俯冲和地幔对流等诸多地球深部过程。基于此，火星探测的一项重大科学命题是在火星上寻找水。

1996 年，美国发射了"火星全球勘测者号"（Mars Global Surveyor，MGS）探测器，开展了火星的高精度全球高程测量和高分辨率成像，以及全球磁场测量。首次获得了高精度和高分辨率的全球地形地貌图[6]以及全球磁场分布图。根据地貌和地形高程特征，推断火星北半球曾经存在古海洋[7]。2001 年，美国"火星奥德赛号"（Mars Odyssey）环绕器携带的高能中子探测器，发现了与水活动相关的硫酸盐等矿物。2003 年欧洲空间局发射的"火星快车"（Mars Express，MEX）探测器和 2006 年美国发射的"火星勘测轨道器"（Mars Reconnaissance Orbiter，MRO），也先后揭示了火星两极冰层的沉积结构及其变化[8,9]，提示在极区附近的冰下可能存在液态水湖泊[10]；此外，还发现火星南半球中纬度区存在地下冰层的证据[11]。广泛分布的硫酸盐、绿脱石、绿泥石、蒙皂石、高岭石等矿物，以及少量的碳酸盐岩等，指示火星表面曾经有过强烈的水活动和次生改造作用[12-14]。

火星车在地面的实地探测则进一步证实了火星早期存在大量液态水形成的沉积环境[15]。2003 年，美国发射的"勇气号"（MER Spirit）火星车和"机遇号"（MER Opportunity）火星车，分别于 2004 年 1 月在火星南半球低纬地区古谢夫撞击坑（Gusev Crater）和子午线高原（Meridiani Planum）成功着陆，地面巡视探测证实了黄钾铁钒、蒙脱石、赤铁矿、碳酸盐等与水相关矿物的存在，发现了交错层理、结核、多边形构造、裂隙充填物、岩皮构造等大量液态水活动的证据[16]。2008 年降落于北半球高纬度地区的美国"凤凰号"着陆器（Phoenix Lander）探测器，原位观测也发现了高氯酸盐和碳酸盐沉积物的证据[17,18]。

结合撞击坑计数定年数据、遥感和就位探测的结果推断，火星在诺亚纪（距今 41 亿～37 亿年）曾出现过大量液态水，当时火星表面环境可能有利于生命存在，在西方纪（距今 37 亿～30 亿年）开始逐渐变干旱。研究人员推断，火星表面大量的液态水的逃逸可能是由于磁场保护的消失，以及火星重力束缚弱，易遭受太阳风的剥蚀所致[19]。最新计算模拟显示，相当多的早期大量液态水也可能通过岩石水化作用而进入火星地壳之中[20]。与地球比较而言，火星缺乏板块构造活动，全球地质活动的改造作用弱，因而它保存了太阳系行星早期演化的重要地质记录。

3. 2011~2020 年：火星宜居性及演化

宜居性是指行星具有或演化出适宜生命产生和繁衍的潜力，宜居环境则是指适宜生命起源和生存的环境。过去 10 年，火星探测的科学主题从寻找液态水，开始转向火星宜居性及演化的探索，即探寻火星宜居性演变规律和驱动要素，搜寻生命线索。

2012 年，美国的"好奇号"（Curiosity）火星车成功降落于火星南半球低纬地区古老的盖尔撞击坑（Gale Crater），该撞击坑的年龄为 35 亿~38 亿年[21]。"好奇号"火星车携带的科学载荷包括阿尔法粒子 X 射线分光计、化学与摄像机仪器（ChemCam）、化学与矿物学分析仪（CheMin）、火星样本分析仪、辐射评估探测器、火星车环境监测站等。在过去 9 年里，"好奇号"对盖尔撞击坑开展了十分详细的就位科学探测，在火星早期地质演化记录和宜居性研究方面，获得了一系列重要科学发现，包括发现诺亚纪-早西方纪古老冲积-湖泊沉积体系、地球化学元素和黏土矿物变化、泥岩中挥发分和有机质等，还首次探测到了现今地表季节性甲烷气体变化[22-31]。2021 年 2 月 19 日，美国"毅力号"火星巡视车成功着陆于火星耶泽罗撞击坑（Jezero Crater）内，这里是一个诺亚纪河流入湖三角洲沉积地质单元。"毅力号"将开展早期火星宜居性研究，搜寻古老生命线索[32]。"毅力号"首次实施火星岩石土壤样本的采集工作，并将它们密封在特制的样品管中，计划在 2030 年左右送回地球供科学家研究。

火星大气和挥发分演化是宜居性研究的另一项重要内容。现今火星的大气密度仅为地球大气的 6%~7%，大气中二氧化碳占 96%，氩气和氮气各占 1.9%，其他成分占 0.2%。2013 年美国发射了专门用于研究火星大气和挥发分的探测器 MAVEN（Mars Atmosphere and Volatile Evolution），并开展了电离层和磁层探测。2016 年，欧洲空间局与俄罗斯航空局还联合发射了火星探测计划-痕量气体环绕探测器（Exo-Mars TGO），主要用于分析火星高层大气逃逸、太阳风能量粒子、磁层动力学过程、电离层性质、地下水冰等及其变化。该探测器观测到了火星大气中的粒子逃逸、拾取和沉降、漫射极光、轨道高度的沙尘等现象[33,34]，首次发现火星沙尘可以将地表水汽带至高层大气而加剧水丢失[35]，这加深了对于火星大气和水如何逃逸的认识。2021年，专门用于研究火星大气动力学的阿联酋的"希望号"环绕器已抵达火星，开展火星低层和高层大气的循环过程和火星天气观测研究。

二、我国首次火星探测任务"天问一号"的科学研究进展

1. "天问一号"的科学目标、任务和载荷配置

我国首次火星探测任务也是围绕火星地质、环境和宜居性等前沿重大科学问题而展开。"天问一号"火星探测任务的科学目标包括：①火星形貌与地质构造特征；②火星表面土壤特征与水冰分布；③火星表面物质组成；④火星大气、电离层及表面环境特征；⑤火星物理场与内部结构[36-38]。"天问一号"火星探测器由环绕器和着陆巡视器组成，着陆巡视器包含进入舱和火星车[39]，环绕器开展火星全球性多手段遥感详查探测。"天问一号"的主要探测任务包括：大气电离层及行星际环境、表面和地下水冰、火壤分布和结构探测、地形地貌特征及其变化、表面物质成分的调查和分析。为此，环绕器配置有中分辨率相机、高分辨率相机、环绕器次表层探测雷达、矿物光谱分析仪、磁强计、离子与中性粒子分析仪、能量粒子分析仪总共7套有效科学载荷。"祝融号"火星车开展巡视区的多手段精细探测，主要探测任务包括：地形地貌和地质构造，火壤结构（剖面）和地下水冰探查，表层物质的元素、矿物和岩石类型，磁场和气象探测等。为此，火星车配置了表面成分探测仪、多光谱相机、地形相机、车载次表层探测雷达、表面磁场探测仪、火星气象测量仪总共6套有效载荷[40]。

2. "天问一号"火星探测器的阶段性进展

"天问一号"探测器于2020年7月23日发射；2021年2月10日进入环火轨道，对着陆区开展科学探测；5月15日实施着陆巡视器和环绕器分离，并实现着陆巡视器软着陆于乌托邦平原南部预选着陆区（109.925°E，25.066°N，高程−4099.4m）；5月22日，"祝融号"火星车驶达火星表面；8月15日，"祝融号"完成了90个火星日（约92个地球日）的巡视探测任务（图2），获取了巡视区域地形地貌、雷达、磁场、气象，以及特定岩石、土壤典型目标的光谱数据等约10 GB的原始数据。之后，"祝融号"继续开展拓展巡视探测，截至2022年5月7日累计行驶1921 m。环绕器已开展火星全球遥感探测，同时为火星车提供中继通信。

"天问一号"所有科学载荷获取的数据，经过地面应用系统科研人员相关处理和质量验证，制作成标准的数据产品后，通过中国探月与深空探测网行星探测工程科学数据发布系统（http://www.clep.org.cn/）、月球与行星数据发布系统（https://clpds.bao.ac.cn），以月为周期面向科学研究团队开放数据申请。为了高效和高水平

图 2 火星巡视车在乌托邦平原南部的行驶路径（a）、火星车与着陆器合影（b）、
周围的火壤物质和沙丘地貌照片（c）

资料来源：国家航天局．http://www.clep.org.cn/n6189350/n6760313/index.html

推进"天问一号"数据科学研究，首次火星探测任务总师系统和科学研究专家委员会
精心组织科学家队伍，开展科学研究以及学术研讨。同时，中国科学院等多家单位已
部署项目，用于支持"天问一号"火星科学数据研究。来自中国科学院以及高校和行
业的数十家研究机构的科学家已牵头申请了"天问一号"所采集的数据，并开展相关
科学研究，分析乌托邦平原南部着陆巡视区的地形地貌、水冰作用、地下分层结构和
磁场、空间物理等科学问题，已取得初步科学研究成果[41]。

三、2021年重要进展及未来研究展望

1. 科学研究进展

2021 年，是火星探测历史上不寻常的一年，在科学研究和探测方面均取得了不少
收获。

在科学研究方面，一是获得了火星内部结构新认识。行星地球物理学家基于 2018

年美国发射的"洞察号"（InSight）着陆探测器获得的火震数据，揭示出了火星具有薄的壳（20 km 或 37 km）[42]、较冷的幔（约 1530 km）[43] 和较大的熔融核（1830 km）[44]，这是科学家第一次利用火震数据获得火星的内部结构，也标志着火星内部结构演化研究向前迈出了重要一步（图 3）。二是在火星天气、气候和环境研究等方面取得了新进展，建立了火星高时空分辨率的全球大气数据库，为研究年际、季节和日变化的火星气候变化（包括沙尘、水冰云、气温等）以及完善全球大气环流模型提供了重要依据[45]；利用 ExoMars TGO 环绕器获得的大量 10～120 km 高度大气一氧化碳（CO）实测数据[46]，观测到火星全球尺度上氘原子（D）和氢原子（H）的比值（D/H 比值）随高度变化[47]，获得火星大气循环的新认识。2021 年 12 月 6 日，ExoMars TGO 环绕器荷载的高精度超热中子探测仪在水手峡谷（Valles Marineris）的观测显示，地下可能有大量水的证据，主要以冰或者含水矿物的形式存在。

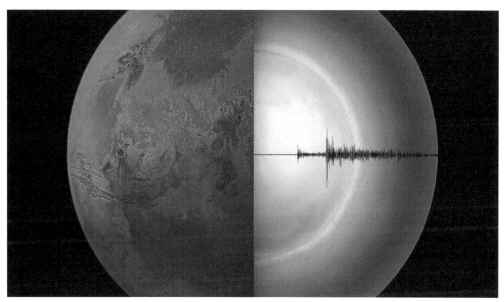

图 3　"洞察号"地震观测揭示的火星内部结构示意图，从内到外分别为火星的核、幔和壳

在探测方面，"希望号"和"天问一号"这两个环绕探测器均已开始科学探测，大大增强了火星大气和空间环境探测能力，"毅力号"和"祝融号"火星车也分别开启了对耶泽罗撞击坑古湖泊三角洲沉积区和乌托邦平原南部盆地的巡视科学探测。作为美国国家航空航天局和欧洲空间局联合制订的 2031 年火星采样返回计划的第一步，"毅力号"已开始在耶泽罗撞击坑钻取地质样品，这是人类首次在火星上开展采样工作，样品预计在 10 年后取回地球。此外，"毅力号"还首次实现了在火星上无人直升

机飞行，获得原位利用二氧化碳制备氧气试验的成功。

2. 研究展望

火星探测的重点已从寻找水的踪迹转变为地外生命探寻。火星宜居性研究和生命信号搜寻依然是未来火星探测的前沿。目前在火星上有多台探测器运行，包括在轨的Mars Odyssey、MEX、MRO、Mangalyaan（印度）、MAVEN、ExoMars TGO、"天问一号"等环绕探测器，地面的"好奇号""毅力号""祝融号"火星车和"洞察号"着陆探测器，具备了天-地联合、多点组网观测能力，正在催生火星立体探测新模式。火星科学研究已进入新发展阶段。2022年有欧洲空间局和俄罗斯联邦航天局联合研制的ExoMars-2022号火星探测器发射，预计2023年抵达火星开展探测。

近20年所获得的火星环境连续观测资料，使年际、季节和日尺度环境和气候变化规律研究已成为可能。在未来几年，有望在火星气候演变、甲烷气体来源、极地冰在火星气候演化中的作用、中低纬度地下水/冰探测、大型沟壑成因、生命线索等火星宜居性研究的若干科学重要问题方面取得突破。

行星内部是行星动力学演化的引擎，也是制约行星宜居性演变的重要因素。虽然内部研究面临很大挑战，但是"洞察号"的最新科学进展令人鼓舞，期待未来在火震及内部结构研究方面有更多探测和科学发现。火星内部发电机（即火星核产生磁场的磁流体动力学过程）停止时间、驱动机制是待解决的重大科学问题。"祝融号"火星车正在获取乌托邦平原南部表面磁场数据，这是人类首次在火星表面开展磁场高精度测量。未来几年，结合在轨探测器和在地面着陆器和巡视车的精确磁场测量，在火星内部发电机研究方面有望取得突破。

随着数据量和复杂度增加，大数据人工智能和机器学习将在火星科学研究领域发挥越来越重要作用。另外，地球上的火星类比环境研究、实验室模拟实验及天体生物学研究等，正在逐步成为火星科学研究新的生长点。

最为令人期待是火星采样返回任务。美国和中国都制订了未来10年火星采样返回计划，"毅力号"已在耶泽罗撞击坑实施样品采集。可以预料，像采样返回为月球研究带来的重大突破一样，火星采样返回任务的实施将为火星研究带来科学和技术的革命。

我国火星科学研究将进入快速发展时期，为此提出以下建议。

（1）以我国行星科学学科建设为契机[48-50]，以"天问一号"首次火星探测任务为抓手，加快行星科学的优秀青年人才培养和队伍建设，及早补齐科学研究的短板。

（2）以火星宜居性及演化重大科学问题理论创新和火星采样返回科学研究为导

向，加大专门研究经费支持力度，倾斜支持科学目标牵引下的科学载荷研发，提升重大科学发现和原始创新能力。

（3）加强火星科学研究的国际合作、科普文化建设，共建开放、合作的健康学术生态，形成创新研究新局面。

希望我们能够坚持科学与技术双轮驱动，不断获得原始科学发现、做出重要创新性成果，更好地服务国家深空战略发展需求，为人类文明进步贡献智慧和力量。

参考文献

[1] Leighton R B, Murray B C, Sharp R P, et al. Mariner 4 photography of Mars-initial results. Science, 1965, 149(3684):627-630.

[2] Carr M H, Masursky H, Saunders R S. Generalized geologic map of Mars. Journal of Geophysical Research, 1973, 78(20):4031-4036.

[3] Mutch T A, Binder A B, Huck F O, et al. Surface of Mars-view from Viking-1 lander. Science, 1976, 193(4255):791-801.

[4] Biemann K, Oro J, Toulmin P, et al. Search for organic and volatile inorganic-compounds in 2 surface samples from Chryse-Planitia region of Mars. Science, 1976, 194(4260):72-76.

[5] Riedler W, Mohlmann D, Oraevsky V N, et al. Magnetic-fields near Mars: First results. Nature, 1989, 341(6243):604-607.

[6] Smith D E, Zuber M T, Solomon S C, et al. The global topography of Mars and implications for surface evolution. Science, 1999, 284(5419):1495-1503.

[7] Head J W, Hiesinger H, Ivanov M A, et al. Possible ancient oceans on Mars: Evidence from Mars Orbiter Laser Altimeter data. Science, 1999, 286(5447):2134-2137.

[8] Seu R, Phillips R J, Alberti G, et al. Accumulation and erosion of Mars' south polar layered deposits. Science, 2007, 317(5845):1715-1718.

[9] Phillips R J, Zuber M T, Smrekar S E, et al. Mars north polar deposits: Stratigraphy, age, and geodynamical response. Science, 2008, 320(5880):1182-1185.

[10] Orosei R, Lauro S E, Pettinelli E, et al. Radar evidence of subglacial liquid water on Mars. Science, 2018, 361(6401):490-493.

[11] Holt J W, Safaeinili A, Plaut J J, et al. Radarsounding evidence for buried glaciers in the southern mid-latitudes of Mars. Science, 2008, 322(5905):1235-1238.

[12] Ehlmann B L, Mustard J F, Murchie S L, et al. Orbitalidentification of carbonate-bearing rocks on Mars. Science, 2008, 322(5909):1828-1832.

[13] Mustard J F, Murchie S L, Pelkey S M, et al. Hydrated silicate minerals on mars observed by the Mars reconnaissance orbiter CRISM instrument. Nature, 2008, 454(7202):305-309.

[14] Wray J J, Murchie S L, Squyres S W, et al. Diverse aqueous environments on ancient Mars revealed in the southern highlands. Geology, 2009, 37(11):1043-1046.

［15］ Ehlmann B L,Edwards C S. Mineralogy of the Martian surface. Annual Review of Earth and Planetary Sciences,2014,42:291-315.

［16］ Squyres S W,Knoll A H,Arvidson R E,et al. Two years at Meridiani Planum:Results from the Opportunity Rover. Science,2006,313(5792):1403-1407.

［17］ Hecht M H,Kounaves S P,Quinn R C,et al. Detection ofperchlorate and the soluble chemistry of Martian soil at the Phoenix lander site. Science,2009,325(5936):64-67.

［18］ Boynton W V,Ming D W,Kounaves S P,et al. Evidence forcalcium carbonate at the Mars Phoenix landing site. Science,2009,325(5936):61-64.

［19］ Villanueva G L,Mumma M J,Novak R E,et al. Strong water isotopic anomalies in the martian atmosphere:Probing current and ancient reservoirs. Science,2015,348(6231):218-221.

［20］ Scheller E L,Ehlmann B L,Hu R,et al. Long-term drying of Mars by sequestration of ocean-scale volumes of water in the crust. Science,2021,372(6537):56-62.

［21］ Hartmann W K, Neukum G. Cratering chronology and the evolution of Mars. Space Science Reviews,2001,96(1-4):165-194.

［22］ Bristow T F,Rampe E B,Achilles C N,et al. Clay mineral diversity and abundance in sedimentary rocks of Gale crater,Mars. Science Advances,2018,4(6):eaar3330.

［23］ Eigenbrode J L,Summons R E,Steele A,et al. Organic matter preserved in 3-billion-year-old mudstones at Gale crater,Mars. Science,2018,360(6393):1096-1100.

［24］ Freissinet C,Glavin D P,Mahaffy P R,et al. Organic molecules in the Sheepbed Mudstone,Gale Crater,Mars. Journal of Geophysical Research-Planets,2015,120(3):495-514.

［25］ Grotzinger J P,Gupta S,Malin M C,et al. Deposition,exhumation,and paleoclimate of an ancient lake deposit,Gale crater,Mars. Science,2015,350(6257):AAC7575.

［26］ Grotzinger J P, Sumner D Y, Kah L C, et al. A habitable fluvio-lacustrine environment at Yellowknife Bay,Gale Crater,Mars. Science,2014,343(6169):1242777.

［27］ McLennan S M,Anderson R B,Bell J F,et al. Elemental geochemistry of sedimentary rocks at Yellowknife Bay,Gale Crater,Mars. Science,2014,343(6169):1244734.

［28］ Ming D W,Archer P D,Glavin D P,et al. Volatile and organic compositions of sedimentary rocks in Yellowknife Bay,Gale Crater,Mars. Science,2014,343(6169):1245267.

［29］ Moores J E,Gough R V,Martinez G M,et al. Methane seasonal cycle at Gale Crater on Mars consistent with regolith adsorption and diffusion. Nature Geoscience,2019,12(5):321-325.

［30］ Webster C R, Mahaffy P R, Atreya S K, et al. Mars methane detection and variability at Galecrater. Science,2015,347(6220):415-417.

［31］ Webster C R,Mahaffy P R,Atreya S K,et al. Background levels of methane in Mars' atmosphere show strong seasonal variations. Science,2018,360(6393):1093-1096.

［32］ 赵宇鴳. 寻找火星远古水世界的潜在生命线索:耶泽罗探测记(下). 科学新闻,2021,23(5):30-33.

[33] Andersson L, Weber T D, Malaspina D, et al. Dust observations at orbital altitudes surrounding Mars. Science, 2015, 350(6261): aad0398.

[34] Schneider N M, Deighan J I, Jain S K, et al. Discovery of diffuse aurora on Mars. Science, 2015, 350(6261): aad0313.

[35] Fedorova A A, Montmessin F, Korablev O, et al. Stormy water on Mars: The distribution and saturation of atmospheric water during the dusty season. Science, 2020, 367(6475): 297-300.

[36] Wan W X, Wang C, Li C L, et al. China's first mission to Mars. Nature Astronomy, 2020, 4(7): 721.

[37] Li C L, Zhang R Q, Yu D Y, et al. China's Mars exploration mission and science investigation. Space Science Reviews, 2021, 217(4): 57.

[38] 欧阳自远, 邹永廖. 火星科学概论. 上海: 上海科技教育出版社, 2015.

[39] Ye P J, Sun Z Z, Rao W, et al. Mission overview and key technologies of the first Mars probe of China. Science China: Technological Sciences, 2017, 60(5): 649-657.

[40] Zou Y L, Zhu Y, Bai Y F, et al. Scientific objectives and payloads of Tianwen-1, China's first Mars exploration mission. Advances in Space Research, 2021, 67(2): 812-823.

[41] Liu J J, Li C L, Zhang R Q, et al. Geomorphic contexts and science focus of the Zhurong landing site on Mars. Nature Astronomy, 2022, 6: 65-71.

[42] Knapmeyer-Endrun B, Panning M P, Bissig F, et al. Thickness and structure of the martian crust from InSight seismic data. Science, 2021, 373(6553): 438-443.

[43] Khan A, Ceylan S, van Driel M, et al. Upper mantle structure of Mars from InSight seismic data. Science, 2021, 373(6553): 434-438.

[44] Stahler S C, Khan A, Banerdt W B, et al. Seismic detection of the martian core. Science, 2021, 373(6553): 443-448.

[45] Giuranna M, Wolkenberg P, Grassi D, et al. The current weather and climate of Mars: 12 years of atmospheric monitoring by the Planetary Fourier Spectrometer on Mars Express. Icarus, 2021, 353(2021): 113406.

[46] Olsen K S, Lefevre F, Montmessin F, et al. The vertical structure of CO in the Martian atmosphere from the ExoMars Trace Gas Orbiter. Nature Geoscience, 2021, 14(2): 67-71.

[47] Villanueva G L, Liuzzi G, Crismani M M J, et al. Water heavily fractionated as it ascends on Mars as revealed by ExoMars/NOMAD. Science Advances, 2021, 7(7): eabc8843.

[48] 万卫星, 魏勇, 郭正堂, 等. 从深空探测大国迈向行星科学强国. 中国科学院院刊, 2019, 34(7): 748-755.

[49] 魏勇, 朱日祥. 行星科学前沿与国家战略. 中国科学院院刊, 2019, 34(7): 749-756.

[50] 潘永信, 王赤. 国家深空探测战略可持续发展需求: 行星科学研究. 中国科学基金, 2021, (2): 181-185.

Science Advances in Mars Exploration

Pan Yongxin, Zhang Rongqiao, Wang Chi, Li Chunlai

Martian habitability as well as searching for life is the main science theme of Mars exploration over the last decade. As a neighbor planet, the red planet's exploration will be beneficial to better understanding of our home planet. The year of 2021 is an unusually busy year for the Mars exploration. The Tianwen-1 mission from China, the Mars Hope Probe from the United Arab Emirates, and the Mars 2020 Perseverance from the United States of America started their scientific exploration activities on Mars, respectively. The Tianwen-1's Zhurong rover and orbiter start investigations of the Martian surface and subsurface geology, magnetic field and space environments and so forth. The Hope probe is probing the Martian atmosphere and its layers. The Perseverance rover has conducted by far the first sampling on Mars, as the first step towards a Mars sample return mission planned jointly by NASA and ESA. In the coming years, possible multiple probes' joint scientific observation and inclusive international cooperation will enhance the ability and efficiency of Mars exploration. Here we briefly overview the main scientific questions, major research advances and perspectives of Mars exploration.

2.2　超快电子源的研究进展与展望

戴　庆　李　驰

（国家纳米科学中心）

电真空器件在国防、民用和科研装备中扮演着"发动机"的角色，发挥着不可替代的作用。电子源是电真空器件的"心脏"，为电真空器件的运行提供具有特定能量、频率、直径的自由电子束。因此，电子源的性能决定了电真空器件的最高工作频率和最快响应速度。新一代关键装备，例如，高数据率通信、超快电子显微镜（简称超快电镜）、自由电子激光器、深空通信等，要求电真空器件工作频率达到太赫兹（0.1～3 THz）水平。因此，传统电子源技术的纳秒（ns，即 10^{-9} s）级响应速度已经不能满足要求。达到飞秒（fs，即 10^{-15} s）级响应速度的超快电子源成为近年来美欧等发达国家和地区竞相布局抢占的战略高地（图 1）。例如，美国能源部（DOE）发布报告《电子源的未来》，明确了未来 X 射线自由电子激光[1]（X ray free electron laser，XFEL）、超快电子衍射[2]（ultrafast electron diffraction，UED）等装备对超快电子源的重大需求。1999 年度诺贝尔化学奖获得者 Ahmed H. Zewail 等撰写的著作指出超快电子源性能决定了超快电子显微镜（ultrafast electron microscope，UEM）的时空分辨率[3]。美国国防部高级研究计划局（DARPA）发布报告，明确了军用高数据率通信装备采用的真空电子管对超快电子源的重大需求[4]。德国在超快电子源基础研究方面起步较早，并在 UEM 领域的应用方面取得了领跑地位[5,6]。瑞士则重点布局了用于自由电子激光器的超快电子源[7]。

鉴于其广泛而重要的应用前景，超快电子源研究迅速成为材料学、真空电子学、纳米光子学和激光强场物理等多学科交叉融合的前沿热点领域，相关理论模型和表征技术正处于起步阶段。德国和美国相关团队曾利用金属纳米结构在激光的辐照下初步实现了超快（飞秒量级）电子源，但存在能量散度大、电流密度低等性能瓶颈[8,9]。提升电子源性能的关键在于优化电子源材料与结构，不断涌现的新型纳米材料为解决上述挑战提供了新的机遇。本文主要综述超快电子源的研究现状及共性关键科学问题，探讨超快电子源的未来发展方向和新技术路线，以推动我国在高频电子器件领域实现科学和技术的重大创新突破。

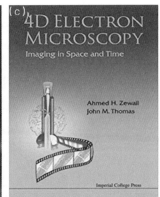
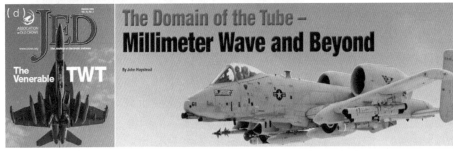

图 1　近年来发布的关于超快电子源及应用的报告

（a）和（b）为 DOE 发布的关于自由电子激光和超快电子衍射的报告；（c）为 Ahmed H. Zewail 等撰写的关于超快电镜的著作；（d）为 DARPA 发布的关于利用超快电子源构筑毫米波器件的报告

一、超快电子源的研究现状

由于电子发射材料的不同，产生电子束的方法有很多，例如热致电子发射、场致电子发射、光致电子发射等。热致电子发射主要是利用电流加热钨灯丝至 2000 ℃ 以上高温，使电子获得足够能量越过真空势垒形成电子发射。热致电子发射作为最早实用化的电子源技术，虽然过去了一百多年，却仍广泛应用于现代电真空装备。虽然热电子源可以提供大功率输出，但由于存在阴极热惯性，其响应速度无法满足 100 GHz 以上的工作频率的要求。场致电子发射主要是利用强电场压缩材料表面真空势垒实现的隧穿发射[10]，同时具有瞬时响应和发散角小的特征，且在发射电流密度、寿命等方面具有明显优势，因此成为 DOE 和 DARPA 重点布局的实现超快电子源的技术路线。在过去的一个世纪里，场致发射电子源电子束发散角小的优势得到了充分利用，推动了高空间分辨电子显微镜、微焦点 X 射线源等装备的巨大发展。然而，其瞬时响应的优势并没有得到充分发挥。现有商业化的场发射装备采用高频电场驱动的方式仅能达到纳秒量级响应速度，无法满足新一代高频电子装备飞秒甚至阿秒（as，即 10^{-18} s）量级响应时间的要求[11]。考虑到电子隧穿的时间尺度在阿秒量级，电子源的响应时间主要由外场调频率的制约，因此必须发展新型超快外场调控技术。

　　近年来发展起来的超快强场激光技术理论上可以突破上述瓶颈问题。采用强激光的光频电场来驱动电子发射是目前已知能够实现更高速率电子发射的唯一技术路线。以超快电子显微镜为例，超快电子源的工作原理如图 2 所示。超快电子源要求电子束要同时具有窄脉宽、高空间相干和高时间相干三个特征[12]。其中，窄脉宽要求电子束发射持续时间短（亚百飞秒）；高空间相干要求电子束发散角度小（小于 5°）；高时间相干要求电子束具有较低的能量散度（小于 0.5 eV）。根据海森伯不确定性原理不难发现，电子束这三个性能特征相互影响，相互制约。虽然可在后期传输过程中通过准直器、单色器等复杂系统来调控电子束的这三个特性，但会大幅牺牲电子束流强度且调控效率有限。如果能通过材料设计调控电子发射行为，直接产生符合要求的电子脉冲，可显著提升电子束的品质并提高器件设计的集成度[13]。

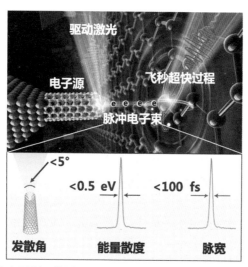

图 2　超快电子源工作原理（以超快电子显微镜为例）及其性能要求

　　目前超快激光驱动的电子源主要有两种工作机制：多光子发射和光场发射。多光子发射是指用强激光照射后，材料表面的电子吸收多个光子获得足够能量克服功函数而逸出[14]。光场发射是利用入射激光的强光场压缩材料表面真空势垒而实现的电子隧穿发射[10]。在光子驱动机制下，对于连续电子结构发射材料（如目前美国和德国主要研究的金属纳米结构）来说，决定能量散度的因素主要是激光光子能量和材料功函数的匹配度[11]。就单色激光来说，很容易通过调整光子能量与材料功函数相匹配来提高单色性。因此，光子驱动是目前常用的激发机制[10]。然而，单色激光的脉宽通常在100 fs以上。由不确定性原理可知，如果想进一步缩短激光脉宽，光子能量将发生严重色散[13]。比如，6 fs的脉冲激光具有约0.8 eV的光子能量散度。而电子的能量散度将在多光子吸收过程中进一步放大。所以，在光子驱动机制下，窄脉宽和低能散是两个相互制约的电子脉冲参数。与光子驱动机制相比，光场驱动的电子脉冲发射与光脉冲相位同步，因此具有阿秒时间分辨尺度，且受光场偏振准直，近年来成为实现高相干超快电子源的另一重要路径[9]。然而，光场驱动机制下，电子在隧穿之后将经历强烈的尾场加速效应，最终导致超大的电子能散（>30 eV）[8]，难以实现时间相干。因此，超快电子源领域的一个关键科学问题就是如何通过发射材料的电子结构设计，实现对电子电离和尾场加速过程的精细控制，进而降低电子能散。

二、碳基相干电子源的研究进展

想要突破上述技术瓶颈，必须发展低维材料超快电子源，利用其几何尺寸优势[14]和能带结构特点[13]实现高相干超快电子发射。在众多的低维电子发射材料当中，碳纳米管（carbon nano-tube，CNT）具有无可比拟的机械强度、长径比、导电性等性能[16]，在超快强激光激发下可以提供稳定的电子发射。同时，碳纳米管量子化的电子发射能级不仅可以提高电子发射的量子效率，还可以限制电子发射的能量展宽，保证较高的时间相干性（单色性）[17]。此外，碳纳米管的一维结构将电子传输路径限制在碳管轴向上，因此可以极大地抑制电子的发散角，获得高空间相干[18]。由此可见，碳纳米管是一种极具潜力的超快电子源材料。

近期研究表明，采用优化制备的尖端半径1 nm的碳纳米管作为光场发射材料，可以在400 nm波段实现超快强场电子隧穿，并获得超低能量散度电子（0.25 eV），不到金属纳米结构的光场发射电子源的1/10，能够满足亚纳米电子显微表征对电子束能散的要求（<0.5 eV）[19]。此外，一维碳纳米管离散的电子能带结构为光场发射电流难以实现相位调制的问题提供了新思路。进一步研究表明，碳纳米管超快电子发射展现出一种极端非线性光电子发射行为[18]。半导体碳纳米管价带顶具有范霍夫奇点，其电子密度远高于导带电子。随着场强增强，来自价带顶的电子在发射电流中将占主导，并出现极端非线性发射电流。利用这种极端非线性效应，碳纳米管光场发射电流具有显著的载波包络相位调制效应，其调制深度达到了100%（是金属的5倍），总调制电流高达2 nA（比金属高两个数量级）。因此，利用碳纳米管作为超快电子源，有望将超快电子源的时间分辨能力推进到亚飞秒量级。

综上所述，利用碳纳米管作为超快电子源，有望同时满足低能量散度，发散角度小和脉宽窄这三个要求。表 1 系统总结了碳纳米管的性能优势。

表 1 光场发射材料性能对比

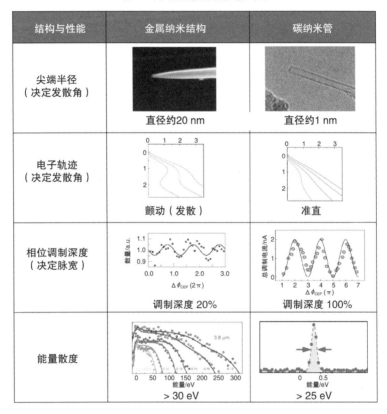

结构与性能	金属纳米结构	碳纳米管
尖端半径 （决定发散角）	直径约 20 nm	直径约 1 nm
电子轨迹 （决定发散角）	颤动（发散）	准直
相位调制深度 （决定脉宽）	调制深度 20%	调制深度 100%
能量散度	> 30 eV	> 25 eV

三、碳基超快电子源研究的未来展望

前期研究工作已经证明，碳纳米管是一种极具潜力的高性能超快电子源材料。然而，其研究也存在着一系列关键难点。

1. 如何进一步完善碳基纳米材料的超快电子发射理论

前期关于超快电子源的研究主要是基于金属纳米结构，其相对简单的电子结构（仅考虑费米面附近的电子发射）有助于研究人员快速建立超快电子发射的物理图像[19]。然而，目前对于超快电子发射全过程的理解并不全面——相较于电子被电离之后的动力学过程，对于电子在碳基纳米材料中的强场电离过程的理解仍处于初级阶段，而后者同样对超快电子发射行为具有决定性作用。但是，碳基纳米材料强场电离过程物理图像复杂，多种机制相互耦合，目前还没有可靠有效的理论计算方法来模拟电子电离过程中材料的电子结构和原子结构对电子渡越、隧穿过程的影响，以及纳米尺度下电子-电子、电子-声子散射等诸多关联因素[22-24]。

为此，下一阶段碳基超快电子源研究要从基础理论出发，发展处理碳基纳米材料与强光场相互作用的第一原理计算方法[25]，从阿秒尺度量子动力学的角度全面、细致地研究典型量子材料对强光场的响应[20]。同时，相关理论研究要建立在实验研究的基础上，可以考虑通过系统研究极端环境（极低温）或条件（超短脉冲）下的超快电子发射行为，来解耦低维体系中的多体相互作用[26]。

2. 如何实现碳基超快电子源材料结构的优化设计

碳纳米管手性结构复杂，难以实现单一手性碳纳米管控制，而不同手性碳纳米管的发射性能之间存在较大差别[25]，这严重制约了碳纳米管在大功率场发射阵列中的应用。同时研究表明，碳纳米管尖端的原子排列对其场发射性能有较大影响，具有完美晶体结构的碳纳米管具有最高的强度、韧性和导电性，但要获得完美晶体结构需要对碳纳米管的生长过程进行精确控制。北京大学和清华大学的研究团队通过调整工艺参数实现对生长过程的精确控制，最终实现了近单一手性超长碳纳米管阵列的合成[28,29]。相关工作为超快电子源的优化设计提供了重要的基础。同时，零维碳材料也为超快电子源的设计提供了更多可能，例如，富勒烯作为电子源比碳纳米管更早进入人们视野[30]。厦门大学研究团队目前已经可以精确灵活地合成富勒烯的同素异形体，甚至可以合成富勒管[31]。这些材料在超快电子源中展现出更加广阔的应用前景。

3. 如何实现碳基超快电子源性能的精确测量

高相干超快电子源在时间和空间维度上都已逼近物理极限，传统的测量表征设备或方法已经不能满足要求。因此，在实验测量方面，需要在对超快电子的发射机制深入理解的基础上，进一步发展在时域与空间域上对光场同时进行调控实现阿秒时间-纳米空间尺度电子源的新方法，研制用于微纳固体超快光电子发射空间和速度分布测量的新型装备[32]，为表征电子发射特性提供新手段。此外，超快电子束的测量与表征还可以借助超快扫描隧道显微镜等新型装备，可在表征发射材料电子结构的同时，实现超快电子发射的调控与探测，最终获得超快电子发射行为与材料电子结构的依赖关系[31]。相关设备的研制将为电子层面研究固体微观动力学等凝聚态物理交叉学科提供崭新的研究手段。

4. 碳基超快电子源的前沿应用

图 3 展示了超快电子源的重要前沿应用领域。

（1）超快电子源是超快电镜的"心脏"，决定了其成像的时间和空间分辨率，受制于电子源的性能瓶颈，当前超快电镜仅能在纳米量级上观测亚皮秒时间尺度的超快过程[34,35]，因此亟待进一步提高碳基超快电子源的时空相干性，以实现原子尺度飞秒超快动力学（如电子运动等）的直接观测。

（2）在自由电子激光器中采用碳基超快电子源有望进一步提升 X 射线自由电子激光的亮度和时间分辨性，为在原子尺度上观测亚飞秒量级超快动力学过程提供关键技术手段[36]。

（3）碳基超快电子源在深空通信技术中也具有重要应用。利用碳纳米管超快电子源发射电流密度大，时空相干性高，光学相位灵敏度高等优势，有望突破当前 X 射线通信在大功率、宽频带等方面的技术瓶颈[37]。

（4）在高数据率传输用的毫米波器件当中采用碳基超快电子源有望实现电子束的预群聚，大幅减小器件的体积和重量[38,39]。

（5）碳基超快电子源还有可能突破电子器件工作频率的瓶颈。在电子芯片当中，电子在晶体管沟道中的传输时间决定了器件的工作频率，但受到当前电子线路驱动方式开关速度瓶颈的制约，当前电子传输时间已经达到了极限[38]。由超快超强激光光场控制的纳米间隙中的超快电子隧穿具有时空相干性和超短的传输时间，因此有望大大提高电子器件的工作频率 5～6 个数量级，达到拍赫兹（PHz，10^{15} Hz）量级[41,42]。

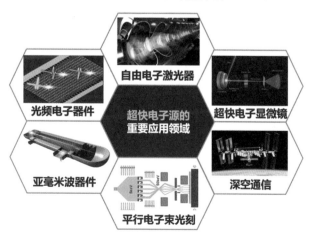

图 3　超快电子源的重要应用领域

然而，要实现以上重大目标，必须应对重大的基础和技术挑战：从深入理解强光场与固体的基本相互作用，到构筑可以用光操作的电子电路；其中，高性能碳基超快电子源是需要重点攻关的核心部件。

四、总结与建议

虽然场致电子发射现象早在一百年前就被发现，然而在长达一个世纪的时间里，理论上具有显著优势的场致发射电子源并没有取代传统热阴极电子源。当前大部分市售电真空装备仍采用热电子源。导致这种现状的原因是场发射瞬时响应（阿秒）的优势没有被发挥出来。回顾历史，展望未来，电子源技术的瓶颈即将被打破，一个新的研究热潮正在悄然兴起——利用场发射原理结合超短强场激光技术以及新材料的原子尺度制造工艺，实现电子源技术的重大突破。此次研究热潮将整体呈现出新材料结构、新理论模型、新器件原理、新测量技术等多层面交叉融合。相信在未来5～10年必将成为国际竞争焦点。

鉴于超快电子源的重要性并基于现有国际研究格局，建议我国对该领域重点布局，在材料、理论、测量、应用四个方面尽快完善碳基超快电子源的研究体系（图4），优先支持具有积累和潜力的团队开展新材料、新技术与新装备的并行研究，使得上述系统早日装备自主可控的高性能超快电子源。

图 4　碳基超快电子源研究体系

参考文献

[1] Wang X, Musumeci P, Lessner E, et al. Future of Electron Sources：The Basic Energy Sciences Workshop held by USDOE Office of Science. 2016. https：//doi. org/10. 2172/1616511.

[2] Hall E, Stemmer S, Zheng H M, et al. Future of Electron Scattering and Diffraction：The Basic Energy Sciences Workshop held by USDOE Office of Science. 2014. https：//doi. org/10. 2172/1287380.

[3] Zewail A H, Thomas J M. 4D Electron Microscopy：Imaging in Space and Time. London：Imperial College Press，2009.

［4］ Haystead J. The venerable TWT. Journal of Electronic Defense,2018,41(3):24-31.

［5］ Wang K,Dahan R,Shentcis M,et al. Coherent interaction between free electrons and a photonic cavity. Nature,2020,582:50-54.

［6］ Kfir O,Lourenço-Martins H,Storeck G,et al. Controlling free electrons with optical whispering-gallery modes. Nature,2020,582:46-49.

［7］ Mustonen A,Beaud P,Kirk E,et al. Five picocoulomb electron bunch generation by ultrafast laser-induced field emission from metallic nano-tip arrays,Applied Physics Letters,2011,99:103504.

［8］ Herink G,Solli D R,Gulde M,et al. Field-driven photoemission from nanostructures quenches the quiver motion. Nature,2012,483:190-193.

［9］ Kruger M,Schenk M,Hommelhoff P. Attosecond control of electrons emitted from a nanoscale metal tip. Nature,2011,475:78-81.

［10］ Li Z,Yang X,He F,et al. High current field emission from individual non-linear resistor ballasted carbon nanotube cluster array. Carbon,2015,89:1-7.

［11］ Vanacore G M,Fitzpatrick A W P,Zewail A H. Four-dimensional electron microscopy:Ultrafast imaging,diffraction and spectroscopy in materials science and biology. Nano Today,2016,11:228-249.

［12］ Zewail A H. Four-dimensional electron microscopy. Science,2010,328:187-193.

［13］ Feist A,Echternkamp K,Schauss J,et al. Quantum coherent optical phase modulation in an ultrafast transmission electron microscope. Nature,2015,521:200-203.

［14］ Deheer W A,Chatelain A,Ugarte D. A carbon nanotube field-emission electron source. Science,1995,270:1179-1180.

［15］ Keimer B,Moore J E. The physics of quantum materials. Nature Physics,2017,13:1045-1055.

［16］ de Jonge N,Lamy Y,Schoots K,et al. High brightness electron beam from a multi-walled carbon nanotube. Nature,2002,420:393-395.

［17］Litovchenko V,Evtukh A,Kryuchenko Yu,et al. Quantum-size resonance tunneling in the field emission phenomenon. Journal of Applied Physics,2004,96:867-877.

［18］ Schmid H,Fink H W. Carbon nanotubes are coherent electron sources. Applied Physics Letters,1997,70:2679-2680.

［19］ Li C,Zhou X,Zhai F,et al. Carbon nanotubes:Carbon nanotubes as an ultrafast emitter with a narrow energy spread at optical frequency. Advanced Materials,2017,29:1-6.

［20］ Li C,Chen K,Guan M et al. Extreme nonlinear strong-field photoemission from carbon nanotubes. Nature Communications,2019,10:4891.

［21］ Park D J,Ahn Y H. Ultrashort field emission in metallic nanostructures and low-dimensional carbon materials. Advances in Physics:X,2020,5:1726207.

［22］ Wang S,Zhao S,Shi Z,et al. Nonlinear Luttinger liquid plasmons in semiconducting single-walled carbon nanotubes. Nature Materials,2020,19:986-991.

[23] Yao F R, Yu W T, Liu C, et al. Complete structural characterization of single carbon nanotubes by Rayleigh scattering circular dichroism. Nature Nanotechnology, 2021, doi: 10. 1038/s41565-021-00953-w

[24] Zhao S H, Wang S, Wu F Q, et al. Correlation of electron tunneling and plasmon propagation in a Luttinger liquid. Physical Review Letters, 2018, 121: 047702.

[25] Zhang Y Y, Zhang Q Y, Schwingenschlögl U. Spin-charge separation in finite length metallic carbon nanotubes. Nano Letters, 2017, 17(11): 6747-6751.

[26] Schlappa J, Wohlfeld K, Zhou K, et al. Spin-orbital separation in the quasi-one-dimensional Mott insulator Sr_2CuO_3. Nature, 2012, 485: 82-85.

[27] Liang S D, Huang N Y, Deng S Z, et al. Chiral and quantum size effects of single-wall carbon nanotubes on field emission. Applied Physics Letters, 2004, 85: 813-815.

[28] Zhang S C, Kang L X, Wang X, et al. Arrays of horizontal carbon nanotubes of controlled chirality grown using designed catalysts. Nature, 2017, 543: 234-238.

[29] Bai Y X, Yue H J, Wang J, et al. Super-durable ultralong carbon nanotubes. Science, 2020, 369 (6507): 1104-1106.

[30] Asaka K, Nakayama T, Miyazawa K, et al. Structures and field emission properties of heat-treated C_{60} fullerene nanowhiskers. Carbon, 2012, 50: 1209-1215.

[31] Koenig R, Tian, H R, Xie S Y, et al. Fullertubes: Cylindrical carbon with half-fullerene end-caps and tubular graphene belts, their chemical enrichment, crystallography of pristine C_{90}-D_{5h} (1) and C_{100}-D_{5d} (1) fullertubes, and isolation of C_{108}, C_{120}, C_{132}, and C_{156} cages of unknown structures. Journal of the American Chemical Society, 2020, 142, 36: 15614-15623.

[32] Li Y L, Sun Q, Zu S, et al., Correlation between near-field enhancement and dephasing time in plasmonic dimers. Physical Review Letters, 2020, 124: 163901.

[33] Garg M, Kern K. Attosecond coherent manipulation of electrons in tunneling microscopy. Science, 2020, 367: 411.

[34] Horstmann, J G, Böckmann H, Wit B, et al. Coherent control of a surface structural phase transition. Nature, 2020, 583: 232-236.

[35] Danz T, Domröse T, Ropers C. Ultrafast nanoimaging of the order parameter in a structural phase transition. Science, 2021, 371: 371-374.

[36] Green M E, Bas D A, Yao H Y, et al. Bright and ultrafast photoelectron emission from aligned single-wall carbon nanotubes through multiphoton exciton resonance. Nano Letters, 2019, 19: 158-164.

[37] Zhao B, Wu C, Sheng LL, et al. Next generation of space wireless communication technology based on X-ray. Acta Photonica Sinica, 2013, 42(7): 801-804.

[38] Basu R, Billa L R, Letizia R, et al. Design of sub-THz traveling wave tubes for high data rate long range wireless links, 2018, Semicond. Sci. Technol. 33 124009. https://doi. org/10. 1088/1361-

6641/aae859.

[39] Gamzina D, Himes L G, Barchfeld R, et al. Nano-CNC machining of sub-THz vacuum electron devices. IEEE Transactions on Electron Devices, 2016, 63: 4067-4073.

[40] Feng J J, Li X H, Hu J N, et al. General vacuum electronics. Journal of Electromagnetic Engineering and Science, 2020, 20(1): 1-8.

[41] Bionta M R, Ritzkowsky F, Turchetti M, et al. On-chip sampling of optical fields with attosecond resolution. Nature Photonics, 2021, 15: 456-460.

[42] Zhou S H, Chen K, Cole M T, et al. Ultrafast electron tunneling devices: from electric-field driven to optical-field driven. Advanced Materials, 2021, 33: 2101449.

Progress and Prospect of Ultrafast Electron Source

Dai Qing, Li Chi

Ultrafast electron source lies at the heart of the next generation high frequency($>$100 GHz) vacuum electronic devices and systems, such as high data rate communication, ultrafast electron microscopy, free electron laser, deep space communication, etc. However, traditional electron source technology with a limited response time up to 10^{-9} s cannot satisfy the high frequency requirement. Therefore, ultrafast electron source with femtosecond response time has become an emerging interdisciplinary research area, involving vacuum electronics, nanophotonic, materials science, strong-field physics. Here, we mainly summarize the present progress and key scientific problems of ultrafast electron sources, and discusses the future development direction and new technology route, so as to promote the achievements and breakthroughs in the field of high frequency electronic devices.

2.3 迈向室温的超导新材料 及可能的应用前景

靳常青 邓 正 望贤成

（中国科学院物理研究所）

超导现象指在某一临界温度（T_c）和磁场下导电材料突然表现为零电阻以及完全的抗磁性。昂内斯（K. H. Onnes）在 1911 年发现了第一个超导体 Hg，并因此于 1913 年荣获诺贝尔物理学奖；1957 年巴丁（J. Bardeen）、库珀（L. V. Cooper）和施里弗（J. R. Schrieffer）提出基于电子-声子耦合作用形成超流库珀电子对的理论（Bardeen-Cooper-Schrieffer theory，简称 BCS 理论），成功解释了常规低温超导现象，这三位科学家因此于 1972 年荣获诺贝尔物理学奖。1986 年，IBM 苏黎世研究实验室的德国物理学家贝德诺尔茨（J. G. Bednorz）与瑞士物理学家米勒（K. Alexander Müller）发现的铜基高温超导材料[1]的 T_c 首次突破了液氮温区（77 K），摆脱了之前超导材料对极为昂贵稀缺的液氦的依赖[2,3]，两位科学家也因这一里程碑式的工作荣获了 1987 年的诺贝尔物理学奖。超导材料被誉为 21 世纪高新战略材料，在能源、交通、国防、信息、电力、空间、环境、健康等几乎所有高技术领域具有重要的应用前景，超导材料和技术研究因此得到产学研界越来越多的青睐。超导材料应用可大致分为强电和弱电两大领域。在强电应用方面，超导材料可用于制作大容量电缆，从而实现电力无损耗传导，有效避免焦耳热；用超导线材绕制磁体，可以达到最高的静态磁场，可用于高能粒子加速器、受控热核反应和核磁共振成像；超导材料的完全抗磁性可用于超导磁悬浮承载，用超导材料构建的磁悬浮列车，实现了目前最高的高铁速度；在弱电应用方面，主要基于约瑟夫森效应（Josephson effect），超导材料可用于精密磁电测量、量子信息存储、量子超快计算等领域，而这些领域是发展变革性信息技术的制高点。根据 T_c，超导材料可大致划分为低温材料（如 NiTi 合金，T_c 在液氮温区）、高温材料（如铜基材料和铁基材料，铜基块材常压可达 130 K、高压 160 K，铁基块材常压 50 K）和近室温材料（如富氢，高压下可达 200 K 以上）几大类，铜基超导材料依然保持常压最高的 T_c（130 K）。目前绝大部分超导应用需要用到液氦冷却，由于液氦极为昂贵且技术复杂，严重制约了超导材料的应用推广。实现无冷却剂的室温超导是科学家长期追求的重大科学目标，如果超导材料能够像随处可见的金属般应用，将为 21 世纪高新技术的跨越式发展提供一揽子解决方案。近来相继有实验报道发现，富氢化合物材料在高压下呈现超过 200 K 的超导性质，T_c 接近室温。如果能进一步实现室温常压超导体的应用，将是超导乃至物质科学领域的一项里程碑式重大突破。

一、富氢超导材料研究进展

氢是宇宙初期产生的第一个元素，也是目前宇宙中已知含量最为丰富的元素，对氢的研究促进了现代量子理论的诞生。氢作为周期表中最轻的元素，其金属化的晶格具有非常高的原子振动频率和电声耦合强度，根据 BCS 理论，金属化的氢极可能具有接近室温的很高的超导转变温度。1968 年，阿什克罗夫特（N. W. Ashcroft）理论预言，金属化的氢可能具有高温超导性质[4]。而早在 1935 年维格纳（E. Wigner）等就指出，氢有可能在很高的压力下实现金属化[5]，引发了对金属氢理论计算的世纪关注[6]。2017 年，美国哈佛大学的西尔韦拉（I. F. Silvera）研究组发文声称在接近 500 GPa 的高压下合成了金属氢[7]，但这一实验结果还存在争议。近期的理论计算表明氢金属化需要的压力在 500 GPa 以上，这超过目前高压实验技术能达到的水平，实现金属氢还需发展先进的超高压技术。

鉴于氢金属化的实现非常困难，20 世纪 70 年代，中国科学院物理研究所的徐济安等人提出用富氢化合物引入化学内压，降低氢的金属化压力的构想[8,9]。2004 年，阿什克罗夫特（N. W. Ashcroft）进一步地明确指出，富氢化合物可以降低氢金属化所需的压强，同时仍保留以氢为主的高温超导属性[10]。该理论一直停留在设想的阶段，直到 2014 年吉林大学马琰铭团队和崔田团队理论预测硫的氢化物 H_2S 和 H_3S 在高压分别具有 T_c 为 80 K[11] 和 204 K[12] 的高温超导性质，相关实验研究才开始取得实质进展。2015 年，德国马克斯-普朗克化学研究所的埃里曼茨（M. I. Eremets）团队实验报道 H_3S 在高压下可达 203 K 超导转变[13]，该发现揭开了近室温超导实验研究的序幕。2019 年，埃里曼茨团队和美国乔治·华盛顿大学赫姆利（R. J. Hemley）团队分别独立报道了 LaH_{10} 在高压下具有高达 250 K 超导转变[14,15]，该结果随后得到中国科学院物理研究所团队的验证[16]。2021 年，中国科学院物理研究所靳常青团队和吉林大学马琰铭团队分别独立发现了 Ca 富氢化物的合成和在高压下具有超过 200 K 的超导性质[17,18]，在材料构成上这是继硫系、镧系之后的第三类超导转变温度在 200 K 以上的富氢超导材料（图 1）。中国科学院物理研究所程金光、靳常青等还相继合成发现了锡基和锆基富氢超导材料[19,20]，这也是首次在过渡金属和Ⅳ族富氢材料发现高温超导，拓展了研制面向应用的富氢超导新材料范围。

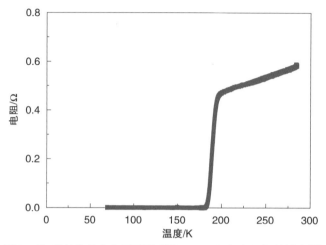

图 1 Ca 富氢化物在高压下呈现的 200 K 以上高温超导转变[18]

此外，美国拉特格斯大学（Rutgers University）的迪亚斯（Ranga P. Dias）研究团队 2020 年宣布在 267 GPa 高压下发现含碳的硫氢化物在 288 K 发生超导转变（图 2)[21]，似乎实现了室温超导。但是该碳-硫-氢体系超导的报道遇到普遍质疑，比如为什么没有超导体通常应有的磁场展宽效应，以及材料的真实结构等。

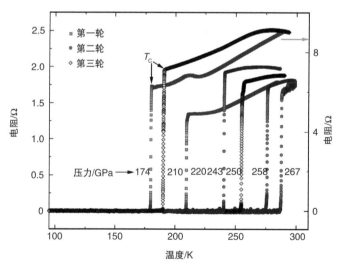

图 2 碳-硫-氢化合物在 267 GPa 高压下在 288 K 发生超导转变[21]
注：第三轮对应的为右纵坐标

二、面向应用的室温超导材料

对于超导材料的应用而言，有三个基本性能参数至关重要，即超导临界温度（T_c）、上临界场（H_c）和临界电流密度（J_c）[22]。根据 BCS 理论基本框架，临界温度和超导能隙呈正相关，高 T_c 通常对应高 H_c 和高 J_c。目前实际应用的超导材料常见的主要包括液氦温区超导材料 NbTi、Nb_3Sn 和 Fe 基材料，以及液氮温区的 Bi 系和 Y 系铜基超导材料等，Cu 系铜基超导材料的高场临界电流密度性能可与 Y 系超导材料相媲美[23]，具有潜在的应用前景[24]。富氢超导材料在高压条件下已经具有很高的 T_c，尽管受限表征技术目前尚无 J_c 数据，但实验已经有一些关于 H_c 的粗略估算。通过富氢超导体在磁场下超导行为研究，实验推测其 H_c 大于 100 T，优于常见的主要超导材料（图 3）。因此，基于优异的超导性质，富氢材料在超导应用方面具有潜在前景。

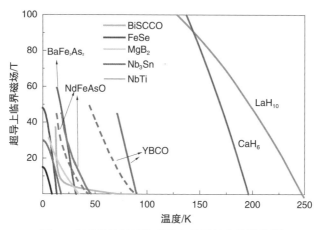

图 3 典型超导材料体系的上临界场参数[22,15,18]

三、超导材料室温应用展望

富氢超导材料虽然具有近室温的超导转变，但需要在很高的压力下才能实现[25]，这将超导从低温极端推向高压极端。为了超导材料在室温环境的应用，首先要实现室温稳定的超导特性，如何降低富氢超导体所需压力是今后努力的重要方向[26-29]。高淼（M. Gao）等理论计算预测，KB_2H_8 可以在 12 GPa 的压力下发生高温超导转变[26]，已经和目前可规模化工业生产的压力相当。卡塔尔多（S. D. Cataldo）等理论预测在 La-H 体系中通过掺杂较小原子半径硼元素引入化学内压，可以使得富氢材料 $LaBH_8$

在较低的压力（40 GPa）下稳定存在，且仍然保持较高的超导转变温度（126 K@40 GPa)[27]。这些理论工作为降低富氢超导体所需压力提供了线索，给出探索低压高温超导体的可能方向。其他轻元素（如 B、C 等）由于原子质量小，德拜频率及电声耦合强度相对较高，其化合物中也有发现近室温超导材料的可能。铜基高温超导材料的 T_c 介于低温超导材料和富氢近室温超导材料之间，且保持了目前常压超导转变温度最高纪录，因此在铜基超导材料上依然存在突破近室温超导的可能[30]，例如，运用先进材料制备技术，通过引入提高 T_c 的新机制，可以实现铜基超导转变温度的进一步的上升。在铜基超导材料中引入 3d（x^2-y^2）和 3d（z^2-r^2）轨道翻转和混合效应，研制发现 Ba_2CuO_{3+y} 具有常压下 T_c 高达 70 K 的超导转变，比基于相同结构的 La_2CuO_4 单层铜基超导材料提高了 80%[31]。如果能够进一步将这个设想推广到 T_c 更高的多层铜氧化物超导体，如目前保持常压最高 T_c 为 130 K 的铜基超导材料，将可能在常压条件实现近室温铜基超导。

由于 BCS 理论并没有明确材料超导转变温度的上限，对于富氢材料或轻元素化合物以及铜基超导材料，都有实现在常压（低压）条件可实用化的室温超导材料的可能。

致谢：中国科学院物理研究所于渌院士、沈保根院士、越忠贤院士等在研究过程中给予鼓励与支持，向涛院士、程金光研究员、胡江平研究员等也在讨论过程中提出了很多建议，谨此致谢！

参考文献

[1] Bednorz J G, Müller K A. Possible high T_c superconductivity in the Ba-La-Cu-O system. Zeitschrift für Physik B Condensed Matter, 1986, 64: 189-193.

[2] Zhao Z X, Chen L Q, Yang Q S, et al. Superconductivity above liquid nitrogen temperature in Ba-Y-Cu oxides. Kexue Tongbao, 1987, 32: 661-664.

[3] Wu M K, Ashburn J, Torng C, et al. Superconductivity at 93 K in a new mixed-phase Y-Ba-Cu-O compound system at ambient pressure. Physical Review Letters, 1987, 58: 908-910.

[4] Ashcroft N W. Metallic hydrogen: A high-temperature superconductor? Physical Review Letters, 1968, 21: 1748-1749.

[5] Wigner E, Huntington H B. On the possibility of a metallic modification of hydrogen. Journal of Chemical Physics, 1935, 3: 764-770.

[6] Mcmahon J M, Ceperley D M. Ground-state structures of atomic metallic hydrogen. Physical Review Letters, 2011, 106(16): 165302.

[7] Dias R P, Silvera I F. Observation of the Wigner-Huntington transition to metallic hydrogen. Science, 2017, 355: 715-718.

[8] 徐济安,朱宰万. 金属氢. 物理,1977,6:296.

[9] 陈良辰. 金属氢研究新进展. 物理,2004,33:261.

[10] Ashcroft N W. Hydrogen dominant metallic alloys: High temperature superconductors? Physical Review Letters,2004,92(18):187002.

[11] Li Y W,Hao J,Liu H Y,et al. The metallization and superconductivity of dense hydrogen sulfide. Journal of Chemical Physics,2014,140(17):174712.

[12] Duan D F,Liu Y X,Tian F B,et al. Pressure-induced metallization of dense(H_2S)$_2 H_2$ with high-T_c superconductivity. Scientific Reports,2014,4:6968.

[13] Drozdov A P,Eremets M I,Troyan I A,et al. Conventional superconductivity at 203 kelvin at high pressures in the sulfur hydride system. Nature,2015,525:73-76.

[14] Drozdov A P,Kong P P,Minkov V S,et al. Superconductivity at 250 K in lanthanum hydride under high pressures. Nature,2019,569:528-531.

[15] Somayazulu M,Ahart M,Mishra A K,et al. Evidence for superconductivity above 260 K in lanthanum superhydride at megabar pressures. Physical Review Letters,2019,122(2):027001.

[16] Hong F,Yang L X,Shan P F,et al. Superconductivity of lanthanum superhydride investigated using the standard four-probe configuration under high pressures. Chinese Physics Letters,2020, 37:107401.

[17] Ma L,Wang K,Xie Y,et al. Experimental observation of superconductivity at 215 K in calcium superhydride under high pressures. 2021,arXiv:2103. 16282.

[18] Li Z W,He X,Zhang C L,et al. Superconductivity above 200K observed in superhydrides of calcium. 2021,arXiv:2103. 16917.

[19] Zhang C L,He X,Li Z. W,et al. Superconductivity in zirconium polyhydrides with T_c above 70K. 2021,arXiv:2112. 14439.

[20] Hong F,Shan P F,Yang L X,et al. Possible superconductivity at～70 K in tin hydride SnHx under high pressure. Materials Today Physics,2022,22:100596.

[21] Snider E,Dasenbrock-Gammon N,Mcbride R,et al. Room-temperature superconductivity in a carbonaceous sulfur hydride. Nature,2020,586:373-377.

[22] Alex G. To use or not to use cool superconductors? Nature Material,2011,10:255-259.

[23] 靳常青. 运用高压技术设计和研制超导材料新体系. 科学通报(纪念液氮温区高温超导发现 30 周年专辑),2017,62:3947.

[24] 赵建发,李文敏,靳常青. 组分简单环境友好的铜基高温超导材料:"铜系". 中国科学:物理学 力学 天文学,2018,48(08):55-60.

[25] Service R F. High-pressure physics at last,room temperature superconductivity achieved but the hydrogen-based material requires high pressure. Science,2020,370:273-274.

[26] Gao M,Yan X W,Lu Z Y,et al. Phonon-mediated high-temperature superconductivity in ternary borohydride KB$_2$H$_8$ around 12 GPa. 2021,arXiv:2106. 07322v1.

[27] Cataldo S D, Christoph H, Wolfgang D, et al. LaBH$_8$: Towards high-T_c low-pressure superconductivity in ternary superhydrides. Physical Review B, 2021, 104: L020511.

[28] Liu H F, Cheng R, Yang K P, et al. Theoretical study on the Y-Ba-H hydrides at high pressure. Physics Letters A, 2021, 390: 127109.

[29] Durajski A P, Szczesniak R. New superconducting superhydride LaC$_2$H$_8$ at relatively low stabilization pressure. Physical Chemistry Chemical Physics, 2021, 23: 25070.

[30] Uchida S. High Temperature Superconductivity. Tokyo: Springer, 2015.

[31] Li W M, Zhao J F, Cao L P, Hu Z et al. Superconductivity in a unique type of copper oxide. Proceedings of the National Academy of Sciences of USA, 2019, 116: 12156.

Approaching Room Temperature Superconductivity & the Applications Prospects

Jin Changqing, Deng Zheng, Wang Xiancheng

Superconductivity(SC) is one of important discoveries in 20th century. Understanding SC mechanism & enhancing materials application greatly advanced the physical sciences, leading to 5 Nobel Prizes on the topic. Recently couples of superhydrides are synthesized while some of them show SC transition above 200K near room temperature. The forthcoming breakthrough of room temperature SC under ambient pressure will be expected when the materials can be put into application.

2.4 光电催化分解水制氢研究进展与展望

李 灿 叶 盛

（中国科学院大连化学物理研究所，洁净能源国家实验室（筹），
催化基础国家重点实验室）

能源危机和环境生态问题是 21 世纪人类面临的最重要的科学挑战。随着全球人口的快速增长，人类对能源的需求急剧增大，化石燃料消耗巨大，而且化石燃料燃烧产生 CO_2 等温室气体，严重影响人类的生存环境。因此，随着人类对可持续能源和环境友好型能源需求的日益增长，相关技术的研发也越来越受重视。在可持续能源技术中，太阳能利用最有前景，一天内照射到地球表面的太阳能高达 173 000 TW，超过地球上所有可再生能源可利用总量的 99%，约为全球能源年消耗量（17.91 TW，以2017 年计算）的 10 000 倍。因而，加大力度开发与利用太阳能，是有效解决能源枯竭和环境生态问题的重要途径。

一、光电催化反应池的组成

光电催化分解水制氢是将太阳光和水转化为化学燃料的有效方法，也是当前最有前景的太阳能利用技术之一[1]。从原理上来说，这种结构类似于自然光合作用体系提供水氧化和生产太阳能燃料的 Z 机制构型，即光生空穴（或电子）迁移到阳极（或阴极）表面，参与水氧化（或还原）反应。由于氧化和还原反应位点在空间上被隔离，水分解过程中的逆反应受到抑制，也不需要再进行人工分离氢气和氧气，降低了制氢装置的成本。

光电催化反应池的基本构造与传统电解槽非常类似，包括阳极、阴极、电解液和隔膜。不同之处在于，电催化分解水的能量完全来自电源供给的电能，而光电催化分解水的能量主要来自光子的光能。因此，具有光响应的光电极是光电催化反应池的核心部件。通常使用半导体作为光电极的捕光材料，其中光阳极选用 n 型半导体，而光阴极则选用 p 型半导体。目前，许多半导体材料可以用于制备光电极，常见的光阳极材料包括 TiO_2、WO_3、$BiWO_4$、$BiVO_4$、$TaON$、Cu_2O、Fe_2O_3、Ta_3N_5、$BaTaO_2N$、

BaNbO$_2$N、n-Si 等[2,3]；常见的光阴极材料包括 CuFeO$_2$、CuBi$_2$O$_4$、CdSe、CdTe、CuInS$_2$、CuInGaSn、InP、InGaAs、GaP、SiC 和 p-Si 等[4,5]。

根据组合方式的不同，光电催化反应池可分为三种类型：单一光阳极体系、单一光阴极体系和光阳极-光阴极耦合体系[6]。顾名思义，单一光阳极体系由光响应的 n 型半导体光阳极（工作电极）和无光响应的铂或碳（对电极）组成；单一光阴极体系由光响应的 p 型半导体光阴极（工作电极）和无光响应的铂或碳（对电极）组成。要实现光电催化分解水过程，这两种单一光电极体系使用的半导体材料需要具有合适的带隙和能带位置，即半导体的导带底要比水的还原电位更负，价带顶要比水的氧化电位更正。然而，目前所开发的大部分半导体材料仅能部分满足能带要求，因此由这些半导体构建的单一光电极体系必须在外加偏压的辅助下才能实现分解水过程。

将上述光阳极和光阴极相耦合，即可实现光电催化全分解水，极大地拓宽了半导体光电极的材料选择范围。根据光照方式的不同，光阳极-光阴极耦合体系可分为平行（parallel）照射模式和串联（tandem）照射模式（图 1）[7]。平行照射模式将每个光电极暴露于整个太阳光谱，两个电极的光吸收过程单独进行，互不干扰。串联照射模式则是将两个光电极按照带隙大小前后排列，使一束光先被宽带隙的半导体光阳极吸收，透过的光再被窄带隙的半导体光阴极吸收，体系能更有效地利用太阳光。此外，相较于平行照射模式，串联照射模式的配置允许通过平面内扩散实现优异的传质，是一种更高效的光电催化分解水途径。

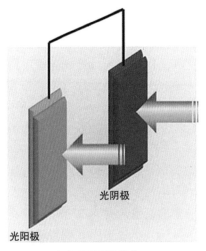
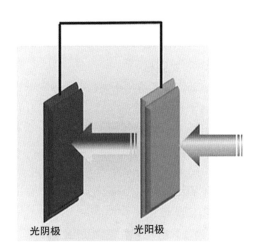

（a）平行照射模式　　　　　　　　（b）串联照射模式

图 1　平行照射模式（a）和串联照射模式（b）照射下的光电催化反应池示意图[7]

二、光电催化分解水制氢研究进展

自 1972 年 A. Fujishima 和 K. Honda 发现 TiO_2 单晶电极可以发生光电催化分解水以来，人们对其产生了极大的兴趣[8]。光电催化制氢的技术经济评估表明，其与化石燃料技术相比具有成本竞争力，但前提是要克服大规模实施光电催化分解水的技术障碍，让太阳能转换（solar-to-hydrogen，STH）效率达到 10%。到目前为止，大多数光电催化分解水的研究都集中在单个光电极（光阳极或者光阴极）上[9,10]，这是由于光阳极和光阴极之间的动力学耦合十分困难。

K. Ohashi 等于 1977 年通过耦合 p-GaP 光阴极和 n-$SrTiO_3$ 光阳极，较早地实现了无偏压光电催化全分解水[11]。自此，研究者们陆续开发了更多的光电催化全分解水体系，包括 p-Si/n-WO_3、p-$CaFe_2O_4$/n-TiO_2、p-CuTiO/n-TiO_2、p-WO_3/n-TiO_2 和 p-Cu_2O/n-WO_3 等[12-16]。然而，这些体系的 STH 效率较低，串联照射模式光电催化反应池因其高效的太阳能转换效率和优异的平面内传质逐渐引起科学家们的关注。

基于此，韩国蔚山科学技术院 Lee 课题组构筑了由 Mo-$BiVO_4$ 光阳极和（Ag，Cu）$GaSe_2$ 光阴极组成的串联照射模式光电催化反应池，该体系的 STH 效率达到 0.67%[17]。此后，该课题组改用活性更高的 Sb_2Se_3 光阴极后，STH 效率进一步提升至 1.5%[18]。瑞士联邦工学院 Grätzel 教授课题组组装了一个串联照射模式全氧化物无偏压太阳能全解水装置，使用 $BiVO_4$ 作为光阳极，Cu_2O/Ga_2O_3 埋置 p-n 结作为光阴极，该体系可以捕获波长 600 nm 以下的太阳光，STH 高达 3%[19]。值得关注的是，日本东京大学 Domen 教授课题组采用 CuInGaSe 光阴极与 $BiVO_4$ 光阳极相耦合，通过调控界面能级，STH 效率达到惊人的 3.7%[20]。近期，中国天津大学巩金龙教授课题组将 Si 光阴极与 $BiVO_4$ 光阳极相耦合，同样实现了 3.7% 的 STH 转化效率[21]。

串联照射模式染料敏化光电催化反应池也逐渐引起科学家们的关注。但到目前为止，只有两例关于该类全解水的研究报道。大连理工大学孙立成教授课题组将钌基光敏剂和水氧化催化剂共吸附在纳米多孔 TiO_2 上用作光阳极，同时将钌基光敏剂和钴基析氢催化剂共同嫁接在纳米多孔 NiO 电极上组成光阴极，实现了可见光驱动的无偏压光电催化全分解水[22]。美国北卡罗来纳大学 Meyer 等采用有机染料光敏剂，同样实现了无偏压光电催化全分解水[23]。

令人欣喜的是，笔者课题组采用多媒介调控策略，成功实现了受自然 Z 机制启发的高效光电催化全分解水过程（图 2）[24]。团队通过将无机氧化物基光阳极和有机聚合物基光阴极相耦合，组装了一个高效的无偏压全分解水光电催化反应池。该体系的 STH 效率达到创纪录的 4.3%。该研究为高效人工光合体系的合理设计和组装提供了新颖的思路和有效的方法。

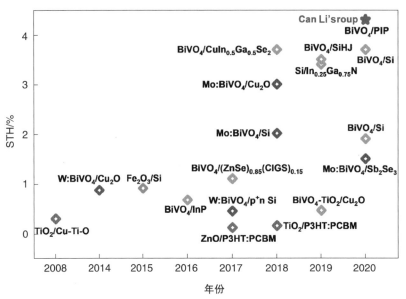

图 2　串联模式光电催化反应池的太阳能转换效率的比较[24]

三、总结与展望

总的来说，Z 机制光电催化全分解水体系的构筑取得了很大的进展，这主要得益于研究者们通过模仿自然光合作用对人工光电器件进行合理的设计和组装。虽然当前光电催化分解水反应池的太阳能转换效率尚无法满足商业应用需求，但是串联模式无偏压全分解水体系因其方便的组装和更有效的太阳能利用，将会是未来商业应用的最有希望的候选者。

为了与全球能源需求相称的规模生产可持续的氢气，应实现高效率和低成本，这两者都高度依赖于光电催化反应池中使用的半导体材料。目前，光阳极的光电性能相对光阴极而言仍然不够高效，这极大地限制了光电催化反应池全分解水的效率。因此，开发廉价、高效的光阳极是人们孜孜以求的研究目标。此外，采用高端的电极材料制备技术，利用先进的光谱表征技术进行相关的机理研究，同时使用仿生策略设计并构建光电催化体系，将给太阳能转化领域带来新的突破。如果能利用太阳能实现高效的光电催化全分解水制氢，改变能源和化工产业对化石资源的过度依赖，将有助于解决气候危机、能源安全等问题，实现经济和地球生态的可持续发展。

参考文献

[1] Pinaud B A,Benck J D,Seitz L C,et al. Technical and economic feasibility of centralized facilities for solar hydrogen production via photocatalysis and photoelectrochemistry. Energy & Environmental Science,2013,6(7):1983-2002.

[2] Ye S,Ding C,Liu M,et al. Water oxidation catalysts for artificial photosynthesis. Advanced Material,2019,31(50):1902069.

[3] Ye S,Ding C,Li C. Artificial photosynthesis systems for catalytic water oxidation. Advacnces in Inorganic Chemistry,2019,74:3-59.

[4] Wu H L,Li X B,Wang Y,et al. Hand-in-hand quantum dot assembly sensitized photocathodes for enhanced photoelectrochemical hydrogen evolution. Journal of Materials Chemistry A,2019,7(45):26098-26104.

[5] Roger I,Shipman M A,Symes M D. Earth-abundant catalysts for electrochemical and photoelectrochemical water splitting. Nature Reviews Chemistry,2017,1(1):0003.

[6] Lianos P. Review of recent trends in photoelectrocatalytic conversion of solar energy to electricity and hydrogen. Applied Catalysis B:Environmental,2017,210:235-254.

[7] Zhang K,Ma M,Li P,et al. Watersplitting progress in tandem devices:moving photolysis beyond electrolysis. Advanced Energy Materials,2016,6(15):1600602.

[8] Fujishima A,Honda K. Electrochemical photolysis of water at a semiconductor electrode. Nature,1972,238(5358):37-38.

[9] Yao T,An X,Han H,et al. Photoelectrocatalytic materials for solar water splitting. Advanced Energy Materials,2018,8(21):1800210.

[10] Kim J H,Jang J W,Jo Y H,et al. Hetero-type dual photoanodes for unbiased solar water splitting with extended light harvesting. Nature Communications,2016,7:13380.

[11] Ohashi K,Mccann J,Bockris J O M. Stable photoelectrochemical cells for the splitting of water. Nature,1977,266:610-611.

[12] Coridan R H,Shaner M,Wiggenhorn C,et al. Electrical and photoelectrochemical properties of WO_3/Si tandem photoelectrodes. The Journal of Physical Chemistry C,2013,117(14):6949-6957.

[13] Ida S,Yamada K,Matsunaga T,et al. Preparation of p-type $CaFe_2O_4$ photocathodes for producing hydrogen from water. Jouranl of the American Chemical Society,2010,132(49):17343-17345.

[14] Mor G K,Varghese O K,Wilke R H T,et al. p-Type Cu-Ti-Onanotube arrays and their use in self-biased heterojunction photoelectrochemical diodes for hydrogen generation. Nano Letters,2008,8(10):3555.

[15] Shaban Y,Khan S U. A Self-driven CM-n-TiO_2/CM-p-WO_3 photoelectrochemical cell for water splitting. ECS Transactions,2008,13(3):65-72.

[16] Lin C Y,Lai Y H,Mersch D,et al. $Cu_2O|NiO_x$ nanocomposite as an inexpensive photocathode in photoelectrochemical water splitting. Chemical Science,2012,3(12):3482-3487.

[17] Kim J H, Kaneko H, Minegishi T, et al. Overallphotoelectrochemical water splitting using tandem cell under simulated sunlight. ChemSusChem, 2016, 9(1):61-66.

[18] Yang W, Kim J H, Hutter O S, et al. Benchmark performance of low-cost Sb$_2$Se$_3$ photocathodes for unassisted solar overall water splitting. Nature Communications, 2020, 11(1):861.

[19] Pan L, Kim J H, Mayer M T, et al. Boosting the performance of Cu$_2$O photocathodes for unassisted solar water splitting devices. Nature Catalysis, 2018, 1(6):412-420.

[20] Kobayashi H, Sato N, Orita M, et al. Development of highly efficient CuIn$_{0.5}$Ga$_{0.5}$Se$_2$-based photocathode and application to overall solar driven water splitting. Energy & Environmental Science, 2018, 11(10):3003-3009.

[21] Liu B, Wang S, Feng S, et al. Double-side Si photoelectrode enabled by chemical passivation for photoelectrochemical hydrogen and oxygen evolution reactions. Advanced Functional Materials, 2020, 31(3):2007222.

[22] Li F, Fan K, Xu B, et al. Organic Dye-sensitized tandem photoelectrochemical cell for light driven total water splitting. Journal of the American Chemical Society, 2015, 137(28):9153-9159.

[23] Shan B, Nayak A, Brennaman M K, et al. Controlling vertical and lateral electron migration using a bifunctional chromophore assembly in dye-sensitized photoelectrosynthesis cells. Journal of the American Chemical Society, 2018, 140(20):6493-6500.

[24] Ye S, Shi W, Liu Y, et al. Unassisted photoelectrochemical cell with multimediator modulation for solar water splitting exceeding 4% solar-to-hydrogen efficiency. Journal of the American Chemical Society, 2021, 143(32):12499-12508.

The Research Progress and Outlook of Photoelectrocatalytic Water Splitting to Produce Hydrogen

Li Can, Ye Sheng

Photoelectrocatalytic (PEC) water splitting to produce hydrogen has been considered as a promising approach for producing chemical energy from solar energy, which is essentially analogous to the Z-scheme of natural photosynthesis with photosystems II and I providing electrochemical potentials to water oxidation and solar fuel production. Here, the recent developments in PEC water splitting are reviewed with special emphasis on unassisted dual-photoelectrode PEC devices. Moreover, we put forward the approaches of achieving large-area, high-performance and highly stable PEC devices for efficient artificial photosynthesis.

2.5　锂金属二次电池——从液态到固态

李　卓　郭　新

（华中科技大学）

锂（Li）电池具有能量密度高、无记忆性等优点，被广泛应用在电动汽车、消费电子设备等领域。但是，目前的锂电池主要使用石墨负极，能量密度几乎达到了其理论极限（约为 300 W·h/kg）。锂负极具有超高的理论比能量 3860 mA·h/g 和最低的还原电势－3.04 V（相对于标准氢电极），因此，锂金属二次电池可以带来更高的能量输出〔如锂氧（Li-O₂）电池的能量密度可高达 3500 W·h/kg〕[1]。但是，自从 20 世纪 70 年代后期锂金属二次电池发明以来，锂金属二次电池一直面临着重大的安全性和循环性的问题（图1）[2]：①充电时不可控的锂枝晶生长造成电池短路；②锂负极与电解质持续的化学反应消耗电解质并生成厚的固体电解质界面层，增加电池的内阻，缩短电池的循环寿命；③锂负极充放电的体积变化使固体电解质界面层破裂，产生游离的"死锂"，加速电解质的消耗。而在 Li-O₂、锂硫（Li-S）二次电池中，由于 O₂ 的溶解和多硫化物的穿梭效应，这些问题愈发严重[3]

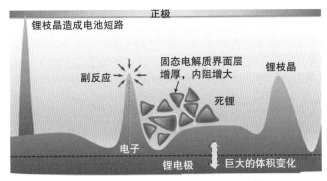

图 1　锂金属二次电池所面临的问题[2]

发展锂金属二次电池需要克服上述问题。抑制锂枝晶生长的关键点主要有：①降低界面上的电场强度，将有效电流减小至极限电流密度之下；②改善界面的不均匀性，实现均匀化的锂沉积[4]。图 2 总结了锂金属负极在固态锂金属二次电池和液态电池中的优缺点和优化办法，并提出了锂金属二次电池从液体到固态的发展策略[5]。液态锂金属二次电池的优化可从有机电解液、界面以及隔膜等方面进行。相比于有机电解液，固体电解质具有安全性更高、电化学窗口更宽等优点，因此，固态化被认为是实现稳定、安全的锂金属二次电池的最终解决方案。针对锂金属负极的问题，固态锂金属二次电池可以从提高固体电解质模量及优化界面接触性来改善。

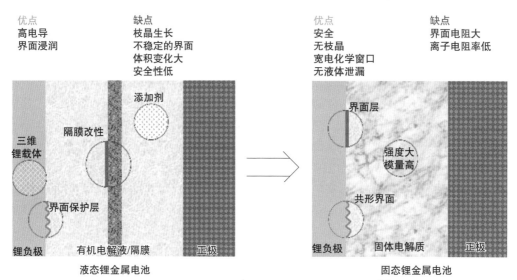

图 2 液态和固态锂金属电池的优缺点及优化策略[5]

一、液态锂金属二次电池的发展现状

（一）电解液优化

电解质优化的主要目的是促使锂负极上原位形成稳定的固体电解质层。稳定的固体电解质层可在长时间的循环中减少电解质与锂负极的直接接触，避免界面的副反应并对锂枝晶的生长进行物理阻碍。此外，在 Li-S 电池中，固体电解质界面层可以抑制多硫化物的穿梭；在 Li-O$_2$ 电池中，固体电解质界面层具有疏氧性，可提高锂负极在氧气环境中的稳定性。

1. 有机溶剂

碳酸酯类的电解质已成功地应用于锂离子电池，但仍然无法在锂金属二次电池中使用；醚类电解质的分解电压低，在高压锂金属二次电池中的应用受限。因此，开发一类新型的有机溶剂对发展锂金属二次电池非常必要。这类有机溶剂应有较低的最高占据分子轨道（highest occupied molecular orbit，HOMO）能级和最低未占分子轨道（lowest unoccupied molecular orbit，LUMO）能级，同时还要有较好的锂离子导电性（图 3）[6]。因此，氟化溶剂具有上述性质，可在锂金属二次电池中得到较好的应用。日本东京大学的山田敦夫教授开发了一种氟化环状磷酸酯有机溶剂，并将其成功应用于锂金属二次电池中[7]。这种氟化环状磷酸酯溶剂具有较低的 HOMO 能级，因此，具有优秀的抗氧化分解性；同时，它也具有较低的 LUMO 能级，有利于在锂负极上生成稳定的固体电解质界面层，可有效地保护锂负极；更重要的是，氟化环状磷酸酯具有阻燃性，大大地提高了锂金属二次电池的安全性。

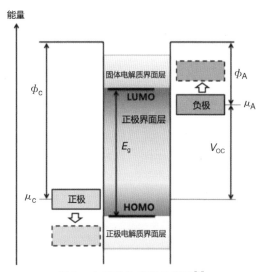

图 3　电解质的能量示意图[6]

2. 锂盐

作为电解液中的重要组成，锂盐是提高锂金属二次电池性能的另一个极为关键的因素。除了需要满足基本要求以外（电化学窗口宽、锂离子电导率高），理想的锂盐还必须对锂负极具有好的电化学/化学稳定性，并能引导形成稳定的固体电解质界面层。尽管目前的一些锂盐（Li［N（SO$_2$F）$_2$］、Li［N（SO$_2$CF$_3$）$_2$］、LiPF$_6$ 等）在锂金属二次电池中具有较好的应用，但是它们在增加锂负极界面的稳定性和抑制枝晶生长方面几乎没有作用。使用双盐（或多盐）体系能平衡电解质的离子电导和黏度，促进锂循环的可逆性；同时，双盐体系增加了竞争机制，使锂盐能参与稳定的界面层的形成。此外，开发新型锂盐也是锂金属二次电池，特别是发展 Li-O$_2$ 和 Li-S 电池的重要课题[8]。

关于锂盐的另一个议题是锂盐的浓度。通常，锂的沉积行为可以通过增加锂盐浓度，开发高浓电解质来改善。首先，相比于标准的锂盐浓度（1 mol/L），高浓度锂盐可以提供足够的锂离子源并减少自由溶剂分子，从而改善锂离子的溶剂化结构。因此，电解质中的阴离子更容易被还原成为固体电解质界面层的主要成分，使界面层更稳定。其次，高浓度锂盐可以提高锂枝晶生长的极限电流密度，从而延缓锂枝晶的生长。在此基础上，美国太平洋西北国家实验室的张继光教授提出了局部高浓电解质的思想[9]。局部高浓电解质是用稀释剂对高浓电解质进行稀释，使高浓度的盐-溶剂簇分散在稀释剂中，且电解质中几乎不存在游离的溶剂分子。在保留高浓电解质优点的同时，局部高浓电解质消除其黏度大、润湿性差、电导率低的缺点。因此，局部高浓电解质增加了锂负极的稳定性，延长了锂金属二次电池的寿命，对锂金属二次电池的发展具有重要意义。

3. 功能添加剂

电解液中的功能添加剂可以改善 Li/电解液的界面、优化固体电解质界面层的性能、消除副反应，从而达到抑制锂枝晶生长的目的。目前常用的功能添加剂有碳酸亚乙烯酯（vinylene carbonate，VC）、氟代碳酸乙烯酯（fluoroethylene carbonate，FEC）和 LiNO$_3$ 等。其中，VC 可以在锂负极界面上开环聚合，形成均匀的界面层；FEC 能促使锂负极表面形成氟化的固体电解质界面层，提高界面锂离子迁移速率；LiNO$_3$ 具有强氧化性，可在锂负极界面形成均匀的钝化层，阻止了锂负极与电解液的直接接触。然而，功能添加剂容易引起界面层的持续增长，增大电池内阻，从而降低电池的循环寿命。

另一类功能添加剂可以降低锂负极界面的局部电场强度，促进锂的均匀沉积，达到消除锂枝晶的目的。如图 4 所示，在锂的沉积过程中，一旦锂枝晶晶芽形成，这类添加剂将聚集在晶芽凸起的尖端，从而形成尖端正电荷聚集，使得尖端的局部电场强度降低。在静电力的作用下，锂的沉积将趋向迁移至晶芽的根部，使得锂的沉积更加均匀，从而消除锂枝晶。这一类添加剂主要包括 Cs$^+$、Rb$^+$ 等金属阳离子[10]，以及一些蛋白质类添加剂[11]。

（二）人工界面层

锂负极表面的固体电解质界面层对稳定锂负极提供直接的物理保护。受此启发，利用界面工程，构建人工界面层，可以有效地保护锂负极。如图 5 所示，理想的人工界面层应该满足以下条件：①力学强度足够高，以抵御锂枝晶的穿透，或者具有柔性

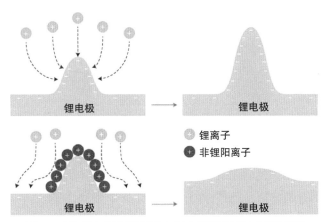

图 4　功能添加剂可降低锂电极表面的局部电场强度[12]

或共形性，能适应循环过程中的表面波动；②实现均匀和快速的锂离子（Li^+）通量，降低界面的局部电流密度，从而消除锂枝晶生长；③完全阻断电解质与锂的接触，直接减少界面副反应，或者涂层与锂之间的反应可控，生成有利的界面层，从而减缓锂的进一步腐蚀[13]。目前大多采用浆料涂敷的办法形成人工界面层。相比之下，先进的制膜技术可以避免制备界面层过程对锂负极的污染，解决锂的低熔点带来的挑战，对镀层厚度、均匀性、符合性和缺陷密度进行精确的控制。同时，先进的制膜技术在人工界面层的结构设计和材料选择上也具有明显的优势。

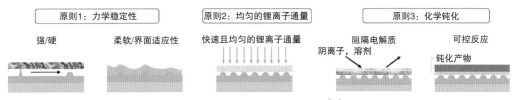

图 5　人工界面层的设计原则[13]

（三）电极结构设计

使用平面结构的锂负极时，锂的形核不均匀，容易导致不均匀的锂沉积和锂枝晶生长。具有特殊几何结构的锂负极具有大的比表面积，可降低局部的电流密度，使锂沉积更为均匀；此外，特殊结构的锂负极可以适应界面的体积变化，对锂金属二次电池具有积极作用。随着纳米技术的发展，纳米结构的锂负极也被应用于锂金属二次电池以改善锂沉积，并取得了明显的效果。

另外，三维集流体也可以用来稳定锂负极表面。用作三维集流体的材料主要有碳基材料和金属框架材料，可形成良好的电子导电网络，从而促进电荷快速转移，使得锂沉积均匀化。此外，这些集流体材料可以在循环过程中保持稳定的体积，避免电池内的应力波动，从而最大限度地改善锂金属二次电池的安全问题。

(四) 隔膜改性

锂枝晶刺穿隔膜造成短路是锂金属二次电池的最大问题，电池发热后这种情况尤为严重。因此，为提高锂金属二次电池的安全性，隔膜改性是一个重要的策略，主要原理有：①改性可提高隔膜的强度，从物理上阻止锂枝晶刺穿隔膜；②改性隔膜表面的涂层材料能与锂枝晶反应，从而消除锂枝晶；③改性隔膜可以通过相变或热交换等方式，消除电池发热/热失控现象，为隔膜创造良好的热环境；④改性的隔膜可提高锂离子迁移数，降低电池的浓差极化，使锂沉积更加均匀。隔膜的改性工艺主要有表面涂层、包覆等，以产生不同的效果。例如，华中科技大学胡先罗教授等将一种相变材料封装在聚丙烯腈的纤维中获得一种新型的隔膜结构，通过相变吸收电池的发热提高了锂金属二次电池的安全性[14]。

二、固态锂金属二次电池

尽管近年来开发了不同的方法来促进锂负极的应用，但是使用电解液的锂金属二次电池仍然面临着难以克服的困难，尤其是有机电解液的可燃性和泄漏风险为锂金属二次电池的应用蒙上阴影。因此，用固体电解质代替有机电解液，开发固态电池，有望成为实现安全的锂金属二次电池的途径之一。开发固态锂金属二次电池的关键是固体电解质，固体电解质应具有以下优点：①不易燃烧，无泄漏风险，安全性好；②室温锂离子电导率足够大；③电化学稳定窗口足够宽，可防止被氧化/还原；④强度足够高，可较大程度阻止锂枝晶穿透。根据材料的组成不同，固体电解质主要分为三类：无机固体电解质、聚合物固体电解质，以及两者组合而成的复合固体电解质。

(一) 无机固体电解质

无机电解质主要有氧化物和硫化物，其室温离子电导率$>10^{-3}$ S/cm，锂离子迁移数约为1，且力学强度高。然而，使用无机固体电解质仍然存在锂枝晶生长的问题（图6）。高弹性模量的固体电解质通常意味着其对金属锂的附着性差，这增加了电极/电解质之间的界面电阻，导致界面的局部电流密度较大，容易引发锂枝晶的生长，通常锂枝晶沿着晶界生长。硫化物电解质的电化学稳定性较差，电化学窗口较窄，较容易被氧化分解，限制了其在锂金属二次电池中的应用。

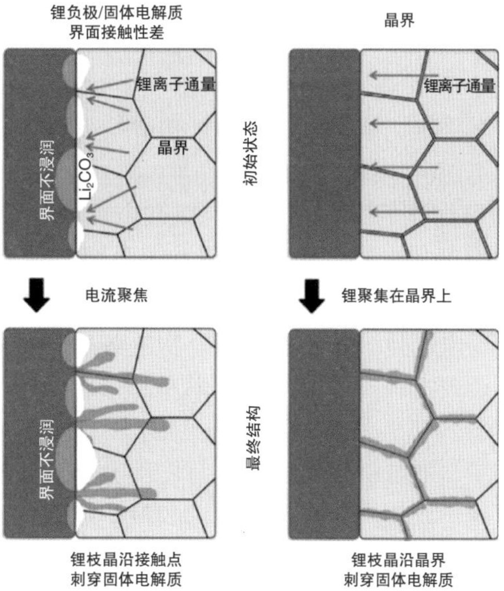

图 6 锂枝晶在无机固体电解质中的生长机理[15]

　　界面改性可以解决锂金属与固体电解质之间浸润性的问题。例如，氧化物电解质 $Li_7La_3Zr_2O_{12}$ 与 Li 之间的浸润性差，一般需要采取沉积界面层（如 Al_2O_3 层）的方法来实现两者之间的浸润[16]。但是，笔者认为 Li 金属天然可浸润 $Li_7La_3Zr_2O_{12}$，实验上表现出的不浸润性主要是由电解质表面的 Li_2CO_3 杂质所引起的[17]。构建三维的界面结构可改善电极/电解质之间的接触性，增加界面接触面积，降低界面电阻，达到降低界面局部电流，抑制锂枝晶生长的目的[18]。此外，采取先进的烧结工艺（如热压烧结）可减少气孔，对抵御锂枝晶的生长具有积极的意义。

　　电极/固体电解质的固-固界面的接触差，导致巨大的界面电阻和不均匀的锂通量，增加了界面的局部电流密度。此外，界面的空隙等缺陷具有较高的表面能，凸起尖端曲率半径较大，使锂枝晶的形核和生长更为容易。在无机固体电解质与金属锂负极之间构建柔性的聚合物电解质缓冲层，可以有效解决界面的接触问题。中国科学院化学研究所的郭玉国研究员团队[19]在 $Li_{1.4}Al_{0.4}Ti_{1.6}(PO_4)_3$ 电解质与锂负极的界面构建了聚氧化乙烯［poly(ethylene oxide)，PEO］的缓冲层，在电解质与正极界面构建了聚丙烯腈（polyacrylonitrile，PAN）的缓冲层。PEO 抗还原性好且对锂金属稳定，避免了电解质与锂负极之间的副反应；PAN 的抗氧化性好，可以搭配高电压的正极材料（如三元正极）。此外，PEO 和 PAN 还具有优秀的柔性，可以改善电极与无机固体电解质的接触性问题。因此，具有缓冲层的无机固体电解质在高电压锂金属二次电池表现出优秀的电化学性能。

（二）聚合物固体电解质

聚合物电解质的力学强度、离子电导率和迁移数均低于无机固体电解质，因此普通的聚合物固体电解质不能阻止锂枝晶的生长。但是，聚合物电解质与锂负极之间的界面相容性好，并且具有优异的柔性，因此，在锂金属二次电池中展现了巨大的潜力。一般来说，聚合物电解质的离子电导率和力学强度是相互矛盾的两个方面，因此，开发聚合物电解质的关键是寻找力学强度和离子电导率之间的平衡。一些拥有特殊结构和官能团的聚合物电解质，如嵌段型聚合物电解质、共聚型聚合物电解质和交联型聚合物电解质等，表现出较高的离子电导率和力学性能，可以在锂金属二次电池中得到较好的应用。

此外，聚合物电解质与锂负极之间可形成共形性界面，从而降低界面的局部电流密度。理想的界面附着性可以通过聚合物电解质的原位固态化技术实现。原位固态化技术是利用电解液对电极的浸润性，在电池注液后再引发电解液原位固化（图7），因此，固态化后聚合物电解质与正、负极形成共形的界面，可修补界面缺陷，从而降低界面的局部电流密度，使得锂沉积更加均匀。笔者团队利用原位固态化技术制备了一系列锂金属二次电池，获得了很好的电池性能[20,21]。未来，原位固态化技术在锂金属二次电池中具有巨大的应用前景。

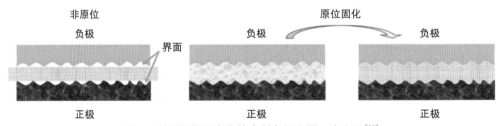

图 7　利用原位固态化技术制备锂金属二次电池[22]

一般来说，聚合物电解质的抗氧化性和抗还原性是相互矛盾的，也就是说，聚合物电解质在高电压下电化学稳定时，通常对金属锂的稳定性差。为了获得更高的能量密度，聚合物电解质往往需要搭配高电压的正极材料且在高电压下工作。在此需求下，诺贝尔化学奖获得者 J. Goodenough 教授的团队[23]开发了一种多层不对称结构的聚合物电解质：正极侧为抗氧化性强的聚（N-甲基-丙二酰胺）[poly（N-methyl-malonic amide），PMA］基聚合物电解质，而负极侧则是抗还原性强的 PEO 基聚合物电解质。因此，所获得的多层聚合物电解质电化学窗口扩宽至 4.75 V，可以应用在

Li ‖ LiCoO₂ 高压固态锂金属二次电池中，并表现出优秀的性能。

在聚合物电解质中引入添加剂可以诱导原位生成稳定的固体电解质界面层，以改善聚合物基锂金属二次电池的性能。浙江工业大学陶新永教授团队[24]在 PEO-LiN（CF₃SO₂）基聚合物电解质中添加了 LiS₂ 添加剂。LiS₂ 的引入加速了锂盐阴离子 N（CF₃SO₂）₂⁻ 的分解，从而促进了 Li/电解质界面上原位生成富 LiF 纳米晶的固体电解质界面层。这种富 LiF 的纳米晶固体电解质界面层抑制了聚合物链上 C—O 键的断裂，阻止了 Li 与 PEO 之间的连续界面副反应，从而提高了聚合物基固态锂金属二次电池的循环寿命。

（三）复合固体电解质

将聚合物电解质与无机陶瓷电解质填料复合可获得复合固体电解质。复合固体电解质保留了聚合物和陶瓷的优点，同时消除了两者的缺点。因此，复合固体电解质具有而于聚合物电解质的力学强度和优于陶瓷电解质的柔性，展现出更好的界面相容性和抵抗锂枝晶的能力。此外，由于空间电荷效应，聚合物与陶瓷电解质的界面可以形成快速的锂离子输运通道，较大程度地提高其离子电导率和迁移数，使得界面的锂通量更加均匀[25]。因此，复合固体电解质对克服锂枝晶的生长有明显的作用，能很好地应用在锂金属二次电池中。

通过复合固体电解质中陶瓷电解质填料几何形状的设计可以增强聚合物电解质，从而提高锂金属二次电池的性能。常用于复合固体电解质中的陶瓷电解质填料主要包括纳米颗粒、纳米纤维、陶瓷颗粒组成的三维网络结构等。如图 8 所示[26]，由于纳米陶瓷颗粒容易团聚的特性，复合固体电解质中导电网络通常是不连续的；纳米纤维较大程度地避免了团聚，可以提供更连续和长程的离子导电网络；三维陶瓷网络结构彻底消除了团聚现象，可以在整个空间范围内提供连续的导电网络。并且，经过烧结的陶瓷电解质本身就具有优秀的离子导电性，可以提供除界面外的额外离子导电通道。相比之下，三维陶瓷电解质网络增强的复合固体电解质在离子电导率和力学强度上均有较为明显的增强，其锂金属二次电池的性能也有明显的提高。锂离子的传导路径较长，然而陶瓷电解质在复合固体电解质中的取向是随机分布的，因此，采用连接电池正负极且方向定向分布的陶瓷电解质纤维，可以有效地缩短锂离子的传导路径和传导时间，进一步提高了聚合物电解质的离子电导率[27]。

三、总结和展望

由于其超高的能量密度，锂金属二次电池获得了广泛的关注。但是受困于不可控

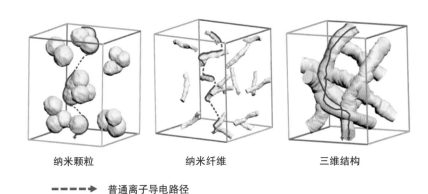

纳米颗粒　　　　　　　　纳米纤维　　　　　　　　三维结构

- - - - → 普通离子导电路径
───── → 快速离子导电路径

图8　使用不同几何结构陶瓷电解质填料的复合固体电解质的离子导电路径的示意图[26]

的锂枝晶生长和锂负极与电解质间持续的副反应等问题，目前仍然无法获得安全、可靠的锂金属二次电池。由于固体电解质具更好的安全性，锂金属二次电池的发展逐渐转向为固态电池。但是，目前固态锂金属二次电池的商业化还面临着上文所述的问题，其发展方向将会多元化。

（1）固体电解质/电极界面巨大的界面电阻是固态锂金属二次电池发展的重大阻碍，界面工程、原位固态化以及特殊结构的设计将促进固态锂金属二次电池的发展。

（2）理解枝晶生长机理是克服锂枝晶的科学基础，需要开发新的物理模型和实验技术来探究锂枝晶生长的热力学和动力学过程，原位表征技术将会在这方面大展身手。

（3）开发性能优异的固体电解质对发展固态锂金属二次电池尤为重要，新的固体电解质应同时满足离子电导率高、力学强度大及界面相容性好的要求。

参考文献

[1] Xu R, Cheng X B, Yan C, et al. Artificial interphases for highly stable lithium metal anode. Matter, 2019,1(2):317-344.

[2] Cheng X B, Zhang R, Zhao C Z, et al. Toward safe lithium metal anode in rechargeable batteries: A review. Chemical Reviews,2017,117(15):10403-10473.

[3] Liu B, Zhang J G, Xu W. Advancing lithium metal batteries. Joule,2018,2(5):833-845.

[4] Gao X, Zhou Y N, Han D, et al. Thermodynamic understanding of Li-dendrite formation. Joule, 2020,4(9):1864-1879.

[5] Yang C, Fu K, Zhang Y, et al. Protected lithium-metal anodes in batteries: From Liquid to solid. Advanced Materials,2017,29(36):1701169.

[6] Zhou Q, Ma J, Dong S, et al. Intermolecular chemistry in solid polymer electrolytes for high-energy-density lithium batteries. Advanced Materials,2019,31(50):1902029.

［7］ Zheng Q, Yamada Y, Shang R, et al. A cyclic phosphate-based battery electrolyte for high voltage and safe operation. Nature Energy, 2020, 5(4):291-298.

［8］ Eshetu G G, Judez X, Li C, et al. Ultrahigh performance all solid-state lithium sulfur batteries: Salt anion's chemistry-induced anomalous synergistic effect. Journal of the American Chemical Society, 2018, 140(31):9921-9933.

［9］ Cao X, Ren X, Zou L, et al. Monolithic solid-electrolyte interphases formed in fluorinated orthoformate-based electrolytes minimize Li depletion and pulverization. Nature Energy, 2019, 4(9):796-805.

［10］ Ding F, Xu W, Graff G L, et al. Dendrite-free lithium deposition via self-healing electrostatic shield mechanism. Journal of the American Chemical Society, 2013, 135(11):4450-4456.

［11］ Wang T, Li Y, Zhang J, et al. Immunizing lithium metal anodes against dendrite growth using protein molecules to achieve high energy batteries. Nature Communication, 2020, 11(1):5429.

［12］ Lin D, Liu Y, Cui Y. Reviving the lithium metal anode for high-energy batteries. Nature Nanotechnology, 2017, 12(3):194-206.

［13］ Yu Z, Cui Y. Bao Z. Design Principles of artificial solid electrolyte interphases for lithium-metal anodes. Cell Reports Physical Science, 2020, 1(7):100119.

［14］ Liu Z, Hu Q, Guo S, et al. Thermoregulating separators based on phase-change materials for safe lithium-ion batteries. Advanced Materials, 2021, 33(15):2008088.

［15］ Liu H, Cheng X B, Huang J Q, et al. Controlling dendrite growth in solid-state electrolytes. ACS Energy Letters, 2020, 5(3):833-843.

［16］ Han X, Gong Y, Fu K K, et al. Negating interfacial impedance in garnet-based solid-state Li metal batteries. Nature Materials, 2017, 16(5):572-579.

［17］ Wu J F, Pu B W, Wang D, et al. *In situ* formed shields enabling Li_2CO_3-free solid electrolytes: A new route to uncover the intrinsic lithiophilicity of garnet electrolytes for dendrite-free Li-metal batteries. ACS Applied Materials & Interfaces, 2019, 11(1):898-905.

［18］ Hitz G T, McOwen D W, Zhang L, et al. High-rate lithium cycling in a scalable trilayer Li-garnet-electrolyte architecture. Materials Today, 2019, 22:50-57.

［19］ Liang J Y, Zeng X X, Zhang X D, et al. Engineering Janus interfaces of ceramic electrolyte via distinct functional polymers for stable high-voltage Li-metal batteries. Journal of the American Chemical Society, 2019, 141(23):9165-9169.

［20］ Li Z, Zhou X Y, Guo X. High-performance lithium metal batteries with ultraconformal interfacial contacts of quasi-solid electrolyte to electrodes. Energy Storage Materials, 2020, 29:149-155.

［21］ Li Z, Xie H X, Zhang X Y, et al. *In situ* thermally polymerized solid composite electrolytes with a broad electrochemical window for all-solid-state lithium metal batteries. Journal of Materials Chemistry A, 2020, 8(7):3892-3900.

［22］ Zhao Q, Liu X, Stalin S, et al. Solid-state polymer electrolytes with in-built fast interfacial transport for secondary lithium batteries. Nature Energy, 2019, 4(5):365-373.

[23] Zhou W,Wang Z,Pu Y,et al. Double-layer polymer electrolyte for high-voltage all-solid-state rechargeable batteries. Advanced Materials,2019,31(4):1805574.

[24] Sheng O,Zheng J,Ju Z,et al. *In situ* construction of a LiF-enriched interface for stable all-solid-state batteries and its origin revealed by cryo-TEM. Advanced Materials,2020:2000223.

[25] Li Z,Huang H M,Zhu J K,et al. Ionic conduction in composite polymer electrolytes:Case of PEO:Ga-LLZO composites. ACS Applied Materials & Interfaces,2019,11(1):784-791.

[26] Li Z,Sha W X,Guo X. Three-dimensional garnet framework-reinforced solid composite electrolytes with high lithium-ion conductivity and excellent stability. ACS Applied Materials & Interfaces,2019,11(30):26920-26927.

[27] Liu W,Lee S W,Lin D,et al. Enhancing ionic conductivity in composite polymer electrolytes with well-aligned ceramic nanowires. Nature Energy,2017,2(5):17035.

Li-Metal Secondary Batteries:From Liquid to Solid

Li Zhuo,Guo Xin

Lithium-metal secondary batteries promise probably the highest energy output, however, to commercialize lithium-metal secondary batteries, several challenges,e. g. ,uncontrolled dendritic Li growth,continuous interfacial reactions and huge volume change, should be solved. Recent studies have shown that the performance and safety of Li-metal anodes can be significantly improved via electrolyte optimization,interface engineering,Li-electrode framework design,separator modification, etc. With mechanical properties superior to liquid electrolytes, solid electrolytes are expected to inhibit the growth of the Li dendrites,which is critical for achieving high safety in Li-metal secondary batteries. However, solid electrolytes suffer from relatively low ionic conductivity and poor Li-electrolyte interfacial contacts; recently significant progresses have been achieved in these aspects. Thanks to the high safety of solid electrolytes, the current trend for Li-metal secondary batteries is moving from liquid-state towards solid-state.

2.6 肿瘤免疫治疗的现状与展望

顾 炎[1] 曹雪涛[1,2,3]

（1. 海军军医大学免疫学研究所暨医学免疫学国家重点实验室；
2. 南开大学免疫学研究所；3. 中国医学科学院免疫治疗研究中心）

肿瘤免疫治疗是指基于机体对肿瘤的免疫监视以及肿瘤免疫逃逸等重要理论，通过对免疫系统的干预，激发或者重塑其对肿瘤的识别与杀伤，从而达到抑制和清除肿瘤的目的的一种治疗方式。肿瘤免疫治疗是近年来肿瘤治疗领域取得的重大突破之一，尤其以免疫检查点阻断疗法和嵌合抗原受体 T 细胞疗法（chimeric antigen receptor T-cell immunotherapy，CAR-T 疗法）为代表的免疫治疗，在临床实践中取得了令人瞩目的疗效，成为继手术、化疗和放疗等传统治疗方法之外，最受关注的治疗手段[1]。

近年来，国内学者瞄准肿瘤等重大疾病防治的需求，勇于挑战与创新，在肿瘤免疫治疗领域取得了一批获得国际同行认可的原创性理论成果和突破性技术创新。尤其是在国家监管部门与研发人员的共同努力下，国产程序性细胞死亡因子 1（programmed cell death 1，PD-1）单抗和 CAR-T 疗法获批上市，细胞治疗临床研究和转化应用的规范化流程逐步建立，这些工作推动了免疫治疗在我国临床应用中的正规化发展。然而，肿瘤免疫治疗也还面临着诸多挑战和问题，例如，如何进一步提高疗效和扩大治疗肿瘤类型、如何筛选特异性个性化的肿瘤抗原、如何发现更多的免疫治疗新靶点、如何更好地预测免疫治疗疗效并解决治疗耐药问题，以及如何探索更多的肿瘤免疫联合治疗策略等。本文将系统总结肿瘤免疫治疗领域的研究现状及挑战，并对未来的发展方向进行展望。

一、肿瘤免疫治疗的现状与研究进展

根据激发机体抗肿瘤免疫应答的机理，肿瘤免疫疗法分为主动免疫治疗（肿瘤疫苗）和被动免疫治疗（包括非特异性免疫激活剂、抗体和过继性细胞疗法等）（图1）。本文重点关注肿瘤疫苗、免疫检查点抑制剂治疗，以及过继性细胞疗法等临床研发及应用重点的肿瘤免疫疗法。

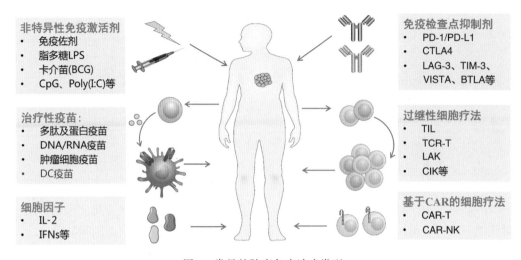

图 1　常见的肿瘤免疫治疗类型

1. 肿瘤疫苗

根据治疗原理，肿瘤疫苗（cancer vaccine）分为预防性疫苗和治疗性疫苗两类。预防性肿瘤疫苗是通过对健康人群注射预防肿瘤发生的病原体疫苗（如乙型肝炎疫苗和人乳头瘤病毒疫苗），预防和降低肿瘤的发生。治疗性肿瘤疫苗主要针对肿瘤患者，将肿瘤抗原（肿瘤细胞、肿瘤相关多肽或蛋白、表达肿瘤抗原的基因或抗原呈递细胞等）导入患者体内，通过激活树突状细胞（dendritic cell，DC），诱导机体产生有效的抗肿瘤免疫应答，这是目前肿瘤疫苗研发的主要方向。治疗性肿瘤疫苗主要分为肿瘤多肽或蛋白疫苗、核酸疫苗，以及细胞疫苗（包括基因修饰的肿瘤细胞疫苗和 DC 疫苗）等（图 2）[2]。

在肿瘤多肽或蛋白疫苗层面，利用肿瘤特定抗原肽作为肿瘤疫苗可诱导针对黑色素瘤、乳腺癌、结肠癌等肿瘤的特异性免疫应答。肿瘤新生抗原（neoantigen）的鉴定是肿瘤个体化治疗的关键，也是该领域研究的热点与难点。新抗原的发现很大程度上取决于肿瘤的突变负荷（tumor mutational burden，TMB），通常用基因编码区每百万个碱基所含的突变数来衡量。除了 TMB 外，还取决于抗原的免疫原性、主要组织相容性复合体（major histocompatibility complex，MHC）分子，以及 T 细胞受体的亲和力等[3]。

在核酸疫苗层面，核酸疫苗（尤其是 RNA 疫苗）取得了令人鼓舞的临床疗效。利用脂质载体将编码新抗原的 RNA 转运至淋巴器官，诱导高效的 T 细胞应答，能够有效控制肿瘤的进展。在临床实践中，基于肿瘤突变产生的新抗原的 RNA 疫苗被用于治疗黑色素瘤，通过超声引导注射至淋巴结内，直接激活淋巴结内的 DC 细胞，诱导黑色素瘤的消退[4]。

在细胞疫苗层面，除了基因修饰的肿瘤细胞疫苗外，备受关注的是 DC 疫苗。目前 DC 疫苗大都采用合成的肿瘤抗原肽加佐剂、肿瘤细胞提取物或凋亡的肿瘤细胞体外致敏 DC 细胞；或将编码肿瘤抗原的基因导入 DC，直接被 DC 呈递；或者将编码细胞因子、MHC 分子、共刺激分子的基因导入 DC，增加 DC 的共刺激能力。2010 年美国食品药品监督管理局（FDA）批准了首个以 DC 细胞和肿瘤融合蛋白 PA2024 为基础的肿瘤疫苗（Sipuleucel-T），用于去势抵抗性前列腺癌的治疗[5]。由医学免疫学国家重点实验室主持研究的国内第一个经中国国家食品药品监督管理总局（CFDA）批准的抗原体外致敏 DC 回输进行肿瘤治疗的项目已进入Ⅲ期临床。

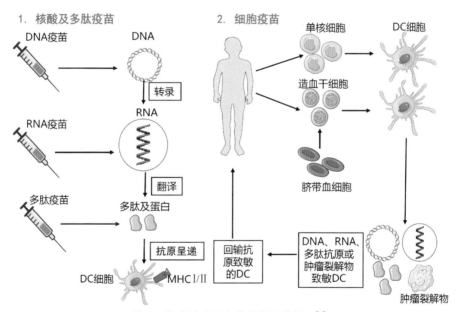

图 2　肿瘤治疗性疫苗种类及其原理[2]

肿瘤的疫苗治疗还面临着诸多挑战，例如：①如何有效地预测并筛选特异性及个体化的肿瘤新生抗原？②如何通过载体在体内外高效表达肿瘤抗原，诱导有效的抗肿瘤免疫应答？③如何克服肿瘤细胞内在的耐药性和肿瘤免疫抑制对肿瘤疫苗疗效的损伤？针对上述问题，通过大规模的基因测序以及大数据信息化处理预测新的肿瘤抗原，并通过高通量的检测平台验证并预测候选肿瘤疫苗的效应，最后通过与其他免疫疗法联合（如免疫检查点抑制剂），有望提高肿瘤疫苗的治疗效应。

2. 免疫检查点抑制剂

免疫检查点（immune checkpoint）是免疫系统的"刹车"分子，最具代表的是细胞毒性 T 淋巴细胞相关蛋白 4（cytotoxic T lymphocyte-associated protein 4，CTLA-4）和 PD-1。利用免疫检查点抑制剂（immune checkpoint inhibitor，ICI）阻断免疫检查点的效应，能有效激发抗肿瘤 T 细胞免疫应答。2018 年，诺贝尔生理学或医学奖授予詹姆斯·艾莉森（James P. Allison）与本庶佑（Tasuku Honjo），以表彰他们在发现靶向 PD-1 和 CTLA-4 的新型肿瘤免疫治疗方法中所做出的卓越贡献。目前，抗 PD-1 /PD-L1 抗体已经在肿瘤免疫治疗中得到广泛使用，以单一药物或与其他疗法联合使用的形式，被列入多数常见肿瘤的一线或二线治疗方案。

然而，包含 ICI 疗法在内的肿瘤免疫治疗仅对部分肿瘤的部分患者有效。因此，研究调控免疫检测点分子的表达机制、寻找预测和提高 ICI 疗效的新策略，并寻找更多的免疫检查点分子和联合治疗方案，是目前研究的主要方向。其中，寻找影响 ICI 疗效的因素是临床实践中迫切需要解决的问题。ICI 疗效受多种因素的影响，包括肿瘤基因组学、宿主遗传学、PD-1/PD-L1 表达水平、肿瘤微环境及肠道微生物等[6]。临床上免疫组化检测肿瘤的 PD-L1 表达是最直观的预测 ICI 疗效的方法之一。另一获得极大关注的预测疗效的指标是肿瘤突变负荷（TMB），肿瘤中 TMB 越高，意味着能够被 T 细胞识别的新抗原产生也越多，ICI 疗效可能就越好。抗 PD-1 抗体也被批准用于高度微卫星的不稳定性（MSI-H）或错配修复缺陷（dMMR）的高突变肿瘤的治疗，是首个基于肿瘤生物学机制而不是肿瘤类型被批准的药物[7]。然而，TMB 并不是完善的预测疗效的靶标，PD-L1 表达和 TMB 在大多数肿瘤中没有显著相关性，TMB 也并不总是与 ICI 反应性相关[8]。因此，还需要增加包括更多关键变量的复合预测疗效因子，如 MHC 和 T 细胞受体库等。

寻找更多的免疫检查点分子，解析 ICI 疗法耐受的机制，并将 ICI 疗法与其他检查点抑制剂联合应用，目前也受到极大关注。此外，ICI 疗法与其他肿瘤治疗方法的联合应用是目前肿瘤免疫治疗的趋势（图3），这部分内容将在展望部分详述。

ICI 疗法中还需注意的是免疫治疗相关不良反应。在 ICI 治疗过程中出现免疫系统过度激活，造成自身的组织器官被免疫细胞攻击，出现了相应的不良反应（如皮疹、结肠炎、免疫性肝炎或心肌炎等），与免疫治疗相关的死亡发生相关。因此，必须谨慎评估引起严重或永久性不良影响的可能性，对于免疫治疗相关不良反应的早期发现、识别和预判是关键。

3. 过继性细胞疗法

过继性细胞疗法（adoptive cell transfer，ACT）是指将患者体内来源的免疫细胞

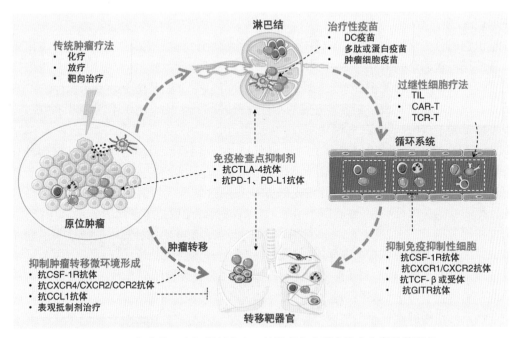

图 3 以免疫检查点抑制剂为中心的肿瘤免疫联合治疗方案及其原理

在体外进行富集、扩增或基因修饰后，回输至患者体内，以直接靶向杀伤肿瘤或激活有效的抗肿瘤免疫。目前，应用较多的是基于 T 细胞的过继回输疗法，如肿瘤浸润性淋巴细胞（tumor-infiltrating lymphocyte，TIL）、T 细胞受体修饰（TCR-modified T cell，TCR-T）和嵌合型抗原修饰的 T 细胞（Chimeric antigen receptor T cell，CAR-T）。

TIL 是存在于肿瘤组织内部具有高度异质性的淋巴细胞，临床研究证实经体外扩增激活的 TIL 可以显著提高临床肿瘤治疗的反应性[9]。早在 20 世纪 80 年代，本研究团队就在国内率先开展了 TIL 分离培养、活化及体内外杀伤能力的基础研究。2019年，TIL 细胞治疗产品 LN-145 获得美国 FDA 授权的突破性疗法认定，这是首个获得该项认定的用于实体肿瘤治疗的细胞免疫疗法。如何筛选和扩增肿瘤抗原特异性的 TIL 细胞克隆，达到精准识别肿瘤细胞，是研究的热点与难点。高通量测序方法极大地推动了肿瘤突变抗原以及抗原特异性 TIL 细胞的筛选和鉴定。例如，采用外显子测序技术鉴别肿瘤突变抗原，并用亲和力算法进行模拟预测评估，合成候选的抗原表位并鉴别出能被 TIL 细胞识别的突变抗原，通过大量扩增该 TIL 细胞克隆进行回输治疗[10]。

TCR-T 疗法是利用基因工程方法让 T 细胞表面表达能够识别特定肿瘤抗原的

TCR，使得 TCR-T 具有特异性杀伤肿瘤的能力，其关键在于治疗靶点的选择。例如，靶向癌症-睾丸抗原家族蛋白 NY-ESO-1 的 TCR-T 应用于治疗神经母细胞瘤、靶向 T 细胞识别黑色素瘤抗原 1（MART-1）或黑色素瘤相关抗原 Gp100 的 TCR-T 应用于转移性黑色素瘤，均取得了较好的疗效[11,12]。特异性抗原的选择，可以提高 TCR-T 的疗效，降低因脱靶效应带来的副作用。利用测序技术（如肿瘤全外显子测序）结合抗原表位新的预测算法，鉴定出有效的个性化的肿瘤新生抗原，用于 TCR-T 的构建，极大地推动了 TCR-T 疗法的应用。

CAR-T 疗法是指利用基因工程方法把 CAR 结构转染至 T 细胞，使得 T 细胞具有既能特异性识别肿瘤抗原又能自身活化的能力。CAR 结构是由抗体的抗原结合部（识别肿瘤抗原）以及 TCRζ 链的胞内部分和共刺激信号（传递活化信号）等偶联组成。靶向 CD19 的 CAR-T 细胞用于治疗 B 细胞恶性肿瘤取得了前所未有的临床疗效，率先获得美国 FDA 批准上市[13]。然而，CAR-T 在大部分实体肿瘤中的疗效不尽如人意，原因是实体肿瘤具有高度异质性及缺乏相对特异的肿瘤抗原，此外肿瘤内免疫抑制环境也抑制了 CAR-T 对于肿瘤的杀伤。因此，寻找肿瘤特异性抗原及通过引入不同的共刺激分子改进 CAR 结构，可以增强 CAR-T 细胞抗肿瘤活性。为提高 CAR 的靶向性及防止因肿瘤抗原丢失而导致的肿瘤复发，双特异性 CAR-T 细胞被设计用于同时识别两种肿瘤抗原，如靶向 CD19 /CD20 抗原的双特异性 CAR-T 在复发性 B 细胞恶性肿瘤治疗中取得了较好的疗效[14]。2019 年，研究人员将 CAR-T 技术与双特异性 T 细胞衔接器（BiTE）技术相结合，利用 BiTE 同时识别肿瘤相关抗原和 T 细胞受体相关分子 CD3，提高了 CAR-T 细胞与肿瘤细胞空间上的相互接触[15]。

尽管过继性细胞疗法具有较好的临床疗效，但依然面临着肿瘤抗原下调、T 细胞衰竭和脱靶毒性等问题。因此，抗原的选择极为关键，高亲和力和特异性的抗原能够降低对非肿瘤组织脱靶毒性。此外，过继性细胞疗法还需注意的主要并发症是细胞因子释放综合征，是由多种促炎细胞因子（包括 IL-6、TNF-α 等）不受控制地释放导致，临床上可表现为高热、低血压和多器官衰竭等。

4. 其他肿瘤免疫治疗策略

其他肿瘤免疫治疗策略包括非特异性免疫刺激剂、外源性重组细胞因子、溶瘤病毒等，都可以增强内源性抗肿瘤免疫反应。此外，靶向肿瘤微环境中的免疫抑制性细胞〔如骨髓来源的抑制性细胞（myeloid-derived suppressor cells，MDSC）、肿瘤相关巨噬细胞（tumor associated macrophage，TAM）〕等也是常见的肿瘤免疫治疗策略。该内容本文不做详述。

二、肿瘤免疫治疗的前景与展望

1. 以肿瘤免疫基础研究为根基的"新的免疫治疗靶标"的鉴定与应用

肿瘤免疫治疗的蓬勃发展是肿瘤基础研究转化应用于临床实践的成功案例。因此，对肿瘤生物学行为以及机体微环境等问题的进一步探索与研究，发现"新抗原"、"新分子"及"新细胞"等重要靶点，将有助于肿瘤免疫治疗瓶颈的突破。

随着抗原理论和测序等技术的发展，肿瘤新生抗原预测和评估体系得以完善。目前，肿瘤新生抗原主要包括非同义单核苷酸变异（non-synonymous single nucleotide variation，nsSNV）、突变框架移位、剪接变异体、基因融合、内源性逆转录病毒（endogenous retro-virus，ERV）等形成的抗原[16]。其中，SNV 新生抗原最受关注，但其作为抗原靶点可能仅适用于一些免疫原性高的肿瘤（如转移性黑色素瘤）[17]。在某些 SNV 低的肿瘤中也含有高特异性的抗原表达，因其高表达移码突变新抗原和 ERV 抗原等，对 ICI 等免疫治疗敏感。肿瘤细胞 ERV 的表达近年来受到关注，其可激活机体的天然免疫系统，用于抗肿瘤细胞治疗和治疗性疫苗，也与肿瘤免疫治疗应答有较强的相关性[18]。因此，不断革新抗原理论和技术，寻找特异性的肿瘤新生抗原，完善有效性评估体系，是做好肿瘤抗原筛选和预测的关键。

除经典免疫检查点分子 CTLA-4 和 PD-1 外，随着肿瘤的免疫抑制以及抗原呈递机制的逐步解析，T 细胞上新的免疫检查点，如淋巴细胞活化基因 3（lymphocyte activation gene 3，LAG-3）、T 细胞免疫球蛋白黏蛋白-3（T cell immunoglobulin and mucin domain 3，TIM-3）和 T 细胞免疫球蛋白和 ITIM 结构域蛋白（T cell Ig and ITIM domain，TIGIT）以及 B7 家族的配体（B7-H3、B7-H4 和 B7-H5），陆续被作为 ICI 治疗的靶点应用于临床试验中。近期研究表明，CD161 及其受体作为潜在的免疫检查点调控 T 细胞免疫应答[19]。抑制表达于 $CD8^+$ T 细胞的免疫检查点分子自然杀伤细胞受体家族 2A（natural killer cell group 2A，NKG2A）可以增强肿瘤疫苗诱导的肿瘤杀伤效应[20]。除 T 细胞外，其他细胞表达的免疫检查点分子也受到关注，如 SIRPa/CD47 可以抑制巨噬细胞介导的吞噬作用。因此，发现更多的免疫检查点分子将进一步拓展 ICI 疗法的临床应用。

肿瘤免疫及免疫治疗离不开肿瘤微环境的协同参与，不少免疫治疗的效果不佳都与肿瘤内部免疫抑制性微环境密切相关。免疫细胞具有功能的高度可塑性，在肿瘤持续"驯化"下，发生功能极化或表型改变，参与免疫抑制形成。近年来备受关注的中性粒细胞既能促进免疫系统对肿瘤的清除，也能在肿瘤的"驯化"下参与免疫抑制并直接促进肿瘤生长转移[21]。除了免疫细胞外，其他造血系细胞（如红细胞、血小板等）和基质细胞（如成纤维细胞、上皮细胞、内皮细胞等）也发挥了重要的免疫调节功能。本研究团队发现，肿瘤晚期患者脾脏中新型红细胞样细胞亚群（Ter 细胞）能够分泌大量的神经营养因子 artemin 进入血液而促进癌症的转移和恶化[22]。因此，解析肿瘤微环境中免疫抑制性细胞及分子对于提高肿瘤免疫治疗效果和作为肿瘤治疗的靶点均具有重要的意义。

2. 以前沿技术应用为动力的"新的疗效预测因素"的发现与应用

组学（如基因组学、转录组学、蛋白质组学、代谢组学等）技术以及单细胞技术的不断研发及应用，使得肿瘤免疫研究既可以进行全景式的大规模筛选，也可以进行单细胞水平的检测，为免疫治疗靶点及疗效预测因素的筛选带来了前所未有的机遇。

利用单细胞测序技术解析肿瘤及肿瘤微环境中免疫细胞的异质性，对于开发针对免疫细胞的靶向治疗及预测 ICI 疗效具有重要的指导意义。针对肿瘤细胞，单细胞测序能够揭示肿瘤内及个体间的细胞异质性。研究人员通过对不同肿瘤细胞基因表达特征的聚类分析，发现其具有独特进化轨迹，通过阐述不同肿瘤类型和细胞异质性，可以发现与免疫治疗应答及治疗抵抗相关的转录特征和细胞亚类[23]。针对微环境中的免疫细胞，国内学者利用单细胞转录组测序（scRNA-seq）绘制了肺癌和结肠癌等肿瘤中的 T 细胞免疫图谱，揭示了 T 细胞的亚群分类、组织分布特征、肿瘤内群体异质性及药物靶基因的表达情况[24]。scRNA-seq 和单细胞 TCR 测序（scTCR-seq）的联合应用，可以将不同个体、同一个体不同 T 细胞克隆间的表型及功能变化完整解析[25]。

此外，CRISPR/Cas9 基因编辑技术的发展使得全基因组水平的筛选得以实现。利用 CRISPR/Cas9 筛选系统可以筛选肿瘤免疫治疗中的关键基因和靶点。利用该技术筛选免疫治疗效果预测因素具有得天独厚的优势，可以在全基因组水平对影响免疫治疗效果的调控因子进行大规模筛选。目前肿瘤细胞内 PD-L1 调控分子、抗原呈递相关分子、干扰素信号通路及其相关分子、表观调控分子及 T 细胞内表观及代谢调控分子等被筛选出与肿瘤免疫治疗的响应性密切相关[26,27]。

肿瘤微环境的复杂性以及肿瘤的异质性，决定了不能用单一的生物标志物来预测肿瘤免疫治疗的疗效。未来有可能通过对大数据的深度挖掘（如肿瘤基因组测序数据、RNA 测序数据、肿瘤患者遗传多态性的测序数据，以及 PD-L1 表达、其他免疫检查点分子表达和反映肿瘤微环境特征的免疫组化检测等），结合基于深度学习的人工智能方法，构建新的疗效预测模型，来预测特定肿瘤患者的肿瘤微环境特征和对肿瘤免疫治疗的响应性[28]。

3. 以临床实践与个体化治疗为引导的"新的联合治疗方案"的制定

鉴于肿瘤的异质性、肿瘤与宿主间复杂的相互作用以及患者的个体差异，没有任何单一或者普适性的肿瘤免疫治疗方法可以攻克所有肿瘤并惠及所有患者。因此，使用肿瘤联合治疗策略是目前的趋势所在。多种联合策略（如免疫疗法间的组合或与传统肿瘤疗法的联合）正在被临床广泛应用。例如，ICI疗法与化疗、放疗、靶向治疗或免疫调节剂之间的联合应用是最常见的联合策略。除了直接致死作用外，传统治疗方法可能会将低免疫浸润性"冷"肿瘤转变为"热"肿瘤，从而形成有利于免疫治疗的肿瘤微环境[29]。此外，新的协同策略正在研究或临床试验中，如多个免疫检查点抑制剂（如LAG-3、TIM-3、VISTA和BTLA等）之间的联合，或与免疫激活剂（如天然免疫激活剂、肿瘤疫苗、细胞因子和溶瘤病毒等）的联合[30]。

然而，综合治疗不是简单的疗法相加，而是要如何取得"1+1＞2"的疗效。因此，首先要对患者进行个性化和系统性的评估，如上文所述，新的高通量和单细胞测序技术以及大规模筛选技术的发展，在评估治疗效果和提供个性化治疗的最佳选择方面显示出巨大的潜力。其次，要选择最优的给药组合和给药顺序。以新辅助治疗为例，免疫治疗通常用于新辅助阶段，以增强对肿瘤的系统免疫，配合后续的手术治疗，实现最佳的临床疗效，最终实现肿瘤的精准化、个体化治疗。

三、总 结

综上所述，肿瘤免疫理论逐步完善作为免疫治疗靶点开发的"根基"、新技术体系不断研发与应用作为治疗技术革新的"动力"以及临床与个体化治疗的强劲需求和稳步推进作为"引导"，肿瘤免疫治疗必将获得更大的理论和技术突破，取得更令人瞩目的临床疗效。我们相信，肿瘤免疫治疗虽面临种种问题和挑战，但未来可期，必将为更多的晚期肿瘤患者治疗带来福音。

参考文献

[1] Waldman A D,Fritz J M,Lenardo M J. A guide to cancer immunotherapy:from T cell basic science to clinical practice. Nature Reviews Immunology,2020,20:651-668.

[2] Saxena M,van der Burg S H,Melief C J M,et al. Therapeutic cancer vaccines. Nature Reviews Cancer,2021,21:360-378.

[3] Bulik-Sullivan B,Busby J,Palmer C D,et al. Deep learning using tumor HLA peptide mass spectrometry datasets improves neoantigen identification. Nature Biotechnology,2019,37:55-63.

[4] Sahin U,Derhovanessian E,Miller M,et al. Personalized RNA mutanome vaccines mobilize poly-

specific therapeutic immunity against cancer. Nature,2017,547:222-226.

[5] Kantoff P W, Higano C S, Shore N D, et al. Sipuleucel-T immunotherapy for castration-resistant prostate cancer. The New England Journal of Medicine,2010,363:411-422.

[6] Havel J J,Chowell D,Chan T A. The evolving landscape of biomarkers for checkpoint inhibitor immunotherapy. Nature Reviews Cancer 2019,19:133-150.

[7] Andre T,Shiu K K,Kim T W,et al. Pembrolizumab in microsatellite-instability-high advanced colorectal cancer. The New England Journal of Medicine,2020,383:2207-2218.

[8] Jardim D L,Goodman A,de Melo Gagliato D,et al. The challenges of tumor mutational burden as an immunotherapy biomarker. Cancer Cell,2021,39(2):154-173.

[9] Knochelmann H M,Rivera-Reyes A M,Wyatt M M,et al. Modeling *ex vivo* tumor-infiltrating lymphocyte expansion from established solid malignancies. Oncoimmunology,2021,10:1959101.

[10] Bentham R,Litchfield K,Watkins T B K,et al. Using DNA sequencing data to quantify T cell fraction and therapy response. Nature,2021,597:555-560.

[11] Singh N,Kulikovskaya I,Barrett D M,et al. T cells targeting NY-ESO-1 demonstrate efficacy against disseminated neuroblastoma. Oncoimmunology,2016,5(1):e1040216.

[12] Simon B,Harrer D C,Thirion C,et al. Enhancing lentiviral transduction to generate melanoma-specific human T cells for cancer immunotherapy. Journal of Immunological Methods,2019,472:55-64.

[13] Sadelain M. CD19 CAR T cells. Cell,2017,171:1471.

[14] Tong C,Zhang Y,Liu Y,et al. Optimized tandem CD19/CD20 CAR-engineered T cells in refractory/relapsed B-cell lymphoma. Blood,2020,136:1632-1644.

[15] Choi B D,Yu X,Castano A P,et al. CAR-T cells secreting BiTEs circumvent antigen escape without detectable toxicity. Nature Biotechnology,2019,37:1049-1058.

[16] Smith C C,Selitsky S R,Chai S,et al. Alternative tumour-specific antigens. Nature Reviews Cancer,2019,19:465-478.

[17] Turajlic S,Litchfield K,Xu H,et al. Insertion-and-deletion-derived tumour-specific neoantigens and the immunogenic phenotype:A pan-cancer analysis. The Lancet Oncology,2017,18:1009-1021.

[18] Canadas I,Thummalapalli R,Kim J W,et al. Tumor innate immunity primed by specific interferon-stimulated endogenous retroviruses. Nature Medicine,2018,24:1143-1150.

[19] Mathewson N D,Ashenberg O,Tirosh I,et al. Inhibitory CD161 receptor identified in glioma-infiltrating T cells by single-cell analysis. Cell,2021,184:1281-1298,e1226.

[20] van Montfoort N,Borst L,Korrer M J,et al. NKG2A blockade potentiates CD8 T cell immunity induced by cancer vaccines. Cell,2018,175:1744-1755,e1715.

[21] Yang L,Liu Q,Zhang X,et al. DNA of neutrophil extracellular traps promotes cancer metastasis via CCDC25. Nature,2020,583:133-138.

[22] Han Y,Liu Q,Hou J,et al. Tumor-induced generation of splenic erythroblast-like Ter-cells promotes tumor progression. Cell,2018,173:634-648,e612.

［23］ Maynard A,McCoach C E,Rotow J K,et al. Therapy-induced evolution of human lung cancer revealed by single-cell RNA sequencing. Cell,2020,182(5):1232-1251,e1222.

［24］ Guo X,Zhang Y,Zheng L,et al. Global characterization of T cells in non-small-cell lung cancer by single-cell sequencing. Nature Medicine,2018,24:978-985.

［25］ Krishna C,DiNatale R G,Kuo F,et al. Single-cell sequencing links multiregional immune landscapes and tissue-resident T cells in ccRCC to tumor topology and therapy efficacy. Cancer Cell,2021,39:662-677,e666.

［26］ Griffin G K,Wu J,Iracheta-Vellve A,et al. Epigenetic silencing by SETDB1 suppresses tumour intrinsic immunogenicity. Nature,2021,595:309-314.

［27］ Patel S J,Sanjana N E,Kishton R J,et al. Identification of essential genes for cancer immunotherapy. Nature,2017,548:537-542.

［28］ Ciccolini J,Benzekry S,Barlesi F. Deciphering the response and resistance to immune-checkpoint inhibitors in lung cancer with artificial intelligence-based analysis:When PIONeeR meets QUANTIC. British Journal of Cancer,2020, 123:337-338.

［29］ Heinhuis K M,Ros W,Kok M,et al. Enhancing antitumor response by combining immune checkpoint inhibitors with chemotherapy in solid tumors. Annals of Oncology,2019,30:219-235.

［30］ Kubli S P,Berger T,Araujo D V,et al. Beyond immune checkpoint blockade:Emerging immunological strategies. Nature Reviews Drug Discovery,2021,20:899-919.

The Present and Future of Cancer Immunotherapy

Gu Yan ,Cao Xuetao

With the innovation of theory and breakthrough of methodology in the field of cancer immunology,cancer immunotherapy has achieved great advances both in basic research and also clinical application in recent years. However,there are still many challenges in this field,such as how to improve the therapeutic efficacy in clinic,identify biomarker predictors for clinical efficacy,screen specific tumor antigens,discover new immunotherapy targets,and explore combined strategies with increased immunotherapy efficacy. Here we focus on summarizing the current research and application of cancer vaccine,immune checkpoint inhibitor and adoptive cell therapy,and also the challenges limiting the development of cancer immunotherapy. We discuss the future prospects of cancer immunotherapy.

2.7　气候变化的检测归因和预估研究进展

周天军[1,2]　张文霞[1]　胡　帅[1]　陈晓龙[1]　江　洁[1]　巫明娜[1]

（1. 中国科学院大气物理研究所，大气科学和地球流体力学数值
模拟国家重点实验室；2. 中国科学院大学地球与行星学院）

一、引　　言

气候系统由大气圈、水圈、冰冻圈、生物圈和岩石圈表层这五个圈层组成，后四个圈层通过与大气圈相互作用影响着地球的气候状态。气候变化表现为气候系统五个圈层及其相互作用的变化。气候变化是指统计上可识别的持续较长时间（典型的为几十年或更长）的气候状态的变化，包括气候平均值、极值和变率的变化。气候变化的原因可能是自然的内部过程，或是自然外部强迫（如太阳活动、火山爆发），或是人为外强迫（如持续改变包括温室气体和气溶胶在内的大气成分和土地利用的形式等）。我们关于气候变化科学的认知能力，取决于观测资料、观测技术和气候模式的改进，以及物理认识上的提高。本文主要针对工业化以来的气候变化、检测归因和预估研究，总结最新国际进展，提出未来研究方向。

二、主要科学进展

1. 支撑检测归因和预估研究的气候模式发展

来自联合国政府间气候变化专门委员会（Intergovernmental Panel on Climate Change，IPCC）第六次评估报告（AR6）最终发布版的数据显示，目前开展全球模拟和预估的模式研发中心或联盟的数量较以往显著增多，达到 28 家；同时完成了第六次国际耦合模式比较计划（CMIP6）核心试验、历史气候模拟试验和情景预估试验等三组试验，并及时完成了数据交汇，模式数据被 IPCC AR6 正式采用的模式版本有 39 个，其中包括 11 个考虑了碳循环过程的地球系统模式。CMIP6 模式的大气平均分

辨率接近 100 km，海洋分辨率接近 75 km。这些模式支撑了包括 IPCC AR6 编写在内的当前国际上关于历史气候变化归因和未来气候变化预估的大部分工作[1]。

气候模式研发的另一重要进展是高分辨率海气耦合模式在气候变化研究中的应用。国际上有 12 家研究机构的高分辨率模拟结果被 IPCC AR6 引用，其中大气分辨率最高达到 20 km，海洋分辨率最高达到 10 km。这类模式目前主要用于过程和机理研究，因其需要的超算资源过高，尚没有大规模用于气候变化检测归因和预估研究中。

2. 气候变化的检测和归因

近年来气候变化检测归因领域的进展来源于多个方面。基于更丰富、更可靠的观测资料和古气候记录，采用更科学的分析方法和改进的检测归因技术，得益于模式模拟能力和物理认知水平的提高，加之人为气候变化的信号逐渐增强，目前检测归因研究已扩展到更多圈层，并且针对区域尺度的气候变化给出了更高信度的归因结论[2-3]。针对全球增温，IPCC AR6 给出了更长时间尺度、更准确的估算，指出过去 10 年（2011～2020 年）全球平均气温较工业化前水平（1850～1900 年平均气温）增暖了1.09（0.95～1.20)℃。并首次给出确凿的检测归因结论，指出"毋庸置疑，人为影响正在使得大气、海洋和陆地变暖。2010～2019 年相对于 1850～1900 年，人为造成的全球表面温度上升幅度的可能范围是 0.8～1.3℃，最佳估计是 1.07℃"[2]。

针对极端事件归因，目前科学界已能对区域尺度、更多类型的极端事件给出更高信度的检测归因结论，并且从长期变化的检测归因发展到事件归因，能够在极端事件发生后迅速给出归因结论。综合已有证据表明，"人类活动造成的气候变化已经影响到全球每个区域的许多极端天气气候事件"，包括极端温度、强降水、干旱、热带气旋、复合型极端事件等[2-3]；其中，与温度相关的极端事件的归因信度显著高于与降水相关的极端事件[4]。例如，人为温室气体排放使得类似 2020 年梅雨期长江流域持续性强降水事件的发生概率提高，而人为气溶胶排放则降低了其发生概率[5]；中国东部城市区域复合型高温事件的增加可归因于温室气体排放和城市化[6]。

3. 年代际近期气候预测问题

面向未来 20 年（2021～2040 年）的近期气候变化预测信息，对于制定应对气候变化的科学决策具有重要参考价值。IPCC AR6 针对近期气候变化的情景评估结果表明，相对于 1995～2014 年，2021～2040 年全球年平均地表温度上升幅度的可能范围是 0.4～1.0℃，CMIP6 各模式的预估结果之间存在较大差异[2]。仅考虑外强迫变化情景的全球年平均地表温度近期预估不确定性大约一半来自内部变率，另一半来自模式不确定性；而预测结果对排放情景的依赖性较小，这与 IPCC 第五次评估报告（AR5）的结果一致[7]。近期预测不确定性的客观存在极大地制约了源于模式气候变化信息的适用性。基于观测资料初始化的年代际气候预测与基于大样本集合模拟的年代际内部变率约束是减少近期气候预测不确定性的两类主要方法。

年代际气候预测是指对未来 2～10 年的气候进行预测，属于初值问题与边值问题的结合[8]。基于气候模式的年代际预测试验需要通过同化观测数据对模式进行初始化，同时考虑与历史气候模拟或气候预估试验类似的外强迫的影响[9]。自 2007 年 Doug M. Smith 等首次利用耦合模式尝试进行年代际预测以来[10]，该领域取得了显著发展。最近的研究表明，当前的年代际预测系统对于中高纬度大气环流的预测普遍存在信噪比相对现实世界显著偏低的问题，这令气候模式中的可预测信号被较强的、不可预测的气候噪音所掩盖，使得年代际预测的准确性降低[11]。基于多模式、大样本年代际预测试验，通过剔除不可预测噪音、提取可预测信号，并对模式的预报结果采用"方差订正"技术，可在很大程度上解决可预测信号偏弱问题。针对我国区域性气候的集合预测研究表明，位于青藏高原主体的羌塘高原夏季降水具有一定的年代际可预报性，"方差订正"技术能够减少约 50% 的年代际预测不确定性，实时年代际预测表明，相对 1986～2005 年平均，2020～2027 年羌塘高原夏季降水将增加约 12.8%[12]。

在 10～30 年的时间尺度上，气候系统年代际变率是造成近期气候预估不确定性的主要因子。有效分离外强迫和内部变率对近期气候预估的影响，有利于提高未来气候变化信息的准确度。近年来，基于单一模式、包含大量样本的大样本集合模拟是估算内部变率影响的重要工具。随着大样本集合模拟试验的发展，已有研究结合观测资料和多套大样本集合模拟数据，通过约束内部变率因子来约束近期气候预测不确定性。例如研究指出，在充分考虑太平洋年代际振荡的位相变化后，能够有效降低南亚夏季降水及极端旱涝发生概率预估结果的不确定性[13]。研究发现，在约束太平洋年代际振荡的位相变化后，未来太平洋沃克环流将减弱，这将引起南亚季风区和海洋大陆地区降水减少及亚马孙地区西北部的气候变干[14]（图 1）。

4. 气候敏感度估算

平衡态气候敏感度（equilibrium climate sensitivity，ECS）是指当 CO_2 浓度升高到工业化前的 2 倍时，气候系统的响应达到新的平衡态时全球平均地表气温的变化。升温越高，说明气候对 CO_2 的增加越敏感。气候敏感度有多大一直是气候变化研究的核心问题之一。IPCC AR5 认为 ECS 的可能范围（66%）为 1.5～4.5℃[7]，这与 30 多年前 J. Charney 报告推测的结果相同[15]。ECS 估算结果不确定性过大给气候变化预估和应对气候变化的决策造成困难。IPCC AR5 以来的研究考虑了辐射强迫与反馈

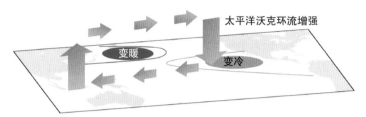

历史变化：正位相→负位相

太平洋沃克环流增强

变暖　　变冷

（a）1980~2015年IPO位相由正转负主导了太平洋沃克环流的增强

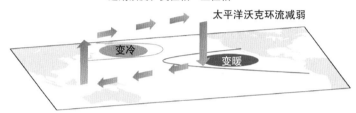

近期预测：负位相→正位相

太平洋沃克环流减弱

变冷　　变暖

（b）至2050年，伴随着IPO正位相的恢复，太平洋沃克环流将减弱

图 1　IPO位相转变影响太平洋沃克环流强度变化的动力学机制示意图

参数之间的负相关关系，可将 ECS 的范围缩小 14%[16]。在此基础上，综合各种器测数据和不同冷暖期的古气候代用记录，以及采用基于观测资料约束模式气候敏感度的方法，IPCC AR6 给出 ECS 的最佳估计值为 3 ℃，可能范围为 2.5～4 ℃，相对于AR5，不确定性显著降低。

5. 气候预估的观测约束

气候模式存在偏差，给气候预估结果带来不确定性。为了得到更加准确的预估结果，IPCC AR6 首次基于多种来源的证据对全球平均地表气温（global mean surface air temperature，GSAT）的未来变化进行评估。一方面采用不同社会经济路径驱动下、基于观测升温约束后的 CMIP6 模式模拟结果，其温升明显弱于无约束的 CMIP6结果；另一方面基于最新评估的 ECS 和瞬态气候响应（transient climate response，TCR）范围，采用两层能量平衡模型的气候仿真器，在不同社会经济路径对应的辐射强迫驱动下产生未来 GAST 的可能变化；最后综合观测约束后的 CMIP6 模拟与基于仿真器的计算，得出各种情景下 GSAT 的预估结果。该方法基本排除了内部变率的影响，充分体现了对外强迫响应的信号[17]。以 SSP5-8.5 情景下长期预估结果为例，无约束预估的近期、中期和长期相对于 1850～1900 年的增温幅度分别为（1.7±0.6）℃、

(2.6±0.9)℃和（4.8±1.7)℃，而约束后的相应统计数据为（1.6±0.3)℃、（2.4±0.5)℃和（4.4±1.2)℃[18]。

为更好地应对气候变化，精准的区域尺度信息非常重要，因此约束区域温度、降水变化是当下的研究前沿。最近的研究发展了多种不同的观测约束方法。例如基于最优指纹法的检测归因技术，利用基于历史变化估算得到的比例因子来调整未来预估中模式响应的幅度。该归因研究发现，CMIP5 多模式整体低估了青藏高原地区的人为气候增暖，经归因约束后，RCP4.5 情景中青藏高原在 21 世纪末相较 1986~2005 年升温约 3℃，比未约束时高 0.32 ℃[19]。再如，还可通过"萌现约束"技术，即建立模式预测不确定性与其对当前气候模拟不确定性之间的可靠联系，利用当前的观测数据约束未来预测结果。将这一方法应用于西北太平洋副热带高压的预估中，与约束前相比，RCP8.5 情景下西太平洋副热带高压显著增强，对应东亚夏季风降水多增加 30%以上，同时模式不确定性下降了 45%[20]。各类观测约束方法已广泛应用于不同气候区和气候指标中[21-23]。

6. 极端事件的预估

针对极端天气气候事件的预估，目前的国际研究热点一方面关注区域尺度的预估信息，另一方面关注更多类型的极端事件。IPCC AR6 强调指出，随着未来全球增温，极端高温事件和强降水事件的频率将随增温幅度而加速增长，越极端的事件，其发生频率的增长百分比越大。针对干旱事件，IPCC AR6 指出未来更多区域将受到农业生态干旱加重的影响[2]。除了传统的单一要素极端事件，复合型极端事件因其影响更大而越来越受到关注。诸多模拟证据表明，未来全球许多区域将面临更高的复合型极端事件风险，如高温-干旱复合事件、与野火有关的复合气象条件、沿海地区复合型洪涝事件[2-3,24]。

极端事件的变化与天气气候变率（即波动性）的变化密切相关。全球增暖背景下的天气气候变率变化是气候变化领域的前沿问题。研究发现，随着未来增温，从天气尺度到月、季节和年际尺度，全球湿润区（主要包括热带、季风区、中高纬地区）降水变率将增强，意味着干湿时期之间的波动将更为剧烈[25]。

三、结论与展望

（1）气候模式是支撑气候变化检测、归因和预估研究不可或缺的手段，其未来发展方向一是对流分辨模式，二是地球系统模式。检测和归因研究依赖于气候模式和大集合模拟试验，当前气候模式对类似 2021 年郑州暴雨这样的极端降水等极端事件的模拟能力不够，公里尺度对流分辨（又称作"风暴解析"）的超高分辨率模式在区域乃至全球气候预估中的应用，将是未来有望取得突破的一个方向。

（2）气候变化的检测和归因结果受到观测资料的质量、模式性能和归因方法的影响。如何在类似东亚季风区这样的区域尺度对极端降水等变化进行归因是一个挑战性问题，需要资料、模式和归因方法三者并进。

（3）面向未来 10～30 年的年代际近期气候预估具有重大决策需求，但是从可预报性的机理到预报系统的构建，当前的能力都需要提高，该方向将是未来的增长点。

（4）气候敏感度研究是一个前沿话题，其准确估算不仅影响到未来气候预估的可靠性，还将直接影响《巴黎协定》温升阈值的出现时间估算，例如，IPCC AR6 第四章指出，如果气候敏感度位于估算范围的较小值处，则在 SSP1-1.9 和 SSP1-2.6 情景下，可以避免 21 世纪气温上升水平超过 1.5℃ 的阈值（中等信度）。

（5）包括东亚在内的区域尺度气候预估，特别是温度和降水变化，如何基于"萌现约束"、发展空间型标度法（pattern scaling）等技术，来提高区域气候预估可靠性是一个前沿问题。对复合型极端事件的预估是需要重点关注的问题。

参考文献

[1] 钱诚,张文霞.CMIP6 检测归因模式比较计划(DAMIP)概况与评述.气候变化研究进展,2019,15(5):469-475.

[2] IPCC. Summary for policymakers. In:Masson-Delmotte V, Zhai P, Pirani A, et al. Climate Change 2021:The Physical Science Basis. Contribution of Working Group I to the Sixth Assessment Report of the Intergovernmental Panel on Climate Change. Cambridge:Cambridge University Press,2021.

[3] Seneviratne S I, Zhang X, Adnan M, et al. Weather and climate extreme events in a changing climate. In:Masson-Delmotte V, Zhai P, Pirani A, et al. Climate Change 2021:The Physical Science Basis. Contribution of Working Group I to the Sixth Assessment Report of the Intergovernmental Panel on Climate Change. Cambridge:Cambridge University Press,2021.

[4] Sun Y, Zhang X, Ding Y, et al. Understanding human influence on climate change in China, National Science Review,2021,nwab113,doi:10.1093/nsr/nwab113.

[5] 周天军,任俐文,张文霞.2020 年梅雨期极端降水的归因探讨和未来风险预估研究.中国科学:地球科学,2021,51(10):1637-1649.

[6] Wang J, ChenY, Liao W, et al. Anthropogenic emissions and urbanization increase risk of compound hot extremes in cities. Nature Climate Change,2021,11(12):1084-1089.

[7] IPCC. Summary for Policymakers. In:Stocker T F, Qin D, Plattner G K, et al. Climate Change 2013:The Physical Science Basis. Contribution of Working Group I to the Fifth Assessment Report of the Intergovernmental Panel on Climate Change. Cambridge:Cambridge University Press,2013.

[8] 周天军,吴波.年代际气候预测问题:科学前沿与挑战.地球科学进展,2017,32(4):331-341.

[9] Meehl G A, Richter J H, Teng H, et al. Initialized Earth System prediction from subseasonal to dec-

adal timescales. Nature Reviews Earth & Environment, 2021, 2(5): 340-357.

[10] Smith D M, Cusack S, ColmanA W, et al. Improved surface temperature prediction for the coming decade from a global climate model. Science, 2007, 317(5839): 796-799.

[11] Smith D M, Scaife A A, Eade R, et al. North Atlantic climate far more predictable than models imply. Nature, 2020, 583(7818): 796-800.

[12] Hu S, Zhou T. Skillful prediction of summer rainfall in the Tibetan Plateau on multiyear time scales. Science Advances, 2021, 7(24): eabf9395.

[13] Huang X, Zhou T, Dai A, et al. South Asian summer monsoon projections constrained by the interdecadal Pacific oscillation. Sci Adv, 2020, 6(11): eaay6546.

[14] Wu M, Zhou T, Li C, et al. A very likely weakening of Pacific Walker Circulation in constrained near-future projections. Nature Communications, 2021, 12: 6502.

[15] Charney J, Arakawa A, Baker D J, et al. Carbon dioxide and climate: A scientific assessment. In: Report of an Ad Hoc Study Group on Carbon Dioxide and Climate. Washington DC: National Academy of Sciences Press, 1979.

[16] Forster P, Storelvmo T, Armour K, et al. The earth's energy budget, climate feedbacks, and climate sensitivity. In: Masson-Delmotte V, Zhai P, Pirani A, et al. Climate Change 2021: The Physical Science Basis. Contribution of Working Group I to the Sixth Assessment Report of the Intergovernmental Panel on Climate Change. Cambridge: Cambridge University Press, 2021.

[17] Lee J Y, Marotzke J, Bala G, et al. Future global climate: Scenario-based projections and near-term information. In: Masson-Delmotte V, Zhai P, Pirani A, et al. Climate Change 2021: The Physical Science Basis. Contribution of Working Group I to the Sixth Assessment Report of the Intergovernmental Panel on Climate Change. Cambridge: Cambridge University Press, 2021, In Press.

[18] Ribes A, Qasmi S, Gillett N P. Making climate projections conditional on historical observations. Science Advances, 2021, 7(4): eabc0671.

[19] Zhou T, Zhang W. Anthropogenic warming of Tibetan Plateau and constrained future projection. Environmental Research Letters, 2021, 16(4): 044039.

[20] Chen X, Zhou T, Wu P, et al. Emergent constraints on future projections of the western North Pacific Subtropical High. Nat. Commun. , 2020, 11(1): 2802, doi: 10. 1038/s41467-020-16631-9.

[21] Li G, Xie S-P, He C, et al. Western Pacific emergent constraint lowers projected increase in Indian summer monsoon rainfall. Nature Climate Change, 2017, 7(10): 708-712, doi: 10. 1038/nclimate3387.

[22] Yan Y, Lu R, Li C. Relationship between the future projections of Sahel rainfall and the simulation biases of present South Asian and Western North Pacific rainfall in summer. Journal of Climate, 2019, 32(4): 1327-1343.

[23] Wang T, Zhao Y, Xu C, et al. Atmospheric dynamic constraints on Tibetan Plateau freshwater under Paris climate targets. Nature Climate Change, 2021, 11(3): 219-225.

[24] Zscheischler J, Westra S, van den Hurk B J J M, t al. Future climate risk from compound events,

Nature Climate Change,2018,8(6):469-477.

[25] Zhang W,Furtado K,Wu P,et al. Increasing precipitation variability on daily-to-multiyear timescales in a warmer world. Science Advances,2021,7(31):eabf8021.

Progresses in Climate Change Detection, Attribution and Projection

Zhou Tianjun,Zhang Wenxia,Hu Shuai,
Chen Xiaolong,Jiang Jie,Wu Mingna

The observed climate change is contributed by both natural variability and anthropogenic forcing. Climate change detection, attribution and projection has been the research foci of climate research community. Here, we present a review on the progresses of international community and highlight the recent achievements in climate model development, the detection and attribution of climate changes with focus on extreme precipitation and extreme heat events, near-term decadal climate prediction, the estimation of climate sensitivity, the constrained projection of future climate and the projection of extreme events including its variability. Key issues calling for further research are also discussed.

2020年中国科研代表性成果

Representative Achievements of Chinese Scientific Research in 2020

3.1 凯勒-里奇流研究的突破

陈秀雄 王 兵

（中国科学技术大学几何与物理研究中心）

自然界中星系的爆破坍塌、热气球的膨胀收缩、肥皂泡的缘起缘灭，如此种种都是空间形状的演化。不同机制的空间演化由不同方程所主导。里奇流是众多空间演化方程中的一种，其本质上是一种热方程，由美国数学家哈密顿（Hamilton）开创[1]。在自然界中，温度分布的演化满足热方程。让 $u=u(\vec{x}, t)$ 代表在时间 t 时刻、在位置 $\vec{x}=(x, y, z)$ 处的温度，在合理假定下，$u(\vec{x}, t)$ 满足如下的偏微分方程：

$$\frac{\partial}{\partial t}u = \Delta u(\vec{x}, t).$$

其中 $\Delta=\partial_{xx}^2+\partial_{yy}^2+\partial_{zz}^2$ 是拉普拉斯（Laplace）算子。在时间充分大之后，u 将趋于稳态，不再随着时间演化。此时上述方程化为

$$0 = \Delta u(\vec{x}, \infty).$$

这意味着 $u(\cdot, \infty)$ 是调和函数。上述找到调和函数的过程称为热流。当我们把函数换成黎曼（Riemann）度量 g，则相应的热流就是里奇流。事实上，给定一个背景空间 M（也叫流形），黎曼度量 g 可以直观理解成一种计算距离的光滑标尺。好的标尺可以清晰测量空间全貌。理论物理告诉我们，最佳标尺是满足如下方程的爱因斯坦（Einstein）度量：

$$Ric = kg. \tag{1}$$

这里 Ric 代表里奇曲率，是一种描述空间弯曲程度的量，而 k 是爱因斯坦常数。如果初始度量 g_0 不是爱因斯坦度量，我们期望用类似热流的方法使其越来越接近爱因斯坦度量。这个热流的对应就是里奇流。它是关于度量 g 的热流，方程可以写为：

$$\frac{\partial}{\partial t}g = -2Ric + rg. \tag{2}$$

这里 r 是一个依赖于时间的数。纯粹类比常规热方程，人们会问是否方程（2）最终会到达一种稳态，使得 g 不再随时间变化？由于里奇曲率中各种量高度纠缠在一起，

所以上述方程的特性要比普通热方程复杂得多。方程（2）一般不会在通常意义下收敛，总会碰到奇点。当空间维数是 3 时，由于哈密顿和佩雷尔曼等人的工作，里奇流的奇点可以被清晰地解构[2]。当空间的维数更高时，里奇流奇点会变得更加复杂。所以人们一般只能研究特殊流形上的里奇流。其中非常重要的一类叫作法诺流形，它是一种特殊凯勒流形。

我们发展了哈密尔顿和佩雷尔曼的分析，将法诺流形上凯勒里奇流的奇点分析清楚，从而证明了此类里奇流的收敛性质[3]，也称为哈密顿-田猜想[2,4]。自 20 世纪 80 年代初曹怀东证明长时间存在性[5]起，该收敛问题就一直是里奇流领域内的核心问题。它和诸多重要问题紧密联系。譬如，基于此猜想的解决，我们率先证明了田［刚偏零阶估计］猜想，也可以得到丘成桐稳定性猜想的一个简洁证明[6]。另外，此猜想的证明也诱导了平均曲率流延拓猜想的解决[7]。

此项研究关键处在于转化视角。原本哈密顿-田猜想关注的是空间本身的紧性，我们观察到研究时空整体紧性才是更加自然的问题。在研究过程中，我们引入了若干新概念，譬如典则半径、锥形等。我们的定理指出，法诺流形上里奇流的极限是锥形上的孤立子。近年来里奇流的发展越来越明显地表明，我们定义的锥形在里奇流研究中具有普遍意义：非坍缩极限里奇流的切流似乎都是发生在锥形上的。相关概念和研究方法在高维里奇流研究中必将起到重要作用。

参考文献

[1] Hamilton R S. Three-manifolds with positive Ricci curvature. Journal of Differential Geometry, 1982,17(2):255-306.

[2] Perelman G. The entropy formula for the Ricci flow and its geometric applications. arXiv: math. DG/0211159.

[3] Chen X X, Wang B. Space of Ricci flows(II)-part B: weak compactness of the flows. Journal of Differential Geometry,2020,116:1-123.

[4] Tian G. Kahler-Einstein metrics with positive scalar curvature. Inventiones Mathematicae, 1997, 130:1-39.

[5] Cao H D, Deformation of Kahler metrics to Kahler-Einstein metrics on compact Kahler manifolds. Inventiones Mathematicae,1985,81(2):359-372.

[6] Chen X X, Sun S, Wang B. Kahler-Ricci flow, Kahler-Einstein metric and K-stability. Geometry and Topology,2018,22(6):3145-3173.

[7] Li H Z, Wang B. The extension problem of the mean curvature flow (I). Inventiones Mathematicae, 2019,218(3):721-777.

On the Convergence of the Kahler-Ricci Flow on Fano Manifold

Chen Xiuxiong，Wang Bing

Ricci flow is a family of Riemannian metrics $g(t)$，satisfying the equation

$$\frac{\partial}{\partial t}g = -2Ric + rg.$$

When the underlying manifold is a Fano manifold and the initial metric is in a class proportional to the first Chern class，the flow is called the Kähler-Ricci flow on Fano manifold. The convergence of the flow is a fundamental problem in the field and it is deeply connected to many other important problems in differential geometry. It was conjectured to hold by Hamilton-Tian. Based on the compactness of the moduli of non-collapsed Calabi-Yau spaces with mild singularities，the authors set up a structure theory for polarized Kähler Ricci flows with proper geometric bounds. This theory is a generalization of the structure theory of non-collapsed Kähler Einstein manifolds. As applications，the authors prove the Hamilton-Tian conjecture and the partial-C_0-conjecture of Tian in full generality.

3.2 "中国天眼"探索快速射电暴的起源

李柯伽

（北京大学/中国科学院国家天文台）

快速射电暴（fast radio burst，FRB）是宇宙中偶发的无线电爆发事件。在几毫秒的时间内，它们所释放的无线电波段的能量，相当于世界当前总发电量累计几百亿年的总和。目前 FRB 研究的两个核心问题是：①快速射电暴来源于何种天体；②高强度射电辐射的辐射机制是什么。关于这两个问题，有很多猜测，对 FRB 中心引擎猜测已经基本集中在中子星、黑洞相关的致密星图像上，但是辐射产生的机制却包含"激波产生"和"磁层产生"两类可能性[1]。

依托于我国自主研制的国家重大科技基础设施"中国天眼"500m 口径球面射电望远镜（FAST），本研究团队开展了对 FRB 的深入观测。借助于 FAST 极高的灵敏度和优异的偏振响应保真度，我们团队得以首次系统地研究了 FRB 重复爆发的偏振多样性并结合高能数据对快速射电暴的脉冲触发机制给出了限制。

快速射电暴相干辐射可能来自中心天体磁层内的物理过程，也可能来自远离中心天体的相对论性激波的作用过程。2019 年，本课题组和合作者监测到快速射电暴 FRB180301 的 15 次重复爆发。在其中 7 次爆发中探测到射电辐射的偏振多样性（图 1）。这个观测支持磁层辐射起源，却不支持传播机制的激波辐射模型。这项研究首次确定了磁层是快速射电暴的产生地，成果于 2020 年 10 月 28 日发表在国际学术期刊《自然》上[2]。

虽然 FAST 的观测结果说明快速射电暴的辐射产生于磁层内，但是其中心天体仍旧未知。2020 年 4 月 28 日，国际上两台射电望远镜（位于加拿大的 CHIME 和位于美国的 STARE2）分别独立发现了一例快速射电暴（FRB200428），并确认其来自银河系内正处于活跃期的磁星 SGR J1935+2154。在 FAST 监测期间，磁星 SGR J1935+2154 释放出 29 个 X 射线爆发，课题组和来自拉斯维加斯大学、北京师范大学、中国科学院高能物理研究所、中国科学院紫金山天文台、南京大学等国内外研究机构的合作团队在对应的射电数据中排除了来自磁星 SGR J1935+2154 的显著信号。FAST 的观测对磁星的射电辐射给出了最严格的流量限制，同时说明快速射电暴与磁星高能爆发的相关性较弱。该研究对磁星产生快速射电暴物理机制的研究有非常强的限制和推动作用，成果于 2020 年 11 月 4 日发表在《自然》[3]上。

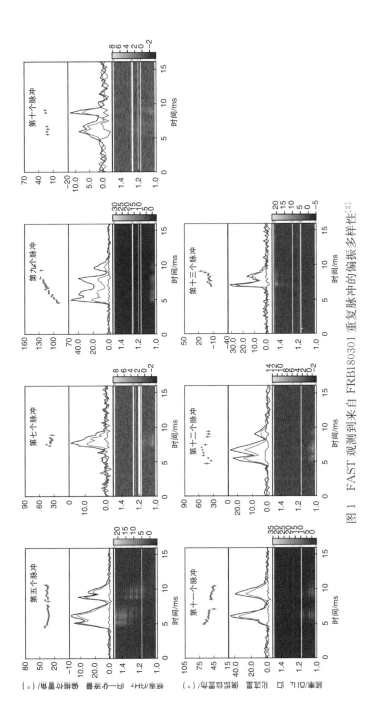

图 1　FAST 观测到来自 FRB180301 重复脉冲的偏振多样性[2]

在这两项研究中，FAST 的高灵敏度使得我们能够探测到来自 FRB 的较弱的脉冲信号或者对辐射强度给出最佳限制。这些观测资料无论是从更大样本的角度看还是从更高灵敏度的参数空间角度看，都充分展示了未来基于 FAST 深入研究 FRB 的广阔前景。

参考文献

[1] Zhang B. The physical mechanisms of fast radio bursts. Nature, 2020, 587: 45-53.

[2] Luo R, Wang B J, Men Y P, et al. Diverse polarization angle swings from a repeating fast radio burst source. Nature, 2020, 586: 693-696.

[3] Lin L, Zhang C F, Wang P, Gao, et al. No pulsed radio emission during a bursting phase of a Galactic magnetar. Nature, 2020, 587: 63-65.

Explore the Origin of Fast Radio Bursts Using FAST

Li Kejia

Fast Radio Bursts (FRB) are millisecond-duration radio burst occurring randomly in the universe. The energy of radio wave released by individual burst is equivalent to the total amount of global electricity generated for tens of billions of years. The two cutting-edge questions are: 1) what kind of celestial body the fast radio burst originates from, and 2) what is the radiation mechanism of radio burst. Using observation from FAST radio telescope, we carried out in-depth studies with FRBs. We confirmed that: 1) magnetospheric rather than shock wave-like environments are preferred for FRB and 2) provided evidence that the radio bursts is not directly associated with high-energy processes.

3.3 具有超高压电性能的透明铁电单晶

徐 卓 李 飞 邱超锐

（西安交通大学）

弛豫铁电单晶（relaxor ferroelectric single crystal）是压电性能最为优异的一类无机非金属材料，具有高压电常数、大应变量、高机电耦合系数和高储能密度，被认为是发明压电陶瓷半个世纪以来的重大突破。它的压电系数可达 2000 pC/N，比锆钛酸铅（PZT）压电陶瓷高 4～6 倍，应变量（1%～2%）比 PZT 压电陶瓷高一个数量级，储能密度（约为 130 J/cm³）比 PZT 陶瓷高 2 个数量级。凭借优异的压电性能，弛豫铁电单晶材料自 20 世纪 90 年代问世以来即成为铁电压电领域研究的热点材料，并为诸多压电器件（尤其是医疗超声探头和水声换能器）的性能提升带来了一次重大的机遇[1]。

然而，随着人们对压电器件性能要求的日益提高，如何进一步优化弛豫铁电单晶材料的性能成为当前急需解决的重要科学问题。自弛豫铁电单晶发明 20 多年以来，其压电性能就再没有过突破，急需新的设计方法来进一步提升其性能。此外，铁电畴壁的存在导致弛豫铁电单晶透光率低，因此无法满足当前压电器件小型化、高灵敏和多功能化的发展需求。

通过相场的方法模拟 PMN-PT 晶体内部的 71°铁电畴在交流电场作用下的演变过程（图 1），我们发现，在电场极性反转的过程中，71°畴壁在（011）和（01$\bar{1}$）平面之间摆动，相邻的 71°铁电畴有相互融合的趋势，导致 71°铁电畴密度降低，铁电畴尺寸增大。而 PMN-PT 晶体内部密集的 71°铁电畴壁正是引起光线散射的"元凶"。由此我们提出，利用交变电场来消除这些对光有散射作用的铁电畴壁，首次在弛豫铁电单晶中同时获得了高压电性（压电系数为 2100 pC /N）和高透光性（图 2），突破了长期以来二者难以共存的瓶颈问题，为研制高性能透明压电材料提供了重要的理论方法。

弛豫铁电单晶的压电性与其内部畴结构有着密切的关系。我们在研究中发现，交流极化 PMN-PT 晶体，改善了其透光性，但并未降低其压电性，相反，其压电性相比传统直流极化的单晶高出 30% 以上。结合相场模拟和原位实验表征，我们发现在 PMN-PT 晶体中，减小畴壁密度（或增大电畴尺寸）会使相应的自由能曲线更为平坦，从而导致晶体压电和介电性能大幅增加。这一新发现颠覆了高畴壁密度产生高压电效应的传统认识，为今后压电材料的设计提供了新思路[2]。

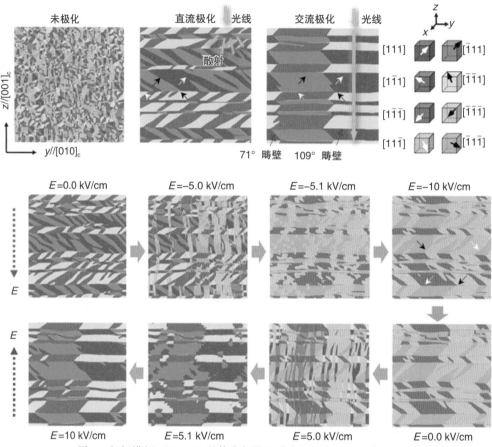

图 1　相场模拟 PMN-PT 晶体内部铁电畴在电场作用下的演变情况

E：极化电场强度；x，y，z：坐标轴方向

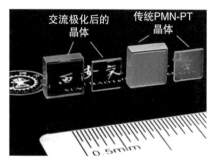

图 2　透明弛豫铁电单晶材料

该成果于 2020 年 1 月发表在《自然》期刊上，并入选"2020 年度中国科学十大进展"。文章发表后，《科学》期刊官网以"自发电透明晶体让隐身机器人和自供电触摸屏成为可能"为题刊文评论[3]。《自然》期刊评论："这项成果将高性能压电晶体列入了光学器件行列，打开了声-光-电器件设计的大门。"[4]

参考文献

[1] Zhang S,Li F,Jiang X,et al. Advantages and challenges of relaxor-PbTiO$_3$ ferroelectric crystals for electroacoustic transducers:A review. Progress in materials science,2015,68:1-66.

[2] Qiu C,Wang B,Zhang N,et al. Transparent ferroelectric crystals with ultrahigh piezoelectricity. Nature,2020,577(7790):350-354.

[3] Service R. Transparent,power-producing crystals could lead to invisible robots,self-powered touch screens. Science,2020. doi:10. 1126/science. aba9180.

[4] Koruza J. Transparent crystals with ultrahigh piezoelectricity. Nature,2020. 577:325-326.

Transparent Ferroelectric Crystals with Ultrahigh Piezoelectricity

Xu Zhuo,*Li Fei*,*Qiu Chaorui*

Relaxor ferroelectric single crystal possesses excellent piezoelectric effect,but its poor transparency greatly restricts the development of photoacoustic technology and electro-optic technology. Through phase-field simulations and extensive experiments,we developed a relatively simple method of using an alternating-current electric field to engineer the domain structures of originally opaque rhombohedral Pb(Mg$_{1/3}$Nb$_{2/3}$) O$_3$-PbTiO$_3$(PMN-PT) crystals to simultaneously generate near-perfect transparency,an ultrahigh piezoelectric coefficient d_{33} ($>$2,100 pC/N),an excellent electromechanical coupling factor k_{33} (about 94 %) and a large electro-optical coefficient γ_{33} ($\sim$220 pm/V). We find that increasing the domain size leads to a higher d_{33} value for the [001]-oriented rhombohedral PMN-PT crystals, challenging the conventional wisdom that decreasing the domain size always results in higher piezoelectricity. This work presents a paradigm for achieving high transparency and piezoelectricity by ferroelectric domain engineering,and provides a brand-new key material for the development of high-performance photoacoustic imaging transducers,electro-optic modulators and optical phased arrays.

3.4　单个超冷分子的相干合成

何晓东　詹明生

（中国科学院精密测量科学与技术创新研究院）

原子-分子是物质组成的一个基本层次，如何从原子可控地生成功能更加强大的分子一直是物理学和化学关心的基础前沿问题[1]。一方面，具有复杂能级结构的分子是化学、凝聚物质和生命本身等更复杂现象出现的物质单元；另一方面，即使是由少量原子组成的简单分子，其系统复杂性也已达到了从量子力学波动方程出发来完全理解其特性的上限。因此，精确控制单分子态和相互作用的能力不仅能够促进人们对于最基本的量子力学基础上的化学过程的认识，而且还可用于凝聚态体系的量子模拟、检验基本物理规律的精确测量、量子信息处理等领域的前沿科学研究[2,3]。

实现对单个分子完全控制意味着既要求分子在实验室坐标系下运动极其缓慢，即实现对单个分子的冷却，又要实现对单分子内部自由度的相干操控。通向单分子完全操控的有效途径之一是借助光阱中完全操控单原子的能力[4]。以双原子分子为例，首先在单个光阱中制备原子样品，进而通过操控原子间的相互作用将原子合成单个超冷弱束缚分子（图1）。

图1　光镊中单分子合成

在光阱中将两个原子合成单个分子，圆球代表不同的原子，波浪线代表电磁场

沿着这一前景广阔但又极具挑战的研究方向，国际上包括中国科学院精密测量科学与技术创新研究院、美国哈佛大学、英国杜伦大学等在内的科研机构的若干研究组开展了将光镊中的超冷双原子样品合成单个双原子分子的工作。在通常的原子合成分子的方案中即传统的费希巴赫共振（Feshbach resonance）以及光缔合这两种方法，前者只适用于特定的原子-分子系统[5]，后者虽较为普适，但是常伴有强的难以克服的

退相干效应，难以实现单分子的相干合成[6,7]。

笔者团队为了克服合成单分子过程中的退相干问题，巧妙地应用了光镊中固有的偏振梯度场对原子产生的等效梯度磁场效应[8]，实现了原子自旋与双原子相对运动的耦合，即所谓的 SMC 机制，并选用了微波而非激光合成分子。相对于激光，微波不会对原子产生自发辐射等退相干效应。最终，该团队在单个光阱中实现了一对超冷异核原子（^{85}Rb 和^{87}Rb）到单个分子（$^{85}Rb^{87}Rb$）之间的相干超冷化学反应，在国际上首次实现了单个分子的相干合成，通过调整微波脉冲的长度，团队观察到了光阱中双原子与单个分子之间长寿命的拉比振荡（图2），即实现了双原子量子体系中原子态与分子态的可控相干叠加与相互转化，奏响了一首优美的双原子"圆舞曲"[9]。

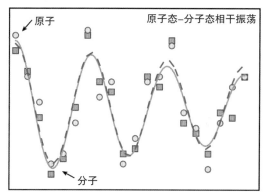

图2 原子态-分子态之间的相干拉比振荡[9]

这项工作从科学上证明了，基于原子之间核间距自由度的相干控制可用于实现分子合成，是纯净的分子态操控方法，具有优越的相干性。这是一种全新的分子合成范式，从物理机制上有别于传统的光缔合和磁缔合的分子合成方案。该新范式兼顾了以上两种方案的优点：既有与基于费希巴赫共振一样的相干特性，又适用于更多的原子-分子系统，如重要的碱金属-碱土金属混合双原子分子系统。这项工作开启了原子-分子体系所有自由度全面相干操控的研究大门。为基元化学反应过程相干控制、量子少体束缚态的相干合成及其量子调控提供了可能性，具有重要的科学价值。

参考文献

[1] Feynman R P. There's plenty of room at the bottom. Caltech's Engineering and Science Magazine，1960,23:22.

[2] Bohn J L,Rey A M,Ye J. Cold molecules:Progress in quantum engineering of chemistry and quantum matter. Science,2017,357:1002.

[3] Lu B, Wang D J. Ultracold dipolar molecules. Acta Physica Sinica, 2019, 68:043301.

[4] Xu P, Yang J H, Liu M, et al. Interaction-induced decay of a heteronuclear two-atom system. Nature Communications, 2015, 6:7803.

[5] Zhang J T, Yu Y C, Cairncross W B, et al. Forming a single molecule by magnetoassociation in an optical tweezer. Physical Review Letters, 2020, 124:253401.

[6] Liu L R, Hood J D, Yu Y, et al. Building one molecule from a reservoir of two atoms. Science, 2018, 360:900.

[7] Liu L R, Hood J D, Yu Y, et al. Molecular Assembly of ground-state cooled single atoms. Physical Review X, 2019, 9:021039.

[8] Wang K P, Zhuang J, He X D, et al. High-fidelity manipulation of the quantized motion of a single atom via Stern-Gerlach splitting. Chinese Physics Letters, 2020, 37:044209.

[9] He X D, Wang K P, Zhuang J, et al. Coherently forming a single molecule in an optical trap. Science, 2020, 370:331-335.

Coherently Forming an Ultracold Single Molecule

He Xiaodong, Zhan Mingsheng

Atom-molecules are a fundamental level of matter composition. The formation of ultracold molecules has had a profound impact on many research areas of physics, ultracold chemistry, precision measurements, quantum simulation, and quantum computation. However, conventional methods of producing such molecules are attainable only for a limited number of systems or they suffer for strong dephasing. The CAS team led by Mingsheng Zhan reported an alternative route to coherently bind two atoms into a weakly bound molecule by coupling atomic spins to their two-body relative motion in a strongly focused laser with inherent polarization gradients. They realized a successful assembly of an ultracold ^{87}Rb-^{85}Rb molecule in an optical tweezer and observed coherent, long-lived atom-molecule Rabi oscillations. They further demonstrated the full control of the internal and external degrees of freedom in the atom-molecule system.

3.5 中国科学家实现超越经典算力的光量子计算

钟翰森[1,2] 陆朝阳[1,2] 潘建伟[1,2]

（1. 中国科学技术大学微尺度物质科学研究中心和近代物理系
2. 量子信息与量子科技前沿协同创新中心）

相比经典计算机，量子计算机有望通过特定算法，在解决一些具有重大社会和经济价值的问题时，实现指数级别的加速，比如密码破译[1]、大数据搜索[2]、量子化学[3]等。当前，研制量子计算机已成为世界科技前沿的最大挑战之一，成为欧美各发达国家角逐的焦点。但是，受限于当前的实验技术水平，搭建有实用价值的大规模容错量子计算机仍然需要长期的努力。如果能基于现有的实验技术，在特定问题上，证明专用量子设备能超越经典超级计算机，或称为量子计算优越性，那么将会给量子计算的进一步发展奠定坚实的基础。

近年来，随着量子技术的发展，实验演示量子计算优越性已经成为可能。玻色采样由阿伦森（Aaronson）和阿尔希波夫（Arkhipov）于 2011 年提出，可用于实现光量子计算优越性[4]，并在 2017 年由哈密顿（Hamilton）等提出了其修改版——高斯玻色采样[5]。基于高斯玻色采样，中国研究者们在 2020 年完成了"九章"光量子计算原型机[6]。

为了完成"九章"实验，笔者团队研发了平均纯度达到 0.938，收集效率为 0.628，压缩系数达到约 1.5 的量子光源；实现了三维结构的 100×100 模式干涉仪，其通过率约为 97.7%，波包重合度优于 99.5%；设计了主动锁相系统和被动稳相系统来保持光路相位稳定；光子从干涉仪输出后，使用平均探测效率超过 80% 的单光子探测器进行探测，并产生最终的样本。

然而，不同于大数分解等问题很容易就能验证结果的正确性，验证采样的正确性并不容易。笔者团队将实验采样分为三个部分：①2～4 个光子的简单部分，可以计算出实验分布和理论分布的相似度；②5～40 个光子的稀疏部分，态空间远比样本量大，无法得出实验分布，但可以通过统计学的方法排除各种潜在的经典易于计算的假冒样本；③40 个光子以上的困难部分，这部分是量子优越性的区域，无法直接对结果做验证，可以测量统计量进行侧面的验证。

首先，只选择 3 个光源作为输入，这个时候实验处于简单区域；一共对 23 种不同的光源组合进行实验，计算出实验样本和理论分布的相似度平均值为 0.990（1）[1]，距离平均值为 0.103（1），验证了少光子数时实验装置的正确性；随后使用了全部的 25 个光源，对完整系统进行检验。对于稀疏部分，使用高权重样本生成（HOG）方法进行验证，排除了最常见的热态分布；对于完整的样本，包括困难区域的样本，计算了样本的两点关联，发现和理论值非常吻合，同时排除了可分辨光子的假设和热态的假设。这些证据，有力地佐证了采样结果的正确性。

为了确认量子计算优越性的存在，同时也为了对稀疏采样区域的结果做验证，笔者团队在神威·太湖之光超级计算机上实现了高斯玻色采样算法，并对可以计算的 40 个光子以下的部分进行了验证。据估算，"九章"光量子计算原型机的采样速度比运行现有采样算法的超级计算机快 14 个数量级。

在"九章"实验中，笔者团队基于高斯玻色采样，在光学体系中实现了量子计算优越性。2020 年 12 月 18 日，《科学》期刊以封面标题的形式发表了此研究成果。这是国际上首次用光子实现量子计算优越性（图 1）。2021 年，团队在"九章"光量子计算原型机的基础上，进行了一系列概念和技术创新，进一步研发"九章二号"量子计算原型机。首先，受到激光（受激辐射光放大）概念的启发，研究人员设计并实现了受激双模量子压缩光源，显著提高了量子光源的产率、品质和收集效率。其次，通过三维集成和收集光路的紧凑设计，多光子量子干涉线路增加到了 144 维。由此，"九章二号"探测到的光子数增加到了 113 个，输出态空间维度达到了 10^{43}。进一步，通过动态调节压缩光的相位，研究人员实现了对高斯玻色取样矩阵的重新配置，演示了"九章二号"可用于求解不同参数数学问题的编程能力。根据目前已正式发表的最优化经典算法，"九章二号"在高斯玻色取样这个问题上的处理速度比最快的超级计算机快亿亿亿倍（即 10^{24} 倍）[7]。

在将来的实验中，笔者团队还将进一步优化升级实验装置，实现可完全编程的高斯玻色采样，同时排除各种目前实验可能存在的漏洞。此外，学界当前正在探索基于现有的量子设备，实现实际应用的可能方案。基于"九章"实验平台，笔者团队将尝试应用到其他有现实意义的问题，比如分子对接[8]、图同构[9]、量子近似优化[10]、分子振动谱[11]等，而不再仅局限于采样。更进一步地，笔者团队还将基于现有的实验技术，探索光学容错量子计算路线，包括 Gottesman-Kitaev-Preskill（GKP）编码[12]、猫态编码[13]等量子纠错方案，为实现真正有实用意义的大规模通用量子计算机奠定基础。

---

① （1）中的 1 表示误差，对应前面数字的最后一位，例如 0.990（1）表示 0.990＋/－0.001。

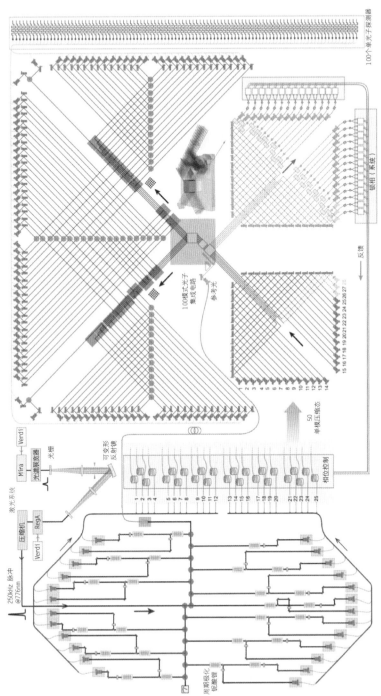

图 1 "九章"光量子计算原型机的原理图

"九章"光量子计算原型机的系统共由四部分组成：压缩光源、锁相系统、干涉仪和单光子探测器。首先，将总功率为 1.4 W 的变极限单脉冲激光分为 13 路，泵浦 25 个 PPKTP 晶体；每路产生的双模压缩光分别被收集到单模光纤中，然后将其注入 100 模式干涉仪；继而使用锁相系统稳定每个压缩光源的相位；最后，输出态由 100 个超导纳米线单光子探测器探测

　　此项研究工作由中国科学技术大学、量子信息与量子科技前沿协同创新中心、中国科学院上海微系统与信息技术研究所、清华大学和北京信息科学与技术国家研究中心的研究团队合作完成。

参考文献

[1] Shor P W. Algorithms for quantum computation: Discrete logarithms and factoring. In: IEEE. Proceedings 35th Annual Symposium on Foundations of Computer Science. 1994:124-134.

[2] Grover L K. A fast quantum mechanical algorithm for database search. In: Association for Computing Machinery. Proceedings of the Twenty-Eighth Annual ACM Symposium on Theory of Computing. STOC'96. 1996:212-219.

[3] Lanyon B P, Whitfield J D, Gillett G G, et al. Towards quantum chemistry on a quantum computer. Nature Chemistry, 2010, 2(2):106-111.

[4] Aaronson S, Arkhipov A. The computational complexity of linear optics. In: ACM. Proceedings of the Forty-Third Annual ACM Symposium on Theory of Computing, 2011:333-342.

[5] Hamilton C S, Kruse R, Sansoni L, et al. Gaussian boson sampling. Physical Review Letters, 2017, 119(17):170501.

[6] Zhong H S, Wang H, Deng Y H, et al. Quantum computational advantage using photons. Science, 2020, 370(6523):1460-1463.

[7] Zhong H S, Deng Y H, Qin J, et al. Phase-programmable Gaussian boson sampling using stimulated squeezed light. Physical Review Letters, 2021, 127(18):180502.

[8] Banchi L, Fingerhuth M, Babej T, et al. Molecular docking with Gaussian boson sampling. Science Advances, 2020, 6(23):eaax1950.

[9] Brádler K, Friedland S, Izaac J, et al. Graph isomorphism and Gaussian boson sampling. Special Matrices, 2021, 9(1):166-196.

[10] Arrazola J M, Bromley T R, Rebentrost P. Quantum approximate optimization with Gaussian boson sampling. Physical Review A, 2018, 98(1):012322.

[11] Huh J, Guerreschi G G, Peropadre B, et al. Boson sampling for molecular vibronic spectra. Nature Photonics, 2015, 9(9):615-620.

[12] Gottesman D, Kitaev A, Preskill J. Encoding a qubit in an oscillator. Physical Review A, 2001, 64(1):012310.

[13] Cochrane P T, Milburn G J, Munro W J. Macroscopically distinct quantum-superposition states as a bosonic code for amplitude damping. Physical Review A, 1999, 59(4):2631-2634.

Chinese Scientists Achieve Quantum Computational Advantage

Zhong Hansen，Lu Chaoyang，Pan Jianwei

Quantum computers are expected to achieve exponential speedup when solving some important problems，such as password cracking，big data searching and quantum chemistry. However，building a large-scale fault-tolerant quantum computer still requires long-term efforts. Quantum computational advantage，that is，based on existing technologies to prove that special quantum devices can surpass classic supercomputers when we solve some certain problem，is a milestone in the area of quantum computing. Recently，Chinese scientists build a prototype photonic quantum computer named *Jiuzhang*，and demonstrated the 76-photon and 113-photon Gaussian Boson sampling experiments. The sampling rate is faster than classical supercomputer using the state-of-the-art algorithm 14 and 24 orders of magnitude，respectively. This means that China has become the second country that achieves quantum computational advantage. Related research results were published in the academic journals *Science* and *Physical Review Letters*.

3.6 "分子围栏"材料催化甲烷
低温氧化制甲醇

王 亮 肖丰收

（浙江大学，化学工程与生物工程学院）

催化在国民经济生产中扮演着关键的角色，超过80%的化学反应过程涉及催化。但是当前多数催化剂依然存在活性或选择性不足的问题，尤其针对惰性分子活化转化等极具挑战的反应过程。高效新颖催化剂的设计合成是催化研究永恒的主题，多通过对催化活性物种本身进行调变以提升性能。不同于这一传统思维，浙江大学肖丰收教授团队创新性地提出控制扩散提高催化性能的新策略，首次提出了"分子围栏"的催化剂设计理念，据此设计合成了相应的催化剂结构来控制关键物种的扩散，在这一重要研究领域取得了国际领先的研究成果。

该研究团队发现氢气和氧气原位生成的过氧化氢物种在反应体系中的空间分布是关键控制因素，因此致力于在微观尺度上控制过氧化氢物种的扩散行为，实现局部高浓度富集以提升其氧化能力。根据这一理念，研究团队原创性地提出"分子围栏"催化剂设计理论和方法。团队通过将AuPd合金纳米颗粒固定在铝硅酸盐沸石晶体内，然后用有机硅烷对沸石外表面进行改性，合成了新型沸石分子筛催化材料。在低温甲烷氧化反应中，由于疏水壳层的"围栏"效应，过氧化氢物种被富集在沸石晶体内部形成局部高浓度，较传统反应体系浓度提升了五个数量级（图1）[1,2]。得益于该富集效应，在70℃下，突破性地得到17.2%的甲烷转化率并且保持甲醇选择性在92%，甲醇的生成速率达到91.6 mmol/（g_{AuPd}·h）这是同类反应体系中的最高水平，相关成果发表于《科学》期刊[1]，在甲烷转化等领域内产生了重要影响。该方法通过反应过程中过氧化氢的富集，实现了催化过程中扩散与反应的高效协同，获得同体系中迄今的最高甲醇产率，从而解决甲烷低温难以活化的巨大挑战，以及高甲烷转化率和高甲醇选择性不可兼得的领域难题。该研究不仅为甲烷的低温氧化提供了新的催化体系，也为多相催化剂的设计提供了新的思路。

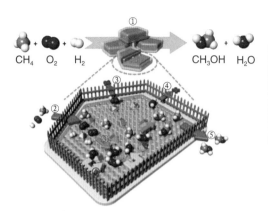

反应过程

①沸石封装纳米金属，外表面修饰疏水壳层
②CH₄、O₂、H₂等分子进入沸石内部
③H₂+ O₂形成H₂O₂
④疏水壳层限制H₂O₂扩散，实现富集
⑤时间延长，产物扩散

图 1　在"分子围栏"催化剂上甲烷氧化的反应过程

参考文献

[1] Jin Z,Wang L,Zuidema E,et al. Hydrophobic zeolite modification for *in situ* peroxide formation in methane oxidation to methanol. Science,2020,367:193-197.

[2] Jin Z,Liu Y,Wang L,et al. Direct synthesis of pure aqueous H₂O₂ solution within aluminosilicate zeolite crystals. ACS Catalysis,2021,11:1946-1951.

Low-Temperature Oxidation of Methane to Methanol over a Molecular-Fence Catalyst

Wang Liang ,Xiao Fengshou

Catalysis plays a key role in the national economy. More than 80% of chemical reaction processes involve catalysis. However, most catalysts still have the problem of insufficient activity or selectivity, especially for the challenging reaction processes such as activation and conversion of inert molecules. The design and synthesis of efficient and novel catalysts is an eternal theme of catalytic research. For example, selective partial oxidation of methane to methanol suffers from low efficiency. Here, we report a heterogeneous catalyst system for enhanced methanol productivity in methane oxidation by *in situ* generated hydrogen peroxide at mild temperature(70℃). The catalyst was synthesized by fixation of

AuPd alloy nanoparticles within aluminosilicate zeolite crystals, followed by modification of the external surface of the zeolite with organosilanes. The silanes appear to allow diffusion of hydrogen, oxygen, and methane to the catalyst active sites, while confining the generated peroxide there to enhance its reaction probability. At 17.3% conversion of methane, methanol selectivity reached 92%, corresponding to methanol productivity up to 91.6 millimoles per gram of AuPd per hour.

3.7　黑磷复合材料"界面重构"实现高倍率高容量锂存储

季恒星

（中国科学技术大学化学与材料科学学院）

　　电动汽车和5G通信等多领域技术的进步对具有快充能力高能量密度电池的需求日益迫切。2020年10月获国务院批准的《新能源汽车产业发展规划（2021—2035年）》也着重提出了对快充电池技术的需求[1]。然而，高能量密度和快充能力难以兼得，这主要受限于负极材料的倍率性能[2]。快速充电时，大电流导致负极极化加剧并迅速达到截止电压，致使电极容量无法发挥，电池能量密度降低；另一方面，负极极化容易导致金属锂析出，产生安全隐患[3]。如何突破电极极化对快速充电时电池能量密度和安全性的限制是目前电池技术面临的一个关键科学难题。在产生电极极化的诸多因素中，锂离子扩散过程，尤其是锂离子跨界面扩散过程缓慢是主要因素。因此，设计具有高界面离子扩散能力，同时兼具高容量和循环稳定的负极材料体系，并系统认识界面结构对倍率性能的影响因素是目前研究面临的重大挑战。

　　对此，笔者研究团队以黑磷这种具有高理论容量和高离子扩散能力的负极材料为基础，从电极界面设计的角度出发，提出了一种崭新的"固体界面共价键合引发界面重构"的设计策略（图1）。在该设计中，一方面，黑磷和石墨边界处的磷、碳原子通过共价键连接，使黑磷边界磷原子重排（图2），获得界面结构稳定、锂离子传导顺畅

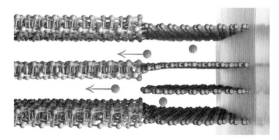

图1　"化学键合"的黑磷-石墨复合电极材料结构模型示意图

图中紫色、黑色、蓝色和浅蓝色分别代表磷、碳、锂和聚苯胺

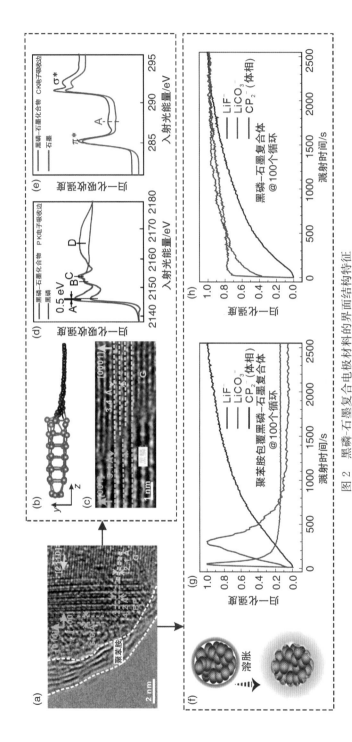

图 2 黑磷-石墨复合电极材料的界面结构特征

(a) 黑磷-石墨复合材料的透射电镜图像；(b) 和 (c) 黑磷-石墨界面结构；(d) 和 (e) X 射线吸收谱验证磷-碳键的形成；(f)～(h) 表面修饰对固态电解质界面膜的改善作用

的黑磷-石墨复合材料。锂离子扩散势垒降低67%。另一方面，通过聚合物包覆，在黑磷-石墨颗粒表面形成富含有机组分的固态电解质界面膜，有利于锂离子的快速传导。通过"界面工程"对锂离子传导能力进行改善，该复合材料仅需充电9分钟即可恢复约80%的电量，2000次充放电循环后仍可保持90%的容量。该负极材料为生产出具有高于350 Wh/kg能量密度并具备快充能力的锂离子电池带来了希望。与此同时，研究人员结合多种原位和非原位的表征技术，系统研究了界面结构对锂离子传输和材料电化学性能的作用规律。

相关系列工作从复合材料界面重构出发降低电极反应的离子电阻，且利用即时产生的层状结构边界配位不饱和原子实现材料制备，为实现兼具高能量密度、功率密度和循环寿命的锂二次电池电极材料提供了一种崭新的设计思路和化学途径。磷—碳共价相连的磷/碳复合材料，为利用原位电化学谱学技术分析固体界面结合方式对电化学性质的影响提供了材料基础。材料有望继商业化石墨和准商业化硅/碳复合负极材料之后，成为具有应用价值的负极材料。部分研究成果于2020年10月发表在《科学》期刊上[4]，并入选教育部2020年度"中国高校十大科技进展"。

参考文献

[1] Liu Y, Zhu Y, Cui Y. Challenges and opportunities towards fast-charging battery materials. Nature Energy, 2019, 4: 540-550.

[2] Berg E J, Villevieille C, Streich D, et al. Rechargeable batteries: grasping for the limits of chemistry. Journal of the Electrochemical Society, 2015, 162: A2468-A2475.

[3] Waldmann T, Hogg B I, Wohlfahrt-Mehrens M. Li plating as unwanted side reaction in commercial Li-ion cells-A review. Journal of Power Sources, 2018, 384: 107-124.

[4] Jin H, Xin S, Chuang C, et al. Black phosphorus composites with engineered interfaces for high-rate high-capacity lithium storage. Science, 2020, 370: 192-197.

Black Phosphorus Composites with Engineered Interfaces for High-Rate High-Capacity Lithium Storage

Ji Hengxing

In response to the difficulty of increasing the power density and energy density of the Li-ion batteries at the same time, Professor Ji Hengxing's group proposed a new design strategy of "composite material with interface reconstruction" to improve the lithium ion diffusion in the anode material，and elucidated the constitutive relationship between the reconstructed interface and lithium ion diffusion. The research results are expected to improve the energy density of Li-ion batteries to over 350 Wh/kg and to achieve fast charging capability. The research results were published in *Science* in 2020，which is the first time that a domestic research institution has published a research paper in this journal as the first unit in the field of lithium-ion battery research. This research result was elected as one of the "Top Ten Scientific and Technological Advances of China Universities" by the Ministry of Education in 2020.

3.8　元 DNA：概念、设计与组装

樊春海

（上海交通大学，化学化工学院与变革性分子前沿科学中心）

1953 年，沃森（James Watson）和克里克（Francis Crick）发现了 DNA 双螺旋的结构[1]，从此开启了分子生物学时代。1982 年，西曼（Nadrian Seeman）首次提出 DNA 纳米技术的概念[2]，即利用核酸分子的双螺旋结构特性来构建可操控的纳米尺度结构，掀开了 DNA 纳米技术的序幕。自此以后，全球科学家致力于利用 DNA 的双螺旋结构特性来设计和构建复杂的纳米结构。这些具有纳米尺度精确可寻址能力的 DNA 结构在纳米生物传感与诊疗、纳米光子学等领域展现了巨大的应用潜力。然而，组装具有可编程特性的更大尺寸（微米到毫米级别）的 DNA 结构，仍然面临挑战。如何将 DNA 的精确自组装能力扩展到更大的尺度范围，是该领域面临的一个关键科学问题。

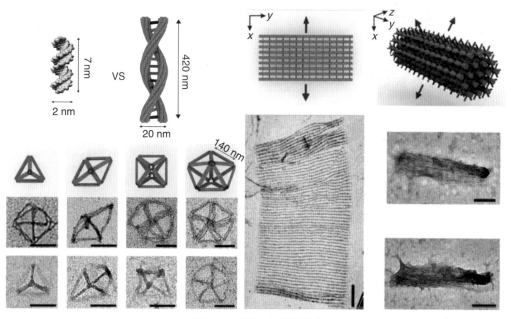

图 1　元 DNA 自组装策略

鉴于此，上海交通大学姚广保、樊春海等人开发了一种通用的"元DNA"（meta-DNA）策略。元DNA源于DNA，并具有类似DNA的结构和杂交特性，其单体长度可达400 nm，是DNA的百倍。该策略允许各种亚微米到微米大小的DNA结构以类似于经典的DNA短链的方式进行自组装。仅通过改变单个元DNA的局部柔性及其相互作用，就能够构建出一维到三维的一系列亚微米或微米尺度的DNA结构，比如元多臂结、元瓦块、四面体、八面体、棱柱体以及多种紧密排列的晶格结构。图1展示了基于元DNA组装策略构建的典型DNA结构。

元DNA策略也可以用于构建动态的反应网络。通过元碱基的多级链置换，能够实现元DNA的取代反应，从而将静态结构转变为动态结构，为构建具有可重构和环境响应能力的微米级复杂结构奠定了基础（图2）。

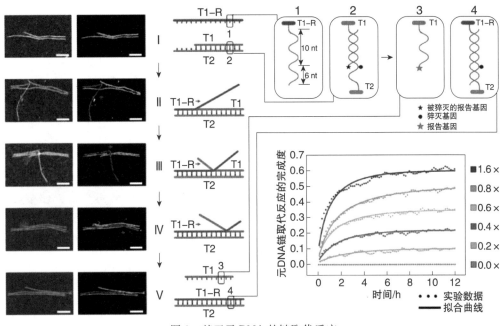

图2　基于元DNA的链取代反应

元DNA策略将DNA纳米技术的精确构建能力从纳米尺度扩展到微米尺度。通过合理的设计策略，完全有可能将尺度扩展到毫米量级、厘米量级等肉眼可见的宏观尺度。结合元DNA链置换反应，有望设计合成出具有纳米尺度分辨率的宏观复杂电路和动态器件，并应用于生物传感和芯片制造等领域。相关研究成果于2020年以"Meta-DNA Structures"为题发表于国际学术期刊《自然-化学》（*Nature Chemistry*）上。

参考文献

[1] Watson J D, Crick F H. Molecular structure of nucleic acids: A structure for deoxyribose nucleic acid. Nature, 1953, 171: 737-738.

[2] Seeman N C. Nucleic acid junctions and lattices. Journal of Theoretical Biology, 1982, 99: 237-247.

[3] Yao G, Zhang F, Wang F, et al. Meta-DNA structures. Nature Chemistry, 2020, 12: 1067-1075.

Meta DNA: Concept, Design and Assembly

Fan Chunhai

DNA has emerged as a highly programmable material to construct customized objects and functional devices at the 10-100 nm scale. Scaling up the size of the DNA structure would enable applications including construction of metamaterial and surface-based biosensors. Here we demonstrated that a six-helix bundle DNA origami nanostructure at the submicrometre scale, termed as meta-DNA, could be utilized as a magnified analogue of single-stranded DNA. By mimicking the molecular behaviours and assembly strategies of natural DNA strands, we realized fabrication of diverse complex structures on the micrometre scale employing meta-DNA building blocks, such as meta-multi-arm junctions, three-dimensional polyhedrons, and various 2D/3D lattices. We also demonstrated a hierarchical strand-displacement reaction on meta-DNA to transfer the dynamic features of DNA into the meta-DNA.

3.9 "原子晕"催化剂设计策略及其低碳烃转化应用

王 熙

（北京交通大学）

根据 2019 年化学协会国际理事会（International Council of Chemical Associations）发布的分析报告[1]，化学工业占全球 GDP 的贡献率为 7 %（5.7 万亿美元），超过 90% 化工产品的生产过程涉及催化剂。而基于催化反应（如 O—O 键活化、C—H 键活化等）设计高效、稳定的催化剂则是化学工业的最重要环节之一。一般而言，常用催化剂的设计策略有两个：策略一，从反应物的化学键属性出发进行设计，一般如均相催化剂合成；策略二，从研究催化剂的活性位入手，再设计合适的反应物（分子），进而开发高性能催化剂。与此不同，受我国催化先驱张大煜院士提出的"表面键理论"的启发和指引，笔者团队（北京交通大学王熙课题组和中国科学院化学研究所姚建年课题组）从二者之间的相互作用的角度出发，提出"原子晕"催化剂设计策略：将二者之间形成的有效域视为整体以促进反应物化学键的开合，从而根据反应物化学键和催化剂活性位的演变机制来设计催化剂（图 1）。

据此新策略我们设计出多种 d 电子和 sp 电子型催化剂并成功应用于诸多催化反应中[2-5]。比如，在 O—O 键活化中，利用 Pt 掺杂的 TiO_2，构筑 Pt—Ti 活性位，实现了 Pt-d 轨道 z 方向的特定暴露，d_z^2，d_{xz} 和 d_{yz} 轨道暴露从传统的 60 %（简并，各 20 %）提升到 99 %（非简并），从而优化其与 O_2-π^* 轨道的耦合，形成有效的 Pt—O—O—Ti 键合域 [图 1 (d)、(e)]。相比传统体系，氧化还原反应（Oxygen Reduction Reaction）效率提升 5 倍[2,3]；在 N≡N 键活化中，通过单原子合金化策略设计 Re_1—Cu (111) 催化剂，窄化的 Re 原子 d 轨道可以与 N_2-π^* 轨道耦合形成有效的 Cu—Re—N≡N 键合域，促进 N≡N 开合，法拉第效率比纯 Cu 提高了 6 倍。

基于该策略活化化学键的普适性，笔者团队进一步建立了"原子晕设计策略—催

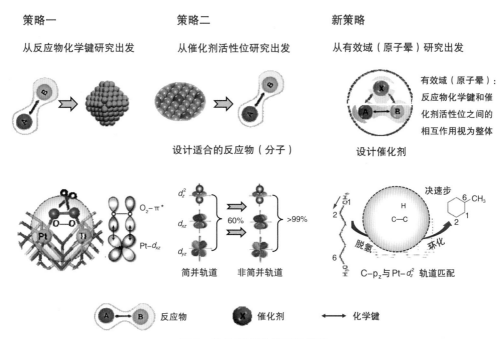

策略一
从反应物化学键研究出发

策略二
从催化剂活性位研究出发

新策略
从有效域（原子晕）研究出发

有效域（原子晕）：
反应物化学键和催
化剂活性位之间的
相互作用视为整体

设计适合的反应物（分子）

设计催化剂

60% >99%

简并轨道   非简并轨道

决速步

C—p_z 与 Pt-d_z^2 轨道匹配

A—B 反应物   X 催化剂   ←→ 化学键

图 1   催化剂设计的三种策略

化剂可控制备—催化剂工程与反应工艺优化"的研究路线，并将其推广到新型芳构化催化剂设计中。针对抽余油芳构化反应，伯碳上氢转移被视为其速率限制步骤。为实现 C—H 键的有效活化，笔者团队制备的抽余油芳构化 Pt/KL 催化剂实现了 Pt-d_z^2 轨道有效暴露，优化其与 C-p_z 轨道耦合，形成有效的 Pt—C—H 键合域［图 1（f）］，芳烃选择性为 80 wt%，是传统 Pt/Al_2O_3 催化剂的 2 倍。基于"原子晕"策略设计的新一代芳构化铂基催化剂打破了美国雪佛龙（Chevron）石油公司和 UOP 公司的垄断。新型抽余油芳构化催化剂已应用于十万吨级抽余油芳构化生产示范（图 2）：①在催化剂载体工程设计方面，通过优化"双分散孔构筑"制备工艺将载体比表面大幅度提高，提高了催化传质效率，并在中石化催化剂长岭分公司进行千吨级工业试生产，实现了高端 Al_2O_3 载体国产化；②在催化剂工程放大及应用示范方面，采用"简化积炭重整"反应优化工艺，催化剂稳定运行超 8000 h 以上。基于此，与中海油惠州炼化公司、中石化胜利油田、成都晟源石化、四川云汇石化等公司分别开展了 43 万 t/a、19 万 t/a、15 万 t/a、15 万 t/a 抽余油芳构化技术开发项目，预期经济效益可达 50 亿元以上。

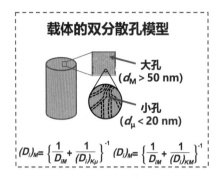

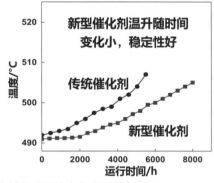

图 2　新型抽余油芳构化催化剂的工程放大与应用示范

参考文献

[1] The International Council of Chemical Associations(ICCA). The global chemical industry：catalyzing growth and addressing our world's sustainability challenges. Oxford Economics，2019，1-29.

[2] Lu F，Wang X，Yao J，et al. Engineering platinum-oxygen dual catalytic sites via charge transfer towards highly efficient hydrogen evolution. Angewandte Chemie International Edition，2020，59(40)：17712-17718.

[3] Lu F，Wang X，Yao J，et al. Regulation of oxygen reduction reaction by the magnetic effect of L_{10}-PtFe alloy. Applied Catalysis B：Environmental，2020，278：119332.

[4] Yi D，Wang X，Yao J，et al. Regulating charge transfer of lattice oxygen in single-atom-doped titania for hydrogen evolution. Angewandte Chemie International Edition，2020，59(37)：15855-15859.

[5] Zhou M，Yi D，Wang X，et al. Modulating 3d orbitals of Ni atoms on Ni-Pt edge sites enables highly-efficient alkaline hydrogen evolution. Advanced Energy Materials，2021，11(36)：2101789.

Atomic-Realm Strategy for Catalyst Design and Its Application in Low-Carbon Hydrocarbon Conversion

Wang Xi

It is well recognized that more than 90% of industrial chemical processes involve at least one catalytic step. To solve the "bottleneck" problems on catalyst design, "Atom-Realm Strategy", mainly focusing on the coupling of orbitals between reactant and the active site of catalyst, was proposed. For instance, the d orbital of Pt atom along the z axis (d_z^2, d_{xz} and d_{yz}) was optimized in Pt_1/TiO_2 to couple with π^* orbital of O_2 molecules, leading to enhanced oxygen reduction reaction (ORR) activity. A research-application route of "atom-realm effect— precise design of catalysts—catalysts engineering and reaction process optimization" was also built and applied in low-carbon hydrocarbon conversion, such as catalytic reforming. We carried out some chemical engineering process optimization including "simplifying coking reforming", realized over 1000 tons/ year domestic production of advanced catalyst supports, and is developing the 2nd generation reforming catalysts for aromatization of raffinate oil.

3.10 群聚信息素 4-乙烯基苯甲醚
诱发蝗虫群聚成灾

郭晓娇　康　乐

（中国科学院动物研究所）

历史上，蝗灾被认为是伴随人类生产生活的主要"瘟疫"之一。时至今日，世界范围内的蝗灾仍然是农业生产、经济发展和环境的重要威胁。其中，飞蝗是世界上分布最广泛的蝗虫，据我国 2000 多年的历史记载，由飞蝗引发的大规模蝗灾就发生过 800 多次。尽管蝗灾与人类长期相伴，但是真正在科学上对蝗灾成因的认识刚满百年。国际著名昆虫学家和蝗虫学之父尤瓦洛夫（Uvarov）1921 年发现，飞蝗能够成灾是因为其可以从无害的低密度的散居型向能够大规模移动或迁飞的高密度的群居型转变。其中，群聚信息素被认为是蝗虫聚集的最关键因素之一。经过 50 多年的研究，虽然有几种化合物被报道为蝗虫的群聚信息素，但是这些化合物中没有一个能符合群聚信息素的所有标准，特别是没有野外种群验证的证据。

针对这一问题，笔者研究团队分析发现，在 35 种群居型和散居型飞蝗的体表和粪便挥发化合物[1]中有 6 种是群居型显著高于散居型。利用气味双选行为仪对这 6 种化合物诱发的行为反应进行检测，发现飞蝗只对 4-乙烯基苯甲醚（4-vinylanisole，4VA）表现出了吸引反应。在此基础上，笔者团队通过一系列行为实验确定 4VA 对群居型和散居型飞蝗的不同发育阶段和性别都有很强的吸引力。化学分析证明，4VA 是一种由群居型蝗虫特异性挥发、释放量低但生物活性非常高的化合物。4VA 的释放量随着种群密度增加而增加，甚至它的产生可由 4～5 只散居型飞蝗聚集而触发，具有很低的诱发阈值。在飞蝗触角上的四种主要感器类型中，笔者团队发现 4VA 能够特异地引起锥形感器的反应，且群散飞蝗触角对 4VA 的反应均具有典型的浓度依赖效应。在飞蝗的上百个嗅觉受体中[2,3]，笔者团队发现，定位在锥形感器中的嗅觉受体 OR35 是 4VA 的特异性受体。使用基因编辑技术 CRISPR/Cas9[4] 敲除 OR35 后，飞蝗突变体的触角与锥形感器神经电生理反应显著降低，同时丧失了对 4VA 的响应行为。

此外，人工草坪上的行为双选和诱捕实验证明，4VA 对蝗虫实验室种群在户外也具有很强的吸引力。在飞蝗的自然生境——天津大港湿地保护区进行的大范围的区块实验进一步证明，4VA 不仅能吸引野外种群，而且不受自然环境中蝗虫背景密度的影响。综上所述，笔者团队鉴定并证明了 4VA 是飞蝗的群聚信息素[5]（图 1）。

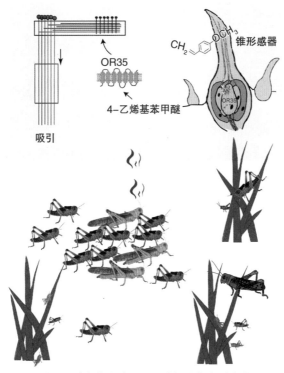

图 1　群聚信息素 4VA 诱发蝗虫聚群成灾

　　该研究不仅揭示了蝗虫群居的奥秘，而且使蝗虫的可持续防控成为可能。这项研究将从多个方面改变人们控制蝗灾的理念和方法：利用人工合成的信息素可以在田间长期监测蝗虫种群动态，为预测预报服务；利用人工合成的信息素可以设计诱集带诱集蝗虫，并在诱集带集中使用化学农药或生物制剂将其消灭，从而极大地减少化学农药的使用；根据 4VA 的结构设计拮抗剂，阻止蝗虫的聚集；嗅觉受体 OR35 的发现，让使用基因编辑技术建立 4VA 反应缺失的突变体成为可能，这种突变体长期释放到野外就可能在重灾区建立起不能群居的蝗虫种群，既在野外维持了一定数量的蝗虫，又达到可持续控制的目的，将环境保护与害虫控制有机地结合起来。因此，4VA 和它的受体的发现将极大地改变防治蝗虫的对策和技术。

　　研究成果于《自然》期刊上线发表的 24 小时内，全球 300 余家媒体予以报道，《自然》期刊附编者按与专文评述该文，文章评论"用科学战胜蝗灾"。F1000Prime 给予该项工作"杰出"级别的评分和推荐。此外，该项工作入选 2020 年度"中国生命科学十大进展"与中国科学院"2020 年度科技创新亮点成果"。

参考文献

［1］ Wei J N,Shao W B,Wang X H,et al. Composition and emission dynamics of migratory locust volatiles in response to changes in developmental stages and population density. Insect Science,2017,24:60-72.

［2］ Wang X,Fang X,Yang P,et al. The locust genome provides insight into swarm formation and long-distance flight. Nature Communications,2014,5:2957.

［3］ Wang Z F,Yang P C,Chen D F,et al. Identification and functional analysis of olfactory receptor family reveal unusual characteristics of the olfactory system in the migratory locust. Cellular and Molecular Life Sciences,2015,72:4429-4443.

［4］ Li Y,Zhang J,Chen D F,et al. CRISPR/Cas9 in locusts:Successful establishment of an olfactory deficiency line by targeting the mutagenesis of an odorant receptor co-receptor(Orco). Insect Biochemistry and Molecular Biology,2016,79:27-31.

［5］ Guo X J,Yu Q Q,Chen DF,et al. 4-Vinylanisole is an aggregation pheromone in locusts. Nature,2020,584:584-588.

Aggregation Pheromone 4-Vinylanisole(4VA) Induces Locust Outbreaks

Guo Xiaojiao,Kang Le

In human history,locust plagues,drought and flood were considered as three major natural disasters,which caused serious agricultural and economic losses all over the world. Aggregation pheromones play a crucial role in the transition of locusts from a solitary form to the devastating gregarious form and the formation of large-scale swarms. However,none of the candidate compounds reported meet all the criteria for a locust aggregation pheromone. Recently,Prof. Le Kang's group in Chinese Academy of Sciences identified and verified 4-vinylanisole (4VA) (also known as 4-methoxystyrene) which is an aggregation pheromone of the migratory locust(*Locusta migratoria*), through chemical analysis, behavioral assays, electrophysiological recording, olfactory receptor identification, gene knockout, field verification, etc. This study may contribute largely to end the controversy on the aggregation pheromone of migratory locust for over 50 years. These findings reveal the mechanism of locust aggregation, provide crucial clues for the development of novel locust control strategies and may lead to new strategies to control locust outbreaks.

3.11　首个新冠病毒蛋白质三维结构的解析及两个临床候选药物的发现

杨海涛[1]　蒋华良[1,3]　饶子和[1,2]

（1. 上海科技大学；2. 清华大学；3. 中国科学院上海药物研究所）

截至 2021 年 9 月 23 日，全球累计确诊新冠肺炎病例 2.3 亿例，死亡人数超过 470 万[1]。新型冠状病毒（SARS-CoV-2，简称新冠病毒）与严重急性呼吸综合征冠状病毒（SARS-CoV）和中东呼吸综合征冠状病毒（MERS-CoV）具有较近的亲缘关系。由于缺乏特效药，针对新型冠状病毒的药物靶点研究及新药研发迫在眉睫。

在疫情暴发伊始，上海科技大学免疫化学研究所、中国科学院上海药物研究所、中国人民解放军军事科学院军事研究院以及中国科学院武汉病毒研究所等多个课题组立即成立联合攻关团队。通过深入分析，攻关团队发现新型冠状病毒异常"狡猾"，在入侵细胞后，会立即利用细胞内的物质合成自身复制必需的两条超长复制酶多肽（pp1a 和 pp1ab）。这两条复制酶多肽需要先被剪切成多个"零件"（如 RNA 依赖的 RNA 聚合酶、解旋酶等）（图 1）；这些"零件"再进一步组装成一台庞大的复制转录机器，然后病毒才能启动自身遗传物质的大量复制[2]。由于两条复制酶多肽的剪切要求异常精确，因此病毒往往需要自身编码蛋白酶来完成该项任务。主蛋白酶（M^{pro}）就是新冠病毒自身编码的一种必需蛋白酶，该酶就像一把神奇的"魔剪"，它在病毒复制酶多肽上至少有 11 个切点[3]，只有当这些位点被正常切割后，与病毒复制相关的"零件"才能从复制酶多肽上释放出来，进而顺利组装成一台高效的复制转录机器，启动病毒的复制。主蛋白酶这把"魔剪"在病毒复制过程中起到了至关重要的作用，且人体中并无类似的蛋白质。因此主蛋白酶可以作为一个抗新冠病毒的关键药靶[4]。

攻关团队率先解析了 2.1Å 的"主蛋白酶-N3"的高分辨率复合物结构（随后又提高至 1.7Å）[5]。高分辨率的晶体结构揭示了抑制剂 N3 与病毒主蛋白酶的精细相互作用，并阐明了其精确靶向蛋白酶活性中心，以及抑制主蛋白酶活性的分子机制（图 2）。这也是世界上第一个被解析的新冠病毒蛋白质的三维空间结构，为后续抗病毒小分子药物的设计和开发奠定了重要的基础。该结构被蛋白质结构数据库（PDB）选为 2020 年 2 月的明星分子（February Molecule of the Month），并被 PDB 撰文报道。

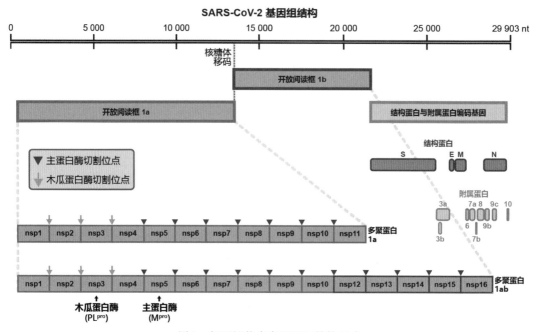

图 1　新型冠状病毒基因组结构组成

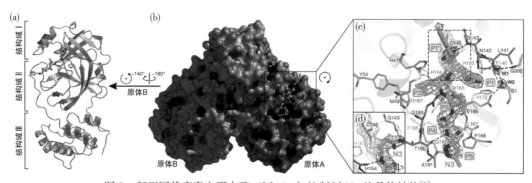

图 2　新型冠状病毒主蛋白酶（M^pro）与抑制剂 N3 的晶体结构[5]

（a）新冠病毒主蛋白酶同源二聚体中一个蛋白分子与 N3 形成复合物的示意图；（b）新冠主蛋白酶同源二聚体表面示意图；（c）新冠病毒主蛋白酶底物结合口袋放大图，其中绿色分子为 N3；（d）碳硫共价键示意图

除此之外，攻关团队运用虚拟筛选和高通量筛选策略相结合的方式，对 1 万多个老药、临床药物以及天然活性产物进行筛选，发现了数种对新冠病毒主蛋白酶有显著抑制作用的先导药物，其中包括双硫仑（disulfiram）、卡莫氟（carmofur）、依布硒（ebselen）等[5]。目前，依布硒已被美国 FDA 批准进入 Ⅱ 期临床试验，用于抗新冠肺炎的治疗；双硫仑则同时被美国 FDA 和英国批准进入 Ⅱ 期临床试验。该项研究为今

后人类在面临新发重大传染病疫情时，如何迅速发现具有临床潜力的治疗性药物提供了一个具有重要借鉴意义的有效应对策略。

本研究成果于 2020 年 4 月 9 日，经国际权威学术刊物《自然》特邀投稿，以"Structure of M^{pro} from SARS-CoV-2 and discovery of its inhibitors"为题在线发表。该项成果入选了 2020 年度"富林沃斯（Falling Walls）生命科学领域十项进展"和2020 年度"中国生命科学十大进展"。

参考文献

[1] Johns Hopkins Coronavirus Resource Center. COVID-19 Dashboard by the Center for Systems Science and Engineering (CSSE) at Johns Hopkins University (JHU). https://coronavirus. jhu. edu/map. html.

[2] Thiel V, Ivanov K A, Putics Á, et al. Mechanisms and enzymes involved in SARS coronavirus genome expression. The Journal of General Virology, 2003, 84 (Pt 9): 2305-2315.

[3] Moustaqil M, Ollivier E, Chiu H P, et al. SARS-CoV-2 proteases PLpro and 3CLpro cleave IRF3 and critical modulators of inflammatory pathways (NLRP12 and TAB1): implications for disease presentation across species. Emerging Microbes & Infections, 2021, 10 (1): 178-195.

[4] Pillaiyar T, Manickam M, Namasivayam V, et al. An overview of severe acute respiratory syndrome-coronavirus (SARS-CoV) 3CL protease inhibitors: Peptidomimetics and small molecule chemotherapy. Journal of Medicinal Chemistry, 2016, 59 (14): 6595-6628.

[5] Jin Z, Du X, Xu Y, et al. Structure of M^{pro} from SARS-CoV-2 and discovery of its inhibitors. Nature, 2020, 582 (7811): 289-293.

Determination of the First Structure of Any Protein from SARS-CoV-2 and the Discovery of Two Clinical Drug Candidates

Yang Haitao, Jiang Hualiang, Rao Zihe

A new coronavirus, known as severe acute respiratory syndrome coronavirus 2 (SARS-CoV-2), is the aetiological agent responsible for the 2019—2020 viral pneumonia outbreak of coronavirus disease 2019 (COVID-19). This programme focused on identifying drug leads that target main protease (M^{pro}) of SARS-CoV-2.

We identified a mechanism-based inhibitor (N3) by computer-aided drug design, and then determined the crystal structure of M^{pro} of SARS-CoV-2 in complex with this compound. Through a combination of structure-based virtual and high-throughput screening, we assayed more than 10,000 compounds—including approved drugs, drug candidates in clinical trials and other pharmacologically active compounds—as inhibitors of M^{pro}. One of these compounds (ebselen) also exhibited promising antiviral activity in cell-based assays. Our results demonstrate the efficacy of our screening strategy, which can lead to the rapid discovery of drug leads with clinical potential in response to new infectious diseases for which no specific drugs or vaccines are available.

3.12　探索器官衰老的秘密

李静宜[1,2,3]　马　帅[1,2,3]　王　思[4]　张维绮[5]
曲　静[1,2,3]　刘光慧[1,2,3]

[1. 中国科学院动物研究所；2. 中国科学院干细胞与再生医学创新研究院；
3. 北京干细胞与再生医学研究院；4. 首都医科大学宣武医院；
5. 中国科学院北京基因组学研究所（中国国家生物信息中心）]

人口老龄化是中国乃至世界面临的重大社会科学问题，伴随年龄增长而发生的组织器官功能衰退，是神经退行性疾病、心血管疾病、糖尿病等衰老相关疾病的关键诱因。随着科学技术的不断发展，人们发现各种组织器官，甚至不同细胞类型均有其特有的衰老特征和规律。因此，探究不同器官衰老的分子机制、鉴定特异性衰老生物标志物及关键调控靶标、研发针对特定器官的衰老与稳态失衡的干预技术，是实现预警及干预器官衰老乃至机体衰老的重要基础，对2021年《政府工作报告》中提出的"全面推进健康中国建设"，"人均预期寿命再提高1岁"的目标具有重要的促进意义①。

围绕"器官衰老及向退行变化演变的基本属性、内外环境因素互作及其对疾病发生发展的调控模式"这一重要科学问题，中国科学院动物研究所刘光慧团队合作利用高通量与高精度单细胞测序技术并联合系统生物学方法，阐明了器官衰老及向退行性变化演变过程中的关键易感细胞类型，并揭示了多器官衰老的生物标志物和潜在调控靶标。

本研究团队以灵长类动物为模型，在衰老的组织、细胞、分子等多个层次开展研究，结合高通量高精度单细胞转录组测序技术，系统性地绘制了卵巢、胰岛、血管、心肺、视网膜等组织器官衰老的细胞和分子图谱，由此揭示多种灵长类器官衰老的特异性机制——氧化还原调控网络失衡是卵巢衰老的关键特征[1]；蛋白稳态失衡是胰岛β细胞衰老的分子驱动力[2]；长寿基因 FOXO3A 的表达下调是血管衰老的主要因

---

① 李克强. 政府工作报告——2021年3月5日在第十三届全国人民代表大会第四次会议上. http://www.gov.cn/guowuyuan/zfgzbg.htm.

素[3]；新型冠状病毒受体 ACE2 在衰老灵长类动物心肺系统中的表达上调是老年个体易感新冠肺炎的潜在机理之一[4]；视网膜色素上皮细胞中的免疫炎性增加是视网膜衰老的主要分子特征[5]。此外，研究人员以大鼠为模型，绘制了哺乳动物衰老与节食状态下 9 种组织器官共计 20 多万个单细胞转录组图谱，从不同维度系统评估了衰老和节食对机体多个类型组织细胞的影响，揭示了热量限制通过调节多种组织的免疫炎症通路进而延缓衰老的新型分子机制[6]。上述系列研究成果为预警和干预衰老及相关疾病提供了新型生物标志物及潜在的干预靶标。

上述研究受到国际学术界的广泛关注，先后入选《细胞》（Cell）期刊封面文章（图 1）、F1000（very good）、"2020 年度中国科学院杰出科技成就奖"、"2020 年度中国科学十大进展"，以及"2020 年度中国生命科学十大进展"（图 2）。其中，大鼠多器官的研究领先于国际其他两项发表于《自然》上的关于小鼠系统衰老的研究报道[7,8]4 月余，并受到《自然》和《科学-转化医学》（Science Translational Medicine）的亮点评述[9]。

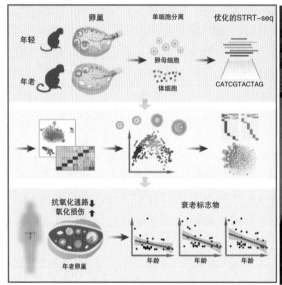
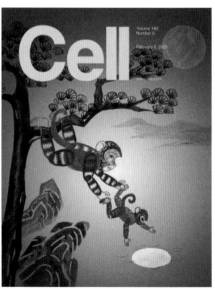

图 1　高精度转录组图谱研究揭示灵长类卵巢衰老机制（左），
该研究被选为《细胞》杂志封面故事——《猴子捞月》（右）

该研究是国际上首次报道的非人灵长类动物器官衰老的高精度单细胞转录组图谱，揭示了细胞类型特异性的氧化还原调控网络失衡是包括人类在内的灵长类动物卵巢衰老的共性分子机制。相关研究成果于 2020 年 2 月在《细胞》杂志发表（封面故事）。

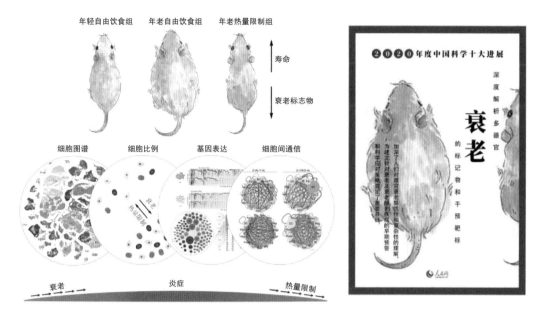

图 2　深度解析卡路里限制延缓大鼠多器官衰老的细胞分子机制

　　该研究系统解析了机体多组织器官衰老的细胞和分子变化规律，发现了慢性炎症是哺乳动物多种器官衰老的共性特征，证明了热量限制在炎症免疫调控中的重要作用，揭示了代谢干预、机体免疫反应与健康状况、寿命之间的密切关联。相关研究成果于 2020 年 3 月在《细胞》期刊上发表。

参考文献

［1］Wang S，Zheng Y X，Li J Y，et al. Single-cell transcriptomic atlas of primate ovarian aging. Cell，2020，180：585-600. e19.

［2］Li J Y，Zheng Y X，Yan P Z，et al. A single-cell transcriptomic atlas of primate pancreatic islet aging. National Science Review，2021，8：nwaa127.

［3］Zhang W Q，Zhang S，Yan P Z，et al. A single-cell transcriptomic landscape of primate arterial aging. Nature Communications，2020，11：1-13.

［4］Ma S，Sun S H，Li J M，et al. Single-cell transcriptomic atlas of primate cardiopulmonary aging. Cell Research，2021，31：415-432.

［5］Wang S，Zheng Y X，Li Q Q，et al. Deciphering primate retinal aging at single-cell resolution. Protein&Cell，2021，12：889-898.

［6］Ma S，Sun S，Geng L L，et al. Resource caloric restriction reprograms the single-cell transcriptional landscape of rattus norvegicus aging. Cell，2020，80(5)：984-1001.

[7] Almanzar N, Antony J, Baghel A S, et al. A single-cell transcriptomic atlas characterizes ageing tissues in the mouse. Nature,2020,583:590-595.

[8] Schaum N, Lehallier B, Hahn O, et al. Ageing hallmarks exhibit organ-specific temporal signatures. Nature,2020,583:596-602.

[9] Su X Y. "You are what you eat!" Science Translational Medicine,2020,12:eabb2771.

Deciphering Organ Aging

Li Jingyi, Ma Shuai, Wang Si, Zhang Weiqi,
Qu Jing, Liu Guanghui

The increase in aging population, coupled with the high incidence of aging-related diseases, is a major social and scientific issue worldwide. The degeneration of tissue and organ function that accompanies aging is a leading factor of age-related diseases such as neurodegenerative diseases, cardiovascular diseases, and diabetes. Thus, it is important to dissect the effect of aging on organ homeostasis, identify organ-specific aging biomarkers and key regulatory targets, and to explore intervention strategies for organ-specific aging and aging-related diseases. This lays the foundation for the early warning and intervention for the aging of the organ and even the body. To understand the aging changes in organs, Prof. Guang-Hui Liu at the Institute of Zoology (Chinese Academy of Sciences) and his colleagues used high-throughput and high-resolution single-cell sequencing technology combined with systems biology methods to clarify the key susceptible cell types in the process of organ aging, and revealed the biological markers and potential regulatory targets of multi-organ aging.

3.13　人类海马体胚胎发育的动态转录组及表观遗传组图谱

钟穗娟　吴　倩　王晓群

（中国科学院生物物理研究所）

人类大脑在进化过程中获得的记忆、思辨、推理决策等高级认知功能赋予了人类与众不同的智慧和创造力。大脑中的海马体与进化程度最高的新皮层不同，是古皮层的一部分，属于进化上相对保守的脑结构，由端脑内侧区域发育而成，负责大脑的信息编码、记忆的形成和存储及空间导航功能，与癫痫、智力障碍和阿尔茨海默病等多种病症的发病机制密切相关[1,2]。深入理解人类海马体的发育模式和进化机制是探索人类高级记忆功能及深入研究相关脑疾病致病机理的关键。然而，目前大量关于海马体的研究都集中在啮齿类动物海马体的结构及功能上，无法真正解析人类海马体发育的独特动态过程及进化机制。

为深入解决该问题，本研究团队利用高通量单细胞转录组测序技术及染色质开放区域测序手段（assay for targeting accessible-chromatin with high-throughout sequencing，ATAC-seq）绘制了人类海马体在胚胎期16～27周的发育动态转录组及表观遗传组图谱。通过生物信息学技术的深入挖掘，本研究团队注释了海马体胚胎发育关键阶段包括神经干细胞、兴奋性神经元、抑制性神经元和各类胶质细胞在内的11个主要细胞类型及47个细胞亚型，展示了海马体复杂的细胞组成及多样性。海马体的细胞组成不仅复杂，而且具有区域独特性。例如，海马体与前额叶皮层区域相邻并且协同发育，环路功能上也密不可分，但是同在胚胎发育过程中，我们却发现在海马体细胞内会特异性地表达 SOX4、SOX11 等可以有效促进海马体神经干细胞分化为神经元的 SOX 家族基因[3,4]，并且海马体的少突胶质细胞的出现时间要远远早于前额叶皮层，部分的兴奋性神经元的成熟度也比前额叶皮层要高（图1）。我们发现在人类海马体具有两种重要的神经干细胞类型，分别是表达 PAX6 或者 HOPX 的神经干细胞，这两类干细胞从两个不同方向汇入海马体，并负责海马体的神经发生及胶质发生，因此我们推测海马体发育的独特性与其神经干细胞的增殖分化是息息相关的。值得一提的是，我们发现表达 HOPX 的神经干细胞只在人类海马体内存在，而在小鼠海马体内是不存在的（图2）[5-7]。于是，我们提出一个假设，尽管海马体在进化过程中相对保守，但仍产生了差异。于是我们结合人类和小鼠海马体的单细胞转录组数据进行挖掘，发现 *NBPF1* 这个灵长类特异性基因能够有效地促进海马体细胞数量增加，促使人类海马体的结构更为复杂，甚至与人类高级的思维记忆功能密切相关[8]。

本研究团队的工作从分子特质、细胞组分、组织形态及环路功能等多个角度充分展示了人类海马体发育的区域独特性，鉴定了海马体胚胎发育的关键时间节点及其重要细胞活动，更好地为揭示海马体相关的认知功能调控机制、了解阿尔茨海默病的致病机理及治疗方案提供重要的理论基础。相关工作于 2020 年发表于《自然》期刊[9]，并荣获中国科学技术协会生命科学学会联合体评选的"中国生命科学十大进展"。

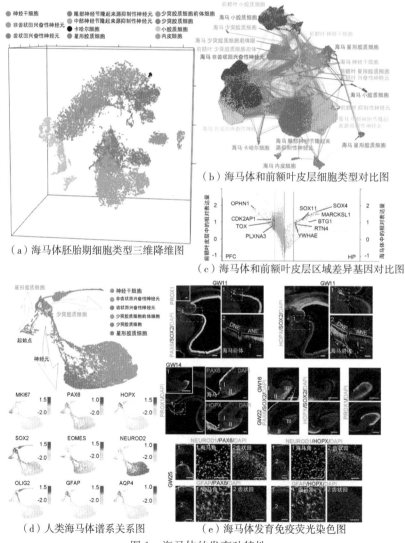

（a）海马体胚胎期细胞类型三维降维图

（b）海马体和前额叶皮层细胞类型对比图

（c）海马体和前额叶皮层区域差异基因对比图

（d）人类海马体谱系关系图

（e）海马体发育免疫荧光染色图

图 1　海马体的发育独特性

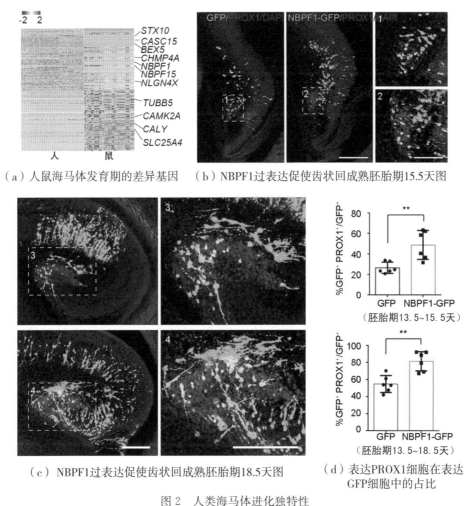

（a）人鼠海马体发育期的差异基因　（b）NBPF1过表达促使齿状回成熟胚胎期15.5天图

（c）NBPF1过表达促使齿状回成熟胚胎期18.5天图　（d）表达PROX1细胞在表达GFP细胞中的占比

图2　人类海马体进化独特性

参考文献

[1] Miller A M，Vedder L C，Law L M，et al. Cues，context，and long-term memory：the role of the retrosplenial cortex in spatial cognition. Frontiers in Human Neuroscience，2014，8：586.

[2] Bird C M，Burgess N. The hippocampus and memory：insights from spatial processing. Nature Reviews Neuroscience，2008，9(3)：182-194.

[3] Mu L，Berti L，Masserdotti G，et al. SoxC transcription factors are required for neuronal differentiation in adult hippocampal neurogenesis. Journal of Neuroscience，2012，32(9)：3067-3080.

[4] Wang Y，Lin L，Lai H L，et al. Transcription factor Sox11 is essential for both embryonic and adult

neurogenesis. Developmental Dynamics,2013,242(6):638-653.

[5] Pollen A A,Nowakowski T J,Chen J,et al. Molecular identity of human outer radial glia during cortical development. Cell,2015,163(1):55-67.

[6] Karalay O,Doberauer K,Vadodaria K C,et al. Prospero-related homeobox 1 gene(Prox1)is regulated by canonical Wnt signaling and has a stage-specific role in adult hippocampal neurogenesis. Proceedings of the National Academy of Sciences of the United States of America,2011,108(14): 5807-5812.

[7] Bandler R C,Mayer C,Fishell G. Cortical interneuron specification:the juncture of genes,time and geometry. Current Opinion in Neurobiology,2017,42:17-24.

[8] Obleness M,Searles V B,Varki A,et al. Evolution of genetic and genomic features unique to the human lineage. Nature Reviews Genetics,2012,13(12):853-866.

[9] Zhong S,Ding W,Sun L,et al. Decoding the development of the human hippocampus. Nature,2020, 577(7791):531-536.

Characterization of the Transcriptional Landscape of Human Developing Hippocampus by scRNA-seq and ATAC-seq

Zhong Suijuan,Wu Qian,Wang Xiaoqun

The hippocampus is an important part of the limbic system in the human brain that plays essential roles in the consolidation of information from short-term memory to long-term memory and spatial navigation. In Alzheimer's disease, the hippocampus is one of the first brain regions to be affected,resulting in memory loss and disorientation as early symptoms. Although the hippocampus is an evolutionarily conserved organ of the brain in vertebrates, knowledge of the cellular and molecular features of the developing human hippocampus is still lacking. We use scRNA-seq and ATAC-seq to illustrate the cell types,cell linage, molecular features and transcriptional regulation of the developing human hippocampus. Our multiomic information provides a blueprint for understanding human hippocampal development and a powerful tool for investigating the mechanisms behind relative diseases.

3.14　神经系统对免疫系统的双向调控

祁　海

（清华大学）

人体是一个非常精妙而复杂的多个系统的集合，每个系统各司其职，以维持人体正常的运转。在诸多系统中，有两个系统尤为特殊：神经系统与免疫系统。通俗地描述，神经系统就是人体的"司令部"，处理各种信息并下达指令；而免疫系统则是"警察局"，保卫机体免遭外来微生物（如病毒、细菌等）的感染。

这两个部门看似并无联系，但实际上却互相影响。我们往往有这样一种经验性的常识：适度并且规律的运动会让人心情愉悦，机体抵抗力也显著增强；但如果过度运动体力透支，则更容易患上感冒等病症。这背后的科学机理是什么，我们如何定义"适度运动"与"过度运动"，这些规律能否为科学指导健身强体提供依据？我们将在下文中进行探讨。

神经系统通过两种方式调控人体：神经元的电信号—化学信号传导以及激素分泌两种途径，前者短期而快速，后者则长期而缓慢。神经元的电信号—化学信号传导是通过神经细胞（也就是神经元）直接实现的。然而遗憾的是，截至本文所描述的论文[1]之前，尚未有明确的任何直接证据表明神经元形成的神经通路可以直接影响免疫系统应答。本研究组与合作团队一起，确立了第一条中枢神经增进适应性免疫功能的脑—脾神经通路，开辟了研究行为增强免疫的新范式[1]。

脾是小鼠中最大的免疫器官，笔者团队新开发了一种在小鼠脾内去除脾神经的手术，使得神经通路的信号无法直接传递到脾脏。脾神经被去除的小鼠在接种疫苗后，产生可分泌抗体的浆细胞的数量明显减少。如上文所述，这一结果提示脾神经信号冲动对 B 细胞应答过程中浆细胞的形成，具有一定的促进作用。接着，研究人员敲除 B 细胞上表达的乙酰胆碱 α9 受体，发现这种促进作用被严重削弱；通过体内 T 细胞剔除实验，发现在肾上腺素能的脾神经和需要感知乙酰胆碱的 B 细胞之间，有一类新近发现的可以感受去甲肾上腺素而分泌乙酰胆碱的 T 细胞，在其中起到了不可或缺的作用[2,3]。利用伪狂犬病毒可以沿着神经元信号传导逆行向上追踪的能力[4]，发现脾神经与大脑中的室旁核（paraventricular nucleus of hypothalamus，PVN）、中央杏仁核（central amygdaloid nucleus，CeA）两个区域有连接，且这两个区域的功能与应激、恐惧反应紧密相关[5]。同时，已知这两处区域共有的一类神经元是分泌促肾上腺皮质激素释放激素（corticotropin releasing hormone，CRH）的神经元[6]。这类 CRH 神经元是掌控垂体—肾上腺轴的上游神经元，其激活可以促使肾上腺大量释放糖皮质激素，从而调整机体应激以抑制免疫系统活动[7,8]。

用光遗传学手段直接刺激 CeA/PVN 这两处的 CRH 神经元，仅仅几秒钟的时间，研究者们就观测到了脾神经的电信号明显增强（图 1）。这一结果证实了大脑 CeA/PVN 区域与脾神经具有直接的联系。这是不是意味着，通过适度的外界刺激，比如运动、情绪变化等，可以增强机体免疫力？

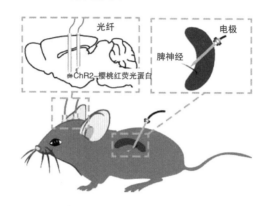

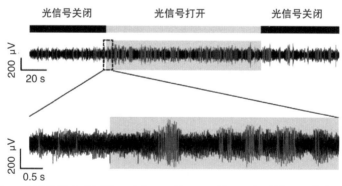

图 1　光遗传学实验证明 CeA/PVN 的 CRH 神经元与脾神经有直接连接

研究者们开发出一种新的行为学实验体系，即"孤立高台站立"（图 2）。让小鼠站立在很高的、它无法跳下的台子上，这种站立行为可以刺激并激活小鼠 CeA/PVN 两个区域中的 CRH 神经元。在每天两次的孤立高台站立实验的影响下，小鼠在接种

NP-KLH 抗原[①]后第二周里，与对照组相比，体内对抗原特异的抗体水平就可以增加约 70%。

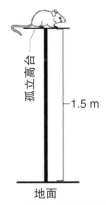

图 2　孤立高台站立模式图

至此，研究者们证明了一条对适应性免疫具有增强功能的脑—脾神经轴，揭示了 CRH 神经元的双重免疫调节功能。研究者们猜测，对特定神经元、神经环路水平做出定量描述，进而评价不同锻炼方式、不同躯体运动形式乃至冥想、禅修等过程对免疫系统的影响，给我们正确选择锻炼或其他修行方式提供更明确的科学依据。这也是图 3"勤動"（勤动）所表达的愿景。

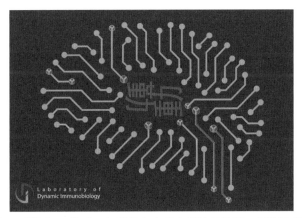

图 3　"勤動"（勤动）与增强免疫的中枢神经核团、环路示意图

此为本研究团队设计的图片：Y 代表抗体，蓝线和圆球代表神经元细胞，整体环绕成大脑的形状；中间两个字是"勤動"（勤动），意思是加强运动促进神经系统活性进而促进免疫力。

---

①　为免疫研究中的模式抗原，是一种蛋白抗原。

参考文献

[1] Zhang X, Lei B, Yuan Y, et al. Brain control of humoral immune responses amenable to behavioural modulation. Nature, 2020, 581: 204-208.

[2] Mauricio R-B, Reardon C, Tracey K J, et al. Acetylcholine-synthesizing T cells relay neural signals in a vagus nerve circuit. Science, 2011, 334: 98-101.

[3] Rinner I, Schauenstein K. Detection of choline-acetyltransferase activity in lymphocytes. Journal of Neuroscience Research, 1993, 35: 188-191.

[4] Saalmüller A, Mettenleiter T C. Rapid identification and quantitation of cells infected by recombinant herpesvirus(pseudorabies virus)using a fluorescence-based β-galactosidase assay and flow cytometry. Journal of Virological Methods, 1993, 44: 99-108.

[5] Keifer O P, Hurt R C, Ressler K J, et al. The physiology of fear: reconceptualizing the role of the central amygdala in fear learning. Physiology, 2015, 30: 389-401.

[6] Peng J, Long B, Yuan J, et al. A quantitative analysis of the distribution of CRH neurons in whole mouse brain. Frontiers in Neuroanatomy, 2017, 11: 63.

[7] Kadmiel M, Cidlowski J A. Glucocorticoid receptor signaling in health and disease. Trends in Pharmacological Sciences, 2013, 34: 518-530.

[8] Vandevyver S, Dejager L, Tuckermann J, et al. New insights into the anti-inflammatory mechanisms of glucocorticoids: an emerging role for glucocorticoid-receptor-mediated transactivation. Endocrinology, 2013, 154: 993-1007.

The Dual-regulation to Immune System by Nervous System

Qi Hai

It has been speculated that brain activities might directly control adaptive immune responses in lymphoid organs, we show that splenic denervation in mice specifically compromises the formation of plasma cells during a T cell-dependent but not T cell-independent immune response. Splenic nerve activity enhances plasma cell production in a manner that requires B-cell responsiveness to acetylcholine mediated by the α9 nicotinic receptor, and T cells that express choline acetyl transferase probably act as a relay between the noradrenergic nerve and acetylcholine-responding B cells. Neurons in the central nucleus of the amygdala

(CeA) and the paraventricular nucleus （PVN） that express corticotropin-releasing hormone （CRH） are connected to the splenic nerve, pharmacogenetic activation of these neurons increases plasma cell abundance after immunization. In a newly developed behaviour regimen, mice are made to stand on an elevated platform, leading to activation of CeA and PVN CRH neurons and increased plasma cell formation. By identifying a specific brain-spleen neural connection that autonomically enhances humoral responses and demonstrating immune stimulation by a bodily behaviour, our study reveals brain control of adaptive immunity and suggests the possibility to enhance immunocompetency by behavioural intervention.

3.15 进食诱导胆固醇合成的机制 及降脂新药靶发现

芦小艺　宋保亮

（武汉大学生命科学学院）

　　胆固醇（cholesterol）是一类具有环戊烷多氢菲基本结构的甾醇类小分子，广泛存在于多种生物体内。它是生物膜的关键组成成分，维持细胞亚结构及调控膜的流动性和相变，同时也是合成胆汁酸和甾醇类激素的前体。胆固醇还可以共价修饰Hedgehog 和 Smoothened 蛋白，调控信号转导、胚胎发育和肿瘤发生[1]（图 1）。

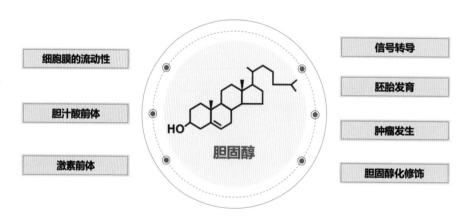

细胞膜的流动性

胆汁酸前体

激素前体

信号转导

胚胎发育

肿瘤发生

胆固醇化修饰

HO

胆固醇

图 1　胆固醇分子结构和生物学功能[1]

　　人体获得胆固醇主要通过两个途径：一是胆固醇的从头合成，细胞利用乙酰辅酶A 为原料，经 30 多步酶促反应合成胆固醇，其中肝脏是胆固醇合成最主要的器官；二是小肠直接从食物中吸收胆固醇。然而，过量的胆固醇会导致高胆固醇血症，进而诱发动脉粥样硬化等心脑血管疾病，严重威胁人类健康。我国有约 2 亿人血脂异常，目前的降胆固醇药物并不能完全达到控制血脂的目的。因此，揭示胆固醇代谢调控机制对于代谢性疾病的防治和创新药物的研发具有重要意义。

　　在漫长的演化过程中，人类的食物也在不断地改变：摄入碳水化合物为主的食物，其中缺乏胆固醇；获取动物性食物，从中吸收外源的胆固醇。人类也不断经历饥饿和饱食的循环。胆固醇合成需要消耗很多营养物质和能量，其中 3-羟基-3-甲基戊二酰辅酶 A 还原酶（HMG-CoA reductase，HMGCR）是胆固醇合成途径限速酶[2]。人体在进食碳水化合物后，胆固醇的合成上调，而饥饿时该过程则受到抑制[3]。然而，机体如何感知自身的营养与能量状态，进而调节胆固醇合成的分子机制尚不清楚。

本团队的研究起始于进食后胆固醇合成升高这一个常见而又非常重要的日常现象。通过表达筛选，发现去泛素化酶 20（ubiquitin specific peptidase 20，USP20）能够稳定 HMGCR 蛋白。肝脏特异性敲除 Usp20 基因的小鼠，HMGCR 蛋白水平不再受进食所诱导，同时进食后胆固醇的合成也比对照小鼠显著下降，表明 USP20 是进食诱导胆固醇合成增加的重要分子。

我们进一步通过精细的生化手段证明，进食通过胰岛素-AKT（Insulin-AKT）、葡萄糖-AMP 依赖的蛋白激酶（Glucose-AMPK）信号通路协同激活雷帕霉素靶蛋白复合体 1（mammalian target of rapamycin complex 1，mTORC1）的活性，并且磷酸化 USP20 第 132 位和 134 位丝氨酸，从而增加 HMGCR 蛋白水平，促进胆固醇合成。

有趣的是，我们还发现抑制 USP20 的活性不仅降低了胆固醇合成，还促进了胆固醇合成中间物 3-羟基-3-甲基戊二酸单酰辅酶 A（3-hydroxy-3-methylglutaryl coenzyme A，HMG-CoA）向琥珀酸（succinate）的转化，而琥珀酸则诱导机体产热，将多余营养物质燃烧掉；USP20 的抑制剂 GSK2643943A 在小鼠中能够显著降低进食后胆固醇合成与高脂食物诱导的体重增加，同时具有提高机体胰岛素敏感性、改善代谢综合征等治疗效果，这为后续药物研发提供重要理论依据（图 2）。相关结果[4]于 2020 年 11 月发表于《自然》期刊，并且因"进食诱导胆固醇合成的机制及降脂新药靶发现"入选了由中国科学技术协会生命科学学会联合体评选的"2020 年度中国生命科学十大进展"。

参考文献

[1] Goldstein J L, Brown M S. Regulation of the mevalonate pathway. Nature, 1990, 343(6257): 425-430.

[2] Chen L G, Chen X W, Huang X, et al. Regulation of glucose and lipid metabolism in health and disease. Science China Life Sciences, 2019, 62(11): 1420-1458.

[3] Luo J, Yang H Y, Song B L. Mechanisms and regulation of cholesterol homeostasis. Nat Nature Reviews Molecular Cell Biology, 2020, 21(4): 225-245.

[4] Lu X Y, Shi X J, Hu A, et al. Feeding induces cholesterol biosynthesis via the mTORC1-USP20-HMGCR axis. Nature, 2020, 588(7838): 479-484.

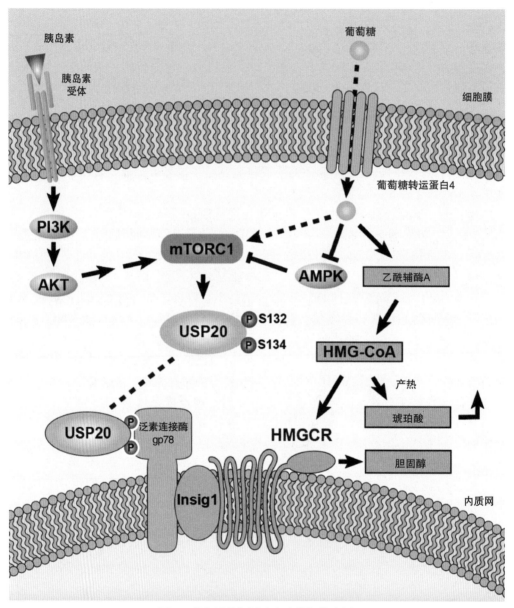

图 2　进食诱导胆固醇合成增加模式图

AMPK：AMP 激活的蛋白激酶；Insig1：胰岛素诱导基因 1；PI3K：磷脂酰肌醇 3 激酶

Mechanism of Feeding-Induced Cholesterol Biosynthesis and Identification of the Novel Lipid-Lowering Drug Target

Lu Xiaoyi，Song Baoliang

Cholesterol is an important structure and signaling molecule in mammals. The humans obtain cholesterol through de novo biosynthesis and intestinal absorption. Cholesterol biosynthesis is nutritionally and energetically costly. It is inhibited in starvation condition and elevated after taking carbohydrate food. Our results reveal that the deubiquitinase USP20 plays a key role in diet-induced cholesterol biosynthesis. In the feeding condition，insulin and glucose synergistically activate mTORC1 that phosphorylates USP20，which then binds HMGCR complex and stabilizes HMGCR，the rate limiting enzyme in cholesterol biosynthesis. Therefore，the liver makes more cholesterol. Genetic ablation or pharmacological inhibition of USP20 decreases lipid synthesis，increases heat production and improves insulin sensitivity in mice. Together，this study suggests that USP20 inhibitors could be used to lower cholesterol levels and treat relevant metabolic diseases.

3.16 水稻氮高效和高产协同调控的分子机制

傅向东

（中国科学院遗传与发育生物学研究所）

氮肥是不可或缺的农业生产资料，在保障农作物生产方面发挥着不可替代的作用。20世纪初，德国化学家弗里茨·哈伯（Fritz Haber）和卡尔·博世（Carl Bosch）发明了利用空气中的氮气合成氨的技术，这项人工固氮技术在第二次世界大战结束后快速在全球普及。20世纪60年代，为了解决因大量施肥导致农作物倒伏和减产的难题，育种家成功地培育并推广了半矮秆农作物品种，提高了农作物的耐高肥、抗倒伏能力，使全世界水稻和小麦产量翻了一番，缓解了因人口快速增长而引发的世界粮食危机。这场农业技术改革在世界农业史上被称为"绿色革命"。

但是，长期以来，高肥料投入条件下选育的农作物品种其产量对于氮肥投入的依赖性很大，这不仅增加了种植成本和能源消耗，而且加剧了全球的温室效应、水体富营养化和土壤酸化等环境问题。面对粮食安全和生态环境安全的双重挑战，提高农作物品种的氮肥利用效率已成为"少投入、多产出、保护环境"的可持续农业发展中一个亟待突破的育种瓶颈。

直到2000年前后，随着植物分子生物学和基因组学的快速发展，科学家逐渐发现"绿色革命"半矮化育种的关键控制点是DELLA蛋白，它是植物激素赤霉素信号途径中关键的植物生长阻遏因子。DELLA蛋白高水平积累，赋予了农作物半矮化、抗倒伏和高产等优良特性。笔者研究团队2018年的一项研究首次在分子水平上揭示了DELLA蛋白积累是导致农作物品种氮肥利用效率下降的关键因素；而通过提高生长调节因子GRF4的基因表达水平或蛋白活性，能够抵消DELLA蛋白积累对氮肥利用效率的负效应[1]。该研究入选了"2018年度中国科学十大进展"。笔者研究团队的后续研究发现，氮肥促进水稻增产需要DELLA蛋白，同时还需要一个能够响应土壤氮肥水平变化的NGR5蛋白。NGR5蛋白通过DNA甲基修饰途径调控水稻分蘖相关基因的表达，最终促进水稻分蘖和增产。DELLA蛋白通过与NGR5竞争性结合赤霉素受体蛋白GID1，使NGR5蛋白免于降解（图1）。笔者研究团队从水稻种质资源材料中挖掘到一个NGR5基因的优异等位变异，通过分子设计育种，将NGR5和GRF4基因的优异等位变异同时导入当前生产上的高产水稻品种中，可以在适当少施氮肥的条件下获得更高产量,实现氮肥利用效率与产量的协同提升[2]。该研究成果以封面文章形式发表在 2020 年《科学》期刊上,并入选"2020 年度中国生命科学十大进展"。

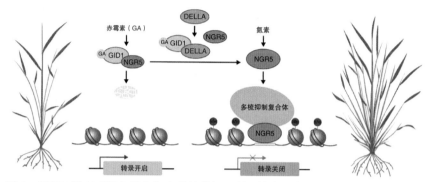

图 1 *NGR5* 提高"绿色革命"水稻品种氮肥利用效率和产量的表观遗传调控机制

近年来,农作物产量和氮肥利用效率等复杂农艺性状形成的分子基础研究进展迅速,但是在生产上能够应用于氮高效和高产协同改良的优异等位基因还非常有限。利用 DELLA-GRF4-NGR5 分子模块改良现有高产农作物品种的氮肥吸收利用能力,可实现农作物氮高效和高产的协同改良[3],这不仅有助于解决农作物品种氮肥依赖的弊端,而且为"少投入、多产出、保护环境"的可持续农业发展和新"绿色革命"提供了新策略(图 2)。

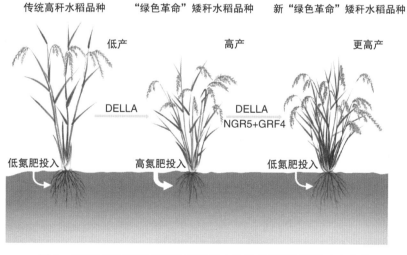

图 2 *GRF4* 和 *NGR5* 基因协同提升农作物的氮肥利用效率和产量

参考文献

[1] Li S,Tian Y,Wu K,et al. Modulating plant growth-metabolism coordination for sustainable agriculture. Nature,2018,560:595-600.

[2] Wu K,Wang S,Song W,et al. Enhanced sustainable green revolution yield via nitrogen-responsive chromatin modulation in rice. Science,2020,367:641.

[3] Wu K,Xu H,Gao X,et al. New insights into gibberellin signaling in regulating plant growth-metabolic coordination. Current Opinion in Plant Biology,2021,63:102074.

Modulating Gibberellin Signaling for Enhanced Sustainable Green Revolution Yield

Fu Xiangdong

The green revolution of the 1960s boosted crop yields through widespread adoption of semi-dwarf plant varieties and nitrogen fertilizer use. The accumulation of growth-repressing DELLA proteins in Green Revolution varieties (GRVs) is associated with lodging resistance,however,GRVs have a poor nitrogen use efficiency. Here we show that increasing nitrogen supply increases NGR5 (NITROGEN-MEDIATED TILLER GROWTH RESPONSE 5) abundance,which in turn recruits polycomb repressive complex 2(PRC2) to repress branching-inhibitory genes via H3K27me3 modification, thereby promoting rice tillering. NGR5 is a target of gibberellin-GID1-promoted proteasomal destruction. DELLA proteins competitively inhibit the GID1-NGR5 interaction, stabilizing NGR5 and increasing tillering of rice GRVs. We also show that increased abundances of both NGR5 and GRF4 (GROWTH-REGULATING FACTOR 4) enhance grain yield and nitrogen use efficiency through modulating plant growth-metabolic coordination, and that provides new breeding strategies to reduce nitrogen fertilizer use while boosting grain yield above what is currently sustainably achievable.

3.17 小麦抗赤霉病基因 *Fhb7* 的克隆
和调控机理解析及育种利用

孔令让 王宏伟 孙思龙 葛文扬

（山东农业大学农学院，作物生物学国家重点实验室）

小麦是全球种植面积最大和分布最广的重要粮食作物，全世界超过 1/3 的人口以小麦作为淀粉、蛋白质、矿物质及维生素等人体所需必要物质的主要来源。但是，冬小麦是跨年度生长的作物，从发芽到收获需要经历"秋冬春夏"四季，可能会受到各种生物或非生物胁迫的威胁。非生物胁迫主要包括干旱、盐碱及温度胁迫等；生物胁迫主要有小麦病害（如赤霉病、锈病、白粉病等）、虫害及杂草危害等。克服这些难关，获得高产优质的小麦品种是科研工作者一直以来努力的目标。

由禾谷镰孢菌（*Fusarium graminearum*）引起的小麦赤霉病是众多胁迫中对小麦产量和品质危害最严重的病害之一[1]，故有小麦"癌症"之称（图 1）。我国是小麦赤霉病受害面积最大的国家，其流行暴发，不仅造成大面积产量损失，更严重的是病原菌分泌的呕吐毒素、玉米赤霉烯酮、T2、HT2 等霉菌毒素严重污染小麦籽粒，危害人畜健康，已成为威胁我国粮食生产和食品安全的重大难题。2020 年 9 月 15 日，小麦赤霉病被我国农业农村部列入《一类农作物病害名录》[2]。然而，目前赤霉病的有效抗病基因资源并不多，因此挖掘新的抗赤霉病基因，对于解决这一世界性难题，保障国家粮食和食品安全具有重要意义。

本研究团队历经 20 年，在发现小麦抗赤霉病基因 *Fhb7* 的基础上，完成了目标基因的初步定位和精细定位[3-6]；随后，利用多种测序技术组装了高质量的二倍体长穗偃麦草（*Thinopyrum elongatum*，EE，$2n=2x=14$）参考基因组，并利用异代换系材料构建多个定位群体，结合 BAC 文库筛选搭建物理图谱，最后完成了主效抗赤霉病基因 *Fhb7* 的图位克隆。进一步研究发现，该基因编码的一种谷胱甘肽巯基转移酶（Glutathione S-transferases，GSTs），对包括脱氧雪腐镰刀菌烯醇（deoxynivalenol，DON）在内的许多主要的单端孢霉烯族毒素（如 NIV、T-2、HT-2 等）都具有广谱的解毒功能；并从分子层面上揭示了 *Fhb7* 编码的蛋白可以打开 DON 毒素的环氧基团，催化其形成谷胱甘肽加合物（DON-GSH），从而产生解毒效应的抗病机制

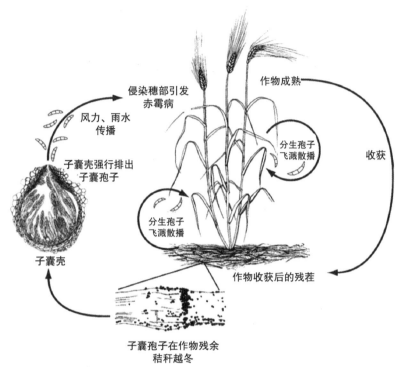

图1 小麦赤霉病发生及病害循环[1]

（图2）。令人惊奇的是，*Fhb7* 和禾本科植物共生的香柱菌（*Epichloë*）基因组序列同源性高达 97%，通过基因组比较分析推测，早期不同生物之间可能通过水平基因转移（horizontal gene transfer，HGT）将 *Epichloë Fhb7* 的 DNA 序列整合到长穗偃麦草或者其更原始的宿主亲本的基因组中，从而进化出抵御镰孢菌属病原菌侵染的功能。本研究首次提供了真核生物核基因组 DNA 水平转移的功能性证据，为研究植物与真菌互作机制和植物基因组进化提供了新思路。同时，育种利用研究表明，在多种不同的遗传背景下，*Fhb7* 基因均显著提高小麦赤霉病和茎基腐病抗性，而对产量无显著的影响。目前，已通过常规育种和分子标记辅助选择技术相结合，选育并审定了携带 *Fhb7* 的高产、优质、综合抗性突出的小麦新品种"山农-48"，为解决小麦抗赤霉病"卡脖子"难题提供了有效基因资源。相关科研成果于 2020 年 5 月 22 日作为封面文章发表在国际学术期刊《科学》[6]。该成果在多个科学层面取得关键性突破，是我国小麦研究领域的首篇《科学》文章，引发全球范围内同行专家的广泛关注，入选由两院院士评选的"2020 年中国十大科技进展新闻"、2020 年度"中国生命科学十大进展"、2020 年"中国高校十大科技进展"、山东省 2020 年"十大科技成果"和"2021中国农业科学重大进展"。

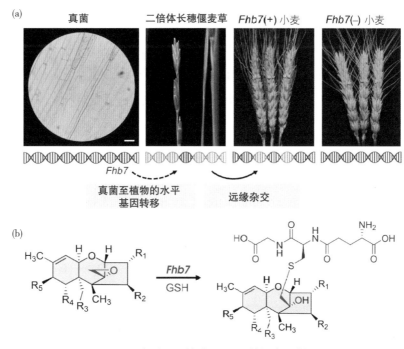

图 2　*Fhb7* 水平基因转移（a）及其解毒机制（b）

参考文献

［1］Dean R，Van Kan J A L，Pretorius Z A，et al. The Top 10 fungal pathogens in molecular plant pathology. Molecular Plant Pathology，2012，13(4)：414-430.

［2］农业农村部．一类农作物病虫害名录．中华人民共和国农业农村部公告，第 333 号．2020 年 9 月 15 日．

［3］Shen X，Kong L，Ohm H. Fusarium head blight resistance in hexaploid wheat(*Triticum aestivum*)-*Lophopyrum* genetic lines and tagging of the alien chromatin by PCR markers. Theoretical and Applied Genetics，2004，108(5)：808-813.

［4］Zhang X，Shen X，Hao Y，et al. A genetic map of Lophopyrum ponticum chromosome 7E，harboring resistance genes to Fusarium head blight and leaf rust. Theoretical and Applied Genetics，2011，122(2)：263-270.

［5］Guo J，Zhang X，Hou Y，et al. High-density mapping of the major FHB resistance gene *Fhb7* derived from *Thinopyrum ponticum* and its pyramiding with *Fhb1* by marker-assisted selection. Theoretical and Applied Genetics,2015,128(11):2301-2316.

[6] Wang H,Sun S,Ge W,et al. Horizontal gene transfer of *Fhb7* from fungus underlies Fusarium head blight resistance in wheat. Science,2020,368(6493):eaba5435.

Horizontal Gene Transfer of *Fhb7* from a Fungus Underlies *Fusarium* Head Blight Resistance in Wheat

Kong Lingrang ,Wang Hongwei ,Sun Silong ,Ge Wenyang

Fusarium head blight(FHB)has devastated wheat production globally. Mycotoxin, especially trichothecenes,produced by the *Fusarium* pathogen is a major safety concern for human health and animal production. *Fhb7* discovered from a wheat relative,*Thinopyrum ponticum*,shows a stable major effect on FHB resistance. Here, we cloned the *Fhb7* by assembling the Triticeae E genome of *Th. elongatum*,a wheatgrass species used in wheat distant hybridization breeding. *Fhb7* encodes a glutathione S-transferase(GST)and confers broad resistance to *Fusarium* species by detoxifying trichothecenes via de-epoxidation. Presence of GST homologs in *Epichloë* fungal species but not in the sequenced plants supports that *Th. elongatum* might gain GST via horizontal gene transfer(HGT)from an endophytic *Epichloë* that established symbiosis with temperate grasses. When introgressed into wheat, GST confers resistance to both FHB and crown rot in diverse wheat backgrounds without yield penalty, providing an excellent *Fusarium* resistant source for breeding.

3.18　创新质谱技术驱动生命科学的新发现

黄超兰

（中国医学科学院北京协和医院，疑难重症及罕见病国家重点实验室）

　　T 细胞主要通过 T 细胞受体（T cell receptor，TCR）来特异性识别抗原，继而引发对入侵病原物或者肿瘤细胞的适应性免疫反应。TCR 与抗原结合后不能直接活化 T 细胞，需依赖其邻近的 CD3 分子向细胞内传递活化信息。TCR-CD3 复合物由多种亚基组成，其中 TCR 的 α、β 亚基负责与抗原结合，CD3 的 γ，δ，ε，ζ 亚基负责完成细胞内信号转导。免疫受体酪氨酸活化基序（immunorecepter tyrosine-based activation motif，ITAM）作为激活 T 细胞的重要元件，在所有 CD3 亚基上均有分布。此前科学家们对 CD3ζ 亚基研究得最为详细，嵌合抗原受体（chimeric antigen receptor，CAR）的设计构建中通常利用 CD3ζ 传递信号。但迄今为止，其他 CD3 亚基的作用尚未明晰。

　　基于此，本团队与中国科学院分子细胞科学卓越创新中心许琛琦团队、美国加州大学圣选戈分校惠恩夫团队开展深度合作，并提出了 3 个关键性生物学问题：

　　（1）在不同抗原刺激下，CD3 亚基 ITAM 磷酸化模式是否存在差异？

　　（2）是否可以定量考察不同 ITAM 磷酸化结果？

　　（3）如果上述差异存在，它会怎样影响 TCR 信号转导？

　　深入探索 ITAM 的动态磷酸化模式将为全面揭示不同 CD3 亚基的功能提供核心信息。然而，一个完整的 TCR-CD3 复合物分布着 10 个 ITAM，共 20 个磷酸化位点，如何在同一时间实现对全部磷酸化位点的定量分析是个极具挑战性的技术难题。为此，本团队开发了一种新颖的绝对定量方法——基于靶向免疫沉淀多重轻标绝对定量质谱法（targeted-IP-multiplex-light-absolute-quantitative mass spectrometry，TIMLAQ-MS）。区别于此前报道的定量手段，TIMLAQ-MS 利用稳定的靶向免疫沉淀体系，成功绕过了同位素重标记肽这一步骤，既节约了成本，又有效降低了方法的复杂性和数据采集误差，同时又显著提高了定量的准确性，最终实现了在单次检测中对 TCR-CD3 复合物酪氨酸动态磷酸化修饰全貌的描绘（图 1）。

　　出乎意料的是，除了观察到 CD3 亚基常见的双磷酸化修饰现象，我们还首次鉴定到了 CD3ε 亚基的单磷酸化修饰模式。正是这一特殊的新发现促使我们深入探索 CD3ε

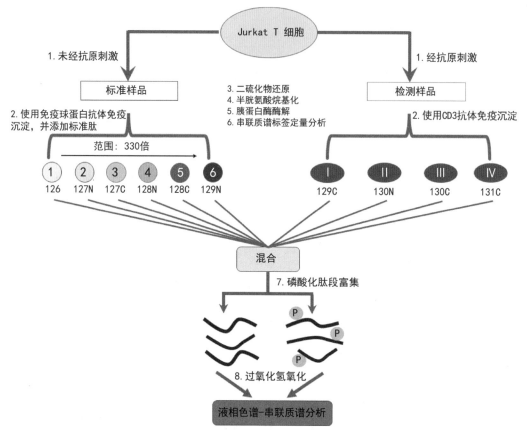

图 1 基于 TIMLAQ-MS 法的 CD3 ITAM 磷酸化修饰鉴定流程图

在 TCR 信号转导通路中潜在的全新多重信号转导功能。结果显示，CD3ε 可以同时招募抑制性信号分子 C-SRC 酪氨酸激酶（Csk）和活化性信号分子磷脂酰肌醇 3-激酶（PI3K）。将 CD3ε 胞内区加入临床使用的第二代 CAR 序列中，可使得 CAR-T 细胞持续性更好，抗肿瘤功能更强，并显著降低发生细胞因子释放综合征的风险（图 2）。

从重要的基础生物学问题开始，为解决问题而开发创新的技术方法，得到新发现，再深入探索生物学功能，最后有望贡献在治疗方法上，本研究完美地演绎了不同交叉领域共同合作而产生的精彩结果。相关成果于 2020 年 7 月发表在国际顶级期刊《细胞》上[1]，由于在"抗原受体信号转导机制及其在 CAR-T 治疗中的应用"方面做出了原创性贡献，成功入选"2020 年度中国生命科学十大进展"和"2020 年度中国医学科学院重要医学进展"。

本团队始终坚持以一切具有重要意义的科学和临床问题为起源来开发质谱和蛋白

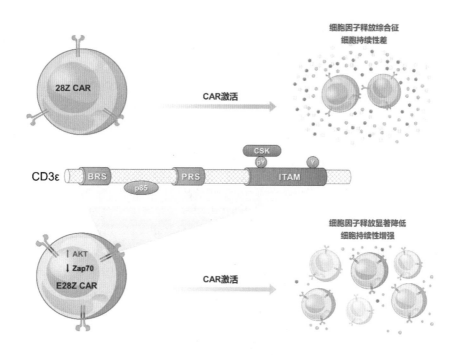

图 2 嵌合抗原受体的信号元件改造

质组学的创新方法，探究和揭示生命科学领域的未知，并以真正高标准高质量的蛋白质组学数据，有效建立临床生物标志物开发和验证的高质量标准化程序，得到能贡献生命科学和人类健康的真正产出。我们还于 2020 年 11 月在国际知名期刊《自然-通讯》（*Nature Communications*）上发表深度尿液蛋白组揭示早期新型冠状病毒感染主要为免疫抑制并或存在"两阶段"机制的重要成果，该文章成功入选期刊年度热门阅读榜单[2]。

参考文献

[1] Wu W, Zhou Q, Masubuchi T, et al. Multiple Signaling Roles of CD3ε and Its Application in CAR-T Cell Therapy. Cell, 2020, 182(4): 855-871.

[2] Tian W, Zhang N, Jin R, et al. Immune suppression in the early stage of COVID-19 disease. Nature Communications, 2020, 11(1): 5859.

Innovative Mass Spectrometry Technology Drives New Discoveries in Life Sciences

Catherine C. L. Wong

T cells trigger an adaptive immune response through specific recognition of antigens by T cell receptor (TCR). The phosphorylation of immunoreceptor tyrosine-based activation motif (ITAM) on CD3 chains plays an important role in TCR-mediated signaling. However, whether different phosphorylation pattern exists among distinct CD3 chains remains elusive. In order to unravel the mystery, we developed a novel absolute quantitative mass spectrometry-based method, which is called Targeted-IP-Multiplex-Light-Absolute-Quantitative Mass Spectrometry (TIMLAQ-MS). Thus, we achieved a comprehensive mapping of all CD3 ITAM phosphorylation patterns in a single test, leading to a discovery of novel multiple signaling roles of CD3ε mono-phosphorylation, and also the development of next-generation chimeric antigen receptor-T (CAR-T) cell therapy with reduced cytokine production, enhanced cell persistence and improved antitumor activity. Starting from an essential scientific question, we used innovative mass spectrometry technology to drive new discoveries in life sciences, and finally contributed to the design of new treatment regimens.

3.19　全球二氧化碳施肥效应时空变化格局

王松寒　张永光　居为民　陈镜明

（南京大学国际地球系统科学研究所）

　　未来全球变暖的速率以及陆地生态系统对全球变暖的响应是国际学术期刊《科学》2018 年列出的未来 25 年需要解决的 125 个重大科学问题之一。工业革命以来，人类活动造成大气中二氧化碳（CO_2）的浓度持续上升。CO_2 浓度的不断增加，在通过温室效应导致全球变暖的同时，也可以提升植被的光合作用速率（即 CO_2 施肥效应），增加陆地生态系统吸收大气 CO_2 的能力（即碳汇能力），从而减缓全球变暖的速率。研究表明，CO_2 施肥效应是近几十年来全球陆地生态系统碳汇显著增加的决定性因素[1,2]，也是全球变绿的主要驱动因子之一[3]。因此，在全球尺度上定量化评估 CO_2 施肥效应，并分析其时空变化格局，对于准确评估全球陆地生态系统的固碳能力以及其变化趋势、降低未来气候变化预测的不确定性十分重要。

　　尽管基于控制实验已经可以在叶片和冠层尺度对 CO_2 施肥效应的机理进行较为深入的研究[4]，但受限于控制实验的数量、空间分布、物种代表性等，全球尺度 CO_2 施肥效应的时空变化格局尚不清楚。长时间序列遥感观测则为全球 CO_2 施肥效应研究提供了连续、准确的基础数据集。因此，笔者团队的研究首先基于 AVHRR①、MODIS② 等系列卫星传感器的观测数据，生成 1982～2015 年全球新型植被指数（NIRv）数据，并验证了将其作为全球植被总初级生产力（gross primary production，GPP）指代的可行性；在此基础上，构建了准确评估全球 CO_2 施肥效应的检测-归因模型，揭示了近 40 年全球 CO_2 施肥效应的时空变化特征；最后，结合欧洲地区的叶片氮磷观测和全球陆地水储量等数据，鉴别了全球 CO_2 施肥效应时空变化趋势的可能原因。

　　研究结果表明，基于长期遥感数据观测，全球 CO_2 施肥效应在近 40 年存在显著

---

　　① 全称为 Advanced Very High Resolution Radiometer，是 NOAA 系列气象卫星上搭载的传感器。

　　② 全称为 Moderate-resolution Imaging Spectroradiometer，是 TERRA 和 AQUA 卫星上搭载的传感器。

的下降的趋势（图 1）[5]；相较于前一个时间区间（1982~1996 年），最后一个时间区间（2001~2015 年）的全球 CO_2 施肥效应显著降低（图 1）。考虑了原始数据和检测-归因模型的不确定性之后，该下降趋势仍然是显著的；改变滑动窗口的长度等分析进一步证明了该下降趋势的鲁棒性。笔者研究团队采用了基于光能利用率模型的全球 GPP 产品等其他多种数据源对该结果进行了验证。

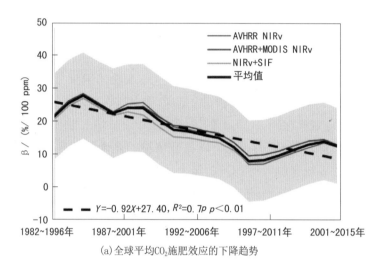

(a)全球平均CO_2施肥效应的下降趋势

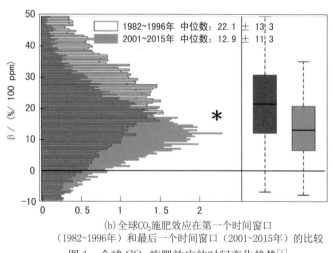

(b)全球CO_2施肥效应在第一个时间窗口
（1982~1996 年）和最后一个时间窗口（2001~2015 年）的比较

图 1　全球 CO_2 施肥效应的时间变化趋势[5]

　　笔者研究团队对全球 CO_2 施肥效应下降趋势的空间分布进行了分析，结果表明全球超过 70%~80% 的陆地植被区域存在着施肥效应下降的趋势，下降趋势较大的地区包括欧洲、西伯利亚、南美洲和非洲大部以及澳大利亚西部地区；在少部分地区施肥效应存在着上升的趋势，例如东南亚部分地区和澳大利亚东部地区。多个生态系统模型同样能够模拟出全球 CO_2 施肥效应的下降趋势，但其下降趋势显著低于基于遥感数据的结果。

进一步分析造成全球 CO_2 施肥效应下降的可能原因。基于欧洲 ICP Forests 等机构的支持，笔者研究团队获取了欧洲地区超过 3 万余条森林叶片氮磷浓度观测数据，并分析了其和 CO_2 施肥效应下降的因果关系。结果表明，欧洲地区植被叶片氮磷浓度有显著的下降趋势，且该趋势与 CO_2 施肥效应的下降呈显著的正相关关系。研究团队还分析了水分供应状况的变化对施肥效应的影响。以陆地水储量作为水分供应的指标，研究发现遥感 GPP 产品对水分供应的敏感性存在显著的增加趋势，而生态系统模型的结果则几乎没有显著的变化趋势。这表明植被对水分胁迫的响应更为敏感，因此水分供应状况的变化是 CO_2 施肥效应下降的可能原因之一。

该研究将植被遥感观测结果与全球变化生态学研究相结合，首次对近 40 年全球 CO_2 施肥效应的时空变化格局进行了定量化评估。该研究成果对更为深入地理解全球变化背景下陆地生态系统的响应机制、更为准确地评估全球陆地生态系统碳收支、更为精准地预测未来气候变暖的速率均具有十分重要的意义。2020 年 12 月 11 日，以上的研究成果[5]在国际学术期刊《科学》发表，获得了地理学和全球变化生态学领域的高度关注，并入选了 2020 年度"中国地理科学十大研究进展"。

参考文献

[1] Schimel D, Stephens B B, Fisher J B. Effect of increasing CO_2 on the terrestrial carbon cycle. Proceedings of the National Academy of Sciences, 2015, 112: 436-441.

[2] Fernández-Martínez M, Sardans J, Chevallier F, et al., Global trends in carbon sinks and their relationships with CO_2 and temperature. Nature Climate Change, 2019, 9: 73-79.

[3] Zhu Z, Piao S L, Myneni R B, et al. Greening of the Earth and its drivers. Nature Climate Change, 2016, 6: 791-795.

[4] Ainsworth E A, Rogers A. The response of photosynthesis and stomatal conductance to rising [CO_2]: mechanisms and environmental interactions. Plant, Cell & Environment, 2007, 30: 258-270.

[5] Wang S H, Zhang Y G, Jun W M, et al., Recent global decline of CO_2 fertilization effects on vegetation photosynthesis. Science, 2020, 370: 1295-1300.

Global Spatio-Temporal Patterns of the CO$_2$ Fertilization Effects on Vegetation Photosynthesis

Wang Songhan, Zhang Yongguang, Ju Weimin, Chen Jingming

The enhanced vegetation productivity driven by increased levels of carbon dioxide (CO$_2$) (the CO$_2$ fertilization effect, CFE) sustains an important negative feedback on climate warming, but the temporal dynamics of CFE remain unclear. Using multiple long-term satellite- and ground-based datasets, our results indicated that global CFE has declined across most terrestrial regions of the globe during 1982—2015, correlating well with changing nutrient concentrations and availability of soil water. Current carbon-cycle models also demonstrate a declining CFE trend, albeit substantially weaker than that from the Earth observations. This declining trend in the forcing of terrestrial carbon sinks by increasing levels of atmospheric CO$_2$ implies a weakening negative feedback on the climatic system and increases the society's dependence on future strategies to mitigate climate warming.

3.20　印度大陆俯冲板片在高原裂谷下方撕裂的图像证据

田小波[1,2]　梁晓峰[1,2]　陈　赟[1,2]

（1. 中国科学院地质与地球物理研究所岩石圈演化国家重点实验室；
2. 中国科学院地球科学研究院）

板块构造理论主要阐述了大洋岩石圈在洋中脊的扩张和汇聚边界的深俯冲，大洋岩石圈的运动携带着大陆漂移，洋盆关闭最终导致大陆碰撞。由于大陆地壳（岩石圈）明显厚于大洋地壳（岩石圈），并且具有明显的流变分层结构，因此大陆岩石圈不易于深俯冲，而且会发生显著的板内形变（陆内形变）。已有的研究显示，陆-陆汇聚的造山及其后的高原演化包含了复杂的地壳（岩石圈）构造活动和变形过程，作为目前人类认识显生宙（大约 5.7 亿年以来）地球演化的主要理论——板块构造理论，急需发展大陆动力学对其进行完善。作为陆-陆汇聚的典型实例，青藏高原是目前地球上最大、最高的高原，是新生代（65 百万年）以来陆内变形作为广泛和强烈的区域，因此被认为是发展大陆动力学的最佳天然实验室。

早新生代，新特提斯洋闭合和印度大陆与欧亚大陆的陆-陆碰撞，近五十几个百万年（Ma）来两个大陆一直在持续汇聚，地壳（岩石圈）汇聚形成了具有巨厚地壳的青藏高原。然而，中新世中期（20～10 Ma）以来高原中南部开始发育数条南北走向的裂谷，代表高原中部地壳由南北向挤压到东西向拉张或伸展的转变。部分学者认为这是高原中-下地壳软弱，在重力势能的作用下高原垮塌的表现，高原中部的垮塌促进了同期（准同期）高原东缘和东南缘的高原侧向加速生长，但这一设想不好解释高原自中新世以来向北（祁连山）的持续扩展。

在印度大陆与欧亚大陆碰撞后的五十几个百万年来，印度大陆岩石圈板片持续向北在青藏高原下方俯冲，在近南北向至少数百千米长的印度大陆岩石圈俯冲进入上地幔。地震学研究显示青藏高原下方印度大陆俯冲板片俯冲倾角由西向东增大[1,2]。根据高原南北向裂谷空间分布特征的动力学分析[3]、火山岩成分[4,5]、地震波走时[6,7]和剪切波分裂[8]等多项研究推测，南北向裂谷的发育与印度大陆俯冲板片的南北向撕裂有关，热的软流圈物质上涌至高原岩石圈底部后东-西向分流，对高原的岩石圈的东西向拉张促使南北向裂谷的发育。然而，这种认识一直缺乏高分辨的地震学图像来证实俯冲板片的撕裂，及其与裂谷的空间对应关系。

为此，本研究团队利用青藏高原中部布设的两个宽频带流动台阵（图 1）记录的天然远震事件的波形记录提取 S 波接收函数，通过大量接收函数的共转换点叠加成像获得了高原中部岩石圈底界面和印度大陆俯冲板片的起伏形态（图 2）。结果显示：在亚东-谷露裂谷下方，印度大陆俯冲板片底界面发生了明显的东西向错断，两侧板片深度相差70 km；在裂谷的两侧，青藏高原岩石圈底界面在东西向也存在 30 km 的深度差异。

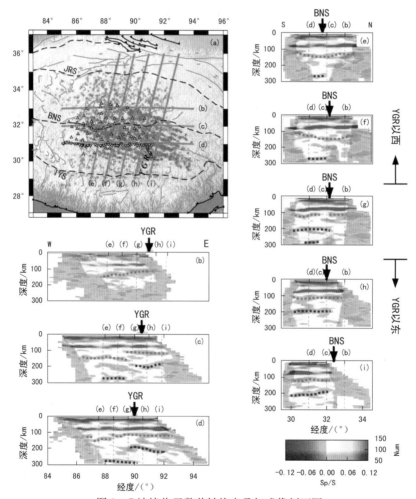

图 1 S 波接收函数共转换点叠加成像剖面图

（a）表示剖面位置和 150 km、200 km 深度平面 S-to-P 转换点分布情况；（b）~（i）蓝色虚线表示岩石圈底界面深部分布；BNS：班公湖-怒江缝合带，IYS：雅鲁藏布江缝合带，JRS：金沙江缝合带，YGR：亚东-谷露裂谷。

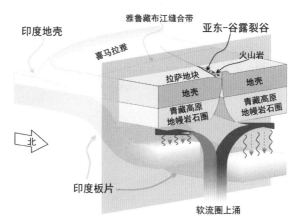

图 2　印度大陆俯冲板片撕裂示意图（阴影遮蔽的南侧区域表示本项研究没有约束）

　　我们的成像获得了印度大陆俯冲板片南北向撕裂的直接证据（图 1），同时结果也表明其上方的高原岩石圈发生相同方式的错断，并且这两个不同深度层次的撕裂或错断都位于亚东-谷露裂谷正下方。这些高分辨的地震学图像为建立印度大陆俯冲板片撕裂与高原裂谷发育的动力学联系提供了关键的证据。综合前人的研究，我们认为（图 2）：青藏高原在印度大陆的北向推挤下，岩石圈缩短增厚，在南北挤压力和重力共同作用下，高原中部物质发生东西向的伸展；由于印度大陆俯冲板片发生撕裂，形成板片窗为深部的软流圈物质上升提供了通道；热的软流圈物质侵蚀、破坏上方东西向伸展的高原岩石圈，促使南北向裂谷的发育。

　　该研究成果[9]于 2020 年以题为 "Complex structure of upper mantle beneath the Yadong-Gulu rift in Tibet revealed by S-to-P converted waves" 的论文发表在国际权威地学期刊《地球与行星科学通讯》（*Earth and Planetary Science Letters*）上。

参考文献

[1] Li C, Van der Hilst R D, Meltzer A S, et al. Subduction of the Indian lithosphere beneath the Tibetan Plateau and Burma. Earth and Planetary Science Letters, 2008, 274(1-2): 157-168.

[2] Zhao J M, Yuan X H, Liu H B, et al. The boundary between the Indian and Asian tectonic plates below Tibet. Proceedings of the National Academy of Sciences of the United States of America, 2010, 107(25): 11229-11233.

[3] Yin A. Mode of Cenozoic east-west extension in Tibet suggesting a common origin of rifts in Asia during the Indo-Asian collision. Journal of Geophysical Research-Solid Earth, 2000, 105 (B9): 21745-21759.

[4] 侯增谦, 赵志丹, 高永丰等. 印度大陆板片前缘撕裂与分段俯冲: 来自冈底斯新生代火山-岩浆作

用证据 . 岩石学报,2006,22(4):761-774.

[5] 丁林,岳雅慧,蔡福龙等 . 西藏拉萨地块高镁超钾质火山岩及对南北向裂谷形成时间和切割深度的制约 . 地质学报,2006,80(9):1252-1261.

[6] Liang X F,Sandvol E,Chen Y J,et al. A complex Tibetan upper mantle:A fragmented Indian slab and no south-verging subduction of Eurasian lithosphere. Earth and Planetary Science Letters,2012,333:101-111.

[7] Li J T,Song X D. Tearing of Indian mantle lithosphere from high-resolution seismic images and its implications for lithosphere coupling in southern Tibet. Proceedings of the National Academy of Sciences of the United States of America,2018,115(33):8296-8300.

[8] Chen Y,Li W,Yuan X,et al. Tearing of the Indian lithospheric slab beneath southern Tibet revealed by SKS-wave measurements. Earth and Planetary Science Letters,2015,413:13-24.

[9] Liu Z,Tian X B,Yuan X H,et al. Complex structure of upper mantle beneath the Yadong-Gulu rift in Tibet revealed by S-to-P converted waves. Earth and Planetary Science Letters,2020,531,115954,DOI:10. 1016/j. epsl. 2019. 115954.

Seismic Images of Tearing the Lndian Subducting Slab Beneath Rift in Tibet

Tian Xiaobo,Liang Xiaofeng,Chen Yun

The continental convergence between the Indian and Eurasian plates has produced the Tibetan plateau since about 50 Ma. The pattern of the Indian mantle lithosphere subducting beneath the plateau and the deformation of the Tibetan mantle lithosphere are the keys for understanding the continental collision process and the evolution of plateau. Based on the teleseismic S-wave receiver function image,our study provides solid evidences that there are offsets of the lithosphere-asthenosphere boundary in both the Tibetan and Indian lithosphere beneath the north-south trending Yadong-Gulu rift. Our study suggests that the thickness of the Tibetan lithosphere and the depth of the underlying Indian mantle slab are saltus across the rift in different scales. The abrupt changes imply that the subducting Indian mantle slab has been torn and provided an upwelling channel for the asthenosphere contributing to the development of the rift.

3.21　千年气候事件研究的进展和趋势

程　海　张海伟　赵景耀　董西瑀

（西安交通大学全球变化研究院）

　　千年尺度的全球气候事件是最显著的地球气候系统的波动之一，尤以过去6万年中"新仙女木"（Younger Dryas，YD）事件、丹斯加德-奥施格（Dansgaard-Oeschger，D/O）旋回和海因里希冰阶（Heinrich Stadial，HS）最为典型。对这些事件的研究有助于深刻理解全球气候快速变化的驱动-传动机制和高-低纬的动力学关联，并有助于气候变化预测。以发生在距今（以公元1950年为截止时间）12 900～11 700年前的快速变冷事件YD为例，本研究团队利用新一代高分辨率、精确定年的石笋记录精确校准了格陵兰冰芯年代，将其年龄误差由原来的100～140年降低为20～40年［图1（A）］；建立了不同气候系统的相关关系，进而发现南极YD事件的开始时间晚于格陵兰地区约100年，而其结束时间早于格陵兰地区约200年；热带低纬地区和南半球提前于北半球大西洋和亚洲季风-西风区数百年便表现出向YD结束阶段转变的气候信号，表明热带地区或南半球可能首先引发了YD事件的结束[1]［图1（A）］。本团队的新一代高精度石笋记录进一步表明千年事件内部南美季风区变得更干旱的转变过程比北半球季风增强和格陵兰升温的时间更早[2-3]，南美季风区变得更干旱对应着亚马逊河径流量减小以及河口地区海表盐度增加，最终通过一系列的正反馈过程加速了大西洋经向翻转环流的增强并引发了事件的结束[2-3]。研究还发现"陨星撞击"事件可能并不是触发YD事件的原因，因为经石笋校准的格陵兰冰芯记录的YD事件的开始时间要早于陨星撞击时间约50年［图1（A）］，且可能发生的陨星撞击（铂元素异常）在石笋记录中并没有对应的显著异常存在[1]。

　　年代际气候变化研究源于深刻的现实需求，与人类生存环境、经济和社会发展息息相关，也是打通古-今气候研究隔阂和提供古气候研究应用"出口"的重要基础研究。过去约2000年以来的石笋、树轮和湖泊沉积物等记录的亚洲夏季风变化均显示与近-现代气象监测记录相一致的年代际变化特征，以60～80年和20～30年准周期为主导。而目前年代际气候预测的困难在于：这些年代际变化特征是否是地球气候系统所固有的？在多大程度上受到全球变化因子（如大气CO_2浓度、大气CH_4浓度、大气温度、海平面、太阳辐射等）的影响？变率是否变化？与过去约2000年相比，一些特殊时段的气候因子变化巨大（如［图1（B）］所示的早中全新世、YD事件、末次

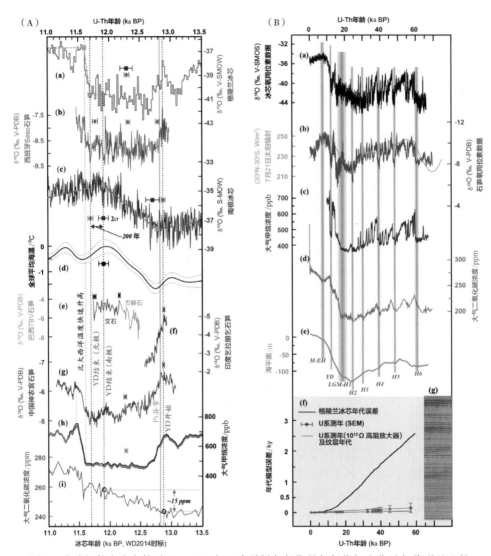

图 1 全球新仙女木事件对比，60 万年以来关键气候期的气候指标变化及年代误差比较

（A）：（a）北极格陵兰冰芯 δ^{18}O 记录[8]；（b）西班牙塞索（Seso）洞穴石笋记录 δ^{18}O[1]；（c）南极冰芯 δ^{18}O[9-10]；（d）全球平均海温曲线[11]；（e）巴西 TBV 洞穴石笋 δ^{18}O[12]；（f）印度乞拉朋乞洞穴石笋 δ^{18}O[1]；（g）中国神农宫洞穴石笋 δ^{18}O[1]；（h）大气甲烷浓度记录[9,13] 和（i）大气二氧化碳浓度记录[14]。（B）：（a）格陵兰冰芯 δ^{18}O 记录[15]；（b）集成的亚洲季风石笋 δ^{18}O 记录[16]；（c）大气甲烷浓度记录[9,13]；（d）大气二氧化碳浓度记录[17]；（e）海平面记录[18]；（f）年代误差比较；（g）激光扫描共聚焦显微镜下的石笋年纹层。

灰色柱为开展高分辨率石笋工作的时段。特殊气候期：Middle-Early Holocene（M-EH）—早中全新世；Younger Dryas（YD）event—新仙女木事件；Last Glacial Maximum（LGM）—末次冰盛期；Heinrich（H）event—海因里希事件。Pee Dee Belemnite（PDB）：美国南卡罗来纳州白垩系皮狄组箭石；Standard Mean Ocean Water（SMOW）：标准平均海洋水；ppm：比率表示，百万分之一；ppb：比率表示，十亿分之一。元素：铂，Platinum（Pt）

冰盛期以及海因里希冰阶等时段），因此这些重要时段的新一代高分辨率石笋记录的研究将能够为回答上述科学问题提供新的关键数据。

利用新一代中国石笋记录系统地建立 6 万年以来的千年突变气候事件的精准年代标尺及全球气候对比关系是国际古气候学界的重要目标。目前，本研究团队在这方面的研究工作取得了实质进展，有望近期完成集成工作——"西安交通大学过去 6 万年气候变化集成数据 1.0 版"（XJTU-1.0），以此回应学术界的长期期待[4]（图 1B-f）；如美国基金委斯蒂芬斯博士（Dr. Stephens）20 年前对南京葫芦洞记录的期待"……中国石笋记录将有望取代格陵兰冰芯记录作为校准全球古气候变化的年代学标尺……"；以及牛津大学亨德森教授（Prof. Henderson）2006 年的预期："古气候学研究的过去 20 年是属于冰芯的，而它未来的 20 年将可能属于洞穴石笋"[5]。另外，葫芦洞[14]C 新数据为国际放射性碳校准曲线 2020（IntCal20）较老时段所采用，提供了精准的[14]C 校准曲线[6,7]，对[14]C 年代学以及千年事件中的碳循环研究具有重要意义。

总之，新一代实验分析技术发展（包括 U-Th 年代、激光扫描共聚焦显微镜年层计数、高分辨率采样等）以及相关时段的关键石笋样品（具有年纹层、高铀含量、生长速率快）将提供新一代高分辨率-精确定年的石笋记录，为研究千年尺度的全球气候事件的触发-响应机制、高-低纬驱动关系，不同气候背景下亚洲夏季风的年代际变化，以及古气候年代学方面提供全新的重要支撑，同时使得中国相关研究走向国际学术前沿。

参考文献

[1] Cheng H, Zhang H, Spötl C, et al. Timing and structure of the Younger Dryas event and its underlying climate dynamics. Proceedings of the National Academy of Sciences of the USA, 2020, 117 (38):23408-23417.

[2] Cheng H, Xu Y, Dong X, et al., Onset and termination of Heinrich Stadial 4 and the underlying climate dynamics. Communication Earth & Environment, 2021, 2, 230. https://doi.org/10.1038/s43247-021-00304-6.

[3] Dong, X, Kathayat G, Sune O R, et al., Coupled atmosphere-ice-ocean dynamics during Heinrich Event 2. Proceedings of the National Academy of Sciences of the USA, 2022, In review.

[4] Cheng H, Zhang H, Zhao J, et al., Chinese stalagmite paleoclimate researches: A review and perspective. Science China Earth Sciences, 2019, 62(10):1489-1513.

[5] Henderson G M. Climate caving in to new chronologies. Science, 2006, 313(5787):620-622.

[6] Cheng H, Edwards R L, Southon J, et al. Atmospheric $^{14}C/^{12}C$ changes during the last glacial period from Hulu Cave. Science, 2018, 362(6420):1293-1297.

[7] Reimer P J, Austin W E, Bard E, et al. The IntCal20 Northern Hemisphere radiocarbon age calibra-

tion curve(0-55 cal kBP). Radiocarbon,2020,62(4):725-757.

[8] Rasmussen S O,Andersen K K,Svensson A,et al. A new Greenland ice core chronology for the last glacial termination. Journal of Geophysical Research-Atmospheres, 2006, 111 (D6). https://doi. org/10. 1029/2005JD006079.

[9] Sigl M,Fudge T J,Winstrup M,et al, The WAIS Divide deep ice core WD2014 chronology-Part 2: Annual-layer counting(0-31 ka BP). Climate of the Past,2016,12(3):769-786.

[10] WAIS Divide Project Members. Precise interpolar phasing of abrupt climate change during the last ice age. Nature,2015,520,661-665.

[11] Bereiter B,Shackleton S,Baggenstos D,et al. Mean global ocean temperatures during the last glacial transition. Nature,2018,553(7686):39-44.

[12] Zhang H,Cheng H,Spötl C,et al. Gradual South - North Climate transition in the Atlantic Realm within the Younger Dryas. Geophysical Research Letters,2021,48(8):e2021GL092620.

[13] Rhodes R H,Brook E J,Chiang J C,et al. Enhanced tropical methane production in response to iceberg discharge in the North Atlantic. Science,2015,348(6238):1016-1019.

[14] Marcott S A,Bauska T K,Buizert C,et al. Centennial-scale changes in the global carbon cycle during the last deglaciation. Nature,2014,514(7524):616-619.

[15] Rasmussen S O,Bigler M,Blockley S P,et al. A stratigraphic framework for abrupt climatic changes during the Last Glacial period based on three synchronized Greenland ice-core records: refining and extending the INTIMATE event stratigraphy. Quaternary Science Reviews,2014,106:14-28.

[16] Cheng H,Edwards R L,Sinha A,et al. The Asian monsoon over the past 640,000 years and ice age terminations. Nature,2016,534(7609):640-646.

[17] Bereiter B,Eggleston S,Schmitt J,et al. Revision of the EPICA Dome C CO_2 record from 800 to 600 kyr before present. Geophysical Research Letters,2015,42(2):542-549.

[18] Spratt R M,Lisiecki L E. A Late Pleistocene sea level stack. Climate of the Past,2016,12(4):1079-1092.

Developments in the Research Forefront of Millennial Abrupt Climate Events

ChengHai,Zhang Haiwei,Zhao Jingyao,Dong Xiyu

The millennialabrupt climate events over the past 60 ky had profoundly affected global hydroclimate and ecological environment,however their trigger-propagation mechanisms and influence on the multi-decadal climate variability remain unclear. Recently, a series of new generation speleothem records with unprecedented high-resolution and precise chronology have been established across a set of the millennial events，leading to substantial progresses in establishments of new and precise chronologic benchmarks over the past 60 ky for correlating and calibrating global climate variability，and deciphering high-low latitude forcing mechanisms，and event trigger-propagation dynamics. Prospectively，the developments will also facilitate a groundbreaking in the research forefront of the Asian monsoon multi-decadal variability.

科技领域与
科技战略发展观察

Observations on Development
of Science and Technology

4.1　基础前沿领域发展观察

黄龙光　边文越　张超星　冷伏海

（中国科学院科技战略咨询研究院）

2020 年基础前沿领域取得多项突破，粒子物理、凝聚态物理、量子技术和光学等领域取得重大突破，合成化学自动化、智能化水平不断提高，纳米材料新热点方向显现，在生命健康领域的应用效果更加显著。欧盟布局和推动粒子物理、可持续化学品、石墨烯等前沿领域，美国和日本深入推进量子技术战略，美国调整纳米技术研发布局，以更好地抢占科技前沿和创新发展的制高点。

一、重要研究进展

1. 凝聚态物理、粒子物理、量子技术和光学等领域取得重大突破

室温超导体面世引发全球关注。美国罗切斯特大学等在高压条件下的富氢化合物中观察到室温（15 ℃）下的超导现象[1]，将超导温度纪录提升约 35 ℃。尽管离常压室温超导还有很长距离，但是这是具有里程碑意义的首个室温超导体。美国麻省理工学院等对双层转角石墨烯的超导和绝缘态之间的相互作用进行了多方位研究[2-4]，为双层转角石墨烯超导机制提供了进一步的探索。美国科罗拉多大学博尔德分校等在名为 RM734 的有机分子中首次发现铁电向列液晶相[5]，证实了 100 多年前科学家的预测。中国科学院物理研究所在磁性外尔半金属中首次提出"自旋轨道极化子"概念[6]，为磁性外尔体系中磁序与拓扑性质调控提供了新思路。

粒子物理重大突破频现。意大利格兰萨索国家实验室太阳中微子实验（Borexino）合作组首次探测到源自太阳中碳氮氧循环产生的碳-氮-氧（CNO）中微子[7]，证实了科学家于 20 世纪 30 年代提出的恒星核合成理论，为了解太阳的核心打开了重要的窗口。欧洲核子研究中心（European Organization for Nuclear Research，CERN）的紧凑缪子线圈（CMS）实验首次在质子-质子碰撞中观测到同时产生 3 个 W 玻色子或 Z 玻色子的现象[8]，为基本粒子及其相互作用提供了进一步的理解。美国费米实验室缪子电离冷却实验（MICE）合作组首次观察到缪子电离冷却[9]，向缪子对撞机的建造

迈出了重要一步。日本 T2K 中微子实验首次观测到中微子振荡过程中轻子电荷-宇称（charge parity，CP）破缺的证据[10]，意大利 XENON1T 暗物质实验发现异常信号[11]。这两个发现虽然没有达到物理学新发现的 5 个标准的置信度，但也引发了人们的高度关注。

量子优越性再获实现。中国科学技术大学等构建的 76 个光子的量子计算原型机"九章"[12]，实现了具有实用前景的"高斯玻色取样"任务的快速求解，使中国成为全球第二个实现"量子优越性"的国家。瑞典斯德哥尔摩大学等通过在单个锶离子的实验中拍摄一系列"快照"[13]，揭示了量子系统的测量结果是逐渐将叠加态转换为经典态。澳大利亚新南威尔士大学等在高于 1 K 的温度条件下可以实现硅量子点的单量子比特控制[14]，凸显出对未来量子系统和硅自旋量子比特进行低温控制的潜力。在量子通信方面，中国科学技术大学等利用"墨子号"量子科学实验卫星在国际上首次实现千公里级基于纠缠的量子密钥分发[15]，将以往地面无中继量子保密通信的空间距离提高了一个数量级。英国布里斯托大学等使用量子密钥分发技术建立了一个可扩展的城域量子网络[16]，该网络可以连接布里斯托 17 km 范围内的超过 8 个用户。

光学领域成果丰硕。美国纽约市立大学等通过二维材料的转角来改变光子[17]，将"转角电子学"拓展到纳米光子学，对纳米成像、量子光学、量子计算和低能光学信号处理等领域具有重要意义。荷兰埃因霍芬理工大学等研制出一种直接带隙硅基材料[18]，其将可用于光学通信和光学计算以及开发化学传感器。南京大学等通过结合超构透镜阵列与非线性晶体[19]，成功制备出高维路径纠缠光源和多光子光源。中国科学技术大学等实现了亚分子分辨的单分子光致荧光成像[20]，将成像空间分辨率提升到 0.8 nm 的亚纳米分辨水平。

2. 化学助力各领域发展，智能化水平不断提高

合成化学自动化、智能化水平不断提高。英国利物浦大学制造了一种可以像人类一样自主开展化学实验的机器人。装载了实验操作程序和算法的机器人可以像人类一样在实验室各种仪器装置间移动，进行装样、溶解、密封、催化实验、色谱分析等操作，解放了研究人员而且提高了研究效率[21]。英国格拉斯哥大学模仿人类开展化学研究"读文献—架仪器—做实验"过程，设计了一套基于自然语言处理技术的化学自动合成操作系统，从论文出发自动合成了利多卡因、戴斯-马丁试剂、AlkylFluor 等 12 种有机物，产率与论文报道值相当甚至略高[22]。波兰科学院和美国西北大学等通过引入机器学习算法对化学软件 Chematica 进行升级，使其可以设计出复杂天然产物的全合成路线，并通过了实验验证[23]。

绿色化学助力节能环保。美国密歇根大学研发出一种铜-铁基催化剂，可借助光

将二氧化碳转化为天然气的主要成分甲烷[24]。加拿大多伦多大学设计了一条电化学合成环氧乙烷的新路线，可在室温通过电解反应将乙烯转化为环氧乙烷。该路线可以与电化学二氧化碳制乙烯路线结合，直接将二氧化碳转化为环氧乙烷，也可以用于制备环氧丙烷[25]。美国加利福尼亚大学设计了一种铂/氧化铝催化剂，可在 280 ℃将废弃聚乙烯直接转化为高价值的长链烷基芳烃，且整个过程无需溶剂和氢气[26]。

化学助力生物、医药领域发展。美国伊利诺伊大学使用锰基大位阻催化剂实现含氮、含氧杂环化合物 α 位 $C(sp^3)$—H 键甲基化反应，对于提高药物分子药效具有重要应用价值[27]。美国斯克里普斯研究所发展了一种酶催化远程碳氢键氧化和化学法碳氢键氧化相结合的合成策略，可合成 9 种结构复杂的二萜化合物[28]。加拿大西蒙菲莎大学从简单的非手性原料出发，利用脯氨酸催化的醛基 α 位氟化羟醛缩合、还原、亲核加成、关环等反应，仅需 2～3 步即可以克级产量合成多种核苷类似物[29]。英国剑桥大学的科学家利用可见光介导的自由基策略，将醛、仲胺和烷基卤代物转化为相应的复杂叔烷基胺，反应过程操作简单、条件温和、无需金属催化[30]。

对基础理论、过程的认识不断深入。英国诺丁汉大学和德国乌尔姆大学利用碳纳米管，通过透射电子显微镜，首次在单分子水平上拍摄到化学键形成与断裂的实时动态影像，为人类全面理解化学键提供了全新视角[31]。奥地利格拉茨大学通过扫描透射显微镜，实现两端溴取代的三(9,9-二甲基芴)(DBTF) 单个分子在银(111)表面精准单方向长距离（150 nm）受控移动[32]。瑞士苏黎世联邦理工学院证明在钛硅沸石-1(TS-1) 催化丙烯环氧化反应中，催化活性位点是双核钛而非骨架中孤立的钛原子[33]。韩国科学技术院发现电极可作为一种官能团，通过改变施加电压可调节共价固定在其上的反应物的化学性质[34]。

3. 纳米材料新热点方向显现，在生命健康领域的应用效果更显著

纳米碳材料仍然是热点研究方向。北京理工大学和东京大学利用拓扑结构的全碳多孔纳米阵列制备出无金属的高度灵敏、高生物相容和高可重现的表面增强拉曼光谱术（surface-enhanced Raman spectroscopy，SERS）基底[35]。天津大学采用便捷、可扩大化的热压和热轧工艺实现了类石墨烯结构片层的紧密焊接，进而构筑出连续的三维类石墨烯网络/铜复合材料[36]。中国科学院金属研究所采用在石墨烯表面涂覆低折射率且高透光的有机质子酸四（五氟苯基）硼酸［tetrakis(pentafluorophenyl)boric acid］的纳米涂层的方式同步提高了石墨烯的电导率和透光率，获得了创纪录的高电/光导率[37]。北京大学等设计并开发出碳纳米管的多重分散和排序过程，在 10 cm 硅片上制备出排列整齐的高密度碳纳米管阵列，其密度可调控为每微米 100～200 个碳纳米管[38]。

二维纳米材料成为新的研究热点。韩国科学技术院和美国德雷塞尔大学采用退火处理的方式得到具有异常高的电磁屏蔽能力的二维过渡金属碳氮化物 Ti_3CNT_x（MXene）层状超材料状结构[39]。华中科技大学和美国雷克塞尔大学通过使用特殊设计的打结碳纳米管，首次制备出可以在$-60\,℃$使用 MXene 基的超级电容器[40]。中国科学院上海微系统与信息技术研究所和奥地利维也纳大学利用两步生长法，实现了嵌入二维 h-BN 纳米沟槽中亚 5 nm 宽单层锯齿型和扶手椅型石墨烯纳米带的手性可控制备[41]。中国科学院物理研究所和北京大学采用两步化学气相沉积法，实现了高度非线性的二维材料 MoS_2 在 SiO_2 光纤内壁的直接生长[42]。

抗癌和抗病毒纳米材料进展明显。武汉大学将 Cd274 的 siRNA 整合到商业化的人乳头瘤病毒（human papilloma virus，HPV）L1 蛋白中，制备出可引发强烈免疫刺激的基于疫苗的纳米系统，用于特异性抑制肿瘤 PDL1 表达[43]。南京医科大学等在人体血液中检测到生物形成的铂（Pt）纳米颗粒，证实了其可以被安全地应用于人体，可以抑制对化疗耐药的肿瘤生长[44]。中国科学院生物物理研究所制备出兼具预防作用和治疗作用的铁蛋白纳米粒子疫苗，可以将乙肝病毒大分子表面抗原的前 S1 递送至特定的骨髓细胞[45]。湖南大学设计了一种用于高特异性的癌症成像和光热治疗的双激活纳米探针，可在提高精准度的同时降低癌症治疗过程中出现的严重毒副作用[46]。

制备方法的革新赋予纳米材料更多奇特功能。北京航空航天大学提出在水凝胶/油界面利用剪切-流变诱导排列二维纳米片制备高度有序二维层状纳米纤维结构的通用策略[47]。中国的电子科技大学和德国的亚琛莱布尼茨材料研究所利用脱氧核糖核酸（deoxyribonucleic acid，DNA）作为可编程燃料，以核酸外切酶Ⅲ为能量耗散单元，制备出四维金纳米结构[48]。美国马里兰大学采用高温脉冲方法快速有效地实现了大纳米材料聚集粒子的再分散[49]。上海交通大学利用将钙钛矿材料封装到由分子筛模板衍生的介孔二氧化硅中的方法制备出类陶瓷稳定且高发光的 $CsPbBr_3$ 纳米晶[50]。香港理工大学等提出共格纳米片层合金的设计理念，开发出具有超高强度、高塑性的新型高熵合金材料[51]。

二、重要战略规划

1. 欧洲粒子物理战略聚焦希格斯玻色子和新加速器技术

欧洲核子研究中心发布《2020 年欧洲粒子物理学战略》[52]，提出粒子物理学近期和长期发展愿景。该战略提出两大高度优先的未来计划，一是建造"希格斯工厂"，愿景是先建设一个正负电子对撞机作为"希格斯工厂"，未来再建设预计最大对撞能量高达 100 TeV 的未来环形对撞机。二是开发新的加速器技术，包括高场磁体、高温超导体、等离子体尾波场加速及其他高梯度加速结构、明亮的介子束、能量回收直线加速器。并且该战略指出，粒子物理学界已准备好向更高能量和更小尺度迈出下一步。

2. 美国和日本深入推进量子技术战略

美国白宫国家量子协调办公室发布《美国量子网络战略远景》报告[53]，提出美国将构建量子互联网。美国能源部发布了量子互联网蓝图[54]，计划在 10 年内建成一个全国性的量子互联网，将构建并整合量子网络设备、开发路由技术并找出量子比特跨量子网络传播时的纠错方法。美国能源部宣布在未来 5 年内将提供 6.25 亿美元，建立 2~5 个多学科的量子信息科学研究中心[55]，以支持国家量子计划。美国量子信息科学小组委员会与国家量子协调办公室共同发布"量子前沿"报告[56]，确定了 8 个前沿领域：增大量子技术造福社会的机会、建立量子工程学科、以材料科学为目标的量子技术、通过量子模拟探索量子力学、利用量子信息技术进行精密测量、为新应用生成并分配量子纠缠、表征并减少量子误差、通过量子信息洞悉宇宙。日本统合创新战略推进会议发布《量子技术创新战略最终报告》[57]，指出量子技术创新战略将被确定为一项新的国家战略，将聚焦包括量子计算机与量子模拟、量子测量和量子传感、量子通信和量子密码学、量子材料等主要技术领域，量子人工智能技术、量子生物技术、量子安全技术等量子融合创新领域，以及量子启发技术和准量子技术。

3. 欧盟推动化学品安全可持续发展

欧盟委员会发布《致力于无害环境的可持续化学品战略》[58]，旨在促进化学品向安全和可持续方向发展，也反映了在新冠肺炎疫情影响下欧盟对化学品供应链的思考。战略提出：从设计角度加强安全和可持续性，循环利用安全产品和无毒害材料，实现化学品生产过程绿色和数字化；识别化学品依赖程度并提出降低依赖程度措施，识别事关绿色和数字化转型的化学品价值链，提高欧盟化学品战略预见能力，围绕化学品价值链加强区域合作，利用欧盟资助和投资机制提高具有重要应用的化学品的供应链弹性。

4. 欧盟石墨烯旗舰计划仍然受到高度重视，采用跨国合作方式支持基础创新和应用创新

欧盟石墨烯旗舰计划新增 17 个成员机构和 16 个合作项目[59]，致力于用于量子技

术的纳米石墨烯、石墨烯辐射传感器的检测机制等 9 个基础研究和创新方向，以及用于高级饮用水处理的石墨烯复合材料、溶液法合成钙钛矿/石墨烯自驱动气敏纳米复合材料等 7 个应用创新研究方向。欧盟未来和新兴技术旗舰计划欧洲研究区域网络 (FLAG-ERA) 石墨烯旗舰计划和人脑计划协同研究项目资助联合跨国项目[60]，致力于细菌对石墨烯及其相关材料 (GRMs) 的降解、二维非晶材料的制备与器件集成等 10 个基础研究领域以及用于生物电子药物的 GRMs 神经接口、用于先进金属离子超级电容器的 GRMs 等 8 个应用研究和创新领域。

5. 美国国家纳米技术计划调整研发布局，着力解决国家面临的重大问题

美国国家纳米技术计划 (NNI) 在 2021 财年的预算[61]为 17.23 亿美元。美国国家纳米技术计划对一些项目进行了调整："纳米电子学 2020""未来计算""癌症纳米技术卓越中心"项目在 2020 年底结束，相关研究将由新成立的"半导体领先研发工作组"继续组织实施。"癌症纳米技术卓越中心"项目在 2020 年底结束，而"癌症纳米技术创新研究"项目继续组织实施。2020 年 2 月，国家癌症研究所新设立"癌症纳米技术转化"项目，旨在推动基于纳米技术的癌症疗法从实验室向临床应用转化。塑料废弃物在环境中降解形成的纳米尺度塑料颗粒（即纳米塑料）成为美国国家纳米技术计划关注的新兴研究主题之一。

三、发展启示与建议

1. 加强基础前沿领域重大科学问题的优先布局

凝聚态物理、粒子物理、量子技术和光学等基础前沿领域重大突破不断涌现，高温超导、中微子、暗物质、量子计算和量子通信等重大科学问题持续取得重要进展。这些重大科学问题涵盖了面向世界科技前沿的宇宙演化、物质结构等基础研究的探索，以及面向国家重大需求的基础核心技术。应加强对这些重大科学问题的优先布局，通过在投入、人才等方面提供政策支持，增强原始创新能力，开辟新的前沿生长点，开发新的技术和开拓新的应用。

2. 积极发展数据驱动型化学研发技术

在大数据时代，化学研究正在从传统的反复试错模式向数据驱动模式转型。通过使用机器学习算法，科研人员能够从大量实验数据和历史文献数据中总结实验规律，发现可行性高的研究路线，提高科研效率。而且，这种人工智能可以与自动化技术结合，使化学实验从主要依靠人工转变为机器自动完成，将科研人员从重复性的工作中解放出来，有更多的时间从事创造性研究。建议研究大数据技术对化学科研范式的影响，前瞻部署数据驱动化学研究技术、自动合成机器等前沿科技，牢牢把握住新一轮科技革命带给化学研究的发展机遇。

3. 大力支持利用新兴技术促进纳米技术和材料发展

纳米技术是人工智能、量子信息等前沿领域获得长足发展的重要基础技术，也是下一代无线通信、先进制造等未来产业的重要基础手段，应持续投资纳米技术，调整纳米技术的研究布局，打破以材料本身为出发点的研究模式，倡导以解决重大社会问题为目标的任务导向。纳米技术在人工智能等新兴技术领域的作用已有显现，但人工智能等新兴技术反作用于纳米材料的创新稍显不足，应加大对利用新兴技术促进纳米技术和材料发展等研究方向的部署和支持，在实现纳米技术自身发展的同时，与其他技术一起共同助力我国 2035 年远景目标的稳步实现。

致谢：中国科学院化学研究所张建玲研究员对本文初稿进行了审阅并提出了宝贵的修改意见，特致感谢！

参考文献

[1] Snider E, Dasenbrock-Gammon N, McBride R, et al. Room-temperature superconductivity in a carbonaceous sulfur hydride. Nature, 2020, 586: 373-377.

[2] Uri A, Grover S, Cao Y, et al. Mapping the twist-angle disorder and Landau levels in magic-angle graphene. Nature, 2020, 581: 47-52.

[3] Cao Y, Rodan-Legrain D, Rubies-Bigorda O, et al. Tunable correlated states and spin-polarized phases in twisted bilayer—bilayer graphene. Nature, 2020, 583: 215-220.

[4] Zondiner U, Rozen A, Rodan-Legrain D, et al. Cascade of phase transitions and Dirac revivals in magic-angle graphene. Nature, 2020, 582: 203-208.

[5] Chen X, Korblova E, Dong D P, et al. First-principles experimental demonstration of ferroelectricity in a thermotropic nematic liquid crystal: Polar domains and striking electro-optics. PNAS, 2020, 117 (25): 14021-14031

[6] Xing Y Q, Shen J L, Chen H, et al. Localized spin-orbit polaron in magnetic Weyl semimetal $Co_3Sn_2S_2$. Nature Communications, 2020, 11: 5613.

[7] The Borexino Collaboration. Experimental evidence of neutrinos produced in the CNO fusion cycle in the Sun. Nature, 2020, 587: 577-582.

[8] CMS Collaboration. Observation of the production of three massive gauge bosons at $\sqrt{s}=13$ TeV. Physical Review Letters,2020,125:151802.

[9] MICE Collaboration. Demonstration of cooling by the muon ionization cooling experiment. Nature, 2020,578:53-59.

[10] T2K Collaboration. Constraint on the matter-antimatter symmetry-violating phase in neutrino oscillations. Nature,2020,580:339-344.

[11] XENON Collaboration. Excess electronic recoil events in XENON1T. Physical Review D, 2020, 102:072004.

[12] Zhong H S,Wang H,Deng Y H,et al. Quantum computational advantage using photons. Science, 2020,370(6523):1460-1463.

[13] Pokorny F, Zhang C, Higgins G, et al. Tracking the dynamics of an ideal quantum measurement. Physical Review Letters,2020,124:080401.

[14] Yang C H,Leon R C C,Hwang J C C,et al. Operation of a silicon quantum processor unit cell above one kelvin. Nature,2020,580:350-354.

[15] Yin J,Li Y H,Liao S K,et al. Entanglement-based secure quantum cryptography over 1120 kilometres. Nature,2020,582:501-505.

[16] Joshi S K,Akta D,Wengerowsky S,et al. A trusted node—free eight-user metropolitan quantum communication network. Science Advances,2020,6:eaba0959.

[17] Hu G G,Ou Q D,Si G Y,et al. Topological polaritons and photonic magic angles in twisted α-MoO_3 bilayers. Nature,2020,582:209-213.

[18] Fadaly E M T,Dijkstra A,Suckert J R,et al. Direct-bandgap emission from hexagonal Ge and SiGe alloys. Nature,2020,580:205-209.

[19] Li L,Liu Z X,Ren X F,et al. Metalens-array-based high-dimensional and multiphoton quantum source. Science,2020,368(6498):1487-1490.

[20] Yang B,Chen G,Ghafoor A,et al. Sub-nanometre resolution in single-molecule photoluminescence imaging. Nature Photonics,2020,14:693-699.

[21] Burger B,Maffettone P M,Gusev V V,et al. A mobile robotic chemist. Nature,2020,583(7815): 237-241.

[22] Mehr S H M,Craven M,Leonov A I,et al. A universal system for digitization and automatic execution of the chemical synthesis literature. Science,2020,370(6512):101-108.

[23] Mikulak-Klucznik B,Gołębiowska P,Bayly A A,et al. Computational planning of the synthesis of complex natural products. Nature,2020,588(7836):83-88.

[24] Zhou B,Ou P,Pant N,et al. Highly efficient binary copper-iron catalyst for photoelectrochemical carbon dioxide reduction toward methane. PNAS,2020,117(3):1330-1338.

[25] Leow W R,Lum Y,Ozden A,et al. Chloride-mediated selective electrosynthesis of ethylene and propylene oxides at high current density. Science,2020,368(6496):1228-1233.

[26] Zhang F, Zeng M, Yappert R D, et al. Polyethylene upcycling to long-chain alkylaromatics by tandem hydrogenolysis/aromatization. Science, 2020, 370(6515): 437-441.

[27] Feng K, Quevedo R E, Kohrt J T, et al. Late-stage oxidative C(sp³)—H methylation. Nature, 2020, 580(7805): 621-627.

[28] Zhang X, King-Smith E, Dong L B, et al. Divergent synthesis of complex diterpenes through a hybrid oxidative approach. Science, 2020, 369(6505): 799-806.

[29] Meanwell M, Silverman S M, Lehmann J, et al. A short *de novo* synthesis of nucleoside analogs. Science, 2020, 369(6504): 725-730.

[30] Kumar R, Flodén N J, Whitehurst W G, et al. A general carbonyl alkylativeamination for tertiary amine synthesis. Nature, 2020, 581(7809): 415-420.

[31] Cao K, Skowron S T, Biskupek J, et al. Imaging an unsupported metal—metal bond in dirhenium molecules at the atomic scale. Science Advances, 2020, 6(3): 1-7.

[32] Civita D, Kolmer M, Simpson G J, et al. Control of long-distance motion of single molecules on a surface. Science, 2020, 370(6519): 957-960.

[33] Gordon C P, Engler H, Tragl A S, et al. Efficient epoxidation over dinuclear sites in titanium silicalite-1. Nature, 2020, 586(7831): 708-713.

[34] Heo J, Ahn H, Won J, et al. Electro-inductive effect: electrodes as functional groups with tunable electronic properties. Science, 2020, 370(6513): 214-219.

[35] Chen N, Xiao T H, Luo Z, et al. Porous carbon nanowire array for surfaceenhanced Raman spectroscopy. Nature Communications, 2020, 11: 4772.

[36] Zhang X, Xu Y X, Wang M C, et al. A powder-metallurgy-based strategy toward three-dimensional graphene-like network for reinforcing copper matrix composites, Nature Communications, 2020, 11: 2775.

[37] Laipeng M, Zhongbin W, Lichang Y, et al. Pushing the conductance and transparency limit of monolayer graphene electrodes for flexible organic light-emitting diodes. PNAS, 2020, doi: 10.1073/pnas. 1922521117.

[38] Liu L J, Han J, Xu L, et al. Aligned, high-density semiconducting carbon nanotube arrays for high-performance electronics. Science, 2020, doi: 10.1126/science. aba5980.

[39] Iqbal A, Shahzad F, Hantanasirisakul K, et al. Anomalous absorption of electromagnetic waves by 2D transition metal carbonitride Ti_3CNT_x (MXene). Science, 2020, 369, (6502): 446-450.

[40] Gao X, Du X, Mathis T S, et al. Maximizing ion accessibility in MXene-knotted carbon nanotube composite electrodes for high-rate electrochemical energy storage. Nature Communications, 2020, 11: 6160.

[41] Wang H S, Chen L, Elibol K, et al. Towards chirality control of graphenenanoribbons embedded in hexagonal boron nitride. Nature Materials, doi: 10.1038/s41563-020-00806-2.

[42] Zuo Y G, Yu W T, Liu C, et al. Optical fibres with embedded two-dimensional materials for ultra-

high nonlinearity. Nature Nanotechnology,2020,15:987-991.

[43] Zheng D W,Gao F,Cheng Q,et al. A vaccine-based nanosystem for initiating innate immunity and improving tumor immunotherapy. Nature Communications,2020,doi:10. 1038/s41467-020-15927-0.

[44] Zeng X,Sun J,Li S P,et al. Blood-triggered generation of platinum nanoparticle functions as an anti-cancer agent. Nature Communications,2020,doi:10. 1038/s41467-019-14131-z.

[45] Wang W J,Zhou X X,Bian Y J,et al. Dual-targeting nanoparticle vaccine elicits a therapeutic antibody response against chronic hepatitis B. Nature Nanotechnology,2020,doi:10. 1038/s41565-020-0648-y.

[46] Teng L L,Song G S,Liu Y C,et al. Nitric oxide-activated "dual-key-one-lock" nanoprobe for *in vivo* molecular imaging and high-specificity cancer therapy. Journal of the American Chemical Society,2019,doi:10. 1021/jacs. 9b05901.

[47] Zhao C Q,Zhang P C,Zhou J J,et al. Layered nanocomposites by shear-flow-induced alignment of nanosheets,Nature,2020,doi:10. 1038/s41586-020-2161-8.

[48] Luo M,Xuan M J,Huo S D,et al. 4-Dimensional DNA-gold nanoparticle assemblies. Angewandte Chemie International Edition,2020,59:17250-17255.

[49] Xie H,Hong M,Hitz E M,et al. A high-temperature pulse method for nanoparticle redispersion. Journal of the American Chemical Society,2020,142:17364-17371.

[50] Zhang Q G,Wang B,Zheng W L,et al. Ceramic-like stable $CsPbBr_3$ nanocrystals encapsulated in silica derived from molecular sieve templates. Nature Communications,doi:10. 1038/s41467-019-13881-0.

[51] Fan L,Yang T,Zhao Y L,et al. Ultrahigh strength and ductility in newly developed materials with coherent nanolamellar architectures. https://doi. org/10. 1038/s41467-020-20109-z[2021-03-03].

[52] CERN. 2020 update of the European strategy for particle physics. https://home. cern/sites/home. web. cern. ch/files/2020-06/2020% 20Update% 20European% 20Strategy. pdf[2020-12-10].

[53] The White House National Quantum Coordination Office. A strategic vision for America's quantum networks. https://www. whitehouse. gov/wp-content/uploads/2017/12/A-Strategic-Vision-for-Americas-Quantum-Networks-Feb-2020. pdf[2020-11-20].

[54] U. S. Department of Energy. U. S. Department of Energy unveils blueprint for the quantum internet at 'launch to the future:quantum internet' event. https://www. energy. gov/articles/us-department-energy-unveils-blueprint-quantum-internet-launch-future-quantum-internet[2020-11-20].

[55] U. S. Department of Energy. Department of Energy announces $ 625 million for new quantum centers. https://www. energy. gov/articles/department-energy-announces-625-million-new-quantum-centers[2020-11-20].

[56] The White House National Quantum Coordination Office. Quantumfrontiers report on community input to the nation's strategy for quantum information science. https://www. quantum. gov/wp-content/uploads/2020/10/QuantumFrontiers. pdf[2020-11-20].

[57] 統合イノベーション戦略推進会議. 量子技術イノベーション戦略最終報告 . https：//www8. cao. go. jp/cstp/siryo/haihui048/siryo4-2. pdf[2020-11-20].

[58] European Commission. Chemicals strategy for sustainability towards a toxic-free environment. https：//ec. europa. eu/environment/pdf/chemicals/2020/10/Strategy. pdf[2020-10-14].

[59] European Commission. Associated members and 16 partnering projects join the graphene flagship. http：//graphene-flagship. eu/news/Pages/17-Associated-Members-and-16-Partnering-Projects-join-the-Graphene-Flagship. aspx[2020-03-20].

[60] European Commission. Joint transnational call 2021 for transnational research projects in synergy with the graphene flagship & human brain project. https：//www. flagera. eu/wp-content/up-loads/2020/10/FLAG-ERA_JTC2021_Pre-announcement_20201001. pdf[2020-11-20].

[61] National Nanotechnology Initiative. NNI supplement to the president's 2021 budget. https：//www. nano. gov/2021budgetsupplement[2020-11-20].

Basic Sciences and Frontiers

Huang Longguang ,Bian Wenyue ,Zhang Chaoxing ,Leng Fuhai

A number of breakthroughs have been made in basic and frontier science in 2020. Major breakthroughs have been achieved in particle physics, condensed matter physics, quantum technology and optics. Synthetic chemistry has become more and more automatic and intelligent. New hot topics emerge in nanomaterials and the effect of application of nanomaterials in the field of life and health becomes more remarkable. Research frontiers such as particle physics, sustainable chemicals and graphene, are employed and delivered in Europe. Quantum technology strategies are further implemented in the US and Japan, while nanotechnology research and development has been adjusted in the US. These strategies aim to achieve a leading position in scientific and technological frontier and innovative developmentin the future.

4.2 生命健康与医药领域发展观察

王 玥 许 丽 苏 燕 施慧琳
李祯祺 杨若南 李 伟 徐 萍

（中国科学院上海营养与健康研究所/中国科学院上海生命科学信息中心）

科技创新离不开科技体制机制的保障。随着新一轮规划期到来，多个国家在科技组织、科技管理、科技投入方式等方面进行改革，以重大原始创新和核心技术突破引领全球科技发展，尤其是新冠肺炎疫情对经济社会发展带来的诸多不确定性引发了政府部门对科技体制机制的反思与改革。在这个背景下，人口健康和医药领域也将进入高速发展阶段。与此同时，随着技术的快速革新，以及学科交叉融合的日趋深入，人口健康和医药领域朝着数字化、智能化、工程化的方向不断加快发展，这一全新的发展模式也将进一步促进该领域的知识发现，推进诊疗体系创新，提高健康护理水平。

一、重要研究进展

1. 生命组学技术的持续进步促进基线研究全面展开

生命组学技术仍然是生命科学发展的重要技术驱动力，随着生命组学技术的持续优化，基线研究越来越广泛地开展起来，逐渐成为人口健康和医药领域知识发现的重要模式。

首先，基因组、蛋白质组、代谢组技术不断升级，新一代串联质谱标签系统TMTpro[1]、空间代谢组分析技术 metaFISH[2] 等一系列技术的优化和开发，进一步提升了组学技术的效率、通量、灵敏性和原位分析能力。在此支撑下，人类蛋白质组测序草图[3]、人类蛋白质组互作组图谱[4]、完整的人类 X 染色体序列[5] 等大量高质量基因、蛋白、代谢分析图谱绘制完成，以揭示生命现象、解析疾病发生机制。同时，能够反映空间信息的生命组学分析技术越发受到关注，空间转录组技术被评为《自然-方法》（*Nature Methods*）2020 年度技术。

其次，单细胞分析技术的不断优化成熟，使细胞图谱的绘制逐渐覆盖了人体全系统、全生命周期。2020 年，除了持续拓展绘制细胞图谱的组织器官[6-9]外，浙江大学还首次绘制了跨越胚胎和成年两个时期，涵盖八大系统的人类细胞图谱[10]，为认识生命组成、解析发育、生长、衰老全生命过程奠定了基础。

分子和细胞图谱的绘制助力疾病发生机制[11]的解析和疾病生物标志物[12]的发现。全基因组泛癌分析联盟（Pan-Cancer Analysis of Whole Genomes，PCAWG）发布了迄今最全面的癌症全基因组图谱[13]；美国哈佛医学院对"癌细胞系百科全书"（Cancer Cell Line Encyclopedia，CCLE）中 375 种细胞系的数千种蛋白质进行定量蛋白质组分析，补全了 CCLE 深度蛋白质组学分析数据[14]，为癌症研究提供了重要的数据基础。

2. 工程化技术的深度渗透改变疾病干预方式

对细胞、组织、器官进行"工程化"改造和制造正成为疾病干预的重要方式，以基因编辑技术、组织器官制造、合成生物学为代表的"工程化"技术，正在逐渐使疾病干预实现从"对症治疗"向"解决致病根源问题"转变。

基因编辑技术的持续优化使其应用范围进一步扩大，实现了 C 到 A/G 的单碱基替换[15]、多 DNA 片段的同时编辑[16]、线粒体 DNA 的精准编辑[17]等。同时，其安全性和可控性也逐渐提升，美国约翰斯·霍普金斯大学等相继开发出光控基因编辑技术[18]、病毒抗Ⅲ/Ⅵ型 CRISPR 酶[19,20]可实现对基因编辑的时空控制。技术进步为疾病治疗带来更加广阔的前景。针对晚期难治性骨髓瘤和转移性肉瘤[21]、非小细胞肺癌[22,23]、转甲状腺素蛋白淀粉样变性[24]等的临床试验陆续开展，并且开始取得良好效果。其中，瑞士 CRISPR Therapeutics 公司等利用基因编辑技术首次成功治愈两种遗传性贫血症[25,26]，标志着基因编辑疗法展现良好前景。

组织器官体外制造技术研发快速发展，干细胞、3D 生物打印、类器官等技术优化升级并深度融合，不断探索"器官制造"应用的可行性。肝脏[27]、皮肤[28]等多种人造组织器官已实现在动物体内存活，我国科研人员还实现了利用 3D 技术打印肝脏缓解小鼠肝脏衰竭[29]。此外，韩国浦项大学开发的类组装体技术还为构建更加仿真的类器官带来了新机遇[30]。这些成果进一步推动了人造器官在组织器官替代治疗中的应用。

合成生物学在蛋白质元件的人工设计合成方面取得巨大突破，并在疾病预防与诊疗技术的研发中取得明显成效。研究人员成功利用酵母等生物底盘合成了大麻素[31]；基于合成生物学的糖尿病和白血病药物也已经获批上市[32]；基于酵母的病毒快速合成平台[33]及刺突蛋白变体[34]的设计与量产技术还为传染病的防诊治提供了技术平台。

3. 智能化成为疾病干预的重要发展趋势

脑科学、脑机接口、人工智能的快速发展，推动健康维护和疾病干预的模式朝着智能化的方向不断迈进，为疾病的诊断和治疗带来了全新的路径。

随着对大脑的结构、分子和发育机理的认识日趋深刻，以及成像技术等工具的优化，脑科学与类脑研究持续深入。新型分子探针 FLiCRE[35]、核糖体标记神经胞体成像技术[36]等新技术提高了研究人员对大脑的观察和控制能力；对运动可推动运动技能学习[37]、提高认知记忆能力[38]的解析为相关神经退行性疾病的研究提供了思路。在类脑计算方面，基于多阵列的忆阻器存算一体系统[39]、亿级神经元类脑操作系统 DarwinOS 的推出标志着类脑计算向高效、便携、低功耗发展；脑机接口技术逐步成熟，研究人员利用多巴胺实现了人工神经元和生物神经元[40]的异构融合，利用脑机接口技术也实现了恢复脊髓损伤患者手部运动功能的同时获得触觉反馈[41]；美国 Neuralink 公司开发的 LINK V0.9 系统获美国食品药品监督管理局（Food and Drug Administration，FDA）"突破性设备认定"，通过将电极植入人脑读取脑电波信号，以期治疗神经系统疾病及颅脑外伤患者。该系统的成功研发加速了脑机接口技术的临床应用。

人工智能技术已逐步应用于医学研究与医疗实践，如对蛋白质结构预测、为肿瘤匹配最佳药物组合[42]，用于乳腺癌筛查、预测新冠肺炎病例[43]等，均取得良好的效果。此外，人工智能医疗产品应用加速，2020 年又有多款产品相继获批上市，主要应用于心脏[44]、椎骨压缩性骨折[45]、脑血管闭塞[46]、新冠肺炎[47]等疾病的筛查与预测。

4. 精准医学应用日趋深入，新疗法持续突破

在疾病诊疗方面，一方面精准医学的新范式不断渗透，推动一系列精准诊断和治疗方法（如免疫治疗、基因治疗等）的快速发展；另一方面，RNA、干细胞等新兴疗法正不断展现出在疾病治疗中的巨大应用潜力。

精准医学研究持续推进，研究范式已在医学研究和临床中实践和推广，2020 年，肺腺癌、乳腺癌等更多疾病的精准分型研究取得突破，纳米孔测序、无细胞甲基化 DNA 免疫沉淀、复合成像等技术的应用提高了肾癌[48]、乳腺癌[49]等疾病检测的精准性，助推疾病精准诊断的实现。已有多款精准医学产品落地，首款结合下一代基因测序技术（next-generation sequencing，NGS）和液体活检技术的实体瘤基因组分析产品 Guardant 360 CDx[50]，以及首个泛肿瘤类型的液体活检产品 FoundationOne® Liquid CDx[51]相继获美国食品药品监督管理局批准上市。

免疫细胞治疗在血液肿瘤治疗研发方面持续稳步推进。2020 年，全球第三款 CAR-T 疗法 Tecartus 获美国食品药品监督管理局批准用于治疗套细胞淋巴瘤[52]。免疫细胞治疗实体瘤仍未攻克，研发主要围绕免疫逃逸、开发新的免疫治疗细胞类型

等。美国宾夕法尼亚大学发现 CD8+ 杀伤性 T 细胞通常不会从血液移动到器官和组织中[53]，解释了免疫细胞治疗实体瘤疗效不足的可能原因；美国宾夕法尼亚大学开发了一种可直接作用于肿瘤的嵌合抗原受体巨噬细胞（CAR-M）[54]，利用该细胞开发的疗法已获美国食品药品监督管理局批准开展临床试验。

基因疗法治疗单基因遗传病再迎新药上市，并在遗传病、神经系统疾病和癌症等适应证上取得多项重要进展。2020 年 12 月，英国 Orchard Therapeutics 公司的 Libmeldy 获欧盟批准上市，用于治疗早发型异染性脑白质营养不良[55]。美国哈佛大学等通过重编程视神经节细胞（retinal ganglion cell，RGC）成功逆转因衰老和青光眼引起的视力损失[56]。CRISPR 等基因编辑技术推动了基因疗法快速发展和突破，通过改造自体 CD34+ 细胞、T 细胞等实现了 β-地中海贫血症和镰状细胞贫血[57]、晚期癌症[58]等疾病的治疗。此外，美国 Editas Medicine 公司还对莱伯氏先天性黑蒙症 10 型患者实施了首次体内 CRISPR-Cas9 基因编辑治疗[59]。

核糖核酸（ribonucleic acid，RNA）疗法逐渐进入产业收获期，尤其是反义寡核苷酸（antisense oligonucleotide，ASO）疗法和小干扰 RNA（small interfering RNA，siRNA）疗法的上市进程加速。2020 年，日本新药株式会社的 ASO 药物 Viltepso 获美国食品药品监督管理局批准，用于治疗 53 号外显子跳跃杜氏肌营养不良症[60]；全球第 3 款和第 4 款 siRNA 药物 Oxlumo[61] 和 Leqvio[62] 相继获批上市，分别用于治疗Ⅰ型原发性高草酸尿症（primary hyperoxaluria Ⅰ，PH1）和成人高胆固醇血症及混合性血脂异常。此外，预防性 mRNA 疫苗上市也实现零的突破，德国 BioNTech 公司的 BNT162b2 和美国 Moderna 公司的 mRNA-1273 分别获得美国食品药品监督管理局的紧急使用授权，用于 SARS-CoV-2 病毒的预防[63,64]。

用于细胞修复的干细胞治疗技术日趋成熟，逐渐形成了从基础研究、临床研究、临床转化，到产业发展的全链条体系。干细胞治疗技术的有效性和安全性不断提升，适应证不断增加，治愈潜力不断提高。2020 年，科研人员进一步攻克了糖尿病干细胞疗法的免疫排斥问题[65]，利用皮肤细胞重编程获得的光敏细胞成功使失明小鼠获得了光感[66]。同时，干细胞治疗技术的临床转化和产业进程也不断加快。2020 年启动的相关临床试验数量接近 600 例。Bluebird 公司开发的干细胞疗法进入了欧洲药品管理局（European Medicines Agency，EMA）的加速审评通道，该疗法利用基因修饰的自体造血干细胞治疗肾上腺脑白质营养不良。

人类微生物组是人体不可或缺的重要组成部分，"解码微生物组"被《自然》（Nature）选为 2020 年最值得关注的技术之一。2020 年，微生物组参考序列解析研究持续推进[67]，其中特定人群微生物组特征解析进一步为疾病的诊断[68]和精准治疗[69]提供了新的思路。与此同时，越来越多的人类微生物组功能的因果机制被不断解析，为证实微生物组影响全生命周期健康（包括生长发育[70,71]、疾病发生发展、营养[72]和药物作用）提供了依据。多项临床试验结果相继证实粪菌移植在疾病治疗中发挥关键作用[73,74]，多款微生态药物临床试验迎来新进展，为批准上市铺平道路[75,76]，基于人类微生物组重构的干预方式逐渐成为疾病治疗的另一个重要手段。

二、重大战略行动

1. 新冠肺炎疫情相关研究资助计划

为应对新冠肺炎疫情，各国纷纷推出科技计划，提供资金支持，开展相关研发。例如，2020 年 4 月和 7 月，美国国家过敏症和传染病研究所（National Institute of Allergy and Infectious Diseases，NIAID）和美国国立卫生研究院（National Institutes of Health，NIH）分别发布了《新冠肺炎战略计划》（*Strategic Plan for COVID-19 Research*），以开展新冠病毒研究，提高新冠肺炎及相关疾病的防诊治能力；欧盟也累计投入了 10 亿欧元，支持病毒相关研究及配套设施和技术研发；日本文部科学省通过 2020 年 4 月发布的《紧急经济对策》对建立新冠病毒相关疾病现场诊疗体系及治疗药物和疫苗开发进行重点布局；澳大利亚也发布了《新冠病毒疫苗和治疗战略》支持新冠病毒疫苗研发。与此同时，全球合作也在应对新冠肺炎疫情中发挥了巨大作用。世界卫生组织与欧盟、流行病防范创新联盟（The Coalition for Epidemic Preparedness Innovations，CEPI）、全球疫苗免疫联盟（The Global Alliance for Vaccines and Immunisation，GAVI）于 2020 年 4 月发起了"新冠肺炎工具加速计划"，针对诊断、治疗、疫苗开发开展全球合作。其中，由包括中国在内的 190 个国家和地区参与的新冠病毒疫苗实施计划为提高疫苗研制效率和生产速度，以及改善新冠病毒疫苗获取的公平性和可及性，奠定了坚实基础。

2. 多国/组织在新一轮规划期进行科技体制机制改革

在新一轮规划期，美国、日本、欧盟、英国系统分析了科技发展趋势、本国/区域科技发展现状，制定了新的科技发展目标，并对其科技体制机制进行了改革。在科技管理体系方面，美国在《无尽的前沿法案》提案、英国在《英国研发路线图》中，阐述了将国家科技资助机构参照美国国防部高级研究计划局的运行模式进行改革的举措、建立以项目经理为核心的项目管理制度等，以更好地支持科技创新。在科技规划与资助方式方面，欧盟、英国和日本均制定举措大力推行任务导向型科技规划模式。其中，欧盟通过新一轮"地平线欧洲"框架计划设置以目标为核心的专项任务，英国和日本则都建立了"登月型"研发制度，即以产生颠覆性创新为目标，鼓励科研人员提出大胆的想法，开展具有挑战性的研发的制度。

3. 多国发布新一轮健康战略规划与专项规划

在健康规划方面，2020 年 8 月，美国卫生与公众服务部（United States Department of Health and Human Services，HHS）发布了《健康公民 2030》，提出国家疾病预防和健康促进的十年发展愿景。同时，美国国立卫生研究院也于 2020 年 3 月发布了未来 5 年的战略计划，重点关注生物医学和行为科学研究，提高科研能力，践行科研诚信、公众责任和社会责任。欧盟在"地平线欧洲"框架计划下设立了癌症专项，并于 2020 年 9 月发布了《癌症专项委员会提案报告》，旨在优化癌症预防、诊断和治疗，改善患者生活质量，同时实现资源的公平分配；此外，还于 2020 年 5 月启动了新一轮的健康计划 EU4Health，旨在强化卫生系统，增强其对健康威胁的应对能力。日本健康医疗战略推进本部于 2020 年 3 月发布了新版《健康医疗战略》，核心目标是加强医疗领域研发，发展有利于长寿社会形成的新产业。

在健康各领域的专项规划制定方面，随着学科交叉融合的深入、信息技术和人工智能的快速发展，工程化、数字化和智能化范式正逐渐渗透入人口健康和医药领域的研发中，因此多个国家/组织出台专项规划以响应科技发展趋势。

在推动工程化发展方面，2020 年 2 月，美国国家科学基金会（National Science Foundation，NSF）发布了设计合成细胞的项目招标指南，推进细胞样系统的设计和开发，为生物进化、生物多样性研究以及生物技术的应用提供工具。2020 年 5 月，美国参议院提出《2020 年生物经济研发法案》[77]，提出了美国在工程生物学领域的研发计划，旨在确保美国在工程生物学领域的国际领导地位。此外，美国工程生物学研究联盟（EBRC）继 2019 年发布了《工程生物学研究路线图》之后，于 2020 年和 2021年初再次发布了《微生物组工程：下一代生物经济研究路线图》和《工程生物学与材料科学：跨学科创新研究路线图》。2020 年 9 月，日本科学技术振兴机构（Japan Science and Technology Agency）提出了《设计细胞》的战略提案，旨在实现利用人工修饰细胞实现疾病控制的目标[78]。

在促进数字化发展方面，2020 年，欧盟委员会相继发布《塑造欧洲数字化未来》[79]、《欧洲数据战略》[80]和《人工智能白皮书》[81]，在健康领域突出构建欧洲健康数据空间，促进对健康数据的访问和交换，并为推进人工智能医疗应用和建立监管框架提出建议。英国也陆续推出"英国基因组：医疗的未来"[82]"英国人群研究"[83]等战略计划，促进基因组学、人群队列、健康数据等领域交互融合，最大限度地提高英国生物医学数据的研究与应用效率。日本则成立了数字化促进本部[84]，专门负责推进日本教育和科技各领域的数字化转型。此外，俄罗斯开始着手编制用于临床医学研究的人工智能系统国家标准，澳大利亚[85]和日本[86]也专门针对智能医疗的人才需求制定了培养规划。

三、启示与建议

通过科技创新推动经济、社会发展是科技发展的最根本目标，一套与之相适应的科技体制机制在科技发展中发挥重要的保障和促进作用。在新一轮规划期，美国、欧盟、日本等陆续开展了国家/组织科技管理体系、资助体系、规划体系的改革，旨在突破现有体制机制障碍，为创新营造更有利的政策环境，以提高本国/区域科技竞争力。我国在"十三五"时期着眼于突破制约科技发展的体制机制障碍，深入推进了科技管理体制改革，健全了科技创新治理机制。进一步落实了科技创新和体制机制创新双轮驱动，在关系经济实力和核心竞争力的关键领域加强了部署，进一步突出解决战略性、基础性、前瞻性重大科学问题，以及开发重大共性关键技术和产品的任务导向，同时围绕科技管理体制改革、高效研发体系、技术成果转移转化机制体系等进行了部署。"十四五"时期，我国还需继续加大改革力度，完善政策实施的具体举措，进一步支持重大颠覆性创新，为我国科技自立自强打造最有利的政策与制度体系。

工程化、数字化和智能化已经逐渐成为健康科技和健康产品制造的主要趋势，其在促进知识发现、推进诊疗体系创新，以及提高健康维护水平等方面均具有很强的推动作用。目前，科技强国都在布局支持医药研发的数字化发展，旨在把握科技新趋势带来的发展新机遇。对此，我国也陆续出台了资助计划进行支持，但还需要进一步提高规划层级，从健康科技与应用场景出发，制定覆盖数字化发展全链条的资助计划和扶持政策，扫清数字化发展中的数据质量不高、数据共享不充分、标准化与规范化不足、基础设施不完善等障碍，推动我国在该领域的发展，提升我国健康科技水平和医疗服务水平。

致谢：复旦大学金力院士、上海交通大学医学院陈国强院士在本文的撰写过程中提出了宝贵的意见和建议，在此谨致谢忱！

参考文献

[1] Li J, Vranken J G V, Vaites L P, et al. TMTpro reagents: a set of isobaric labeling mass tags enables simultaneous proteome-wide measurements across 16 samples. Nature Methods, 2020, 17: 399-404.

［2］ Geier B，Sogin E M，Michellod D，et al. Spatial metabolomics of in situ host—microbe interactions at the micrometre scale. Nature Microbiology，2020，5(3)：498-510.

［3］ Adhikari S，Nice E C，Deutsch E W，et al. A high-stringency blueprint of the human proteome. Nature Communications，2020，11：5301.

［4］ Luck K，Kim D K，Lambourne L，et al. A reference map of the human binary protein interactome. Nature，2020，580：402-408.

［5］ Miga K H，Koren S，Rhie A，et al. Telomere-to-telomere assembly of a complete human X chromosome. Nature，2020，585：79-84.

［6］ Litviňuková M，Talavera-López C，Maatz H，et al. Cells of the adult human heart. Nature，2020，588：466-472.

［7］ Park J E，Botting R A，Conde C D，et al. A cell atlas of human thymic development defines T cell repertoire formation. Science，2020，367(6480)：eaay3224.

［8］ Bian Z，Gong Y，Huang T，et al. Deciphering human macrophage development at single-cell resolution. Nature，2020，582(7813)：1-6.

［9］ Zhong S J，Ding W Y，Sun L，et al. Decoding the development of the human hippocampus. Nature，2020，577：531-536.

［10］ Han X，Zhou Z，Fei L，et al. Construction of a human cell landscape at single-cell level. Nature，2020，581(7808)：303-309.

［11］ Jin X，Demere Z，Nair K，et al. A metastasis map of human cancer cell lines. Nature，2020，588(7837)：331-336.

［12］ Hoshino A，Kim H S，Bojmar L，et al. Extracellular vesicle and particle biomarkers define multiple human cancers. Cell，2020，182(4)：1044-1061.

［13］ Cieslik M，Chinnaiyan A M. Global genomics project unravels cancer's complexity at unprecedented scale. https://www. nature. com/articles/d41586-020-00213-2［2020-02-05］.

［14］ Nusinow D P，Szpyt J，Ghandi M，et al. Quantitative proteomics of the cancer cell line encyclopedia. Cell，2020，180(2)：387-402.

［15］ Zhao D，Li J，Li S，et al. Glycosylase base editors enable C-to-A and C-to-G base changes. Nature Biotechnology，2020，39：1-6.

［16］ Thomas G P，Michael A，Brown K R，et al. Genetic interaction mapping and exon-resolution functional genomics with a hybrid Cas9-Cas12a platform. Nature Biotechnology，2020，38：638-648.

［17］ Mok B Y，Moraes M H D，Zeng J，et al. A Bacterial Cytidine Deaminase Toxin Enables CRISPR-free Mitochondrial Base Editing. Nature，2020，583(7817)：631-637.

［18］ Liu Y，Zou R S，He S，et al. Very fast CRISPR on demand. Science，2020，368(6496)：1265-1269.

［19］ Athukoralage J S，Mcmahon S A，Zhang C，et al. An anti-CRISPR viral ring nuclease subverts type Ⅲ CRISPR immunity. Nature，2020，577(7791)：1-4.

［20］ Meeske A J，Jia N，Cassel A K，et al. A phage-encoded anti-CRISPR enables complete evasion of type Ⅵ-A CRISPR-Cas immunity. Science,2020,369(6499):54-59.

[21] Stadtmauer E A,Fraietta J A,Davis M M, et al. CRISPR-engineered T cells in patients with refractory cancer. Science,2020,367(6481):eaba7365.

[22] Lu Y,Xue J,Deng T,et al. Safety and feasibility of CRISPR-edited T cells in patients with refractory non-small-cell lung cancer. Nature Medicine,2020,26(5):732-740.

[23] Dolgin E. First systemic CRISPR agent in humans. Nature Biotechnology,2020,38:1364.

[24] Intellia. Intellia therapeutics receives authorization to initiate phase 1 clinical trial of NTLA-2001 for transthyretin amyloidosis (ATTR). https://ir. intelliatx. com/news-releases/news-release-details/intellia-therapeutics-receives-authorization-initiate-phase-1[2021-02-08].

[25] Frangoul H,Altshuler D,Cappellini M D, et al. CRISPR-Cas9 gene editing for sickle cell disease and β-thalassemia. New England Journal of Medicine,2020,384:252-260.

[26] Esrick E B,Lehmann L E,Biffi A,et al. Post-transcriptional genetic silencing of BCL11A to treat sickle cell disease. New England Journal of Medicine,2020,384(3):205-215.

[27] Takeishi K,I'Hortet A,Wang Y,et al. Assembly and function of a bioengineered human liver for transplantation generated solely from induced pluripotent stem cells. Cell Reports, 2020, 31: 107711.

[28] Lee J,Rabbani C C,Gao H,et al. Hair-bearing human skin generated entirely from pluripotent stem cells. Nature,2020,582(7812):399-404.

[29] Yang H,Sun L,Pang Y,et al. Three-dimensional bioprinted hepatorganoids prolong survival of mice with liver failure. Gut,2021,70(3):567-574.

[30] Kim E,Choi S,Kang B,et al. Creation of bladder assembloids mimicking tissue regeneration and cancer. Nature,2020,588:664-669.

[31] Luo X,Reiter M A,d'Espaux L,et al. Complete biosynthesis of cannabinoids and their unnatural analogues in yeast. Nature,2019,567(7746):123.

[32] Voigt C A. Synthetic biology 2020-2030:six commercially-available products that are changing our world. Nature Communications,2020,11(1):1-6.

[33] Thao T T N,Labroussaa F,Ebert N,et al. Rapid reconstruction of SARS-CoV-2 using a synthetic genomics platform. Nature,2020,582(7813):561-565.

[34] Hsieh C L,Goldsmith J A,Schaub J M,et al. Structure-based design of prefusion-stabilized SARS-CoV-2 spikes. Science,2020,369(6510):1501-1505.

[35] Kim C K,Sanchez M I,Hoerbelt P,et al. A molecular calcium integrator reveals a striatal cell type driving aversion. Cell,2020,183(7):2003-2019.

[36] Chen Y M,Jang H,Spratt P W E,et al. Soma-targeted imaging of neural circuits by ribosome tethering. Neuron,2020,107(3):454-469.

[37] Li H Q,Spitzer N C. Exercise enhances motor skill learning by neurotransmitter switching in the adult midbrain. Nature Communications,2020,11:2195.

[38] Horowitz A M, Fan X L, Bieri G, et al. Blood factors transfer beneficial effects of exercise on neurogenesis and cognition to the aged brain. Science, 2020, 369(6500): 167-173.

[39] Yao P, Wu H Q, Gao B, et al. Fully hardware-implemented memristor convolutional neural network. Nature, 2020, 577: 641-651.

[40] Keene S T, Lubrano C, Kazemzadeh S, et al. A biohybrid synapse with neurotransmitter-mediated plasticity. Nature Materials, 2020, 19: 969-973.

[41] Ganzer P D, Colachis S C, Schwemmer M A, et al. Restoring the sense of touch using a sensorimotor demultiplexing neural interface. Cell, 2020, 181(4): 763-773.

[42] Kuenzi B M, Park J, Fong S H, et al. Predicting drug response and synergy using a deep learning model of human cancer cells. Cancer Cell, 2020, 38(5): 613-615.

[43] Menni C, Valdes A M, Freidin M B, et al. Real-time tracking of self-reported symptoms to predict potential COVID-19. Nature Medicine, 2020, 26: 1037-1040.

[44] FDA. FDA authorizes marketing of first cardiac ultrasound software that uses artificial intelligence to guide user. https://www.fda.gov/news-events/press-announcements/fda-authorizes-marketing-first-cardiac-ultrasound-software-uses-artificial-intelligence-guide-user[2021-02-08].

[45] Wire B. Zebra medical vision secures its 5th FDA clearance, making its vertebral compression fractures AI solution available in the U.S. https://www.businesswire.com/news/home/20200518005487/en/[2021-02-08].

[46] AIDOC. AI saving brain: FDA clears AIDOC's complete AI stroke package. https://www.aidoc.com/blog/news/ai-saving-brain-fda-clears-aidocs-complete-ai-stroke-package/[2021-02-08].

[47] AIDOC. AIDOC to market an ai algorithm under the FDA's enforcement policy for imaging systems. https://www.aidoc.com/blog/news/fda-aidoc-covid19/[2021-02-08].

[48] Nuzzo P V, Berchuck J E, Korthauer K, et al. Detection of renal cell carcinoma using plasma and urine cell-free DNA methylomes. Nature Medicine, 2020, 26(7): 1041-1043.

[49] Jackson H W, Fischer J R, Zanotelli V R T, et al. The single-cell pathology landscape of breast cancer. Nature, 2020, 578(7796): 615-620.

[50] FDA. FDA approves first liquid biopsy next-generation sequencing companion diagnostic test. https://www.fda.gov/news-events/press-announcements/fda-approves-first-liquid-biopsy-next-generation-sequencing-companion-diagnostic-test[2021-02-08].

[51] Foundation Medicine. FDA approves foundation medicine's foundationone®liquid CDx, a comprehensive pan-tumor liquid biopsy test with multiple companion diagnostic indications for patients with advanced cancer. https://www.foundationmedicine.com/press-releases/445c1f9e-6cbb-488b-84ad-5f133612b721[2021-02-08].

[52] FDA. Tecartus(brexucabtagene autoleucel). https://www.fda.gov/vaccines-blood-biologics/cellular-gene-therapy-products/tecartus-brexucabtagene-autoleucel[2020-12-19].

[53] Buggert M, Vella L A, Nguyen S, et al. The identity of human tissue-emigrant CD8$^+$ T cells. Cell,

2020,183:1-16.

[54] Klichinsky M, Ruella M, Shestova O, et al. Human chimeric antigen receptor macrophages for cancer immunotherapy. Nature Biotechnology,2020,38:947-953.

[55] EMA. Libmeldy. https://www. ema. europa. eu/en/medicines/human/EPAR/libmeldy[2021-02-01].

[56] Lu Y, Brommer B, Tian X, et al. Reprogramming to recover youthful epigenetic information and restore vision. Nature,2020,588:124-129.

[57] Frangoul H, Altshuler D, Cappellini M D, et al. CRISPR-Cas9 gene editing for sickle cell disease and β-thalassemia. New England Journal of Medicine,2020,384(3):252-260.

[58] Stadtmauer E A, Fraietta J A, Davis M M, et al. CRISPR-engineered T cells in patients with refractory cancer. Science,2020,367:1-12. eaba7365.

[59] Ledford H. CRISPR treatment inserted directly into the body for first time. Nature, 2020, 579 (7798):185.

[60] FDA. FDA approves targeted treatment for rare duchenne muscular dystrophy mutation. https:// www. fda. gov/news-events/press-announcements/fda-approves-targeted-treatment-rare-duchenne-muscular-dystrophy-mutation[2021-02-06].

[61] FDA. FDA approves first drug to treat rare metabolic disorder. https://www. fda. gov/news-events/press-announcements/fda-approves-first-drug-treat-rare-metabolic-disorder[2020-12-19].

[62] NOVARTIS. Novartis receives EU approval for Leqvio®* (inclisiran), a first-in-class siRNA to lower cholesterol with two doses a year**. https://www. novartis. com/news/media-releases/novartis-receives-eu-approval-leqvio-inclisiran-first-class-sirna-lower-cholesterol-two-doses-year[2020-12-19].

[63] FDA. Pfizer-BioNTech COVID-19 Vaccine. https://www. fda. gov/emergency-preparedness-and-response/coronavirus-disease-2019-covid-19/pfizer-biontech-covid-19-vaccine[2021-02-01].

[64] FDA. Moderna COVID-19 vaccine. https://www. fda. gov/emergency-preparedness-and-response/coronavirus-disease-2019-covid-19/moderna-covid-19-vaccine[2021-02-01].

[65] Yoshihara E, O'Connor C, Gasser E, et al. Immune-evasive human islet-like organoids ameliorate diabetes. Nature,2020,586:606-611.

[66] Mahato B, Kaya K D, Fan Y, et al. Pharmacologic fibroblast reprogramming into photoreceptors restores vision. Nature,581(7806):83-88.

[67] Almeida A, Nayfach S, Boland M, et al. A unified catalog of 204 938 reference genomes from the human gut microbiome. Nature Biotechnology,2021,39:105-114.

[68] Poore G D, Kopylova E, Zhu Q, et al. Microbiome analyses of blood and tissues suggest cancer diagnostic approach. Nature,2020,579:567-574.

[69] Nejman D, Livyatan I, Fuks G, et al. The human tumor microbiome is composed of tumor type—specific intracellular bacteria. Science,2020,368(6494):973-980.

[70] Vuong H E, Pronovost G N, Williams D W, et al. The maternal microbiome modulates fetal neuro-development in mice. Nature,2020,586:281-286.

［71］ Ikuo K,Junki M,Ryuji O K,et al. Maternal gut microbiota in pregnancy influences offspring metabolic phenotype in mice. Science,367(6481):eaaw8429.

［72］ Liou C S,Sirk S J,Diaz C A C,et al. A metabolic pathway for activation of dietary glucosinolates by a human gut symbiont. Cell,2020,180(4):717-728. e19.

［73］ de Groot P,Nicolic T,Pellegrini S,et al. Faecal microbiota transplantation halts progression of human new-onset type 1 diabetes in a randomised controlled trial. Gut,2020,70(1):92-105.

［74］ Barunch E N,Youngster I,Ben-Betzalel G,et al. Fecal microbiota transplant promotes response in immunotherapy-refractory melanoma patients. Science,2020,eabb5920.

［75］ Reiotix. Rebiotix and ferring announce world's first with positive preliminary pivotal phase 3 data for investigational microbiome-based therapy RBX2660. https://www. rebiotix. com/news-media/press-releases/rebiotix-announces-worlds-first-positive-pivotal-phase-3-data-investigational-microbiome-based-therapy-rbx2660/［2020-12-07］.

［76］ Seres. Seres therapeutics announces positive topline results from ser-109 phase 3 ecospor Ⅲ study in recurrent c. difficile infection. https://ir. serestherapeutics. com/news-releases/news-release-details/seres-therapeutics-announces-positive-topline-results-ser-109［2020-12-07］.

［77］ The Digest. Senate committee passes bioeconomy research and development act of 2020. http://www. biofuelsdigest. com/bdigest/2020/05/21/senate-committee-passes-bioeconomy-research-and-development-act-of-2020/［2020-12-07］.

［78］ CRDS. (戦略プロポーザル)『デザイナー細胞』～再生・細胞医療・遺伝子治療の挑戦～/CRDS-FY2020-SP-01. https://www. jst. go. jp/crds/report/report01/CRDS-FY2020-SP-01. html［2020-12-07］.

［79］ European Commission. Shaping Europe's digital future. https://ec. europa. eu/info/sites/info/files/communication-shaping-europes-digital-future-feb2020_en_4. pdf［2021-02-08］.

［80］ European Commission. A European strategy for data. https://ec. europa. eu/info/sites/info/files/communication-european-strategy-data-19feb2020_en. pdf［2021-02-08］.

［81］ European Commission. White paper on artificial intelligence —a European approach to excellence and trust. https://ec. europa. eu/info/sites/info/files/commission-white-paper-artificial-intelligence-feb2020_en. pdf［2021-02-08］.

［82］ HM Government. Landmark strategy launched to cement UK's position as global leader in genomics. https://assets. publishing. service. gov. uk/government/uploads/system/uploads/attachment_data/file/920378/Genome_UK_-_the_future_of_healthcare. pdf［2021-02-08］.

［83］ HDRUK. Launching the development of population research UK. https://www. hdruk. ac. uk/wp-content/uploads/2020/12/161220_Launching-the-development-of-PRUK. pdf［2021-02-08］.

［84］ 文部科学省. 第1回文部科学省デジタル化推進本部開催. https://www. mext. go. jp/b_menu/activity/detail/2020/20200925. html［2021-02-08］.

［85］ Digitalhealth. Gov. Au. National digital health workforce and education roadmap. https://www.

digitalhealth. gov. au/about-the-agency/workforce-and-education[2021-02-08].

［86］文部科学省. 保健医療分野における AI 研究開発加速に向けた人材養成産学協働プロジェクト. https://www. mext. go. jp/a_menu/koutou/iryou/1383121_00004. htm[2021-02-08].

Life Health and Medicine

Wang Yue, Xu Li, Su Yan, Shi Huilin, Li Zhenqi, Yang Ruonan, Li Wei, Xu Ping

Scientific and technological innovation can't leave the guarantee of the management system. With the arrival of a new planning period, many countries have carried out reforms in science and technology management and investment, aiming to lead the global science and technology development with original innovations and breakthroughs in core technologies. In particular, the uncertainties brought by COVID-19 to economic and social development have prompted the government to reflect on and reform the science and technology management system. In this context, the field of human health and medicine will also enter a stage of rapid development. At the same time, with the rapid upgrading of technology, as well as the in depth integration of disciplines, the field of human health and medicine is accelerating towards digital, intelligent, and engineering development. The development of this new model will also further promote the knowledge discovery, promote the innovation in diagnosis and treatment system, and improve the level of health care.

4.3 生物科技领域发展观察

陈 方 丁陈君 郑 颖 吴晓燕 宋 琪

（中国科学院成都文献情报中心）

当前，人类社会迎来百年未有之大变局这一历史进程的重要节点，新一轮科技革命和产业变革正在重塑世界，全球竞争格局和治理体系发生了深刻变革。新冠肺炎疫情这一"黑天鹅"事件的发生，促使人类开始重新反思科技发展、人类活动与自然的关系，也为生物科技及产业的发展带来新的危中之机。2020年，生物科技的发展突飞猛进，在生物资源与生物多样性研究、生物技术前沿研究和应用等方面取得多项进展，各国纷纷加快推动下一代生物经济的战略规划和项目部署，促进新一轮生物科技产业变革。"十四五"时期，我国发展进入新的重要战略机遇期，应进一步加快生物科技自立自强的建设步伐。

一、国际重大研究进展与趋势

1. 生物安全前沿技术在疫情防控中彰显巨大作用

面对新冠肺炎疫情给全球公共卫生安全带来的严峻挑战，生物科学家应用先进的工具和技术快速响应，开发从病毒检测到疫苗、药物等医疗对策的各类产品和应用，发挥了中坚作用。美国哈佛大学-麻省理工学院博德研究所的张锋教授领衔改进基于CRISPR的合成生物学平台SHERLOCK，可用于新冠病毒的简易、灵敏的分子诊断检测[1]。得益于核糖核酸（ribonucleic acid，RNA）修饰及递送技术的发展，人类首次完成信使核糖核酸（mRNA）疫苗设计研发，美国生物技术公司莫德纳（Moderna）公司研制的mRNA新冠疫苗从开发到获批仅用了数月时间。新冠特效药物研发取得进展，多种靶向病毒刺突蛋白的抗体类药物进入临床试验阶段，中国人民解放军军事科学院陈薇院士团队发现首个靶向新冠病毒刺突蛋白N端结构域的高效中和单克隆抗体，这为新冠肺炎治疗药物研发提供新的有效靶标[2]。中国医学科学院的秦川教授等率先建立了新冠病毒感染肺炎的转基因小鼠模型，突破了疫苗、药物从实验室向临床转化的关键技术瓶颈[3]；中国科学院微生物研究所高福院士、严景华研究员领衔开发

了针对 β 冠状病毒感染性疾病的通用疫苗策略，主导研制的具有自主知识产权的新冠病毒重组蛋白疫苗于 2020 年 6 月获批进入临床试验[4]。

2. 生物资源与生物多样性前沿研究取得重要进展

生命组学技术、新一代测序技术等结合大数据分析技术，在生物资源的研究、鉴定和分类，生物多样性监测、保护及生物资源的挖掘利用等方面取得多项成果。美国、中国、法国、英国等国的研究人员对来自不同生态系统的宏基因组数据进行 DNA 测序，在系统发育上定义了人类和其他动物微生物群及海洋、湖泊、沉积物、土壤和建筑环境中巨大噬菌体的主要进化枝[5]。2020 年 7 月，国际"DNA 元件百科全书"（encyclopedia of dna elements，ENCODE）计划发布第三阶段的成果总结，为更好地理解人类和小鼠基因组组织和功能提供了丰富的资源[6]。2020 年 11 月，中国联合美国、丹麦等国的多家机构合作发表了"万种鸟类基因组计划"第二阶段的研究成果，完整地描绘出鸟类物种谱系基因组动态演化图谱，对揭示物种类群分化具有重要意义[7]。

可再生资源的高值化利用驱动面向包括 CO_2 在内的生物质原料向多样化高值产品转化的工业体系转型。北京化工大学的谭天伟教授和瑞典查尔姆斯理工大学的 Jens Nielsen 教授合作提出利用大气 CO_2 及绿色清洁能源（光、废气中的无机化合物、光电、风电等）进行绿色生物制造的"第三代生物炼制"概念[8]；德国马克斯·普朗克科学促进协会陆地微生物学研究所开发了一种人造叶绿体自动化组装平台，实现人类首次在人工叶绿体内将 CO_2 转化为多碳化合物的重大进展[9]；英国剑桥大学的研究人员开发了一种可以将阳光、CO_2 和水转换成氧气和甲酸的方法[10]；日本理化学研究所的研究人员利用海洋光合细菌建立可持续的细胞工厂，成功地稳定大量生产蛛丝蛋白[11]；瑞士伯尔尼应用科技大学等的研究人员利用微生物群落降解木质纤维素，将复杂底物直接转化为短链脂肪酸[12]，相关研究为缓解能源、水和粮食危机，解决固体废弃物和全球变暖问题提供了可行的解决思路。

3. 生物技术前沿研究和颠覆性应用突破层出不穷

基因组测序、基因组编辑等生物技术工具的发展持续迭代提速，推动技术应用向精准高效和规模化的方向发展。美国加利福尼亚大学伯克利分校的研究人员开发了新测序方法 CiBER-seq，突破性地实现了细胞中数百个基因表达同时检测[13]。我国西湖大学的卢培龙研究员与华盛顿大学的 David Baker 团队等合作，首次在世界上实现跨膜孔蛋白的精确从头设计[14]。美国约翰斯·霍普金斯大学的研究人员开发出光诱导控制基因编辑技术 vfCRISPR，将基因编辑的精确度提升至前所未有的高度[15]。美国哈佛大学-麻省理工学院博德研究所的刘如谦团队开发了 DddA 衍生胞嘧啶碱基编辑器 DdCBE，为人类线粒体基因组研究提供了新工具[16]。

生物技术与人工智能、新一代信息技术、自动化技术等交叉融合、相互促进，不断催生前沿重大颠覆性创新。美国伊利诺伊大学的研究人员开发的"打孔卡"DNA 存储方法显著降低写入延迟、增加测读精度，有望实现更低成本、更大容量的 DNA 存储[17]。美国能源部布鲁克海文国家实验室的研究人员使用 DNA 自组装方法成功制造了三维纳米超导体，将在量子计算和传感中发挥重要作用[18]。美国佛蒙特大学和塔夫茨大学的团队联合利用"深绿"超级计算机设计，将青蛙的细胞组装成全新的生命形式，创造了世界上第一个毫米级活体可编程机器人 Xenobots，实现了人类破解"形态学代码"的重要一步[19]。

4. 工程生物研究突破及平台开发推动生物工业发展

工程生物学研究推动计算机辅助设计越来越多地参与修饰生命和创造生命的过程，合成生物技术向多样化应用导向的工程化进一步发展。英国萨里大学的合作团队开发了一种新的计算技术 ReProMin，该技术能够识别在细菌细胞中表达非必需基因的过程，使研究者可以对其选择性去除，将细胞资源分配给其他功能，从而更高效地生产有用产品[20]。美国能源部劳伦斯伯克利国家实验室的研究人员使用计算模型和基因编辑修饰微生物简化代谢重新连接过程，实现"产品-底物配对"，极大地加速了生物制造工艺的研发[21]。美国西北大学开发的体外原型系统 iPROBE 可以快速地发现细胞代谢工程的最佳生物合成途径，识别最佳途径的酶和配比，推动生物制造的产业化[22]。

国内外学者在生物路线合成各类产品方面取得多项成果。美国斯坦福大学的研究揭示了几乎完整的秋水仙素生物合成路径[23]；利用工程酵母以糖和氨基酸为原料合成托品烷生物碱及其衍生物[24]；美国能源部橡树岭国家实验室等机构开发了一种利用微生物生产乙烯的全新方法[25]；韩国科学技术院（KAIST）筛选出了一种迄今生产效率最高的琥珀酸合成菌株[26]；英国爱丁堡大学的科学家开发了利用工程细菌从愈创木酚生产己二酸的方法[27]。

5. 环境友好生物基材料替代及回收涌现成功案例

现代生物技术的可行性、经济性和成熟度不断提高，为解决生态环境领域的重大难题和挑战提供了绿色解决方案。例如，针对全球面临的塑料污染难题，生物技术创新在塑料降解回收、循环利用及可再生材料制造等方面发挥了重要作用。法国图卢兹大学的研究人员设计筛选的新型水解酶突变体可以在 10 h 内水解 90% 的聚对苯二甲酸乙二酯（PET）塑料瓶[28]；英国朴次茅斯大学的研究团队发现并设计了降解 PET 的新型嵌合酶 MHETase-PETase，将 PET 塑料的自然分解速度提高了 6 倍[29]；德国弗劳恩霍夫协会的研究团队等开发了一个全新的聚酰胺家族，初始产物为纤维素生产的副产品（3-蒈烯）[30]；美国加利福尼亚大学圣迭戈分校等的研究团队合作开发了由藻类油制成的可降解聚氨酯泡沫，展示了从生物基原料到产品，再到生物降解为单体的可持续循环利用的完美闭环[31]。

二、国际重大战略规划和政策措施

1. 推动系统变革，加强生物资源与生物多样性保护

在世界范围内，生物多样性保护的总体形势仍在恶化。新冠肺炎疫情引起的危机则进一步提示人类必须推进系统性变革以消除流行病的环境驱动因素。联合国《生物多样性公约》第十五次缔约方大会（COP15）发布了"2020 年后全球生物多样性框架"零草案，提出到 2050 年生物多样性保护的发展愿景并设定了相关的长期目标[32]。欧盟发布《欧盟 2030 年生物多样性战略》，旨在制定一项雄心勃勃的欧盟自然恢复计划[33]。世界自然基金会（World Wide Fund for Nature 或 World Wildlife Fund，WWF）的《新冠肺炎：紧急呼吁保护人与自然》报告提出恢复人类与自然关系的具体建议，目标是建设一个自然健康、碳中和、可持续和公平的社会[34]。

2. 强化顶层设计，推动向下一代生物经济形态过渡

在全球性环境威胁和新冠肺炎疫情的影响下，基于化石原料的线性经济发展模式更加显现出脆弱性和不可持续性，因此向生物经济的过渡变得比以往任何时候都更加关键[35]，主要经济体纷纷加强后疫情时代生物经济战略的顶层设计。美国国家科学院、美国国家工程院和美国国家医学院于 2020 年 1 月联合发布《保卫生物经济 2020》报告，对美国的生物经济进行了定义和现状评估，提出了发展战略要点；2020 年5 月，美国参议院通过《2020 年生物经济研发法案》，旨在推动国家工程生物学研发计划，以确保美国在该领域持续发挥领导作用。2020 年 3 月，欧盟生物基产业联盟（BIC）发布《战略创新与研究议程（SIRA）2030》报告草案[36]，提出了"2050 年循环生物社会"愿景并阐述了主要挑战和路线图，以及至 2030 年的里程碑和关键绩效指标。2020 年 1 月，德国通过新版《国家生物经济战略》[37]，并指定由一个独立的咨询委员会针对其多项目标和实施计划提出具体建议。2020 年 6 月和 2021 年 1 月，日本先后发布新版《生物战略 2020》基本措施版、市场领域措施版，围绕"到 2030 年成

为世界最先进的生物经济社会"的目标提出重点发展技术领域与产业布局。

3. 加强项目部署，促进新一轮生物科技产业变革

围绕发展愿景及目标，各国加强部署生物科技创新项目和配套举措，积极驱动科技产业颠覆性变革。

美国持续加大对生物技术研发的投入，美国能源部分别投入 6800 万美元和 9700 万美元用于生物能源原料作物和生物能源及生物基产品的研发；美国国防部投入 8750 万美元建立生物工业制造和设计生态系统（BioMADE）推动非医药生物工业制造业发展；美国农业部宣布为生物燃料基础设施和相关计划提供大额赠款。欧盟加速布局生物基产业，生物基联合产业（BBI-JU）投入 1.06 亿欧元资助了 22 个年度项目；欧洲投资银行（European Investment Bank，EIB）投资 7 亿欧元启动农业和生物经济领域计划，牵头成立 2.5 亿欧元的欧洲循环生物经济基金（ECBF），扩大资助创新型生物基企业和项目。德国联邦政府投入 36 亿欧元启动生物经济行动计划，加速可持续资源替代化石原料利用的进程。意大利发布未来 5 年生物经济行动计划[38]并做出旗舰项目部署。日本新能源产业技术综合开发机构（New Energy and Industrial Technology Development Organization，NEDO）启动"开发生物基产品生产技术，加速实现碳循环"计划，推动制造业革命并促进经济可持续增长。

三、对我国的启示与建议

2020 年，我国将科技自立自强作为国家发展的战略支撑，以科技助力经济发展，成为全球唯一正增长的主要经济体。步入"十四五"新时期，我国的科技事业发展进入重要战略机遇期。国家发展改革委等部门已在 2020 年 9 月联合印发《关于扩大战略性新兴产业投资 培育壮大新增长点增长极的指导意见》，再次将生物技术列为九大战略性新兴产业之一。在新阶段、新形势下，坚持"四个面向"，加快生物科技创新发展，成为我国科技高水平自立自强、经济社会高质量发展和全面绿色转型的重要任务。

1. 加强生物资源与生物多样性保护体系建设，践行绿色发展理念

全球范围的自然环境和生物多样性正在面临前所未有的破坏和威胁。生物多样性和生态系统保护的任务长期且艰巨，应进一步强化生物资源创新研究能力建设，利用现代生物技术开展生物资源的鉴定、评价和保藏；完善构建生物多样性保护监测平台，加强生物多样性调查、观测和评估；利用动物园、植物园、海洋馆等开展迁地保

护，同时加大公众科普宣传力度，提高公众保护意识，营造有利于生物资源保护和利用的良好社会氛围；加强野生动物管理，推动全健康（One Health）方法研究与实践，强化生物安全前沿技术支撑，完善早期疫情预警机制建设。

2. 重视科学技术交叉融合创新和人才培养，提升颠覆性创新能力

在全球学科之间、科学与技术之间、不同技术之间的交叉融合日益深入的大趋势下，我国应面向世界科技发展前沿，调整、优化学科布局，打造综合交叉学科群，促进学科交叉融合创新，以契合国家重大战略需求和经济社会发展需求。立足生物与信息、材料、工程技术交叉研究领域的重大科学研究，加快培育具有多学科背景的创新型人才，带动我国生物科技水平的整体提升。紧紧围绕攀登战略制高点，前瞻部署关键任务，瞄准支持核心技术研发，培育颠覆性创新能力。同时，完善两用生物技术的法规体系和监管机制建设，加强伦理、社会和安全监督防范。

3. 强化生物制造核心关键技术能力，把握生物经济发展战略主动权

全球生物经济发展的步伐逐渐加快，主要发达国家的生物科技和产业竞争加速，纷纷抢占下一轮经济增长的战略制高点。美国将生物经济视为关系经济繁荣和国家安全的重要议题，将多项前沿关键生物技术纳入对我国的出口管制清单，限制多种生物药品、生物化工制品及相关原料和仪器设备试剂的出口。在此形势下，我国应重视生物技术的自主创新，加大原创性研究成果的保护力度，强化生物制造核心关键技术能力，推动生物技术产品的研发与应用，在促进我国经济高质量发展、满足人民生命健康对科技创新需求的同时，保障我国生物产业安全，把握未来生物经济发展的战略主动权。

致谢：四川大学华西医学中心的王莉教授、中国科学院成都生物研究所的于源研究员等专家对本章节内容提出了宝贵的修改完善建议，特致谢忱！

参考文献

[1] Joung J, Ladha A, Saito M, et al. Detection of SARS-CoV-2 with SHERLOCK One-Pot Testing. the New England Journal of Medicine, 2020, 383(15): 1492-1494.

[2] Chi X Y, Yan R H, Zhang J, et al. A neutralizing human antibody binds to the N-terminal domain of the Spike protein of SARS-CoV-2. Science, 2020, 369(6504): 650-655.

[3] Bao L, Deng W, Huang B, et al. The pathogenicity of SARS-CoV-2 in hACE2 transgenic mice. Nature, 2020, 585: 830-833.

[4] Dai L, Zheng T, Xu K, et al. A universal design of betacoronavirus vaccines against COVID-19,

MERS and SARS. Cell,2020,182(3):722-733.

[5] Al-Shayeb B,Sachdeva R,Chen L X,et al. Clades of huge phage from across Earth's ecosystems. Nature,2020,578:425-431.

[6] The ENCODE project consortium,Snyder M P,Gingeras T R,et al. Perspectives on ENCODE. Nature,2020,583:693-698.

[7] Feng S,Stiller J,Deng Y,et al. Dense sampling of bird diversity increases power of comparative genomics. Nature,2020,587:252-257.

[8] Liu Z,Wang K,Chen Y,et al. Third-generation biorefineries as the means to produce fuels and chemicals from CO_2. Nature Catalysis,2020,3(3):274-288.

[9] Miller T E,T Beneyton T,Schwander T,et al. Light-powered CO_2 fixation in a chloroplast mimic with natural and synthetic parts. Science,368(6491):649-654.

[10] Wang Q,Warnan J,Rodríguez-Jiménez S,et al. Molecularly engineered photocatalyst sheet for scalable solar formate production from carbon dioxide and water. Nature Energy,2020,5:703-710.

[11] Foong C P,Higuchi-Takeuchi M,Malay A D,et al. A marine photosynthetic microbial cell factory as a platform for spider silk production. Communications Biology,2020,3(1):357-364.

[12] Shahab R L,Brethauer S,Davey M P,et al. A heterogeneous microbial consortium producing short-chain fatty acids from lignocellulose. Science,2020,369(6507):1214-1221.

[13] Muller R,Meacham Z A,Ferguson L,et al. CiBER-seq dissects genetic networks by quantitative CRISPRi profiling of expression phenotypes. Science,2020,370(6522):9662.

[14] Xu C,Lu P,El-Din T,et al. Computational design of transmembrane pores. Nature,2020,585(7823):129-134.

[15] Liu Y,Zou R S,He S,et al. Very fast CRISPR on demand. Science,2020,368(6496):1265-1269.

[16] Mok B Y,Moraes M,Zeng J,et al. A bacterial cytidine deaminase toxin enables CRISPR-free mitochondrial base editing. Nature,2020,583:631-637.

[17] Tabatabaei S K,Wang B,Athreya N,et al. DNA punch cards for storing data on native DNA sequences via enzymatic nicking. Nature Communications,2020,11(1):1742-1751.

[18] Shani L,Michelson A N,Minevich B,et al. DNA-assembled superconducting 3D nanoscale architectures. Nature Communications,2020,11(1):5697-5683.

[19] Kriegman S,Blackiston D,Levin M,et al. A scalable pipeline for designing reconfigurable organisms. Proceedings of the National Academy of Sciences,2020,117(4):1853-1859.

[20] Lastiri-Pancardo G,Mercado-Hernández J S,Kim J,et al. A quantitative method for proteome reallocation using minimal regulatory interventions. Nature Chemical Biology,2020,16:1026-1033.

[21] Banerjee D,Eng T,Lau A K,et al. Genome-scale metabolic rewiring improves titers rates and yields of the non-native product indigoidine at scale. Nature Communications,2020,11(1):5385-5395.

[22] Karim A S, Dudley Q M, Juminaga A, et al. In vitro prototyping and rapid optimization of bios ynthetic enzymes for cell design. Nature Chemical Biology, 2020(16):912-919.

[23] Nett R S, Lau W, Sattely E S. Discovery and engineering of colchicine alkaloid biosynthesis. Nature, 2020, 584(7819):148-153.

[24] Srinivasan P, Smolke C D. Biosynthesis of medicinal tropane alkaloids in yeast. Nature, 2020, 585:614-619.

[25] North J A, Narrowe A B, Xiong W, et al. A nitrogenase-like enzyme system catalyzes methionine, ethylene, and methane biogenesis. Science, 2020, 369(6507):1094-1098.

[26] Ahn J H, Seo H, Park W, et al. Enhanced succinic acid production by Mannheimia employing optimal malate dehydrogenase. Nature Communications, 2020, 11(1):1970-1981.

[27] Suitor J T, Varzandeh S, Wallace S. One-pot synthesis of adipic acid from guaiacol in escherichia coli. ACS Synthetic Biology, 2020, 9(9):2472-2476.

[28] Tournier V, Topham C M, Gilles A, et al. An engineered PET depolymerase to break down and recycle plastic bottles. Nature, 2020, 580(7802):216-219.

[29] Graham R. Characterization and engineering of a two-enzyme system for plastics depolymerization. Proceedings of the National Academy of Sciences, 2020, 117(41):25476-25485.

[30] Stockmann P N, Pastoetter D L, Wlbing M, et al. Biobased chiral semi-crystalline or amorphous high-performance polyamides and their scalable stereoselective synthesis. Nature Communications, 2020, 11(1):5724-5735.

[31] Nrg A, Mt B, Acs C, et al. Rapid biodegradation of renewable polyurethane foams with identification of associated microorganisms and decomposition products. Bioresource Technology Reports, 2020, 11:100513-100521.

[32] COP 15. Zero draft of the post-2020 global biodiversity framework. https://www. cbd. int/doc/c/efb0/1f84/a892b98d2982a829962b6371/wg2020-02-03-en. pdf[2020-02-03].

[33] EU. EU Biodiversity Strategy for 2030: Bringing nature back into our lives. https://4post2020bd. net/wp-content/uploads/2020/05/EU-Biodiversity-Strategy. pdf[2020-05-20].

[34] WWF. COVID 19: urgent call to protect people and nature. 2020. https://www. worldwildlife. org/publications/covid19-urgent-call-to-protect-people-and-nature[2020-06-17].

[35] GBS. Expanding the sustainable bioeconomy-vision and way forward. Communiquéof the Global Bioeconomy Summit 2020. https://knowledge4policy. ec. europa. eu/publication/expanding-sustainable-bioeconomy-%E2%80%93-vision-way-forward-communiqu% C3% A9-global-bioeconomy_ en [2020-11-20].

[36] BIC. The strategic innovation and research agenda(SIRA 2030)for a circular bio-based Europe. https://biconsortium. eu/sites/biconsortium. eu/files/documents/Draft% 20SIRA% 202030% 20-% 20 March% 202020. pdf[2020-03-20].

[37] BMBF. Nationale bioökonomiestrategie. https://www. bmbf. de/files/bio% C3% B6konomiestrategie%

20kabinett. pdf[2020-01-15].

[38] NBCG. Implementation action plan (2020-2025) for the Italian bioeconomy strategy BIT Ⅱ. https://enrd. ec. europa. eu/news-events/news/implementation-action-plan-2020-2025-italian-bio-economy-strategy-bit-ii_en[2020-07-19].

Bioscience and Biotechnology

Chen Fang , Ding Chenjun , Zheng Ying , Wu Xiaoyan , Song Qi

At present, our society is at an important point in its historical process of unprecedented changes in the past century, and a new round of technological revolution and industrial change is reshaping the world, the global competition landscape and governance system are undergoing profound changes. While the "black swan" incident of the COVID-19 epidemic has caused mankind to rethink the relationship between technological development, human activities and the nature, which also brings new opportunities in the midst of crisis for biotechnology industrial development. In 2020, there has been rapid progresses in the field of bioscience and biotechnology, with many advances having been made in research on biological resources and biodiversity, cutting-edge research and application of biotechnology, etc. Countries/terroirs have accelerated their strategic planning and deployment of projects to promote the next-generation bioeconomy and a new round of biotechnology industrial transformation. China is also entering an important strategic opportunity period in the 14th Five-Year Plan, accelerating the pace of building a self-sustainable biotechnology industry.

4.4　农业科技领域发展观察

袁建霞　邢　颖　李　超

（中国科学院科技战略咨询研究院）

2020 年是我国决胜全面建成小康社会、决战脱贫攻坚的收官和乡村振兴及"十四五"科技发展规划启动之年。农业生产持续发展，农村经济全面繁荣，农民生活显著改善，其中农业科技发挥了重要作用。本文在全面监测国际农业科技领域研究动态和战略举措的基础上，总结了国内外重要科学研究进展和国际重要战略行动，为我国把握未来农业科技发展态势、制定中长期发展战略提供启示建议。

一、国内外重要科学研究进展

1. 主要作物基因组测序研究取得重大进展

加拿大萨斯喀彻温大学主导的"10＋基因组计划"① 国际团队对世界各地具有代表性的 15 个小麦品种进行基因组测序，绘制出了较全的小麦基因组序列图谱[1]。中国科学院遗传与发育生物学研究所对 2898 份大豆样本进行重测序，选取其中 26 份进行基因组组装，在植物中首次实现了基于图形结构的泛基因组构建[2]。华中农业大学发布了 8 个甘蓝型油菜种质基因组序列，构建了含有 1.5 万个基因、约 1.8 Gb 的泛基因组[3]，并全面系统描绘了 20 个水稻品种的参考表观基因组图谱，注释了覆盖基因组 82% 的功能性 DNA 元件，产生了 500 多套组学数据[4]。中国农业科学院深圳农业基因组研究所首次测序完成杂合二倍体马铃薯基因组，提供了较为完整的杂合马铃薯基因组和较全面的马铃薯单体型比较分析[5]。

2. 作物抗病基因挖掘和病虫害防控机理研究取得重大突破

山东农业大学在小麦近缘植物长穗偃麦草发现了抗赤霉病主效基因 *Fhb7*，并利

---

① "10＋基因组计划"（10＋ Genome Project）是一项由全球 10 个国家的 95 名科学家组成的小麦联合研究计划，目标是破译小麦的数千个基因组序列，包括从小麦野生亲缘种带来的遗传物质。

用该基因培育出抗赤霉病小麦品种。半个多世纪以来，赤霉病研究全球鲜有突破性进展，特别是可用的主效抗赤霉病基因非常稀少，因此 Fhb7 基因的发现是禾谷类作物种质改良和创新难得的抗赤霉病基因[6]。中国科学院遗传与发育生物学研究所和中国农业科学院作物科学研究所合作发现了一个新类型小麦抗白粉病基因 Pm24。该基因编码 WTK3 蛋白。当 WTK3 第 5 外显子缺失 6 个碱基时即可获得抗白粉病功能[7]。中国科学院动物研究所鉴定到一种由群居型蝗虫特异性挥发的气味分子 4-乙烯基苯甲醚（4VA），并从多个层面证明 4VA 是飞蝗群聚信息素。该研究对世界蝗灾的控制和预测具有重要意义[8]。

3. 水稻氮高效利用和豆科作物固氮机理研究取得重大突破

中国科学院遗传与发育生物学研究所和英国牛津大学合作发现了水稻氮肥利用效率和产量协同调控新机制，即赤霉素信号转导途径中的转录因子 NGR5 可以调控植物响应土壤氮素水平的变化，同时实现赤霉素调控植物生长发育。高产水稻品种在氮肥减少的条件下增加 NGR5 表达仍可获得高产[9]。中国科学院上海生命科学研究院植物生理生态研究所揭示了豆科植物与根瘤菌共生固氮的分子机制。该研究发现，豆科植物因皮层细胞获得 SHR-SCR 干细胞分子模块而具有结瘤固氮的能力，使其有别于非豆科植物，加深了对共生固氮的理解，为非豆科植物皮层细胞改造奠定了基础，并为减少作物对氮肥的依赖，实现农业可持续生产提供了新思路[10]。

4. 主要作物驯化和进化研究取得重要进展

广州大学与中国科学院遗传与发育生物学研究所合作揭示了大豆驯化过程中开花的进化和选择机制，发掘出两个长日照条件下控制开花期的关键位点 Tof11 和 Tof12。这两个位点发生了渐进式变异和人工选择，使得栽培品种的开花期和成熟期普遍提前并缩短，提高了栽培大豆的适应性[11]。南京农业大学、得克萨斯大学和哈森阿尔法生物技术研究所合作绘制了所有 5 个异源四倍体棉花的高精度参考基因组图谱，发现这 5 个种的形成是单一起源，并历经了 20 万～60 万年的自然演化。研究还发现，不同棉种在 150 万年的杂交、多倍化和进化过程中基因数量和排列结构并没有非常显著的变化，而在 8000 年左右的人工驯化过程中，陆地棉和海岛棉的纤维长度和品质等发生了显著改变[12]。

5. 作物基因组编辑技术及其育种应用取得多项重要进展

中国科学院遗传与发育生物学研究所取得三项重要成果。一是将胞嘧啶和腺嘌呤单碱基编辑系统"合二为一"，开发出饱和靶向内源基因突变编辑系统，并获得了 3

个全新的水稻除草剂抗性突变位点[13]。二是基于胞嘧啶脱氨及碱基切除修复机理，首次建立了新型可预测多核苷酸删除系统，且在水稻和小麦成功应用[14]。三是建立并优化了适用于植物基因组的引导编辑系统[15]。美国明尼苏达大学开发出两种从头诱导分生组织的双子叶植物基因组编辑方法，打破了传统植物遗传转化技术需经过组织培养的瓶颈[16]。美国加利福尼亚大学戴维斯分校利用 CRISPR/Cas9 基因组编辑技术获得了富含类胡萝卜素的无标记黄金大米[17]。

6. 家畜分子育种与繁殖技术研发取得重要进展

美国华盛顿州立大学和英国爱丁堡大学罗斯林研究所合作，利用基因组编辑和干细胞技术培育出"代孕"种公畜。研究人员首先利用 CRISPR-Cas9 基因编辑技术敲除雄性动物生育能力相关基因 NANOS2，获得不育雄性猪、山羊和牛，然后将分离自同种其他优良雄性动物的精元干细胞移植到这些家畜的睾丸中，使其可以正常产生精子，且这些精子只含有供体动物的遗传物质。这些雄性不育家畜被称为"代孕"种公畜，可以充当优良精子生产机器并帮助雌性受孕。该技术有效解决了传统选择性育种和人工授精方法受时空限制的弊端，可以加快畜种改良进程，提高畜产品生产效率[18]。

7. 作物生物传感器研究取得重大进展

美国麻省理工学院利用单壁碳纳米管，开发出一种可实时监测作物对胁迫（包括受伤、感染和光害等）进行防御反应的纳米传感器。它可以嵌入植物叶片中，追踪损伤诱导的过氧化氢信号。当植物细胞释放过氧化氢时，相邻细胞内会触发钙质释放，刺激这些细胞释放更多的过氧化氢，进而刺激细胞产生次级代谢产物，帮助修复损伤，已经在生菜、芝麻菜、菠菜、草莓、酸模和拟南芥等植物中成功试验[19]。中国香港大学另开发出一种黄色荧光蛋白传感器，可通过检测植物代谢产物还原型辅酶Ⅱ（NADPH）和 NADH/NAD$^+$ 比率动态变化来实时观察植物光合作用能量变化[20]。

二、国际重要战略行动

1. 美国农业部发布 2020～2025 年科学蓝图

美国农业部 2020 年 2 月发布《美国农业部科学蓝图：2020～2025 年科研路线图》[21]，确定了可持续集约化、适应气候变化及食品和营养转化 3 个重点领域及其 6

个战略研究方向：①植物生产、健康和遗传学。加速育种进程，优化资源利用，提高植物产品营养素、代谢物和其他成分的效用和价值。②动物生产、健康和遗传学。研究动物生产和保护的基本知识和工具，加速育种进程，改善动物福利，优化资源利用，评估传感器、数据分析和精准农业增强技术的采用。③景观尺度保护和管理。加强最佳管理实践、创新技术和工具的使用，促进有韧性的农场、森林和牧场，改善生态系统服务。④气候研究和韧性。识别极端事件后粮食和森林系统复原力提高的机制，评估适应气候变化措施的环境影响。⑤食品安全和健康。收集和分析食品安全干预措施成本效率数据，分析病原体基因组数据，开发病原体和污染物检测和干预技术，研究食源性病原体的产生和传播。⑥营养与健康促进。促进健康饮食选择，利用生物活性物质改善现有食品的营养质量，识别并改善可减少肥胖和慢性病的生物与行为因素。

2. 美国国立卫生研究院发布未来十年营养研究战略

美国国立卫生研究院于 2020 年 5 月发布《2020～2030 年 NIH 营养研究战略计划》[22]，响应 2016 年美国《2016～2021 年国家营养研究路线图》，围绕"精准营养研究"核心愿景，提出了 4 个战略目标：通过基础研究促进发现和创新、探索膳食模式和饮食行为对最佳健康状态的影响、确定营养在整个生命周期中的作用及减轻疾病临床负担。与此同时，美国国立卫生研究院还提出了面向战略目标的若干交叉研究领域，包括特定人群健康差异研究，女性健康研究，提高营养研究的严谨性和可再现性，数据科学、系统科学和人工智能研究等。

3. 英国新版农业法案保障粮食生产和环境保护的平衡

英国于 2020 年 1 月出台了新版农业法案[23]。该法案的新内容包括：明确提出为农民保护或改善土壤质量提供资助，开展化肥监管和有机产品监管等。根据该法案，农民和土地管理者能够因提供如下公共产品而获得公共资金的奖励。内容主要包括：保护和改善土地、水和空气；支持植物和野生动物繁衍；减少和防止环境危害；适应和缓解气候变化；维护和修复自然文化遗产并使其融入环境建设；改善动物健康和福利。该法案将实现粮食生产和环境保护的平衡，使英国农业实现转型，在农业创新和环境保护上迈出具有里程碑意义的一步。

4. 欧盟委员会提出建立公平、健康和环保的粮食体系

欧盟委员会于 5 月正式发布《从农场到餐桌战略：建立公平、健康和环保的粮食体系》报告[24]，旨在减少欧盟食品体系的环境和气候足迹，增强防御能力，在气候变

化和生物多样性丧失的情况下确保粮食安全，并带领全球向从农场到餐桌的可持续性发展过渡，开拓新机遇。报告建议在"欧洲地平线"计划下，在食品、生物经济、自然资源、农业、渔业、水产养殖业和环境及数字技术和基于自然的农业食品解决方案的研发上投入100亿欧元。关键研究领域将涉及微生物组、海洋食物、城市食物系统，并增加替代蛋白质（如植物、微生物、海洋和昆虫蛋白及肉类替代品）的可获得性和来源。土壤健康和食品领域的任务是开发恢复土壤健康和功能的解决方案。新知识和创新也将扩大初级生产中的农业生态方法，减少农药、肥料和抗菌剂的使用。为加快创新和知识转移，欧盟委员会将与成员国加强合作，欧洲区域发展基金也将投资食品价值链上的创新和协作。

5. 美国植物科学研究网络发布未来十年植物科学愿景

汇聚美国植物科学领域15个学会专家的植物科学研究网络（PSRN）2020的年9月发布《2020~2030年植物科学十年愿景：重塑植物对健康和可持续未来的潜力》报告[25]，提出4个研究目标和2个技术目标：①利用植物提高地球的适应力。重点创建数字生物圈和预测植物系统的脆弱性。②研究满足多目标的可持续植物生产系统的先进技术。重点改良作物系统、创新作物技术及拓展农业生态学含义。③发展植物科学的新应用，以改善营养、健康和福祉。重点提高植物营养价值及加强与其他学科的交叉合作。④推出描述植物及其组成的交互式可视化数字工具。重点开发透明植物工具①、植物基因组研究和植物模拟。⑤开发可彻底变革研究的新技术。重点是高通量技术、便携式实验室等。⑥管理并实现大数据的潜力。重点是信息基础设施、现场应用的数据通信、机器学习和人工智能。

6. 国际农业研究磋商小组提出食品系统转型重点研究问题

国际农业研究磋商小组（Consultative Group for International Agricultural Research，CGIAR）于2020年6月发布《气候变化情况下食品系统转型行动》（*Actions to Transform Food Systems Under Climate Change*）报告[26]，提出4个优先行动领域的重点研究问题。①更新农业和农村生计模式：提高商品供应链透明度的方法、工具和政策有哪些？哪些因素能支持气候适应实践和技术的快速推广？无法进行农业生产时，农村居民如何发展生计？②消除农户谋生和农场生产中的风险：如何克服气候服务"最后一公里"的挑战？农民如何借助数字化技术改变传统农业发展路

---

① 透明植物工具（the transparent plant tool）是一个交互式的可视化和查询工具，一种描述植物及其组成部分的数字方式，最终支持自主设计植物的创建。

径？③减少食品和农业产业链各环节的碳排放：向健康和可持续饮食转变的最有效机制是什么？减少粮食损失和浪费的瓶颈是什么，以及如何突破？④调整社会运动和创新的政策与金融工具：如何扭转保守的思想观念，促进各国农业和粮食部门改革？食品系统转型的最佳信贷模式是什么？复制和扩大社会变革的行为因素有哪些？支持转型的最佳知识创造实践是什么？

7. 英国研究与创新署大力支持农业尖端技术研发

英国研究与创新署（UK Research and Innovation，UKRI）于 2020 年 7 月宣布投资 2400 万英镑，资助 9 个新的农业技术重大创新项目，以减少碳排放，提高生产力和赢利能力，使食物生产系统到 2040 年实现净零排放，并以更有效、更具弹性和可持续的方式生产食物[27]。9 个项目包括：利用二氧化碳转化为蛋白质的工艺，将来自发电厂的二氧化碳转化为动物饲料用蛋白；提供控制温室气候、灌溉和照明的自主技术；示范机器人技术和自主技术，协助人类采摘和包装水果及防治病虫害；研发下一代成本较低的主要作物经济型自主种植系统；基于藻类生长系统，利用天然海水在沙漠中生产食物；开发可通过过滤病原体来节约用水并保护植物的新材料；研发奶牛精准养殖技术，主要涉及跟踪奶牛行为和营养状况的可穿戴设备等；水果和蔬菜垂直种植系统及马铃薯收获决策支持系统。

三、启示与建议

2020 年中央经济工作会议提出要"解决好种子和耕地问题"，保障粮食安全，落实藏粮于地、藏粮于技战略。要开展种源"卡脖子"技术攻关，打一场种业翻身仗，要实施国家黑土地保护工程、加强农业面源污染治理等。对此，本文结合上述国内外重要科学研究进展和国际重要战略行动，提出如下建议。

1. 识别我国农业育种基础科研优势方向和短板进行分类布局和支持

重大科研进展显示，我国在作物基因组测序、抗病功能基因挖掘、病虫害机理研究、养分高效利用和固氮机理研究、进化和遗传资源挖掘及基因编辑育种等方向表现突出，但在家畜生物育种、农业生物传感器等领域的前沿尖端技术研发方面开创性不足，与美国和英国有较大差距。对于前者，应予以持续支持以争取更大突破，并在此基础上加强成果转化应用研究；对于后者，则需要加强跨学科合作，着力攻关，突破家畜良种培育的科学难题和技术难题，同时实现农业生物传感器高质量开发和应用，以通过精准监测动植物生长发育助力农业育种及生产效率的提高。

2. 在气候变化背景下推动耕地保护和利用

气候变化驱动的粮食和食品系统转型是未来农业发展的目标之一，其中健康的土壤与减少温室气体排放和缓解气候变化密切相关。在耕地保护和利用中需要支持气候适应性技术的快速推广，包括实施沃土工程、测土配方施肥工程，发展保护性耕作及加强农田水利设施建设等。同时，创新知识和技术（包括新型农药和肥料技术、固碳技术和土壤修复技术等），通过保持土壤健康来减少环境足迹和气候足迹。此外，为农民保护或改善土壤质量提供资金支持，进行化肥等农业投入品监管，减少农药、肥料和抗菌剂等对农田土壤有害物质的使用。

3. 以保障人民生命健康为目标促进食品安全、营养和健康协调发展

实施从田间到餐桌战略，以整个食品供应链的视角推进健康的饮食系统。通过改善土壤质量和化肥使用，加强动植物疫病防控，完善食品病原体、污染物的检测和干预技术等，保障食品安全；通过提高植物产品营养素、代谢物和其他成分的效用和价值，发展海洋食物及增加替代蛋白质食品等，生产更有营养价值和有益健康的食品；通过生物信息学、微生物组学等基础研究增加对营养的认识和理解，探索膳食模式和饮食行为对健康的影响，确定营养在整个生命周期中的作用和提高食品的医用价值等，实现精准营养供应。

致谢：中国科学院遗传与发育生物学研究所高彩霞研究员和田志喜研究员、中国农业机械化科学研究院吴海华研究员、中国农业科学院作物科学研究所马有志研究员对本文进行了审阅并提出了宝贵修改意见，特致谢忱！

参考文献

[1] Walkowiak S, Gao L, Monat C, et al. Multiple wheat genomes reveal global variation in modern breeding. Nature, 2020, 588: 277-283. https://doi.org/10.1038/s41586-020-2961-x[2020-11-25].

[2] Liu Y C, Du H L, Li P C, et al. Pan-genome of wild and cultivated soybeans. Cell, 2020, 182(1): 162-176.

[3] Song J M, Guan Z L, Hu J L, et al. Eight high-quality genomes reveal pan-genome architecture and ecotype differentiation of Brassica napus. Nature Plants, 2020, 6: 34-45. https://doi.org/10.1038/s41477-019-0577-7[2020-01-13].

[4] Zhao L, Xie L, Zhang Q, et al. Integrative analysis of reference epigenomes in 20 rice varieties. Nature Communications, 2020, 11: 2658. https://doi.org/10.1038/s41467-020-16457-5[2020-05-27].

［5］ Zhou Q, Tang D, Huang W, et al. Haplotype-resolved genome analyses of a heterozygous diploid potato. Nature Genetics, 2020, 52: 1018-1023. https://doi. org/10. 1038/s41588-020-0699-x［2020-09-28］.

［6］ Wang H, Sun S, Ge W, et al. Horizontal gene transfer of *Fhb7* from fungus underlies *Fusarium* head blight resistance in wheat. Science, 2020. 368 (6493): eaba5435. https://science. sciencemag. org/content/368/6493/eaba5435. full［2020-05-22］.

［7］ Lu P, Guo L, Wang Z, et al. A rare gain of function mutation in a wheat tandem kinase confers resistance to powdery mildew. Nature Communications, 2020, 11: 680. https://doi. org/10. 1038/s41467-020-14294-0［2020-02-03］.

［8］ Guo X, Yu Q, Chen D, et al. 4-vinylanisole is an aggregation pheromone in locusts. Nature, 2020, 584: 584-588. https://doi. org/10. 1038/s41586-020-2610-4［2020-08-12］.

［9］ Wu K, Wang S, Song W, et al. Enhanced sustainable green revolution yield via nitrogen-responsive chromatin modulation in rice. Science, 2020: 367(6478): eaaz2046. https://science. sciencemag. org/content/367/6478/eaaz2046［2020-02-07］.

［10］ Dong W, Zhu Y, Chang H, et al. An SHR-SCR module specifies legume cortical cell fate to enable nodulation. https://www. nature. com/articles/s41586-020-3016-z［2020-12-09］.

［11］ Lu S, Dong L, Fang C, et al. Stepwise selection on homeologous PRR genes controlling flowering and maturity during soybean domestication. Nature Genetics, 2020, 52: 428-436. https://doi. org/10. 1038/s41588-020-0604-7［2020-03-30］.

［12］ Chen Z J, Sreedasyam A, Ando A, et al. Genomic diversifications of five Gossypium allopolyploid species and their impact on cotton improvement. Nature Genetics, 2020, 52: 525-533. https://doi. org/10. 1038/s41588-020-0614-5［2020-04-20］.

［13］ Li C, Zhang R, Meng X, et al. Targeted, random mutagenesis of plant genes with dual cytosine and adenine base editors. Nature Biotechnology, 2020, 38: 875-882. https://doi. org/10. 1038/s41587-019-0393-7［2020-01-13］.

［14］ Wang S, Zong Y, Lin Q, et al. Precise, predictable multi-nucleotide deletions in rice and wheat using APOBEC-Cas9. Nature Biotechnology, 2020, 38: 1460-1465. https://doi. org/10. 1038/s41587-020-0566-4［2020-06-29］.

［15］ Lin Q, Zong, Y, Xue C, et al. Prime genome editing in rice and wheat. Nature Biotechnology, 2020, 38: 582-585. https://doi. org/10. 1038/s41587-020-0455-x［2020-03-16］.

［16］ Maher M F, Nasti R A, Vollbrecht M, et al. Plant gene editing through *de novo* induction of meristems. Nature Biotechnology, 2020, 38: 84-89. https://doi. org/10. 1038/s41587-019-0337-2［2020-12-16］.

［17］ Dong O X, Yu S, Jain R, et al. Marker-free carotenoid-enriched rice generated through targeted gene insertion using CRISPR-Cas9. Nature Communications, 2020, 11: 1178. https://doi. org/10. 1038/s41467-020-14981-y［2020-03-04］.

[18] Ciccarellia M, Giassettia M I, Miao D. et al. Donor-derived spermatogenesis following stem cell transplantation in sterile *NANOS2* knockout males. PNAS, 2020, 117(39): 24195-24204. https://www.pnas.org/content/117/39/24195[2020-09-29].

[19] Lew T T S, Koman V B, Silmore K S, et al. Real-time detection of wound-induced H_2O_2 signalling waves in plants with optical nanosensors. Nature Plants, 2020, 6: 404-415. https://doi.org/10.1038/s41477-020-0632-4[2020-04-15].

[20] Lim S L, Voon C P, Guan X, et al. In planta study of photosynthesis and photorespiration using NADPH and NADH/NAD^+ fluorescent protein sensors. Nature Communications, 2020, 11: 3238. https://doi.org/10.1038/s41467-020-17056-0[2020-06-26].

[21] United States Department of Agriculture. USDA science blueprint: a roadmap for USDA science from 2020 to 2025. https://www.usda.gov/sites/default/files/documents/usda-science-blueprint.pdf[2020-02-08].

[22] National Institutes of Health. NIH releases strategic plan to accelerate nutrition research over next 10 years. 2020-2030 Strategic Plan for NIH Nutrition Research. https://www.nih.gov/news-events/news-releases/nih-releases-strategic-plan-accelerate-nutrition-research-over-next-10-years [2020-05-27].

[23] Farming Online. Agriculture Bill to boost environment & food production. https://www.farming.co.uk/news/agriculture-bill-to-boost-environment—food-production[2020-01-16].

[24] European Commission. A farm to fork strategy for a fair, healthy and environmentally-friendly food system. https://ec.europa.eu/food/sites/food/files/safety/docs/f2f_action-plan_2020_strategy-info_en.pdf[2020-05-20].

[25] Henkhaus N, Bartlett M, Gang D, et al. Plant science decadal vision 2020-2030: reimagining the potential of plants for a healthy and sustainable future. https://doi.org/10.1002/pld3.252[2020-09-01].

[26] Steiner A, Aguilar G, Bomba K, et al. Actions to transform food systems under climate change. agriculture and food security (CCAFS). https://cgspace.cgiar.org/bitstream/handle/10568/108489/Actions%20to%20Transform%20Food%20Systems%20Under%20Climate%20Change.pdf[2020-06-29].

[27] UK-China Sustainable Agriculture Innovation Network. UKRI invests £24 million in cutting-edge technology to transform UK agriculture. UK agriculture brief. http://www.sainonline.org/pages/News/%E8%8B%B1%E5%9B%BD%E5%86%9C%E4%B8%9A%E7%AE%80%E8%AE%AF%20UK%20Agriculture%20Brief%20072020.pdf[2020-07-30].

Agricultural Science and Technology

Yuan Jianxia，Xing Ying，Li Chao

We summarized the important scientific research progress and international strategic actions. Globally, agricultural science and technology achieved great progress and breakthrough in crop genome sequencing, disease-resistance gene mining and pathogenesis, efficient utilization of nutrients, nitrogen fixation mechanism, crop domestication and evolution, crop gene editing and breeding, biological breeding of livestock, and agricultural biosensor during this year. In addition, there were several significant events that deserve attention: "USDA Science Blueprint: A Roadmap For USDA Science From 2020 to 2025" and "2020-2030 Strategic Plan for NIH Nutrition Research" were released in USA, "Agriculture Bill to boost environment & food production" was introduced in UK, European Commission released "A Farm to Fork Strategy for a fair, healthy and environmentally-friendly food system", and CGIAR released "Actions to Transform Food Systems Under Climate Change", et al. Finally, we proposed several suggestions: to identify the advantages and disadvantages of agricultural breeding basic scientific research for making different research strategy, to promote the protection and utilization of arable land in the context of climate change, and to promote the coordinated development of food safety, nutrition and health.

4.5 生态环境科学领域发展观察

廖 琴[1] 曲建升[2] 曾静静[1] 裴惠娟[1]
董利苹[1] 刘燕飞[1] 刘莉娜[1]

（1. 中国科学院西北生态环境资源研究院；
2. 中国科学院成都文献情报中心）

2020 年是全球生态环境领域发展历程中具有里程碑意义的一年，碳中和目标及新冠病毒肺炎疫情暴发后的绿色复苏成为全球新趋势。同时，生物多样性保护、塑料污染治理、可持续发展目标等仍然是全球生态环境领域关注的热点。2020 年，科学界在生态环境领域取得了以下重要发现和突破：①新冠病毒肺炎疫情防控暂时减少空气污染和二氧化碳排放；②气候敏感性范围大幅度缩小；③臭氧层恢复使得南半球环流变化趋势暂缓；④全球生物多样性急剧丧失；⑤卫星图像精准定位大区域树木位置和大小；⑥塑料废弃物治理材料与技术获重要突破。在战略部署方面，各国重点围绕碳中和目标、绿色经济复苏、塑料污染、土壤与生物多样性等方面提出行动计划。此外，澳大利亚还注重森林火灾的相关研究及应对。

一、重要研究进展

1. 新冠病毒肺炎疫情防控暂时减少空气污染和二氧化碳排放

新冠病毒肺炎疫情对全球二氧化碳排放量和空气质量的影响引起了科学界的关注。比利时皇家空间和高层大气物理研究所（Royal Belgian Institute for Space Aeronomy，BIRA-IASB）与荷兰皇家气象研究所（Royal Netherlands Meteorological Institute，KNMI）等机构的研究人员估计了新冠病毒肺炎疫情主要暴发国/地区空气中的二氧化氮（NO_2）含量变化。研究发现，新冠病毒肺炎疫情防控措施使中国、韩国、西欧和美国空气中的 NO_2 含量出现前所未有的下降[1]。德国马克斯·普朗克气象研究所（Max Planck Institute for Meteorology）等机构的研究发现，2020 年 1 月 23 日开始疫情防控以来，中国细颗粒物（$PM_{2.5}$）和 NO_2 在空气中的含量大幅下降，但地表臭氧水平有所增加[2]。英国东英吉利大学与美国斯坦福大学等机构的联合研究显示，截至 2020 年 4 月初，全球 2020 年的二氧化碳日排放量比 2019 年的日平均排放量下降了 17%，疫情对 2020 年的年度排放总量的影响取决于各国控制措施的持续时间[3]。中国清华大学与法国巴黎萨克莱大学等机构的研究人员基于全球近实时监测数据，分析了新冠病毒肺炎大流行对全球二氧化碳排放的影响。研究发现，2020 年上半年，全球二氧化碳排放量比 2019 年同期的全球二氧化碳排放量减少了 8.8%[4]。2020 年 12 月，"全球碳项目"（Global Carbon Project，GCP）发布《2020 年全球碳预算》报告指出，新冠病毒肺炎的全球封锁使 2020 年全球化石燃料二氧化碳排放量比 2019 年减少了约 24 亿吨（减少 7%）[5]。

2. 气候敏感性范围大幅度缩小

地球的气候敏感性，一般是指大气中的二氧化碳浓度增加 1 倍时引起的全球平均温度变化，长期以来被视为了解全球气候变化的起点[6]。40 多年前，气候科学家对二氧化碳引起的气候变化首次进行的全面评估显示，如果大气中的二氧化碳浓度比前工业化时期的浓度增加 1 倍，那么地球最终可能升温 1.5～4.5℃。2020 年，在世界气候研究计划（World Climate Research Programme，WCRP）支持下，来自澳大利亚新南威尔士大学悉尼分校、英国气象局哈德利中心、美国华盛顿大学等机构的 25 名科学家利用多种证据对气候敏感性范围进行的评估表明，大气中的二氧化碳浓度增加 1 倍，可能引起地球升温 2.6～3.9℃[6]。该研究大幅度缩小了气候敏感性的范围，入选《科学》评选的 2020 年度全球十大科学突破[7]。德国阿尔弗雷德韦格纳研究所与加拿大麦吉尔大学的研究人员提出了预测全球变暖的一种新方法。该方法可以通过直接观测估计气候敏感性及其不确定性，而不需要过多的假设。研究发现，全球可能将在 2027～2042 年升温超过 1.5℃，缩小了联合国政府间气候变化专门委员会（Intergovernmental Panel on Climate Change，IPCC）估计的 2020～2052 年范围，且预期的变暖幅度比先前的预测低 10%～15%[8]。

3. 臭氧层恢复暂缓南半球环流变化趋势

人为排放消耗臭氧层物质（ozone-depleting substances，ODS）导致南极平流层臭氧耗竭，从而引起 20 世纪末南半球近地面的环流出现一些变化，包括中纬度急流向极区移动、南半球环状模呈正位相、哈得来环流出现扩张。美国国家海洋和大气管理局（National Oceanic and Atmospheric Administration，NOAA）与科罗拉多大学等机构的联合研究显示，这些环流趋势在 2000 年左右暂停或略有逆转。大气纬向风模式检测和归因分析表明，环流趋势的暂停是由人类活动造成的，而不仅是由气候系统的内部或自然变率引起的。该研究证明，1987 年签署的《蒙特利尔破坏臭氧层物质管制议定书》(简称《蒙特利尔议定书》,*Montreal Protocol on Substances that Deplete the Ozone Layer*)带来的平流层臭氧恢复是这一环流趋势暂停的主要驱动因素。由于2000年以前的环流趋势已经影响到降水,并潜在地影响海洋环流和盐度,因此研究人员预计这些趋势的暂停将对地球系统产生更广泛的影响[9]。这项研究成果入选《自然》评选的2020年度全球十大重要科学发现[10]。

4. 全球生物多样性急剧丧失

美国亚利桑那大学的研究指出,到2070年,气候变化可能使地球1/3的动植物物种面临灭绝。如果成功实现《巴黎协定》目标(即到2100年全球升温幅度控制在1.5℃以内),到2070年可能有16%的物种面临灭绝[11]。奥地利维也纳大学与瑞士弗里堡大学等的联合研究指出,全球外来物种数量增加可能导致生物多样性急剧丧失,甚至对生态系统造成永久性破坏[12]。南非开普敦大学和美国康涅狄格大学等的研究人员探讨了气候变化背景下生物多样性的丧失是立即发生还是随着时间加剧,预计气候变化可能会造成整个21世纪全球范围内的生物多样性突然、潜在的灾难性损失[13]。2020年9月世界自然基金会发布的《2020年地球生命力报告》[14]指出,全球野生动物种群数量在1970~2016年平均下降68%。其中,淡水野生动物种群的下降幅度最大,达到84%。南京大学和中国科学院南京地质与古生物研究所等的研究人员获得了全球首条高精度的古生代3亿多年的生物多样性变化曲线,精确刻画出地质历史中多次重大生物灭绝和辐射事件及其与环境变化的关系[15]。

5. 卫星图像精准定位大区域树木位置和大小

对陆地生态系统的定义在很高程度上取决于其木本植物。因此,掌握生态系统中木本植被结构的准确信息,对理解全球尺度生态学、生物地理学及碳、水和其他营养素组成的生物地球化学循环至关重要。丹麦哥本哈根大学和美国国家航空航天局(National Aeronautics and Space Administration,NASA)戈达德太空飞行中心等机构的研究人员利用亚米级分辨率的卫星图像和深度学习方法,对非洲西北部撒哈拉沙漠、萨赫勒地区和半湿润地区130万km²土地上的树木进行了统计,绘制了逾18亿个树冠的位置和大小[16]。此前,科学家从未在这种大面积土地上绘制过如此精细的树木地图。该研究提出了一种监测全球森林以外的树木的方法,未来有可能在大的区域或全球尺度上绘制每棵树的位置和树冠大小。这一进展将推动人们思考、监测、模拟和管理全球陆地生态系统的方式发生根本性变化[17]。这项研究成果入选《自然》评选的2020年度十大科学发现。

6. 塑料废弃物治理材料与技术获重要突破

法国图卢兹大学等的研究人员改进了一种聚对苯二甲酸乙二醇酯（polyethylene glycol terephthalate，PET）水解酶。改进后的水解酶在 10 h 内可以实现至少 90% 的 PET 解聚为单体。这种高效、优化的酶的性能优于迄今报道的所有 PET 水解酶。研究还表明，通过这种酶解聚的 PET 废料可以再次生产出可生物回收的 PET，并且具有与石化 PET 相同的性能，有助于实现循环 PET 经济[18]。美国麻省理工学院的研究人员使用工业热固性聚合物聚双环戊二烯（polydicyclopentadiene，PDCPD）为模型系统，制备了可裂解的共聚单体，获得可降解、可回收的热固性塑料。该研究表明，引入适量可降解单体可以实现热固性聚合物的可控降解与回收[19]。美国加利福尼亚大学、伊利诺伊大学和康奈尔大学的研究人员提出了将废弃聚乙烯转化为高价值产品的创新方法。他们使用负载铂的 γ-氧化铝催化剂，在 280 ℃ 的温度下基于串联氢解/芳构化反应将废弃聚乙烯转化为高价值的长链烷基芳烃和烷基环烷烃。该方法为最终实现塑料废弃物的循环利用提供了可行的路径[20,21]。

二、重大战略行动

1. 多国相继提出碳中和目标

先进国家通过立法和政策文件的形式相继提出碳中和目标。2019 年 6 月，英国通过《2008 年气候变化法案（2050 年目标修正案）》[22]，提出到 2050 年实现温室气体净零排放的目标，成为全球首个通过净零排放法案的主要经济体。欧盟于 2019 年 12 月公布《欧洲绿色协议》[23]，提出旨在到 2050 年实现气候中和目标的详细路线图和政策框架，并于 2020 年 3 月发布《欧洲气候法》草案[24]，将 2050 年实现气候中和（climate neutrality）的目标纳入法律。2020 年 9 月，中国提出力争于 2060 年前实现碳中和目标。10 月，日本首相宣布到 2050 年实现碳中和、脱碳的社会[25]。12 月，日本政府推出《面向 2050 碳中和的绿色增长战略》[26]，确定了日本到 2050 年实现碳中和的目标，并对海上风电、燃料电池、氢能等 14 个产业提出了具体的发展目标和任务。11 月，加拿大通过《净零排放问责法案》[27]，要求加拿大设定减少温室气体排放的国家目标，以期到 2050 年实现温室气体净零排放。2020 年 12 月，韩国发布《2050 年碳中和战略》[28]，提出到 2050 年实现碳中和的远景。

2. 多国致力于新冠病毒肺炎疫情暴发后的绿色经济复苏计划

新冠病毒肺炎疫情暴发后，多国推出经济刺激计划，积极转向绿色低碳领域。

2020 年 6 月德国政府发布的价值 1300 亿欧元的刺激计划[29]包括三大支柱,其中"对未来友好型德国进行投资"这个支柱被分配了 500 亿欧元的资金,专注于能源转型和可持续交通。7 月,韩国政府启动价值约 114 万亿韩元的新政[30],包括"数字新政"和"绿色新政"两大核心支柱,其中"绿色新政"到 2025 年将投资 73.4 万亿韩元。9 月,法国政府推出 1000 亿欧元的法国复苏计划[31],生态目标是使法国成为欧洲第一个主要的低碳经济体,将提供 300 亿欧元用于建筑翻新、交通运输、农业转型和能源 4 个生态优先领域。9 月,加拿大宣布 100 亿加元的重大基础设施投资计划[32],其中至少有 60 亿加元分配给绿色项目,包括清洁能源项目、零排放公交车和节能建筑改造。11 月,英国发布《绿色工业革命 10 项计划》[33],将提供约 120 亿英镑的政府资金,涵盖清洁能源、交通、建筑、碳捕集、自然和创新技术等领域。12 月,日本发布新经济刺激计划[34],提出要实现数字化和绿色社会,将提供 2 万亿日元资金支持创新技术发展,以实现 2050 年碳中和目标,促进电动汽车和净零能源住宅。

3. 联合国关注可持续发展目标的进展

2020 年 3 月联合国亚洲及太平洋经济社会委员会(United Nations Economic and Social Commission for Asia and the Pacific,ESCAP)发布的《2020 年亚太地区可持续发展目标进度报告》[35]指出,亚太地区在可持续发展目标环境领域的大部分量化指标中均表现不佳,急需在实现可持续发展目标领域加快进展,扭转当前的不良态势。7 月联合国经济和社会事务部(United Nations Department of Economic and Social Affairs,UNDESA)发布的《2020 年可持续发展目标报告》[36]指出,新冠病毒肺炎疫情引发了前所未有的危机,进一步妨碍了可持续发展目标的进展。此外,气候变化的发生速度远超过预期,土地继续退化,大量物种面临灭绝的危险,不可持续的消费和生产方式仍然普遍存在。12 月,联合国开发计划署(The United Nations Development Programme,UNDP)发布《2020 年人类发展报告》[37],首次纳入二氧化碳排放和材料足迹两大衡量指标,呼吁所有国家重新设计发展道路,将人类活动对地球环境的破坏纳入考量。

4. 多国推进塑料污染科学研究及其治理

2020 年 10 月,英国自然环境研究理事会(Natural Environment Research Council,NERC)、新加坡国立研究基金会(National Research Foundation,NRF)以及英国商业、能源与产业战略部(BEIS)共同宣布为创新性国际研究合作提供 600 万英镑资助,以研究如何降低塑料污染对东南亚海洋生态系统的影响[38]。10 月,加拿大环境与气候变化部和卫生部发布《塑料污染科学评估》报告[39]，确定了当前塑料污染研究存在的关键知识差距。为了填补这些知识差距，11 月，加拿大环境与气候变化部和卫生部通过"提高塑料污染知识计划"提供了 225.65 万加元来资助 16 个科学研究项目，以进一步了解塑料污染对自然环境和人类健康造成的影响[40]。2020 年 3 月，欧盟发布新的《循环经济行动计划》[41]，针对再生塑料、微塑料、生物基塑料和一次性塑料制品提出了进一步的治理举措。10 月，美国发布《处理全球海洋垃圾问题的联邦战略》[42]，侧重从 4 个方面解决全球海洋垃圾问题：提升建立更好的废弃物和垃圾管理系统的能力、鼓励全球循环利用、促进研究和开发、清除海洋垃圾。

5. 欧美发布土壤与生物多样性保护相关战略

2019 年 12 月，欧盟在世界土壤日启动"关爱土壤就是关爱生命"的土壤健康与食物计划[43]。该计划的目标是，到 2030 年，确保欧盟 75% 的土壤保持健康，并能够提供基本的生态系统服务，包括停止（新的）土壤硬化、提高城镇土壤的再生利用、土地占用零增长（不再新批土地开发项目）、工程移出土壤的适当归位，以及废弃物的重新循环利用。2020 年 5 月，欧盟发布《2030 年生物多样性战略》[44]，提出到 2030 年欧盟要实现的具体承诺和行动，包括改善和扩大保护区网络，制定欧盟自然恢复计划，以保护自然并扭转生态系统的退化，使欧洲生物多样性在 2030 年前走上恢复之路。2021 年 4 月，欧盟议会通过关于土壤保护的决议[45]，呼吁欧盟设计一个保护和可持续利用土壤的共同法律框架，推动土壤保护的各项行动。2020 年 8 月，美国内政部发布《入侵物种战略计划》草案[46]，提出了美国减少入侵物种风险的战略目标及其实施策略，包括经济有效地防止入侵物种在美国境内的引入和扩散、加强早期监测和快速响应工作、经济有效地控制已入侵物种种群、提高入侵物种数据管理的决策服务能力等。

6. 澳大利亚部署行动应对森林火灾

2019～2020 年，澳大利亚遭遇了 21 世纪最严重的森林火灾。2020 年 1 月，澳大利亚环境部宣布投入 5000 万澳元帮助火灾后的动植物恢复，其中 2500 万澳元用于紧急干预基金和必要的关键干预措施，另外 2500 万澳元用于支持野生动物救援等[47]。1 月与 2 月，澳大利亚环境部先后发布受威胁物种与生态群落的初步清单。2 月，澳大利亚环境部发布临时清单，确定林火之后数周乃至数月内最需要采取紧急干预措施帮助恢复的物种，同时还发布"国家指示性综合火灾数据集"，提供了澳大利亚可能受火灾影响地区的综合信息[48]。3 月，澳大利亚濒危物种科学委员会（TSSC）发布森林火灾应对计划[49]，确定了应对火灾的关键目标与具体行动，包括：防止受

2019～2020年火灾影响的本地物种及生态系统的灭绝和衰退；减少未来火灾的影响；学习与持续改进。

三、启示与建议

1. 推进"十四五"规划中的气候应对行动

我国应进一步将应对气候变化的任务目标纳入"十四五"整体经济社会发展规划，制定应对温室气体排放的长期路线图和具体工作方案，重点包括清洁能源转型和可持续城市发展等行动领域，并与中长期社会经济发展战略保持一致。

2. 制定碳中和目标的实施路径及配套政策

开展气候变化对我国经济与社会的影响评估和分析，研究到2030年和2060年前实现碳达峰与碳中和目标的路径，并制定实现目标的政策路线，重点关注电力、工业、交通和负排放技术等关键领域的减排[50]。提高全国碳排放交易体系的有效性和覆盖范围，建立制度完善的全国碳排放交易市场。

3. 加快新冠病毒肺炎疫情暴发后的绿色经济重建

尽管新冠病毒肺炎疫情防控在短期内减少了空气污染和二氧化碳排放，但新冠病毒肺炎疫情暴发后的政府行动和经济激励措施可能会影响未来全球排放的走向。在经济重建中，应限制高耗能和高排放的大型基础设施项目投资，考虑将绿色标准纳入项目投资，同时利用好新基础设施建设带来的数字化和智能化，实现绿色转型。

4. 落实塑料污染的进一步治理

新冠病毒肺炎疫情导致塑料制品的消费量再次大幅增加[51]。2020年，国家发展改革委、生态环境部等部委先后印发《关于进一步加强塑料污染治理的意见》、《关于进一步加强塑料污染治理近期工作要点的通知》和《关于扎实推进塑料污染治理工作的通知》，对我国塑料污染治理进行了整体部署。各地还需加大政策落实力度，提出有效的推进措施。此外，还需加强塑料回收的科技研发，形成更加完善的塑料循环经济体系。

致谢：中国科学院城市环境研究所朱永官院士、中国科学院南京土壤研究所骆永明研究员、南京农业大学农业资源与生态环境研究所潘根兴教授等审阅了本文并提出了宝贵的修改意见，中国科学院西北生态环境资源研究院王金平、吴秀平、牛艺博、李恒吉等对本文的资料收集和分析工作亦有贡献，在此一并表示感谢。

参考文献

[1] Bauwens M, Compernolle S, Stavrakou T, et al. Impact of coronavirus outbreak on NO_2 pollution assessed using TROPOMI and OMI observations. Geophysical Research Letters, 2020, 47(11): e2020GL087978.

[2] Shi X, Brasseur G P. The response in air quality to the reduction of chinese economic activities during the COVID-19 outbreak. Geophysical Research Letters, 2020, 47(11): e2020GL088070.

[3] Quéré C L, Jackson R B, Jones M W, et al. Temporary reduction in daily global CO_2 emissions during the COVID-19 forced confinement. Nature Climate Change, 2020, 10: 647-653.

[4] Liu Z, Ciais P, Deng Z, et al. Near-real-time monitoring of global CO_2 emissions reveals the effects of the COVID-19 pandemic. Nature Communications, 2020, 11: 5172.

[5] Global Carbon Project. Global Carbon Budget 2020. https://www.globalcarbonproject.org/carbon-budget/20/highlights.htm[2020-12-11].

[6] Sherwood S C, Webb M J, Annan J D, et al. An assessment of earth's climate sensitivity using multiple lines of evidence. Reviews of Geophysics, 2020, 58: e2019RG000678.

[7] Cohen J. 2020 breakthrough of the year. https://vis.sciencemag.org/breakthrough2020/[2020-12-17].

[8] Hébert R, Lovejoy S, Tremblay B. An observation-based scaling model for climate sensitivity estimates and global projections to 2100. Climate Dynamics, 2021, 56: 1105-1129.

[9] Banerjee A, Fyfe J C, Polvani L M, et al. A pause in Southern Hemisphere circulation trends due to the Montreal Protocol. Nature, 2020, 579: 544-548.

[10] Nature News and Views. Viruses, microscopy and fast radio bursts: 10 remarkable discoveries from 2020. Nature, 2020, 588: 596-598.

[11] Román-Palacios C, Wiensa J J. Recent responses to climate change reveal the drivers of species extinction and survival. PNAS, 2020, 117(8): 4211-4217.

[12] Essl F, Lenzner B, Bacher S, et al. Drivers of future alien species impacts: an expert-based assessment. Global Change Biology, 2020, 26(9): 4880-4893.

[13] Trisos C H, Merow C, Pigot A L. The projected timing of abrupt ecological disruption from climate change. Nature, 2020, 580: 496-501.

[14] WWF. living planet report 2020. https://www.wwf.org.uk/sites/default/files/2020-09/LPR20_Full_report.pdf[2020-09-09].

[15] Fan J, Shen S Z, Erwin D H. A high-resolution summary of Cambrian to Early Triassic marine invertebrate biodiversity. Science, 2020, 367(6475): 272-277.

[16] Brandt M, Tucker C J, Kariryaa A, et al. An unexpectedly large count of trees in the West African

Sahara and Sahel. Nature,2020,587:78-82.

[17] Hanan N P,Anchang J Y. Satellites could soon map every tree on Earth. Nature,2020,587:42-43.

[18] Tournier V,Topham C M,Gilles A,et al. An engineered PET depolymerase to break down and recycle plastic bottles. Nature,2020,580(7802):216-219.

[19] Shieh P,Zhang W X,Husted K E L,et al. Cleavable comonomers enable degradable,recyclable thermoset plastics. Nature,2020,583(7817):542-547.

[20] Zhang F,Zeng M H,Yappert R D,et al. Polyethylene up-cycling to long-chain alkylaromatics by tandem hydrogenolysis/aromatization. Science,2020,370(6515):437-441.

[21] Weckhuysen B M. Creating value from plastic waste. Science,2020,370(6515):400-401.

[22] UK Parliament. Legislating for net zero. https://researchbriefings. parliament. uk/ResearchBrief-ing/Summary/CBP-8590♯fullreport[2019-12-16].

[23] European Commission. Communication on the european green deal. https://ec. europa. eu/info/files/communication-european-green-deal_en[2019-12-11].

[24] European Commission. European climate law. https://ec. europa. eu/clima/policies/eu-climate-action/law_en[2020-03-04].

[25] Prime Minister of Japan and His Cabinet. Policy speech by the prime minister to the 203rd session of the diet. https://japan. kantei. go. jp/99_suga/statement/202010/_00006. html[2020-10-28].

[26] 日本経済産業省. 2050 年カーボンニュートラルに伴うグリーン成長戦略を策定しました. https://www. meti. go. jp/press/2020/12/20201225012/20201225012. html[2020-12-25].

[27] House of Commons of Canada. Canadian net-zero emissions accountability act. https://www. parl. ca/LegisInfo/BillDetails. aspx? Mode=1&billId=10959361&Language=E[2020-11-19].

[28] The Government of the Republic of Korea. 2050 carbon neutrality strategy. https://unfccc. int/sites/default/files/resource/LTS1_RKorea. pdf[2020-12-31].

[29] Ergebnis Koalitionsausschuss. Corona-folgen bekämpfen,wohlstand sichern,zukunftsfähigkeit stärken. https://www. bundesfinanzministerium. de/Content/DE/Standardartikel/Themen/Schlaglichter/Konjunkturpaket/2020-06-03-eckpunktepapier. pdf? __blob=publicationFile&v=8[2020-06-03].

[30] Ministry of Economy and Finance. Government releases an english booklet on the Korean new deal. https://english. moef. go. kr/pc/selectTbPressCenterDtl. do? boardCd ＝ N0001&seq ＝ 4948♯fn_download[2020-07-14].

[31] Premier Minister. Plan de relance. https://www. gouvernement. fr/france-relance[2020-09-03].

[32] Canada Infrastructure Bank. Canada infrastructure bank's growth plan. https://cib-bic. ca/en/can-ada-infrastructure-banks-growth-plan-backgrounder/[2020-09-30].

[33] Department for Business,Energy & Industrial Strategy. The ten point plan for a green industrial revolution. https://www. gov. uk/government/publications/the-ten-point-plan-for-a-green-indus-trial-revolution[2020-11-18].

[34] 内阁府. Comprehensive economic measures to secure people's lives and livelihoods toward relief

and hope. https://www5. cao. go. jp/keizai1/keizaitaisaku/keizaitaisaku. html[2020-12-08].

[35] ESCAP. Asia and the pacific SDG progress report 2020. https://www. unescap. org/publications/asia-and-pacific-sdg-progress-report-2020. [2020-03-25].

[36] UNDESA. Sustainable development goals report 2020. https://www. un. org/development/desa/publications/publication/sustainable-development-goals-report-2020[2020-07-07].

[37] UNDP. Human development report 2020. https://www. hdr. undp. org/en/2020-report[2020-12-15].

[38] UK Research and Innovation. Impacts of marine plastic pollution in South-east Asia researched. https://www. ukri. org/news/impacts-of-marine-plastic-pollution-in-south-east-asia-researched/[2020-10-22].

[39] Environment and Climate Change Canada，Health Canada. Science assessment of plastic pollution. https://www. canada. ca/en/environment-climate-change/services/evaluating-existing-substances/science-assessment-plastic-pollution. html♯toc50[2020-10-07].

[40] Environment and Climate Change Canada，Health Canada. The government of canada invests in research on plastic pollution in our environment. https://www. canada. ca/en/environment-climate-change/news/2020/11/the-government-of-canada-invests-in-research-on-plastic-pollution-in-our-environment. html[2020-11-13].

[41] European Commission. Circular economy action plan. https://eur-lex. europa. eu/legal-content/EN/TXT/? qid＝1583933814386&uri＝COM：2020：98：FIN[2020-03-11].

[42] NOAA，EPA. Federal partners announce marine litter strategy. https://oceanservice. noaa. gov/news/oct20/marine-litter-strategy. html[2020-10-28].

[43] European Commission. Asia and the pacific SDG progress report 2020. caring for soil is caring for life. https://op. europa. eu/en/publication-detail/-/publication/4ebd2586-fc85-11ea-b44f-01aa75ed71a1[2019-12-05].

[44] European Commission. EU biodiversity strategy for 2030. https://eur-lex. europa. eu/legal-content/EN/TXT/? qid＝1590574123338&uri＝CELEX：52020DC0380[2020-05-20].

[45] European Parliament. Resolution on soil protection. https://oeil. secure. europarl. europa. eu/oeil/popups/summary. do? id＝1660535&t＝d&l＝en[2021-04-28].

[46] U. S. Department of the Interior. Invasive Species strategic plan. https://www. doi. gov/ppa/doi-invasive-species-strategic-plan♯：～：text＝DOI%20Invasive%20Species%20Strategic%20Plan%20Pursuant%20to%20the，The%20Act%20became%20law%20on%20March%2012C%202019[2020-08-13].

[47] Australian Government. Joint media release：initial commitment of $50 million for emergency wildlife and habitat recovery. https://minister. awe. gov. au/ley/media-releases/additional-funding-for-research-on-bushfire-impacts-on-wildlife[2020-01-13].

[48] Department of Agriculture，Water and the Environment. Wildlife and threatened species bushfire recovery research and resources. http://www. environment. gov. au/biodiversity/bushfire-recovery/

research-and-resources[2020-02-13].

[49] Department of Agriculture, Water and the Environment. The threatened species scientific committee 10-point bushfire response plan. http://www. environment. gov. au/biodiversity/threatened/publications/threatened-species-scientific-committee-bushfire-response-plan[2020-03-02].

[50] WRI. Accelerating the net-zero transition: strategic action for china's 14th five-year plan. https://www. wri. org/publication/accelerating-net-zero-transition-china? downloaded=true[2020-12-02].

[51] Oceans Asia. COVID-19 facemasks & marine plastic pollution. https://oceansasia. org/covid-19-facemasks/[2020-12-07].

Ecology and Environmental Science

Liao Qin, Qu Jiansheng, Zeng Jingjing, Pei Huijuan, Dong Liping, Liu Yanfei, Liu Lina

2020 is a year of milestone for the development of the global ecological environment. Carbon neutrality and green recovery from the COVID-19 pandemic have become new trends around the world. At the same time, biodiversity conservation, plastic pollution control, and sustainable development goals, etc. are still hotspots in this field. In 2020, the scientific community has made the following important discoveries and breakthroughs in the field of ecological environment: ①the outbreak of COVID-19 caused a temporary reduction in major air pollutants and carbon dioxide emissions; ②the range of climate sensitivity has been greatly reduced; ③the recovery of stratospheric ozone has paused the trends of Southern Hemisphere circulation; ④ biodiversity loss was accelerating throughout the world; ⑤ satellite images could accurately locate the location and size of large-scale trees; ⑥new materials and technologies have emerged for the treatment of plastic waste. In terms of strategic deployment, countries put forward action plans focusing on carbon neutral goals, green economic recovery, plastic pollution, soil and biodiversity protection. In addition, Australia also pays attention to forest fire related research and response.

4.6　地球科学领域发展观察

郑军卫[1]　刘文浩[1]　张树良[1]　刘　学[1]　王立伟[1]　翟明国[2]

（1. 中国科学院西北生态环境资源研究院文献情报中心；

2. 中国科学院地质与地球物理研究所）

　　地球科学作为历史悠久的自然科学之一，在解决人类繁荣所需的基本资源和健康环境方面发挥着至关重要的作用。2020 年地球科学领域①在地球演化、关键矿产研究与绿色回收技术、行星与月球构造运动、地震监测、大气监测与观测、地球科学监测平台以及数据集成与利用等方面取得了一系列新的重要进展，一些国家和国际组织亦围绕上述领域进行了研究部署。

一、重要研究进展

1. 地球深部研究新发现推动对地球演化的认识

　　2020 年 4 月，澳大利亚蒙纳士大学联合德国波鸿鲁尔大学等机构[1]研究认为"剥离过程"是地球发展出现代板块构造的起源，并基于数学模型来描绘这种剥离汇聚的动态过程，同时模拟地壳不同部分的压力和温度条件。该研究为中-新太古代全球构造背景的演化提供了新的线索，为生命提供了生存的条件，也为人类提供了赖以生存的资源。5 月，美国伊利诺伊大学香槟分校[2]发布成果称，通过重复地震和新的数据处理方法所产生的详尽的地震数据，发现了能揭示地球磁场形成原理的最好证据。该证据表明，地球内核正在旋转，这使得人类对地球内核-外核边界附近的动力学过程控制地球磁场的发生有了更好的理解。6 月，中国北京大学和美国芝加哥大学等机构[3]联合宣布成功模拟了上地幔条件下盐水中 NaCl 的解离过程。这是人类首次实现对高温高压条件下水中盐的自由能的研究，为理解地幔条件下的化学反应和过程奠定了基础。同月，美国马里兰大学[4]的研究人员使用了一种被称为"序列器"（sequencer）的机器学习算法，分析了数千个地震波沿着核幔边界的地震回波特征，获得了整

---

　　① 本文所指的地球科学主要涉及地质学、地球物理学、地球化学、大气科学、行星科学等学科。

个太平洋区域地震波回波的全景视图，并且发现地球核幔边界存在比之前已知的更广泛的横向异质性，即存在异常密集的热岩区域。这些成果对于未来地球深部探测、板块构造运动机制及地球演化历史的研究至关重要。8月，美国卡内基科学研究院和英国牛津大学[5]的联合研究表明，地幔组分的形成主要依赖于其与俯冲洋壳的相互作用。为了探究亏损地幔的成因，研究人员开发出一种新的地球化学盒子模型。基于该模型的模拟和分析发现，亏损地幔的形成主要与俯冲洋壳在地幔深处的聚集有关，而不是先前人们认为的从地幔抽取岩浆形成大陆壳的结果。

2. 关键矿产（稀土）研究及绿色回收技术取得突破

稀土元素是现代工业技术必不可少的金属元素，素有工业"维生素"之称，被美国、日本等列为21世纪的"战略元素"。为探讨自然界中稀土元素的迁移机制，澳大利亚国立大学和英国埃克塞特大学[6]合作使用轻稀土镧（La）元素和重稀土镝（Dy）元素在1200 ℃和1.5 GPa到200 ℃和0.2 GPa的条件下进行实验，结晶出与方解石、白云石、铁白云石交生的氟磷灰石。结果表明，导致稀土元素可溶的关键成分是碱性元素钠（Na）和钾（K），而不是此前普遍认为的阴离子溶液［氯（Cl）、氟（F）和碳酸根离子］。因此，碱金属络合物对稀土的迁移和成矿具有重要意义。该研究成果有助于人们未来对稀土矿的勘探和开发。美国宾夕法尼亚州立大学[7]的研究人员开发了一种新的酸性矿山排水两阶段处理工艺，能够利用比先前更少的化学物质从煤矿排出的酸性废水中回收更高浓度的稀土元素。新的处理工艺可以降低回收成本，在稀土市场更具竞争力。美国普渡大学[8]的研究人员开发出一种新的稀土萃取提纯工艺——双区配体辅助置换色谱法（ligand-assisted displacement chromatography，LAD），可以生产高纯度（＞99％）和高回收率（＞99％）的金属，而且几乎没有不利的环境影响。

3. 行星地质学在揭示地外天体构造运动方面取得新进展

行星地质学是地球科学的一个重要分支学科[9]，在对地球进行类比研究和探寻宜居星球等方面发挥着重要作用。2018年11月26日，美国NASA"洞察号"火星探测器在火星着陆。作为此次火星探测任务重要内容的内部结构地震实验仪（SEIS）首次开展了火星地下和上层地壳的直接地震测量。2020年2月，美国加州理工学院联合瑞士苏黎世联邦理工学院等[10]在《自然-地球科学》发表"洞察号"火星探测器有关火星地震直接测量的研究结果。研究结果显示，火星的地震活动处于中等水平，介于地球和月球之间。这是迄今人类首次完成对地外行星的直接地球物理测量，第一次真正了解了火星的内部结构和地质过程。4月，瑞士伯尔尼大学和美国布朗大学[11]利用美国 NASA 的月球勘测轨道飞行器（Lunar Reconnaissance Orbiter，LRO）所配备的观测设施对月球表面的温度进行测量，在月球表面狭窄的山脊上发现了 500 多块裸露的基岩，揭示月球上可能存在活跃的构造体系。

4. 利用现有地下通信设施开展地震监测成为新兴研究方向

美国加州理工学院联合西班牙阿拉卡拉大学等[12]通过分析他们在比利时近海分布式声学传感（DAS）阵列上的地震和海浪观测成果，认为利用现有海底光纤装置的分布式声学传感阵列可以提供高价值的地震和海洋学数据产品。该技术具有填补全球地震网络中某些巨大盲点的能力。2020 年 3 月，美国密歇根大学联合墨西哥国立自治大学等以一个长度约 4.8 km 的原型光纤阵列进行的实验表明，利用现有的地下光纤网络来监测和研究地震极具潜力[13]。在该研究中，"光纤地震学"被首次用于建立地下特性的标准测量方法，拓展了之前利用斯坦福光纤测试回路的工作，生成了浅地表的高分辨率地图。研究结果表明，光纤可以用来探测地震波，获得速度模型和地面的共振频率。该研究为将用于建设高速互联网和传输高清视频的光纤作为地震传感器使用的可能性提供了新的证据。

5. 全球大气监测与观测水平取得重大进展

2020 年 5 月，世界气象组织（World Meteorological Organization，WMO）网站报道了一项天气预报和大气监测的重大进展[14]，欧洲气象卫星应用组织（Eumetsat）于 5 月 12 日开始，利用"风神"（Aeolus）卫星对风的垂直分布进行观测，并将风场观测数据公开分发至欧洲国家气象服务部门，以便各国的国家气象服务部门进行天气预报。6 月，欧洲中期天气预报中心（European Centre for Medium-Range Weather Forecasts，ECMWF）与美国橡树岭国家实验室（Oak Ridge National Laboratory，ORNL）的科学家合作研究团队完成了世界上首个以 1 km 网格间距运行的季节尺度全球大气模拟[15]。

6. 地学数据集成与利用助力地球科学快速发展

地球科学领域已建立数据库的数据规模快速增加，同时新类型数据库不断建立，以此为依托的高水平研究也越来越多。2020 年 1 月，全球资源信息数据库阿伦达尔中心[16]（Grid-Arendal）启动了世界上第一个可公开访问的全球尾矿储存设施数据库——全球尾矿门户网站（Global Tailings Portal，GTP）。4 月，美国斯坦福大学[17]的研究人员成功研发了新型北美地震图，将对未来能源勘探与开发、地震预警等工作提供有力支撑。7 月，澳大利亚政府[18]宣布开放一个在线数据门户网站，为勘探者提供 250 个数据集的免费访问。这些数据集是澳大利亚政府 2.25 亿澳元"勘探未来"计划的组成部分。8 月，美国能源部[19]宣布为五个项目提供 850 万美元的资助。这些项目旨在使人工智能模型和数据更易于访问和重用，以加速人工智能研发（R&D）。

二、重大战略行动

1. 六大地学学会强调地学知识对应对全球社会挑战至关重要

地球科学家所拥有的专业知识和技能对于保障人类繁荣所需的基本资源和健康环境至关重要。为了确保矿产、能源和生态资源的可靠供应，满足人类和环境对清洁水、清洁空气和肥沃土壤的需求，管理废物和保护环境，强化公共卫生，并为社会建立起应对一系列自然和人为灾害影响的恢复力，2020 年 5 月 4 日，世界上最大的六个地球科学学会——欧洲地球科学联合会（European Geosciences Union，EGU）、美国地球物理学会（American Geophysical Union，AGU）、亚洲大洋洲地球科学学会（Asia Oceania Geosciences Society，AOGS）、美国地质学会（Geological Society of America，GSA）、日本地球行星科学学会（Japan Geoscience Union，JpGU）和伦敦地质学会（Geological Society of London，GSL），联合宣告国际地球科学界承诺利用其专门知识应对社会挑战，并承诺共同努力支持重要的地球科学研究[20]。未来的行动主要涉及：①为今世及后世制定保护和可持续开发重要资源的战略。②利用科学研究成果提高社会对单一、多重和潜在相互关联威胁的恢复力，以帮助人类预防、更好地准备，进而从地方、区域和全球危机中恢复过来。③公正地分析与自然和人为危害有关的风险，包括个人危险和连锁危险，并为直接解决这些问题制定全面而有前瞻性的方案提供支撑。④促进科学方法、研究和相关成果的广泛推广。⑤提高学术标准、培养责任心，支持公平、多样、包容和透明的资助机制，鼓励学术道德建设。⑥提倡科学自由，并制定促进科学诚信的最佳行动方案。⑦使科学多样化并重视代表性不足的研究人员的声音和观点，包括全球南方的研究人员和初级研究人员。⑧认识和发展专业技能与教育技能，使公共和政治领域的不同受众都能接触到地球科学知识。⑨有效地交流科学方法和研究成果，以提高公众对科学的信任，吸引决策者制定和实施能促进支持全球福祉的科学政策。⑩促进地球科学的学科交叉和学科发展，推动地球科学在解决重大社会问题方面发挥积极作用。

2. 美国 NRC 发布地球科学未来十年愿景

2020 年 5 月，美国国家研究理事会（National Research Council，NRC)[21]发布

《美国国家科学基金会地球科学十年（2020—2030 年）愿景：时域地球》报告，为未来美国地球科学研究优先事项、基础设施和设备、伙伴关系提供重要指南。报告发布了未来 10 年（2020—2030 年）内有可能取得重大进展的优先科学问题。这 12 个引人注目的、高度优先的问题反映了地质时间的重要性、地球表面和内部的联系、地质和生命的共同演化及人类活动的影响，主要包括：①地球内部磁场是如何产生的？②板块构造启动的时间、原因和方式是什么？③关键元素在地球上是如何分布和循环的？④地震的本质和驱动力是什么？⑤什么驱动了火山活动？⑥地形变化的原因和后果是什么？⑦关键带是如何影响气候的？⑧地质历史揭示出什么样的气候系统动力学？⑨地球的水循环是如何变化的？⑩生物地球化学循环是如何演化的？⑪地质过程是如何影响生物多样性的？⑫地球科学研究是如何降低地质灾害的风险和损失的？

3. 美国、欧盟多措并举保障关键矿产资源安全

美国推出多项措施确保其关键矿产资源安全、稳定供应。2020 年 1 月，美国和加拿大制定关键矿产合作行动计划，推进关键矿产的安全供应[22]。2020 年 4 月开始，美国能源部[23]陆续投入大笔资金支持关键矿产的基础研究[24]与关键矿产的提取与加工技术研发[25]；6 月，美国稀土公司（USA Rare Earth）和得克萨斯矿产资源公司（Texas Mineral Resources）[26]宣布，二者合作开发的圆顶（Round Top）重稀土项目位于科罗拉多州麦岭市的稀土试点工厂已获得许可并正式作业。其一旦投产，将是自1999 年以来美国首个可对所有稀土元素进行分离的稀土加工厂。10 月，美国总统特朗普签署了一项行政命令，宣布采矿业处于紧急状态，此举旨在扩大稀土矿在美国国内的生产，以减少对中国的依赖[27]。2020 年 9 月，欧盟委员会发布《关键原材料弹性：寻找出一条更安全和可持续供应的路径》[28]和《欧盟战略技术和行业所需关键原材料：前瞻性研究》[29]等报告，公布了包含稀土、铌、钽、铍等 30 种关键原材料的2020 年欧盟关键原材料清单，其中锂被首次添加到该清单中；针对关键原材料安全和可持续供应面临的挑战，报告提出了 10 项具体行动，包括启动一个以行业为主导的"欧洲原材料联盟"、开展废物处理、先进材料和替代品的关键原材料研究和创新、绘制关键原材料的二次供应潜力图等；前瞻性研究强调了锂离子电池、燃料电池、风电等 9 项技术关键原材料的未来需求与挑战，并提出了欧盟在关键领域开展工作的建议。

4. 澳大利亚扩展"勘探未来"计划以勘探全国资源

2020 年 6 月，澳大利亚联邦政府[30]宣布为"勘探未来"计划追加投资 1.25 亿澳元以使该计划再延长 4 年。7 月，澳大利亚地球科学局[31]发布《勘探未来（2016—

2020年）更新计划》，概述了"勘探未来"计划在澳大利亚北部的勘探活动及2020年7月前已完成的项目。8月，澳大利亚地球科学局[32]宣布为"勘探未来"计划制定两条具有开发新资源潜力的陆地走廊，进行更具针对性的资源勘探。该计划未来4年的目标之一是将从澳大利亚南部收集的勘探大数据与该计划2016～2020年间在澳大利亚北部收集的勘探大数据结合起来，绘制更大规模的资源潜力图。

5. 各国继续加紧部署地学监测设施平台建设

2020年4月，英国自然环境研究理事会（Natural Environment Research Council，NERC）[33]宣布未来3年将资助290万英镑，用于启动大气测量与观测设施中心（AMOF）。该设施中心将为英国大气科学界提供联合服务，利用专家资源规划并开展世界一流的大气观测，进行先进的数据质量评估，并提供开放数据存储。7月，美国能源部[34]宣布投入1900万美元用于资助大气科学领域的31个新项目，旨在提高地球系统模式预测天气和气候的能力。9月，美国国家科学基金会[35]宣布将提供近500万美元资金资助为期5年的"EarthWorks"项目，以支持开发高分辨率地球系统模型，供全球科学家开放使用。

三、启示与建议

1. 持续开展地球深部科学研究力争取得新突破

近年来，依托不断进步的研究方法和测试技术，科学家能够有机会去更深入地开展地球科学尤其是深部地球科学的研究工作。不断涌现的新发现、新结论、新推测也正在推动新一轮地球形成和演化机制的"破与立"。例如，地球对流问题的高级求解器、序列器等全新工具、新模型助力科研人员开展构造板块运动和形变随时间和空间的变化、岩浆流动和地球内部的水循环、深部地球的结构和地表演化等领域的模拟与探索研究。基础研究对于科学技术发展的重大作用更加凸显。我国"十四五"规划明确指出，未来我国应在深地前沿领域开展透视地球等基础科学研究，并且加大对地球深部探测装备的研究。因此，建议我国应在现有成果的基础上，持续开展地球科学研究，把握前沿、提前布局、敢于引领，力争在地球各圈层相互机制和演化历史等方面取得一批具有国际影响力的原创性新认识、新成果。

2. 关键矿产"卡脖子"风险值得关注

近年来，国际贸易摩擦不断加剧，各国对关键矿产资源领域供需安全的担忧空前高涨。特别是在新冠肺炎疫情全球大流行后，新的世界政治和经济格局的变化对全球资源配置产生的影响值得人们关注。澳大利亚、日本、美国、欧盟多国/组织基于国家安全和经济发展考量，纷纷制定和发布新的关键矿产战略和清单，启动关键原材料的重大研究及勘探计划，旨在新一轮矿产资源争夺中抢占战略制高点。因此，建议我国准确掌握国际关键矿产领域的政策战略部署和科学技术研究的整体发展态势，厘清未来中国关键矿产资源领域的研究需求，准确评估中国的优势资源和劣势资源的家底，多措并举保障国家关键矿产资源安全。

3. 新兴技术方法加速地球科学研究关键问题突破

当前，科学研究范式已经进入基于数据驱动的全新阶段，新兴技术集群正在驱动地球科学研究进入迅速发展的快车道。面向地球科学研究关键问题的大尺度观测、模拟、分析及计算等技术方案正在不断推动地球科学问题实现突破。美国国家科学基金会投资超过1亿美元启动国家人工智能研究所联合建设计划，将支持海洋科学、大气科学等多个重要领域人工智能前沿研究。美国能源部、英国自然环境研究理事会等多个机构启动多项资助计划，开展超级计算、人工智能数据管理、无人机观测等领域的新型技术的研发布局。抢占技术先机成为保持地球科学研究优势的关键。未来，我国也应紧抓国际技术前沿，在地球科学核心技术领域早布局、抓落实，争取突破"卡脖子"技术，努力确保在地球科学技术研发方面实现世界领先。

致谢：中国地质大学（武汉）马昌前教授、中国科学院地质与地球物理研究所徐志方研究员、西北大学陈亮副教授等审阅了本文并提出了宝贵的修改意见，中国科学院西北生态环境资源研究院赵纪东、刘燕飞、李小燕、王晓晨等对本文也做出了贡献，参与了本文的部分资料收集与翻译，在此一并表示感谢。

参考文献

[1] Chowdhury P, Chakraborty S, Gerya T V, et al. Peel-back controlled lithospheric convergence explains the secular transitions in Archean metamorphism and magmatism. Earth and Planetary Science Letters,2020,538:116224.

[2] Yang Y, Song X. Origin of temporal changes of inner-core seismic waves. Earth and Planetary Science Letters,2020,541:116267.

[3] Zhang C Z, Giberti F, Sevgen E, et al. Dissociation of salts in water under pressure. Nature Communications,2020,11(1):3037.

[4] Kim D, Lekić V, Ménard B, et al. Sequencing seismograms: a panoptic view of scattering in the core-mantle boundary region. Science,2020,368(6496):1223-1228.

[5] Tucker J M, van Keken P E, Jones R E, et al. A role for subducted oceanic crust in generating the depleted mid-ocean ridge basalt mantle. Geochemistry, Geophysics, Geosystems, 2020, 21 (8): e2020GC009148.

[6] Anenburg M, Mavrogenes J A, Frigo C, et al. Rare earth element mobility in and around carbonatites controlled by sodium, potassium, and silica. Science Advances, 2020, 6(41): eabb6570.

[7] Hassas B V, Rezaee M, Pisupati S V. Precipitation of rare earth elements from acid mine drainage by CO_2 mineralization process. Chemical Engineering Journal, 2020, 399: 125716.

[8] Ding Y, Harvey D, Wang N H L. Two-zone ligand-assisted displacement chromatography for producing high-purity praseodymium, neodymium, and dysprosium with high yield and high productivity from crude mixtures derived from waste magnets. Green Chemistry, 2020, 22(12): 3769-3783.

[9] Rossi A P, van Gasselt S. Planetary Geology. Berlin: Springer International Publishing, 2018: 1-414.

[10] Banerdt W B, Smrekar S E, Banfield D, et al. Initial results from the InSight mission on Mars. Nature Geoscience, 2020, 13(3): 183-189.

[11] Valantinas A, Schultz P H. The origin of neotectonics on the lunar nearside. Geology, 2020, 48 (7): 649-653.

[12] Williams E F, Fernandez-Ruiz M R, Magalhaes R, et al. Distributed sensing of microseisms and teleseisms with submarine dark fibers. Nature Communications, 2019, 10: 5778.

[13] Spica Z J, Perton M, Martin E R, et al. Urban seismic site characterization by fiber-optic seismology. Journal of Geophysical Research: Solid Earth, 2020, 125(3): e2019JB018656.

[14] World Meteorological Organization. Aeolus provides data on Earth's winds. https://public. wmo. int/en/media/news/aeolus-provides-data-earth%E2%80%99s-winds[2020-05-14].

[15] Wedi N. A Baseline for global weather and climate simulations at 1 km resolution. https://www. ecmwf. int/en/about/media-centre/science-blog/2020/baseline-global-weather-and-climate-simulations-1-km[2020-06-22].

[16] Arnoldi M. Global tailings database launched. https://www. miningweekly. com/article/norwegian-enviro-company-launches-global-tailings-database-2020-01-24[2020-01-24].

[17] Tucker D T. Seismic map of North America reveals geologic clues, earthquake hazards. https:// earth. stanford. edu/news/seismic-map-north-america-reveals-geologic-clues-earthquake-hazards#gs. 5gc6m0[2020-04-23].

[18] Pitt K. Delivering data certainty for mining explorers. https://www. minister. industry. gov. au/ministers/pitt/media-releases/delivering-data-certainty-mining-explorers[2020-07-07].

[19] DOE. Department of Energy announces $8. 5 million for fair data to advance artificial intelligence for science. https://www. energy. gov/articles/department-energy-announces-85-million-fair-data-advance-artificial-intelligence-science[2020-08-10].

[20] European Geosciences Union. International declaration: geoscience expertise is crucial for meeting societal challenges. https://www. eurekalert. org/pub_releases/2020-05/egu-idg050320. php[2020-05-04].

［21］National Research Council. A Vision for NSF Earth Sciences 2020-2030：Earth in Time. Washington，DC：The National Academies Press，2020.

［22］Webb M. Canada and US finalise critical mineral collaboration plan. https：//www. miningweekly. com/article/canada-and-us-finalise-critical-mineral-collaboration-plan-2020-01-10［2020-01-10］.

［23］DOE. Department of Energy to provide ＄18 million for research on critical materials. https：//www. energy. gov/articles/department-energy-provide-18-million-research-critical-materials［2020-04-14］.

［24］DOE. DOE awards ＄20 million for research on rare earth elements. https：//www. energy. gov/articles/doe-awards-20-million-research-rare-earth-elements［2020-08-25］.

［25］DOE. Department of Energy announces ＄30 million for innovation in critical materials processing technologies. https：//www. energy. gov/articles/department-energy-announces-30-million-innovation-critical-materials-processing［2020-05-14］.

［26］Mining. com Editor. Rare earths processing facility opens in Colorado. https：//www. mining. com/rare-earths-processing-facility-opens-in-colorado/［2020-06-11］.

［27］Bloomberg. Trump moves to expand rare Earths mining，cites China threat. https：//www. bloomberg. com/news/articles/2020-09-30/trump-moves-to-expand-rare-earths-mining-citing-china-threat［2020-10-01］.

［28］European Commission. Critical raw materials resilience：charting a path towards greater security and sustainability. https：//ec. europa. eu/docsroom/documents/42849/attachments/2/translations/en/renditions/native［2020-09-03］.

［29］European Commission. Critical raw materials for strategic technologies and sectors in the EU—a foresight study. https：//ec. europa. eu/docsroom/documen ts/42881/attachments/1/translations/en/renditions/native［2020-09-03］.

［30］Pitt K. Government moves to strengthen resources exploration. https：//www. minister. industry. gov. au/ministers/pitt/media-releases/government-moves-strengthen-resources-exploration［2020-06-23］.

［31］Geoscience Australia. Exploring for the future program update 2016-2020. https：//www. ga. gov. au/eftf/program-update［2020-07-01］.

［32］Geoscience Australia. Exploring for the future expands across Australia. https：//www. ga. gov. au/news-events/news/latest-news/exploring-for-the-future-expands-across-australia［2020-08-11］.

［33］UK Research and Innovation. New facility for atmospheric measurements launches april 2020. https：//nerc. ukri. org/press/releases/2020/new-facility-for-atmospheric-measurements-launches-april-2020/［2020-04-07］.

［34］DOE. Department of Energy announces ＄19 million for new atmospheric research. https：//www. energy. gov/articles/department-energy-announces-19-million［2020-07-15］.

［35］NCAR，UCAR. CSU，NCAR to develop high-res global model for community use. https：//news. ucar. edu/132760/csu-ncar-develop-high-res-global-model-community-use［2020-09-29］.

Earth Science

Zheng Junwei ,Liu Wenhao ,Zhang Shuliang ,
Liu Xue ,Wang Liwei ,Zhai Mingguo

As one of the natural sciences with a long history，geoscience plays a vital role in solving the basic resources and healthy environment needed for human prosperity. In 2020，a series of new and important progress has been made in the field of geoscience in terms of earth evolution，research and green recycling technology of key minerals，planetary and lunar tectonic movement，seismic monitoring，atmospheric monitoring and observation，geoscience monitoring platform and data integration and utilization. Some countries and international organizations have also carried out research and deployment in the above fields.

4.7　海洋科学领域发展观察

高　峰[1]　王金平[1]　冯志纲[2]　王　凡[2]　吴秀平[1]

（1. 中国科学院西北生态环境资源研究院；2. 中国科学院海洋研究所）

2020 年，全球海洋科学领域成果显著，在物理海洋、海洋生物、海洋地质、海洋环境及海洋技术等领域的研究持续推进；在北极海冰减少、深海探索与绘制、海洋生态系统研究和极地海洋研究等方面取得诸多突破。国际组织和主要海洋国家围绕相关海洋研究方向进行了部署。

一、海洋科学领域重要研究进展

1. 物理海洋研究成果丰硕

在全球变暖的背景下，北极海冰流失速度加快，海冰面积创新低。美国阿拉斯加费尔班克斯大学的研究人员指出，白令海海冰面积达到数千年来的最小值[1]。阿尔弗雷德·魏格纳海洋与极地研究所（Alfred Wegener Institute for Polar and Marine Research，AWI）等机构基于卫星观测数据和建模数据评估，指出格陵兰岛冰盖在 2019 年的总损失量创下新高，达到 5320 亿 t，较上一个创纪录年份（2012 年）的损失量高出 680 亿 t，相当于全球海平面平均上升 1.5 mm[2]。美国纽约州立大学布法罗分校等的研究指出，21 世纪格陵兰岛冰盖的流失速度可能比过去 1.2 万年中的任何时候都快[3]。根据未来的气候情景预测，格陵兰岛冰盖可能会在 1000 年内完全消失。

美国夏威夷大学马诺阿分校的研究人员通过利用全球气候模型对未来海平面进行预测后指出，从季节到年际时间尺度来看，未来海平面变化进一步加剧已成为全球趋势[4]。美国 NASA 和英国国家海洋学中心（National Oceanography Centre，NOC）联合研究了自 1900 年至今海平面上升速度的所有变化，证实了 20 世纪 70 年代以来海平面上升速度的加剧是海洋热膨胀和格陵兰岛海冰流失增加的共同作用所致[5]。

美国夏威夷大学蒙纳分校的研究人员指出，自 1982 年以来，热带气旋在海洋盆地中的移动速度一直在加快。与此同时，北大西洋地区飓风出现频率有所上升，太平洋和大西洋的热带气旋活动均向极地方向转移[6]。韩国蔚山国家科学技术研究院

(Ulsan National Institute of Science and Technology，UNIST）联合美国斯克利普斯海洋研究所（Scripps Institution of Oceanography，SIO）等机构的研究发现，极地气候变化的物理路径可以影响热带信风的强度[7]。美国国家海洋与大气管理局等机构研究人员首次揭示了南大西洋关键深层洋流的日变化率，指出了这些变化与全球气候的关联性[8]。

2. 海洋生物学研究获得新认识

日本海洋与地球科学技术局（Japan Agency for Marine-Earth Science and Technology，JAMSTEC）等机构首次绘制了海洋沉积物的生物多样性分布图并发现海底深处的微生物多样性与地球表面一样丰富[9]。加拿大不列颠哥伦比亚大学等的研究发现，在全世界范围内，一些常见的鱼类和无脊椎动物的数量正在急剧下降，包括罗非鱼、普通章鱼和粉红凤凰螺等[10]。加拿大达尔豪斯大学等的研究发现，世界上许多珊瑚礁中的鲨鱼实际上已经灭绝，其中近20%的珊瑚礁中没有观察到鲨鱼[11]。美国罗格斯大学等的研究指出，受海洋变暖影响，底栖无脊椎动物的生存受到威胁[12]。美国俄勒冈州立大学牵头的研究指出，海洋环境的变化导致斑海雀的食物选择受到限制，其所需的古森林长期丧失，因此面临海洋变暖和古森林流失的双重压力[13]。英国南极调查局（British Antarctic Survey，BAS）的研究显示，南极的帝企鹅聚居地比之前认为的多出近20%。该发现为监测环境变化对这种标志性鸟类数量的影响提供了一个重要基准[14]。

美国蒙特利海湾水族研究所（Monterey Bay Aquarium Research Institute，MBARI）的研究发现，至少两种栉水母自身能够产生腔肠素，从而为深海中的其他物种提供补给。这一发现不仅有助于研究人员认识动物发光的生物化学过程，而且为生物医学研究提供了一个新工具[15]。

3. 海洋地质研究取得新发现

德国地球科学研究中心（German Research Centre for Geosciences）等的研究指出，格陵兰岛和南极的新数据表明，在覆盖地球陆地表面约10%的冰盖下方，流动的微量元素数量比此前预计的更多[16]。由美国夏威夷大学马诺阿分校牵头，来自世界各地的19位海洋科学家联合研究发现，深海采矿不仅对周围海域构成重大威胁，而且还对海底数百至数千英尺①的中层水域生态系统构成重大威胁。通过提出更全面的评估方法，该研究有助于社会和管理者决定是否及如何进行深海采矿[17]。

---

① 1英尺＝0.3048 m。

4. 海洋环境问题研究的新视角

全球每年有超过 800 万 t 的塑料进入海洋。海洋塑料污染已成为当今世界面临的最紧迫的环境问题之一，其中 80 % 的海洋塑料来自陆地，如海滩废弃物和污水排放[18]。英国国家海洋学中心（National Oceanography Centre，NOC）牵头的一项研究指出，大西洋上层水域微塑料的数量巨大，为 1200 万～2100 万 t[19]。英国的利物浦约翰摩尔斯大学、贝尔法斯特女王大学和英国南极调查局（British Antarctic Survey，BAS）的联合研究指出，南极海底的微塑料污染程度与北大西洋及地中海地区的污染程度相同[20]。

日本东京工业大学等提出了首个可量化的塑料摄入对海洋动物所产生影响的机制模型。这一量化手段表明，塑料摄入可能会导致海龟种群数量减少[21]。

海洋保护协会（Ocean Conservancy）等的研究指出，在计入美国的废弃塑料出口量及最新的非法倾倒和随意丢弃的塑料废弃物数量后，就造成海岸塑料污染的全球排名而言，美国位列第三[22]。这项新的研究推翻了之前关于美国对塑料废物进行了充分"管理"的观点。

5. 海洋观测新技术研发与应用

美国斯克里普斯海洋研究所（Scripps Institution of Oceanography，SIO）利用水下虚拟手段开发了虚拟电子珊瑚礁技术。该技术已经以 1 mm 的分辨率绘制了 30 公顷的珊瑚礁，这对于太平洋的凤凰群岛保护区具有重要意义[23]。美国蒙特利海湾水族研究所（Monterey Bay Aquarium Research Institute，MBARI）的研究小组开发了深海粒子成像测速系统（DeepPIV）。通过将其安装在深潜机器人上，该系统可利用激光对自然环境中的胶质动物进行三维扫描[24]。

日本海洋与地球科学技术局牵头开发了一种利用"波浪滑翔器"（Wave Glider），并结合海底地壳运动观测结果开展全球卫星导航系统声波定位的系统。该系统将极大地提高大地震概率预测的可靠性，从而有助于减轻其对海洋设施和沿海社区的影响[25]。

二、海洋科学领域重要研究部署

1. 国际组织

以联合国为代表的国际组织持续关注海洋可持续发展、海洋韧性、蓝色经济等方向。2020 年 6 月，可持续海洋经济高级别小组发布《海洋可再生能源和深海矿产在可持续的未来中将扮演什么角色？》蓝皮书[26]，探讨了海洋可再生能源、深海矿产同海洋可持续发展之间的关系，通过考虑海洋可再生能源需求和潜力，揭示了深海矿产涉及的风险和挑战，同时还提出确保海洋健康和韧性并以可持续发展方式利用海洋可再生能源的路径。8 月，联合国教科文组织（UNESCO）政府间海洋学委员会发布了《联合国海洋科学促进可持续发展十年（2021—2030 年）实施计划摘要》[27]，介绍了"海洋十年"的目标、愿景和挑战等。9 月，可持续海洋经济高级别小组发布《应对新冠肺炎危机的可持续与公平的蓝色复苏》[28]报告，提出了利用海洋进行全球经济复苏的路线图。10 月，全球海洋观测系统观测协调小组发布的《2020 年海洋观测系统报告》[29]介绍了全球海洋观测系统（global ocean observing system，GOOS）的最新状况，包括新冠肺炎疫情暴对海洋观测的影响、观测活动取得的进展及与重要观测服务之间的联系。10 月，国际大洋发现计划（IODP）发布的《2050 年科学框架：通过科学的大洋钻探方式探索地球》提出了 7 大战略目标、5 大旗舰计划和 4 项赋能元素[30]。12 月，联合国教科文组织在线发布的第二版《全球海洋科学报告》在集成第一版的成功经验基础上，另行关注"海洋科学对可持续发展的贡献、科学应用在专利中的体现、对海洋科学人力资源性别平等问题的深入分析、海洋科学能力建设"四个主题[31]。12 月，可持续海洋经济高级别小组的 14 个成员国共同发布的《实现可持续海洋经济的转型：保护、生产和繁荣的愿景》提出各国将通过一系列优先行动在海洋财富、海洋健康、海洋公平、海洋知识与海洋投资这五个领域实现可持续蓝色经济[32]。

2. 美国

美国持续加大对海洋科学领域的投入力度。2020 年 6 月，美国国家海洋和大气管理局、阿拉斯加州和阿拉斯加制图执行委员会（AMEC）共同制定的《阿拉斯加海岸线制图战略》[33]提出在未来 5 年内将获得优先领域海岸线制图数据集。7 月，美国国家海洋和大气管理局发布的《海洋、沿海及大湖区酸化研究计划（2020—2029 年）》从美国国家层面和区域层面对美国海洋酸化的未来研究方向进行了规划[34]。9 月，美国国会研究处（Congressional Research Service，CRS）发布的《北极变化：国会面临的背景与问题》重点介绍了美国参与北极研究的背景、与北极相关的主要国家政策和机构及国会面临的问题[35]。

3. 欧洲

欧洲持续加强其海洋设施能力并关注海洋与人类健康之间的关系。2020 年 3 月，欧洲海洋局（European Marine Board，EMB）牵头发布的《海洋与人类健康战略研究议程（2020—2030 年）》确定了在欧洲建立海洋和人类健康研究能力的目标主题及优先研究领域[36]。4 月，欧洲海洋局发布的《未来科学——海洋科学中的大数据》[37] 概述了大数据支持海洋科学的最新进展、挑战和机遇，制定了增加数字化和在海洋科学中应用大数据的发展目标。5 月，国际"气候变化和欧洲水产资源"项目发布的《气候变化和欧洲水产资源》重点关注最具商业价值的鱼类和贝类，进而发现欧洲渔业和水产养殖业面临的气候变化风险、机遇和不确定性[38]。5 月，欧洲海洋能技术与创新平台（European Technology & Innovation Platform for Ocean Energy，ETIP OCEAN）发布的《海洋能战略研究与创新议程》[39] 提出 2021～2025 年重点研究与创新领域以及未来展望。6 月，英国国家海洋学中心发布的《国家海洋设施（NMF）技术路线图（2020—2021 年）》概述了英国当前的海洋设施能力，并对海洋科学的未来及新技术进行了展望[40]。7 月，英国海洋科学协调委员会（Marine Science Co-ordination Committee，MSCC）发布了《面向海洋可持续性和生产力的海洋科学路线路概要》[41]，旨在助力实现《英国海洋愿景》提出的洁净、健康、安全、极具生产力和生物多样性的海洋这一目标。

4. 其他国家

2020 年 4 月 26 日，澳大利亚南极科学委员会（Australian Antarctic Science Council）发布的为期 10 年的《澳大利亚南极科学战略规划》[42] 提出寻找并研究百万年之久的南极冰芯，这将有助于气候研究取得突破性进展，同时促进对南大洋关键物种的研究。11 月，南极研究科学委员会（Scientific Committee on Antarctic Research，SCAR）制定的"南极与南大洋保护综合科学"（Ant-ICON）[43] 四年计划旨在更深入地认识并解决气候变化、生物入侵、污染、人类活动碳足迹增加等环境威胁的加剧对南极面貌的改变及其影响，从而改善对南极和南大洋生态系统、物种和环境的保护。

三、启示与建议

1. 持续推进海洋生态系统保护，充分释放海洋蓝色碳汇的潜力

在 2020 年 9 月的第七十五届联合国大会上，我国明确要采取更加有力的政策和措施，二氧化碳排放量力争于 2030 年前达到峰值，努力争取 2060 年前实现碳中和。要实现这一目标，需要充分利用海洋的碳吸收潜力。因此，保护修复海洋系统有利于发挥海洋的固碳能力，进而作为自然解决方案而更好地应对气候变化。未来，应在采取有力的海洋保护措施的同时充分利用海洋的生态修复能力和自然资本。

2. 继续深化极地研究，关注海洋系统的连通性

在全球气候变暖日益加剧的背景下，北极海冰面积急剧减少，其作为气候变化关键指标和"放大器"的作用进一步凸显。海洋生态系统具有高度的连通性。研究指出，席卷非洲的巨型沙尘暴"哥斯拉"、北半球中纬地区极端强寒潮事件频发、气溶胶等污染物向青藏高原输送的增多均与北极变暖和海冰减少有关。鉴于极地的战略重要性与应对气候变化的需要，各国纷纷加快了北极地区的研究部署，未来应进一步深化对海洋-海冰-大气相互作用的认识。

3. 借助新兴技术，提升海洋勘探的智能化、数字化水平

随着云计算、大数据和人工智能、数据处理与存储能力的提高，新兴技术越来越多地用于海洋勘测活动。通过创新观测方法，加快天气建模发展，从而创建更精确、更可靠的预测模型。这些新兴技术将为海洋保护、蓝色经济等领域提供支撑。以"奋斗者"号全海深载人潜水器为代表，我国在大深度载人深潜领域取得了瞩目成就，未来应推进海洋勘探的智能化和数字化，利用云计算平台来识别、理解和预测海洋系统的变化。

致谢：中国科学院海洋研究所李超伦研究员、中国海洋大学高会旺教授、自然资源部第一海洋研究所王宗灵研究员、中国海洋大学于华明教授对本报告初稿进行了审阅并提出了宝贵修改意见，在此表示感谢！

参考文献

[1] Jones M C, Berkelhammer M, Keller K J, et al. High sensitivity of bering sea winter sea ice to winter insolation and carbon dioxide over the last 5500 years. Science Advances, 2020, 6(36): EAAZ9588.

[2] Sasgen I, Wouters B, Gardner A S, et al. Return to rapid ice loss in Greenland and record loss in 2019 detected by the GRACE-FO satellites. Communications Earth & Environment, 2020, 1(1): 396-400.

[3] Briner J P, Cuzzone J K, Badgeley J A, et al. Rate of mass loss from the Greenland Ice Sheet will exceed Holocene values this century. Nature, 2020, 586: 70-74.

[4] Widlansky M J, Long X, Schloesser F. Increase in sea level variability with ocean warming associated with the nonlinear thermal expansion of seawater. Communications Earth & Environment, 2020, 1(1): 2022-2025.

[5] Frederikse T, Landerer F, Caron L, et al. The causes of sea-level rise since 1900. Nature, 2020, 584(7821): 393-397.

[6] Kim S H, Moon I-J, Chu P S. An increase in global trends of tropical cyclone translation speed since 1982 and its physical causes. Environmental Research Letters, 2020, 15(9): 84-94.

［7］ Kang S M，Xie S P，Shin Y，et al. Walker circulation response to extratropical radiative forcing. Science Advances，2020，6（47）：2375-2548.

［8］ Kersalé M，Meinen C S，Perez R C，et al. Highly variable upper and abyssal overturning cells in the
South Atlantic. Science Advances，2020，6（32）：eaba7573.

［9］ Hoshino T，Doi H，Uramoto G-I，et al. Global diversity of microbial communities in marine sedi-
ment. Proceedings of the National Academy of Sciences of the United States of America，2020，117
（44）：27587-27597.

［10］ Palomares M L D，Froese R，Derrick B，et al. Fishery biomass trends of exploited fish populations
in marine ecoregions，climatic zones and ocean basins. Estuarine，Coastal and Shelf Science，2020，
243：157-166.

［11］ MacNeil M A，Chapman D D，Heupel M，et al. Global status and conservation potential of reef
sharks. Nature，2020，583：801-806.

［12］ Fuchs H L，Chant R J，Hunter E J，et al. Wrong-way migrations of benthic species driven by ocean
warming and larval transport. Nature Climate Change，2020，10：1052-1056.

［13］ Betts M G，Northrup J M，Rivers J W，et al. Squeezed by a habitat split：Warm ocean conditions
and old-forest loss interact to reduce long-term occupancy of a threatened seabird. Conservation
Letters，2020，13：e12745.

［14］ British Antarctic Survey. Scientists discover new penguin colonies from space. https：//www.
bas. ac. uk/media-post/scientists-discover-new-penguin- colonies-from-space/［2020-08-05］.

［15］ Bessho-Uehara M，Huang W T，Patry W L，et. al. Evidence for *de novo* Biosynthesis of the Luminous
Substrate Coelenterazine in Ctenophores. iScience，https：//doi. org/10. 1016/j. isci. 2020. 101859.

［16］ Hawkings J R，Skidmore M L，Wadham J L，et. al. Enhanced trace element mobilization by Earth's
ice sheets. Proceedings of the National Academy of Sciences，2020，117（50）：31648-31659.

［17］ Drazen J C，Smith C R，Gjerde K M，et. al. Opinion：midwater ecosystems must be considered when
evaluating environmental risks of deep-sea mining. Proceedings of the National Academy of Sciences，
2020，117（30）：17455-17460.

［18］ ECCA Family Trust. Surfacing innovative solutions for reducing marine plastic pollution. https：//
avpn. asia/wp-content/uploads/2019/11/Executive-Summary-Surfacing-Innovative-Solutions-
for-Reducing-Marine-Plastic-Pollution. pdf［2020-01-09］.

［19］ National Oceanography Centre. New study estimates there is at least 10 times more plastic in the
Atlantic than previously thought. https：//noc. ac. uk/news/new-study-estimates-there-least-10-
times-more-plastic-atlantic-previously-thought［2020-08-18］.

［20］ Cunningham E M，Ehlers S M，Dick J T A，et al. High abundances of microplastic pollution in
deep-sea sediments：evidence from antarctica and the southern ocean. Environmental Science &
Technology，2020，54（21）：13661-13671.

［21］ Marn N，Jusup M，Kooijman S A L M，et al. Quantifying impacts of plastic debris on marine wildlife identifies ecological breakpoints. Ecology Letters,2020,10(23):1479-1487.

[22] Law K L,Starr N,Siegler T R,et al. The United States' contribution of plastic waste to land and ocean. Science Advances,2020,44(6):0288.

[23] Science Focus. The technology solving the ocean's greatest mysteries. https://www. sciencefocus. com/planet-earth/the-technology-solving-the-oceans-greatest-mysteries/[2020-08-26].

[24] Katija K,Troni G,Daniels J,et al. Revealing enigmatic mucus structures in the deep sea using DeepPIV. Nature,2020,583:78-82.

[25] Japan Agency for Marine-Earth Science and Technology. Success in multi-point long-term observations of sea floor crustal movements using an unmanned surface vehicle:Major progress in developing a high-temporal-resolution understanding of the current state of Seismogenic zone. http://www. jamstec. go. jp/e/about/press_release/20200930/[2020-09-30].

[26] High Level Panel for a Sustainable Ocean Economy. Whatrole for ocean-based renewable energy and deep-seabed minerals in a sustainable future? https://oceanpanel. org/sites/default/files/ 2020-06/Energy%20and%20Deep-Sea%20Minerals%20Full%20Paper%20Final%20Web. pdf[2020-06-24].

[27] United Nations Decade of Ocean Science for Sustainable Development. Oceandecade:summary of the implementation plan(version 2. 0). https://oceanexpert. org/document/27348[2020-08-10].

[28] High Level Panel for a Sustainable Ocean Economy. Asustainable and equitable blue recovery to the COVID-19 crisis. https://oceanpanel. org/bluerecovery[2020-09-16].

[29] Global Ocean Observing System Observations Coordination Group. Oceanobserving system report card 2020. https://www. ocean-ops. org/reportcard2020/reportcard2020. pdf[2020-10-20].

[30] Integrated Ocean Drilling Program. 2050 science framework:exploring earth by scientific ocean drilling. http://www. iodp. org/iodp-future-doc/1086-2050-science-framework-full-document/file [2020-10-25].

[31] United Nations Educational Scientific and Cultural Organization. Global ocean science report 2020: charting capacity for ocean sustainability. https://unesdoc. unesco. org/ark:/48223/pf0000375147? posInSet=27&queryId=N-776787c5-07ff-41a0-9255-05def1bf21ce[2020-12-14].

[32] High Level Panel for a Sustainable Ocean Economy. Transformations for a Sustainable Ocean Economy A Vision for Protection, Production and Prosperity. https://www. oceanpanel. org/ocean-action/files/transformations-sustainable-ocean-economy-eng. pdf[2020-12-02].

[33] State of Alaska,Alaska Mapping Executive Committee,National Oceanic and Atmospheric Administration, Alaska Ocean Observing System,United States Geological Survey. Mapping the coast of alaska-a 10-year strategy in support of the united states economy,security,and environment. https://iocm. noaa. gov/about/documents/strategic-plans/alaska-mapping-strategy-june2020. pdf[2020-06-11].

[34] NOAA Ocean Acidification Program. Ocean, coastal and great lakes acidification research plan: 2020-2029. https://oceanacidification. noaa. gov/ResearchPlan2020. aspx[2020-07-27].

［35］Congressional Research Service. Changes in the arctic: background and issues for congress. https://fas. org/sgp/crs/misc/R41153. pdf［2020-09-10］.

［36］European Marine Board. Astrategic research agenda for oceans and human health. https://www. marineboard. eu/publications/strategic-research-agenda-oceans-and-human-health［2020-03-10］.

［37］European Marine Board. Big data in marine science. https://www. marineboard. eu/sites/marineboard. eu/files/public/publication/EMB_FSB6_BigData_Web_0. pdf［2020-04-05］.

［38］Climate change and European aquatic resources. CERES project synthesis report. https://ceresproject. eu/wp-content/uploads/2020/05/CERES-Synthesis-Report-18-05-2020_format. pdf［2020-05-13］.

［39］European Technology & Innovation Platform for Ocean Energy. Strategic research and innovation agenda for ocean energy. https://www. oceanenergy-europe. eu/wp-content/uploads/2020/05/ETIP-Ocean-SRIA. pdf［2020-05-28］.

［40］National Oceanography Center, National Marine Facilities, Natural Environment Research Council. National marine facilities technology roadmap 2020/21. https://noc. ac. uk/files/documents/about/ispo/COMMS1155% 20NMF% 20TECHNOLOGY% 20ROADMAP% 20202021% 20V4. pdf［2020-06-29］.

［41］Marine Science Co-ordination Committee. UKmarine science for sustainable and productive seas road map summary. https://assets. publishing. service. gov. uk/government/uploads/system/uploads/attachment_data/file/905452/mscc-road-map-summary. pdf［2020-07-28］.

［42］Australian Antarctic Science Council. Australianantarctic science strategic plan. https://www. antarctica. gov. au/site/assets/files/53908/australian-antarctic-science-strategic-plan. pdf［2020-04-26］.

［43］Australian Antarctic Division. New science to improve antarctic-wide protection. https://www. antarctica. gov. au/news/2020/new-science-to-improve-antarctic-wide-protection/［2020-11-10］.

Marine Science

Gao Feng, Wang Jinping, Feng Zhigang, Wang Fan, Wu Xiuping

In 2020, significant achievements have been made in the field of global oceanography. Research in the fields of physical oceans, marine biology, marine geology, marine environment, and marine technology continues to advance, and many breakthroughs have been made in Arctic sea ice reduction, deep-sea exploration and mapping, marine ecosystem research, and polar research. International organizations and major maritime countries have deployed relevant ocean research directions.

4.8 空间科学领域发展观察

杨 帆 韩 淋 王海名 范唯唯

（中国科学院科技战略咨询研究院）

2020 年，科学家首次发现快速射电暴来源，获得中等质量黑洞存在的最有力证据，首次在轨实现玻色-爱因斯坦凝聚，美国主导的"阿尔忒弥斯"载人月球探索计划持续推进，火星探测再掀新的高潮。我国应该进一步加强研究规划，强化科学目标牵引和原创成果突破，激励空间科学源头创新，重视以我为主的国际空间合作。

一、重要研究进展

1. 首次确定快速射电暴来源

"雨燕伽马射线暴"任务、"费米伽马射线空间望远镜"、"中子星内部构成探测器"、"加拿大氢强度测绘实验"、"瞬变天文射电发射巡天 2"、中国"500 米口径球面射电望远镜"等多个天基和地基天文台幸运地探测到发生在银河系内的快速射电暴事件 FRB 200428 及相关信号。基于以上观测结果，多项独立研究共同指出，磁星 SGR 1935＋2154 是快速射电暴 FRB 200428 的起源[1-3]。这是人类首次确定一个快速射电暴的起源，也是首次在银河系内观测到快速射电暴，观测结果倾向支持快速射电暴起源于致密天体磁层这一理论模型。成果入选《自然》2020 年度十大科学发现及《科学》2020 年度十大科学突破。

2. "慧眼"发现迄今距离黑洞最近的高速喷流

中国科学院高能物理研究所领衔的国际团队利用"慧眼"卫星在高于 200 keV 的能段发现黑洞双星系统的低频准周期振荡，这是迄今发现的能量最高的低频准周期振荡现象。研究表明，此低频振荡应起源于黑洞视界附近的相对论喷流的进动，为解决存在争议的低频准周期振荡物理起源问题提供了重要依据，对研究黑洞附近的广义相对论效应、物质动力学过程和辐射机制等具有重要意义[4]。

3. 多任务助力发现中等质量黑洞存在的最有力证据

基于"XMM 牛顿望远镜"、"钱德拉 X 射线天文台"和"哈勃空间望远镜"的观测数据，研究人员发现 X 射线耀斑 3XMM J215022.4-055108 所在的星团是一个低质量矮星系的核心，在排除了该 X 射线源是中子星的可能性后，最佳解释是 X 射线源来自一个质量是 5 万倍太阳质量的黑洞[5]。这是迄今中等质量黑洞存在的最有力证据，为寻找宇宙中隐藏着的更多中等质量黑洞打开了大门，未来有望回答超大质量黑洞是否由中等质量黑洞发展而成、中等质量黑洞如何形成、密集星团是否是中等质量黑洞最爱的"居所"等更多未解之谜。

4. "光谱-RG"完成首次全天空 X 射线巡天

"光谱-RG"成功完成对全天空的首次巡天，绘制的宇宙地图中包含的天体超过 100 万个，是 X 射线天文学 60 年发展历程中发现的 X 射线源数量的两倍[6]。大部分新发现的 X 射线源是宇宙学距离尺度上的活跃星系核，标志着宇宙时间尺度上巨型黑洞的增长。"光谱-RG"任务还在银河系中发现了约 2 万个星系团和 20 万颗带有热晕的恒星，绘制出反映银河系中吸收辐射的高温气体和冷气体分布情况的弥漫 X 射线图①。"光谱-RG"任务在 6 个月的时间内绘制的硬 X 射线图中发现了超过 600 个辐射源，其中包括数十个未知天体，如被冷气体包围的软 X 射线下不可见的超大质量黑洞。任务计划在未来 3 年内绘制另外 7 张详细的宇宙地图[7]。

5. "嫦娥五号"月球探测任务成功采样返回

"嫦娥五号"搭载"长征五号"成功发射，在经历了地月转移、近月制动、两两分离、平稳落月、钻表取样、月面起飞、交会对接、样品转移、环月等待、月地转移、再入返回、安全着陆等 12 个重大阶段和关键步骤后，共采回 1731 克月球样品。我国由此成为世界上继美国、苏联之后第三个实现月球采样返回的国家，标志着我国具备了地月往返能力，实现了"绕、落、回"三步走规划的完美收官，为我国未来月球与行星探测奠定了坚实基础[8]。成果入选 2020 年度中国科学十大进展。"嫦娥四号"任务团队优秀代表获得国际宇航联合会 2020 年度最高奖——"世界航天奖"，这也是该组织成立 70 年来首次把这一奖项授予中国航天科学家。

6. 火星探测再掀新高潮

阿联酋"希望号"、中国"天问一号"和美国"火星 2020"/"毅力号"先后成功发

---

① 参见：https://www.mpa-garching.mpg.de/845848/news20200622。

射并到达火星，再掀火星探测新高潮。阿联酋"希望号"是阿拉伯国家的首个火星探测器，旨在首次揭示火星大气的完整图景，了解火星空气和水分流失的原因及过程[9]。中国成功实施"天问一号"任务，"祝融号"火星车开展火表巡视探测，我国成为世界上首个通过一次任务实现火星"绕、着、巡"、第二个实现火星着陆和巡视探测的国家[10]。"火星2020"任务将利用漫游车"毅力号"研究火星的宜居性，搜寻生物信号，保存样本待后续取回，测试制氧技术，验证"机智号"火星直升机飞行技术等[11]。此外，美国"好奇号"发布了拍摄到的最高分辨率火星表面全景图[12]，"洞察号"的研究揭示火星是一颗经常地震、刮风并拥有奇特磁脉冲的活跃星球[13]，ESA"火星快车"拍摄的火星北极冰盖图像帮助人们增进对火星极区的认识[14]，ESA和俄罗斯国家航天集团公司（Roscosmos）合作的"火星生命探测计划2016任务"精准测量了水从火星大气逸散的速度并发现该过程比先前认为的要快得多[15]。

7. 日美小行星探测和采样返回任务取得里程碑式进展

日本"隼鸟2号"对小行星"龙宫"的观测结果揭示了"龙宫"母星的撞击史[16]。"隼鸟2号"成功送返小行星样本，完成主任务目标后又踏上新任务的征程，计划在2031年造访小行星1998 KY26[17]。美国NASA首个小行星采样任务"起源、光谱分析、资源识别与安全-风化层探测器"成功触地小行星"贝努"（Bennu），并将足量样本收入样本返回舱，标志着其完成了最具挑战性的科研任务阶段[18]。美国、日本双方已经同意分享各自采回的小行星样本。此外，德国、日本启动"行星际旅程、法厄松飞越和尘埃科学的空间技术验证与实验"/"命运＋号"小行星探测任务[19]。

8. "卫星遥感＋人工智能"助力实现18亿棵树木的精确定位

欧洲和美国多个科研机构的科学家合作，利用人工智能算法分析卫星遥感图像，详细描绘出西非地区130多万 km^2 区域内超过18亿棵树的位置和树冠大小，并发现此前同类研究低估了干旱地区的树木数量[20]。该研究有望根本性地改变我们思考、监测、模拟和管理全球陆地生态系统的方式，未来可能实现对区域乃至全球尺度上重要目标的天基自动化、智能化的全面识别和精确定位。成果入选《自然》2020年度十大科学发现。

9. 在微重力条件下首次实现玻色-爱因斯坦凝聚

美国NASA喷气推进实验室首次在国际空间站冷原子实验室中实现并观测到第五种物质状态——玻色-爱因斯坦凝聚[21,22]。在微重力条件下，玻色-爱因斯坦凝聚持续超过1 s，表明空间微重力环境将帮助人类更好地探索这种奇异物质状态，为研究量子气体和原子干涉创造了新机会。这一重大突破有助于揭示量子力学领域最棘手的一些难题[23]。

10. 国际空间站科研活动亮点纷呈

国际空间站持续产出科研应用成果，除首次在轨实现玻色-爱因斯坦凝聚外，还包括基于纳米颗粒开发新型药物输运系统、诊断航天员太空贫血问题、发现微重力环境对在轨 6 周的航天员的动作控制产生影响、证实空间站环境适宜开展淀粉样蛋白形成机制研究、确认肌肉生长抑制素可有效应对空间环境导致的肌萎缩、证明瞬态发光事件和地球伽马射线闪光具有相关性、首次实现基于小卫星平台观测系外行星凌日、通过观察航天员的微生物"指纹"判断其到达及离开时间等[24,25]。

11. 多个空间科学创新平台成功发射

中国"引力波暴高能电磁对应体全天监测器"[26]，欧洲空间局"宇宙憧憬"计划的第一项中型任务——"太阳轨道器"[27]，韩国环境卫星"静止轨道综合卫星2B"[28]，ESA 和 NASA 合作的"哨兵-6"全球海平面监测卫星[29]，阿联酋、中国、美国的火星探测器等相继成功发射，为空间科学未来取得更多重大突破奠定基础。

二、重大战略行动

1. 美国发布新版国家航天政策和空间科学规划

美国政府发布《国家航天政策》[30]，重申美国在当今空间探索复兴时代的雄心和领导力。NASA 科学任务部发布《科学 2020—2024：卓越科学愿景》[31]，未来 5 年将以探索和科学发现为核心，领导全球合作的科学发现计划，鼓励创新、造福生活、激发灵感，以期发现宇宙奥秘，搜寻地外生命，保护和改善地面生活。NASA 发布《"阿尔忒弥斯-3"任务科学定义组报告》，明确 2024 年载人登月任务的科学优先事项[32,33]。NASA 和美国国家科学院启动 2023～2032 年行星科学和天体生物学 10 年调查任务[34]。NASA 和 ESA 制定火星采样返回行动任务规划[35]。

2. 日本发布新版《宇宙基本计划》

日本内阁府发布新版《宇宙基本计划》[36]，将继续发展"隼鸟号"等广受世界关注的项目，加强国际合作，研发世界顶尖的通用型和创新型技术，积极参与"阿尔忒弥斯"计划，充分发挥国际空间站的作用。日本宇宙航空研究开发机构（Japan Aerospace Exploration Agency，JAXA）发布第二版《国际空间探索任务场景》，梳理了以月球、火星及其卫星为主要探测目标的国际空间探索计划情况，提出日本可能参与的国际任务场景[37]。

3. ESA启动月球着陆器任务征集

ESA确定参与"阿尔忒弥斯"计划[38]，同时将开发欧洲首个月球着陆器——"欧洲大型后勤着陆器"，通过定期着陆月球，为科学研究和机器人任务提供前所未有的机遇[39]。

4. 俄罗斯酝酿制定金星探索计划

俄罗斯国家航天集团公司正在酝酿制定金星探索计划，可能参考已经成型的俄罗斯月球开发计划，并包含《2016—2025联邦航天计划》中的"金星-D"任务[40,41]。研究人员希望将金星探索任务纳入国家预算中，并在2030年前发射一颗金星探测器。

5.《全球探索路线图》更新月表探索场景

国际空间探索协调工作组宣布更新2018年版《全球探索路线图》的月表探索场景部分，以反映当前全球探月新高潮背景下的最新进展情况[42]。

6. 美欧稳步推进空间科学任务部署与实施

美国NASA遴选出宇宙大爆炸及其残留物、临近恒星爆发的耀斑如何影响围绕其运行的行星的大气层等天体物理学概念任务[43]，金星探测、木卫一和海卫一探测任务概念[44]，首个"地球探险接续"机会任务[45]，批准"太阳射电干涉仪空间实验"[46]，"磁层响应日地观测器"、HelioSwarm、"多缝隙太阳探测器"、"极光重构立方体卫星星座"和Solaris空间环境任务概念[47]，以及"极紫外高通量光谱望远镜"和"电喷流塞曼成像探测器"太阳物理任务[48]。

ESA"宇宙憧憬"计划第4个中型任务——"系外行星大气遥感红外大规模巡天"，以及"赫拉"行星防御任务[49]进入建设阶段[50]。

三、发展启示与建议

2020年，世界空间科学领域在黑洞、暗物质、暗能量、太阳系的起源和演化、微重力科学等前沿热点方向持续取得新发现和新突破，我国"慧眼"X射线天文卫星和"嫦娥五号"月球探测任务的卓越成就举世瞩目，多信使、创新概念与国际合作研究已成为空间科学重大突破的关键。多国继续推进未来空间科学部署，加强原始创新、强化独特能力和优势成为与国际合作并重的战略优先事项。美国着力通过"阿尔忒弥斯"计划牵引全球深空载人探索路线重新聚焦月球，力图打造月球"北约"集团，全球在阶段性的调整中酝酿新的组合。我国亟须进一步加强空间科学研究规划，力争在深化人类对基本科学问题的认知方面率先实现突破，立足于基础科学发展和满足国家战略目标的实现，逐步提升我国在空间科学领域国际合作中的影响力和领导力。

致谢：中国科学院国家空间科学中心范全林研究员、中国科学院科技战略咨询研究院张凤研究员对本文的撰写提出许多宝贵的修改意见，特此致谢。

参考文献

［1］CHIME/FRB Collaboration. A bright millisecond-duration radio burst from a Galactic magnetar. Nature,2020,587(7832):54-58.

［2］Bochenek C D,Ravi V,Belov K V,et al. A fast radio burst associated with a Galactic magnetar. Nature,2020,587(7832):59-62.

［3］Lin L,Zhang C F,Wang P,et al. No pulsed radio emission during a bursting phase of a Galactic magnetar. Nature,2020,587(7832):63-65.

［4］Ma X,Tao L,Zhang S N,et al. Discovery of oscillations above 200 keV in a black hole X-ray binary with Insight-HXMT. Nature Astronomy,2021,5(1):94-102.

［5］Lin D,Strader J,Romanowsky A J,et al. Multiwavelength follow-up of the hyperluminous interme-diate-mass black hole candidate 3XMM J215022. 4-055108. The Astrophysical Journal Letters,2020,892(2):L25.

［6］MPG. Our deepest view of the X-ray sky. https://www. mpg. de/14999689/our-deepest-view-of-the-x-ray-sky? c=2249[2020-06-19].

［7］Roscosmos. Вести:《Спектр-РГ》построил лучшую в мире рентгеновскую карту неба. https://www. roscosmos. ru/28713/[2020-06-22].

［8］国家航天局. 嫦娥五号创造五项中国首次. http://www. cnsa. gov. cn/n6758823/n6758844/n6760243/n6760244/c6810929/content. html[2020-12-18].

［9］UAE Space Agency. Emirates Mars mission overview. https://www. emiratesmarsmission. ae/mission/about-emm[2021-10-11].

［10］国家航天局. 中国首次火星探测任务探测器发射成功 迈出中国行星探测第一步. http://www. clep. org. cn/n5982341/c6809884/content. html[2020-07-23].

［11］NASA. Touchdown! NASA's Mars perseverance rover safely lands on red planet. https://www. nasa. gov/press-release/touchdown-nasas-mars-perseverance-rover-safely-lands-on-red-planet[2021-02-19].

［12］NASA. NASA's curiosity Mars rover snaps its highest-resolution panorama yet. https://www. nasa. gov/

feature/jpl/nasas-curiosity-mars-rover-snaps-its-highest-resolution-panorama-yet[2020-03-04].

[13] NASA. A year of surprising science from NASA's insight Mars mission. https://www. nasa. gov/feature/jpl/a-year-of-surprising-science-from-nasas-insight-mars-mission[2020-02-24].

[14] ESA. Rippling ice and storms at Mars' north pole. http://www. esa. int/Science_ Exploration/Space_Science/Mars_Express/Rippling_ice_and_storms_at_Mars_north_pole[2020-01-13].

[15] Fedorova A A,Montmessin F,Korablev O,et al. Stormy water on Mars:the distribution and saturation of atmospheric water during the dusty season. Science,2020,367(6475):297-300.

[16] Tatsumi E,Sugimoto C,Riu L,et al. Collisional history of Ryugu's parent body from bright surface boulders. Nature Astronomy,2021,5(1):39-45.

[17] JAXA. Video for the extended mission. https://www. hayabusa2. jaxa. jp/en/topics/20201116_extMission/[2021-10-11].

[18] NASA. NASA's OSIRIS-REx successfully stows sample of asteroid bennu. https://www. nasa. gov/press-release/nasa-s-osiris-rex-successfully-stows-sample-of-asteroid-bennu[2020-10-30].

[19] DLR. DES-TINY＋-Germany and Japan begin new asteroid mission. https://www. dlr. de/content/en/articles/news/2020/04/20201112 _ destiny-germany-and-japan-begin-new-asteroid-mission. html[2020-11-12].

[20] Brandt M,Tucker C J,Kariryaa A,et al. An unexpectedly large count of trees in the West African Sahara and Sahel. Nature,2020,587(7832):78-82.

[21] PHYS. ORG. Quantum'fifth state of matter'observed in space for first time. https://phys. org/news/2020-06-quantum-state-space. html[2020-06-11].

[22] 中国科学院. 第五种物质状态在国际空间站诞生,为研究量子气体和原子干涉创造新机会. http://www. cas. cn/kj/202006/t20200612_4749765. Shtml[2020-06-12].

[23] Aveline D C,Williams J R,Elliott E R,et al. Observation of bose-einstein condensates in an Earth-orbiting research lab. Nature,2020,582(7811):193-197.

[24] NASA. Annual highlights of results from the international space station. https://www. nasa. gov/sites/default/files/atoms/files/np-2020-11-021-jsc_iss_annual_results_highlights_2020-121720_c. pdf[2020-11-02].

[25] NASA. What we learned this year from space station science. https://www. nasa. gov/mission_pages/station/research/news/what-we-learned-from-space-station-2020[2020-12-29].

[26] 中国科学院. 引力波暴高能电磁对应体全天监测器卫星升空,捕捉稍纵即逝的光. http://www. cas. ac. cn/tt/202012/t20201210_4770258. shtml[2020-12-11].

[27] NASA. New mission will take 1st peek at Sun's poles. https://www. nasa. gov/feature/goddard/2020/new-mission-will-take-first-peek-at-sun-s-poles[2020-01-28].

[28] KARI. 천리안위성 2B 호, 오늘(2 월 19 일) 아침 발사 성공. https://www. kari. re. kr/cop/bbs/BBSMSTR_ 000000000011/selectBoardArticle. do? nttId＝7491&kind＝&mno＝sitemap_02&pageIndex＝1&searchCnd＝&searchWrd＝[2020-02-19].

［29］ NASA. NASA，US and European partners launch mission to monitor global ocean. https：//www. nasa. gov/press-release/nasa-us-and-european-partners-launch-mission-to-monitor-global-ocean［2020-11-22］.

［30］ White House. National space policy of the united states of America. https：//www. whitehouse. gov/wp-content/uploads/2020/12/National-Space-Policy. pdf［2020-12-09］.

［31］ NASA. Explore Science 2020-2024. A vision for science excellence. https：//science. nasa. gov/science-pink/s3fs-public/atoms/files/2020-2024_Science. pdf［2020-05-27］.

［32］ NASA. Artemis Ⅲ science definition report. https：//www. nasa. gov/sites/default/files/atoms/files/artemis-iii-science-definition-report-12042020c. pdf［2021-10-11］.

［33］ NASA. NASA defines science priorities for first crewed artemis landing on moon. https：//www. nasa. gov/press-release/nasa-defines-science-priorities-for-first-crewed-artemis-landing-on-moon［2020-12-08］.

［34］ SpaceNews. Planetary science decadal survey to include astrobiology and planetary defense. https：//spacenews. com/planetary-science-decadal-survey-to-include-astrobiology-and-planetary-defense/［2020-03-18］.

［35］ SpaceNews. NASA and ESA outline cost of Mars sample return. https：//spacenews. com/nasa-and-esa-outline-cost-of-mars-sample-return/［2020-07-29］.

［36］ 日本内阁府. 宇宙基本計画の変更について. https：//www8. cao. go. jp/space/plan/kaitei_fy02/fy02. pdf［2020-06-30］.

［37］ JAXA. 日本の国際宇宙探査シナリオ（案）2019. http：//www. exploration. jaxa. jp/assets/img/news/pdf/scenario/EZA-2019001_% E6% 97% A5% E6% 9C% AC% E3% 81% AE% E5% 9B% BD% E9% 9A% 9B% E5% AE% 87% E5% AE% 99% E6% 8E% A2% E6% 9F% BB% E3% 82% B7% E3% 83% 8A% E3% 83% AA% E3% 82% AA（% E6% A1% 88）2019_Executive% 20Summary. pdf［2021-10-11］.

［38］ NASA. NASA，European space agency formalize artemis gateway partnership. https：//www. nasa. gov/press-release/nasa-european-space-agency-formalize-artemis-gateway-partnership［2020-10-27］.

［39］ ESA. Get your ticket to the Moon：Europe's lunar lander for science and more. http：//www. esa. int/Science_Exploration/Human_and_Robotic_Exploration/Get_your_ticket_to_the_Moon_Europe_s_lunar_lander_for_science_and_more［2020-05-29］.

［40］ РИА. У России появится программа исследования Венеры. https：//ria. ru/20200131/1564116446. html［2020-01-31］.

［41］ ТАСС. Роскосмос поручил сформировать программу исследования Венеры. https：//tass. ru/kosmos/7657083［2020-01-31］.

［42］ ISECG. Global exploration roadmap supplement -lunar surface exploration scenario update. https：//www. globalspaceexploration. org/? p=1049［2020-08-28］.

［43］ NASA. NASA selects proposals to study volatile stars，galaxies，cosmic collisions. https：//www. nasa. gov/press-release/nasa-selects-proposals-to-study-volatile-stars-galaxies-cosmic-collisions

[2020-03-17].

[44] NASA. NASA selects four possible missions to study the secrets of the solar system. https://www. nasa. gov/press-release/nasa-selects-four-possible-missions-to-study-the-secrets-of-the-solar-system[2020-02-14].

[45] NASA. NASA selects new instrument to continue key climate record. https://www. nasa. gov/press-release/nasa-selects-new-instrument-to-continue-key-climate-record[2020-02-27].

[46] NASA. NASA selects mission to study causes of giant solar particle storms. https://www. nasa. gov/press-release/nasa-selects-mission-to-study-causes-of-giant-solar-particle-storms[2020-03-30].

[47] NASA. NASA selects proposals for new space environment missions. https://www. nasa. gov/press-release/nasa-selects-proposals-for-new-space-environment-missions[2020-08-29].

[48] NASA. NASA approves heliophysics missions to explore Sun, Earth'S aurora. https://www. nasa. gov/press-release/nasa-approves-heliophysics-missions-to-explore-sun-earth-s-aurora[2020-12-30].

[49] ESA. Industry starts work on Europe's Hera planetary defence mission. http://www. esa. int/Safety_Security/Hera/Industry_starts_work_on_Europe_s_Hera_planetary_defence_mission[2020-09-15].

[50] ESA. Ariel moves from blueprint to reality. http://www. esa. int/Science_ Exploration/Space_Science/Ariel_moves_from_blueprint_to_reality[2020-11-12].

Space Science

Yang Fan, Han Lin, Wang Haiming, Fan Weiwei

In 2020, the scientific research of space science has attracted much attention, such as the first discovery of the source of fast radio bursts(FRBs), the acquisition of the most convincing evidence for intermediate-mass black holes, the realization of the first on-orbit Bose-Einstein condensation, the continuous advancement of the US-led Artemis manned lunar exploration program, and the new upsurge in Mars exploration. China should further strengthen the scientific research planning, encourage scientific goal-driven researches and original scientific breakthroughs, inspire original innovations in space science, and continue to attach importance to the china-oriented international cooperations in space science.

4.9　信息科技领域发展观察

唐　川　王立娜　张　娟　杨况骏瑜　黄　茹

（中国科学院成都文献情报中心）

2020 年，科技强国继续深化信息科技领域战略布局，推动信息技术创新发展，谋求关键技术重大突破和产业化应用，为数字产业化和产业数字化转型注入新动力，抢占数字经济发展的主动权。本文以 2020 年全球人工智能（artificial intelligence，AI）、半导体、量子信息和高性能计算四个关键领域为对象，重点剖析了领域重要研究进展与各国战略规划，以揭示信息科技领域的战略新动向、技术新趋势、未来关键挑战与机遇。

一、重要研究进展

1. 人工智能

人工智能系统的规模日益扩大，算法性能和效率不断提升，实际应用场景显著拓展，能力大幅提升。

（1）人工智能加速学科领域创新发展。谷歌（Google）公司旗下的深度思考（DeepMind）公司在第 14 届国际蛋白质结构预测竞赛（CASP）上使用 AlphaFold2 人工智能系统，精准预测出蛋白质三维结构[1]，准确性可与使用冷冻电子显微镜、核磁共振或 X 射线晶体学等实验技术解析的三维结构媲美，平均误差约为 0.16 nm。

（2）类脑计算芯片化、系统化趋势明显。英特尔（Intel）公司推出具有 1 亿个神经元的神经形态研究系统 "Pohoiki Springs"，是其截至 2020 年 3 月开发的最大规模神经形态计算系统[2]。来自中国清华大学、北京信息科学与技术国家研究中心和美国特拉华大学的科研团队首次提出了 "类脑计算完备性" 概念[3]。奥地利维也纳大学和美国麻省理工学院组成的国际研究团队开发出一种基于线虫等微小动物大脑的全新人工智能系统[4]。

（3）人工智能模型持续演进和自主进化。谷歌公司提出了一个简单的视觉表示对比学习框架（SimCLR），简化并改进了在图像上进行自监督表示学习的方法[5]。美国

OpenAI 公司开发出一种面向自然语言处理的强大深度学习模型 GPT-3，在诸多实际任务中大幅接近人类水平[6]。

2. 半导体

随着摩尔定律脚步的放缓和新兴技术的加速发展，多样化的下一代集成电路前沿技术百家争鸣，新型计算架构和互补金属氧化物半导体（complementary metal oxide semiconductor，CMOS）微缩的替代方案推陈出新。

（1）5 nm 制程工艺开始量产，3 nm 工艺即将试产。台湾积体电路制造股份有限公司和三星集团纷纷实现 5 nm 极紫外光刻（extreme ultraviolet lithography，EUVL）工艺芯片量产。其中，台湾积体电路制造股份有限公司具有领先优势，5 nm 芯片已经交付苹果（Apple）公司等大客户，所生产的苹果 A14 仿生处理器芯片集成了 118 亿个晶体管，每秒可处理 11 万亿次运算。此外，台湾积体电路制造股份有限公司已于 2021 年试产 3 nm 制程工艺，计划于 2022 年实现量产[7]。

（2）存内计算架构快速演进。清华大学研发出一款基于多个忆阻器阵列的存算一体系统[8]，在处理卷积神经网络时的能效比图形处理器芯片的能效高两个数量级。国际商业机器公司（International Business Machines Corporation，IBM）欧洲研发中心研发出一种基于相变存储器的非冯·诺依曼架构芯片技术[9]，能以超低功耗实现复杂且准确的深度神经网络推理。

（3）碳基电子芯片取得跨越式发展。北京大学的研究人员在碳基半导体制备材料上解决了纯度、面积、密度、顺排等长期无法攻克的难题，批量制备出五阶环形振荡器电路，成品率超过 50%，最高振荡频率为 8.06 GHz，为碳基半导体进入规模化商业应用奠定了基础[10]。美国加利福尼亚大学伯克利分校制造出一种完全由碳材料组成的超窄金属线[11]，克服了全碳基集成电路制造中的一个关键障碍。

3. 量子信息

量子信息技术持续快速发展，新技术层出不穷，量子计算和量子通信的新突破不断，量子网络研究成为新热点。

（1）量子计算研究取得长足进展。澳大利亚新南威尔士大学和荷兰代尔夫特理工大学将硅基量子芯片的运行温度大幅升至最高 1.5 K，推动量子计算机走向实际应用[12]。QuTech 量子计算实验室推出欧洲首个公共量子计算平台 Quantum Inspire，是全球首台使用可扩展自旋量子位组成的量子处理器[13]。美国霍尼韦尔国际公司宣布开发出全球功能最强大的量子计算机，其量子体积是 IBM 公司于 2019 年推出的 53 个量子位量子计算机的 2 倍[14]。

（2）量子通信取得阶段性突破。美国普林斯顿大学在相距 4 mm 的两个硅自旋量子比特间实现了信息交换，为解决量子比特间的互连问题奠定了基础[15]。英国布里斯托尔大学首次实现了两个可编程芯片之间的量子隐形传态，为量子通信的大规模集成铺平了道路[16]。中国科学技术大学首次实现了千公里级量子纠缠密钥分发，为量子卫星通信的规模化、商业化应用奠定了坚实的基础[17]。

（3）量子网络研究持续升温。美国陆军研究实验室研发出一种可同时存储多种量子激励模式的全新量子设备，为创建量子网络开辟了新路径[18]。美国陆军研究实验室与马里兰大学联合建立了混合量子网络前身，扩展了量子网络节点间的基线[19]。法国巴黎索邦大学实现了量子纠缠在两个量子存储器中的高效可逆转移，有助于提升量子互联网的可扩展性[20]。

4. 高性能计算

全球超算研发的竞争日益激烈，中国、欧盟、日本不约而同地在百亿亿次超算中采用 ARM（Advanced RISC Machines）架构的芯片，以期在获得高性能的同时实现关键核心技术自主可控。2020 年 6 月，采用 ARM 架构芯片的日本超级计算机"富岳"首次夺得 Top 500 排行榜桂冠，颠覆了 ARM 架构芯片主要面向移动端、性能相对较差的发展历史。"富岳"同时在运算速度、模拟计算方法、人工智能学习性能、大数据处理性能等项目的测评中荣获第一名，成为全球超级计算机首个"四冠王"，为百亿亿次超算的研发开辟了一条切实可行的新途径。除"富岳"外，多个国家和地区也投入大量超算资源抗击新冠肺炎疫情。

二、重要战略规划

1. 人工智能

各国/组织继续完善人工智能战略布局，深化人工智能生态系统建设，为未来数字化转型奠定基础。同时，人工智能技术监管和标准体系建设也成为关注重点。

（1）美国加强国际合作，加大在人工智能领域的投入。美国参议院发起的《无尽前沿法案》授权未来 5 年内拨款 1000 亿美元支持人工智能等十大关键技术研发[21]。2020 年 9 月，美国和英国签署《人工智能研究与开发合作宣言》[22]，联合开展人工智能研发工作，提议人工智能规划优先事项，协调人工智能领域相关活动，建立人工智能研发生态系统。

（2）欧盟重视人工智能协同发展、技术主权和数据安全。2020 年 2 月，欧盟委员会发布了《人工智能白皮书：通向卓越和信任的欧洲路径》[23]，提出欧洲必须加强各成员国之间的协作，建立更多人工智能联合研发中心和合作网络，确保欧洲成为未来数字化转型的领导者。此外，欧洲先后通过了《通用数据保护条例》《欧洲数据战略》《人工智能白皮书》《数字服务法案》等法律和政策文件，严格监管人工智能等数字技术。

（3）日本系统布局人工智能相关工作。日本先后出台了多项人工智能战略计划，包括《人工智能技术战略》《人工智能技术战略实施计划》《人工智能战略 2019》等，具体部署相关部门的人工智能推进工作及完成期限，以确保政策执行效率。2020 年 7 月，日本发布《统合创新战略 2020》[24]，对人工智能的发展做出了具体的规划，包括开展教育改革、重构研究体系、准备数据连接基础等。

（4）中国建设新一代人工智能标准体系。国家标准化管理委员会、中共中央网络安全和信息化委员会办公室、国家发展和改革委员会、科学技术部、工业和信息化部联合印发《国家新一代人工智能标准体系建设指南》[25]，明确了新一代人工智能标准体系的建设目标，部署了具体建设思路和建设内容，旨在加强人工智能领域标准化顶层设计，推动人工智能产业技术研发和标准制定，促进产业健康可持续发展。

2. 半导体

美国、欧盟、韩国等国/组织积极部署重磅半导体研究战略，巩固现有行业地位，加强先进芯片制造能力，拓展未来半导体产业新兴增长空间。

（1）美国建立微电子学计划，推进半导体制造业复苏。美国国会于 2020 年 12 月通过了 2021 财年《国防授权法案》，其中包括一项半导体相关法案《为美国半导体生产和代工创造有益的激励措施》（S. 4982）[26]。该法案举措包括创建财政补贴项目、建立一个微电子学国家研发网络、设立微电子学小组委员会和咨询委员会、建立国家半导体技术中心等。

（2）欧盟巩固优势领域，建立先进芯片设计能力和制造设施。欧盟中的 18 个成员国于 2020 年 12 月签署了关于处理器和半导体技术的联合声明[27]，旨在加强欧洲的电子和嵌入式系统价值链、完善处理器和半导体生态系统，建立先进的欧洲芯片设计能力和制造设施，以应对关键技术、安全和社会挑战。此外，"欧洲电子组件与系统领先联合计划"（ECSEL JU）和"欧洲处理器计划"也在大力发展欧洲的电子设计与制造能力。

（3）韩国将立足点转向人工智能半导体产业强国战略。韩国科学信息通信技术和未来规划部（MSIT）于 2020 年 10 月发布《人工智能半导体产业发展战略》计划[28]，提出了人工智能半导体产业技术、人才、工业生态系统创新战略，旨在 2030 年实现

"人工智能半导体强国"目标，把人工智能半导体正式培育为"第二个DRAM"产业。

（4）中国将集成电路列入"十四五"规划。《中共中央关于制定国民经济和社会发展第十四个五年规划和二〇三五年远景目标的建议》指出，我国将瞄准集成电路、人工智能、量子信息等前沿领域，实施一批具有前瞻性、战略性的国家重大科技项目[29]。国务院发布《新时期促进集成电路产业和软件产业高质量发展的若干政策》，要求聚焦高端芯片、集成电路装备和工艺技术、集成电路关键材料、集成电路设计工具等关键核心技术领域进行重点研发[30]。国务院学位委员会将"集成电路科学与工程"设为一级学科，加速集成电路相关人才的培养[31]。

3. 量子信息

全球量子信息技术竞赛持续升温，美国、欧盟、俄罗斯等国/组织大力部署量子计算、量子通信和量子互联网研究。

（1）美国建立多个量子研究中心，成立量子联盟指导委员会。2020年1月，美国能源部宣布投资6.25亿美元建立多个量子研究中心[32]；8月，启动5个研究中心建设工作，重点关注量子互联网、量子传感、量子计算等方向，拓展量子信息技术在计算、通信、传感器、化学和传感等方面的应用[33]。9月，美国成立量子经济发展联盟（QED-C）指导委员会，为新兴的量子产业提供支持与统一指导，辅助联邦确定优先事项、设立标准法规、培养人才[34]。

（2）欧盟大力布局量子通信技术。2020年5月，欧盟"量子技术旗舰计划"官网发布《战略研究议程》（SRA）[35]，拟在未来三年推动建设欧洲范围的量子通信网络，完善和扩展现有数字基础设施，为未来的"量子互联网"远景奠定基础，并将推动量子通信与传统的网络基础设施和应用相结合，开发用于全球安全密钥分发的卫星量子密码。

（3）俄罗斯成立专属机构，制定量子通信发展路线图。俄罗斯数字发展委员会于2020年9月批准了由俄罗斯铁路公司与主要专家和科学组织共同制定的俄罗斯量子通信发展路线图[36]。这个路线图涵盖了俄罗斯计划2024年前研发的9项优先技术、15种产品及35项关键性能指标（包括产量和销量、量子网络长度、技术准备水平、人力资源供应等指标），并详述了120多项发展措施和项目。

（4）印度启动新国家量子任务，斥巨资推动量子技术发展。2020年2月，印度科学技术部计划未来5年投入800亿卢比（约合11.2亿美元），推动量子技术的发展，主要投资领域包括量子计算、量子通信和量子密码学[37]。

4. 高性能计算

欧盟和美国相继推出下一代计算研发规划，通过与时俱进的战略部署，维持并扩大在高性能计算领域的领先优势。

（1）美国一直追求引领未来计算研发。2020年11月，美国发布《引领未来先进计算生态系统：战略计划》报告[38]，致力于携手政产学研界打造一个未来先进计算生态系统，研发可持续软件和数据生态系统，推进基础性、应用性和转化研究，并培养多样化、高水平的劳动力队伍，维持在科学工程、经济竞争和国家安全方面的领先优势。

（2）欧盟致力开发自主可控的下一代超算。2020年9月，欧洲高性能计算联合执行体（EuroHPC）发布新章程[39]，拟投资80亿欧元支持以百亿亿次（E级）计算和量子计算为主的新一代超算技术和系统的研究和创新，聚焦基础设施、超算服务的联合、技术、应用、技能发展5大核心领域，为欧洲打造世界级的超算生态系统。12月，欧盟宣布将通过"数字欧洲"计划为超级计算提供22亿欧元的资助[40]，构建并加强欧盟的超算和数据处理能力，尤其是E级和后E级超算能力。

三、发展启示与建议

1. 系统布局国家战略计划，细化具体实施措施

随着"后摩尔"时代的到来，集成电路新型芯片、超级计算机系统新体系结构、人工智能新技术等领域面临跨越式发展的机遇与挑战。我国应在着眼于破解眼前"卡脖子"问题的同时，前瞻性布局潜在颠覆性信息技术，对重大科技问题、重要技术演进开展深入分析，结合自身发展需求和基础，制定中长期国家级战略，从顶层设计层面部署科学化、体系化的信息技术发展举措，编制关键信息技术研发路线图，明确优先发展领域、发展目标、关键挑战和时间节点，"自上而下"逐层进行战略布局，引领未来发展。

2. 建立产学研联合攻关机制，实现关键技术自主可控

针对当前信息科技"卡脖子"问题，建立产学研联合攻关机制，实现关键技术自主可控，逐步摆脱对国际产业链的依赖。针对当前人工智能、半导体芯片、量子信息等颠覆性信息技术企业，改善其投资体制，允许风险投资等资本投资颠覆性技术领域，建设关键信息产业生态环境。

3. 构建高层次复合型人才培养体系

面向当前学科交叉发展趋势，聚焦电子信息、人工智能、量子信息、系统架构、

软件编程、算法设计等多个技术方向，加强学科交叉融合建设，强化科教融合和产教融合模式，打通产学研联合人才培养渠道，推动人才培养、科学研究与产业对接，构建高层次复合型人才培养体系，为我国关键信息技术研究和产业发展奠定人才基础。

致谢：中国科学院自动化研究所孙哲南研究员、中国科学院计算技术研究所洪学海研究员、中国科学技术大学韩永建教授等专家审阅了本文，并提出了宝贵的修改意见，特致感谢！

参考文献

[1] DeepMind. AlphaFold：a solution to a 50-year-old grand challenge in biology. https://deepmind. com/blog/article/alphafold-a-solution-to-a-50-year-old-grand-challenge-in-biology[2020-11-03].

[2] Intel. Intel scales neuromorphic research system to 100 million neurons. https://newsroom. intel. com/news/intel-scales-neuromorphic-research-system-100-million-neurons/#gs. 1ia3e2[2020-03-18].

[3] Zhang Y, Qu P, Ji Y, et al. A system hierarchy for brain-inspired computing. Nature, 2020, 586：378-384.

[4] Lechner M, Hasani R, Amini A, et al. Neural circuit policies enabling auditable autonomy. Nature Machine Intelligence, 2020, 2：642-652.

[5] Chen T. Advancing self-supervised and semi-supervised learning on ImageNet. https://ai. googleblog. com/2020/04/advancing-self-supervised-and-semi. html[2020-04-08].

[6] Tom B, Benjamin M, Nick R, et al. Language models are few-shot learners. arXiv preprint arXiv：2005. 14165, 2020.

[7] 黄晶晶. 台积电最新公布 5nm＋/4nm/3nm 工艺制程进度, 推出超低功耗工艺 N12e. http://www. elecfans. com/d/1280498. html[2020-08-25].

[8] Yao P, Wu H, Gao B, et al. Fully hardware-implemented memristor convolutional neural network. Nature, 2020, 577：641-646.

[9] Joshi V, Le Gallo M, Haefeli S, et al. Accurate deep neural network inference using computational phase-change memory. Nature Communications, 2020, 11：2473.

[10] Liu L, Han J, Xu L, et al. Aligned, high-density semiconducting carbon nanotube arrays for high-performance electronics. Science, 2020, 368(6493)：850-856.

[11] Van Erp R, Soleimanzadeh R, Nela L, et al. Co-designing electronics with microfluidics for more sustainable cooling. Nature, 2020, 585(7824)：211-216.

[12] Yang C H, Leon R C C, Hwang J C C, et al. Operation of a silicon quantum processor unit cell above one kelvin. Nature, 2020, 580(7803)：350-354.

[13] QuTech. Minister Ingrid van Engelshoven and European commissioner Mariya Gabriel launch

quantum inspire. https://qutech. nl/minister-ingrid-van-engelshoven-and-european-commissioner-mariya-gabriel-launch-europes-first-quantum-computer-in-the-cloud-quantum-inspire/[2020-04-20].

[14] IEEE Spectrum. Honeywell claims it has most powerful quantum computer. https://spectrum. ieee. org/tech-talk/computing/hardware/honeywell-claims-it-has-most-powerful-quantum-computer [2020-06-24].

[15] Borjans F,Croot X G,Mi X,et al. Resonant microwave-mediated interactions between distant electron spins. Nature,2020,577(7789):195-198.

[16] First chip-to-chip quantum teleportation harnessing silicon photonic chip fabrication. https://www. bristol. ac. uk/news/2019/december/quantum-teleportation. html[2019-12-23].

[17] Yin J,Li Y H,Liao S K,et al. Entanglement-based secure quantum cryptography over 1,120 kilometres. Nature,2020,582(7813):501-505.

[18] USA Army. New device accelerates development of extraordinary quantum networks. https://www. army. mil/article/231525/new_device_accelerates_development_of_extraordinary_quantum _networks[2020-01-09].

[19] USA Army. Army research brings hybrid quantum computing closer to reality. https://www. army. mil/article/237779/army_research_brings_hybrid_quantum_computing_closer_to_reality[2020-08-03].

[20] Quantum Flagship. World's record entanglement storage sets up a milestone for the European Quantum Internet Alliance. https://qt. eu/about-quantum-flagship/newsroom/5817/[2020-11-05].

[21] United States Congress. Endless Frontier Act. https://www. congress. gov/bill/116th-congress/senate-bill/3832[2020-5-21].

[22] U. S. Department of state. Declaration of the United States of America and the United Kingdom of Great Britain and Northern Ireland on cooperation in artificial intelligence research and development:a shared vision for driving technological breakthroughs in artificial intelligence. https://www. state. gov/declaration-of-the-united-states-of-america-and-the-united-kingdom-of-great-britain-and-northern-ireland-on-cooperation-in-artificial-intelligence-research-and-development-a-shared-vision-for-driving/[2020-09-25].

[23] European Commission. On artificial intelligence—a European approach to excellence and trust. https://ec. europa. eu/info/sites/default/files/commission-white-paper-artificial-intelligence-feb2020_en. pdf[2020-02-19].

[24] 内閣府 . 統合イノベーション戦略 2020. https://www8. cao. go. jp/cstp/tougosenryaku/2020. html[2020-07-17].

[25] 国家标准化管理委员会 中央网信办 国家发展改革委 科技部 工业和信息化部 . 关于印发《国家新一代人工智能标准体系建设指南》的通知 . http://www. gov. cn/zhengce/zhengceku/2020-08/09/content_5533454. htm[2020-07-27].

[26] Congress. S. 4982-creating helpful incentives for producing semiconductors for America and foundries act. https://www. congress. gov/bill/116th-congress/senate-bill/4982/text? __cf_chl_

jschl_tk__ = c202182c4e95809ba2dbbdc0b9daaaa6d606263e-1612423645-0-AflzDvq0sq6K8_Cm-
VkW3XTa4c5RdwgWC0fBr0HnxN2uDwheZ1nuTwbw1FKfFqGRcCykIvirfxcP50ZUnsrJZLsqJRw
fJoCGI3G6YKyjWFviNk0AMT-IC8SVqYi37ov6QrR6oyiuu-9Dkjt5NLvlYWGySgTyobW1EaOJue
UTbOQmOsvIfHaDxKw7w61wVRJ1g4Ndu-9y_LjAMCK5_mcmYuZQxEgoI9C8gkEtkjHPqeBd-
DgCYYjx3J8CrMkfWIf7PwaeR4VANvjiZVky3NS4FBslbOVVSebBZSv23JZvtJs0wWgzdAxP15aK
fD-dMJvC9oQ0LJ-ygQO6P94Lrur_uutRXJfNPuXW1TRZH_6lVc7SZY[2020-12-08].

[27] European Commission. Member states join forces for a European initiative on processors and semiconductor technologies. https://ec. europa. eu/digital-single-market/en/news/member-states-join-
forces-european-initiative-processors-and-semiconductor-technologies[2020-12-07].

[28] MSIT. South Korea wants to develop 50 types of AI chips by 2030. https://www. msit. go. kr/
web/msipContents/contentsView. do? cateId=_policycom2&artId=3137757[2020-10-12].

[29] 新华社 . 中共中央关于制定国民经济和社会发展第十四个五年规划和二〇三五年远景目标的建
议 . http://www. gov. cn/zhengce/2020-11/03/content_5556991. htm[2020-11-03].

[30] 国务院 . 国务院关于印发《新时期促进集成电路产业和软件产业高质量发展若干政策》的通知 .
http://www. gov. cn/zhengce/content/2020-08/04/content_5532370. htm[2020-07-27].

[31] 教育部 . 国务院学位委员会 教育部关于设置"交叉学科"门类、"集成电路科学与工程"和"国家
安全学"一级学科的通知 . http://www. moe. gov. cn/srcsite/A22/yjss_xwgl/xwgl_xwsy/
202101/t20210113_509633. html[2020-12-30].

[32] GCN. Quantum science gets funding infusion. https://gcn. com/articles/2020/01/10/energy-quantum-
research-centers. aspx[2020-01-10].

[33] 光子盒研究院 . 美国宣布斥资 9. 65 亿美元建立五个量子信息科学中心 . https://www. sohu.
com/a/415188396_120762490[2020-08-27].

[34] NIST. Quantum economic development consortium confirms steering committee. https://www.
nist. gov/news-events/news/2020/09/quantum-economic-development-consortium-confirms-steer-
ing-committee[2020-09-16].

[35] 赛智时代 . 全球量子科技战略 . https://www. ciomanage. com/front/article/6157. html[2020-10-
29].

[36] RailNews. Russian railways develop quantum communication roadmap. http://www. railnews. in/
russian-railways-develop-quantum-communication-roadmap/[2020-08-11].

[37] Padma T V. India bets big on quantum technology. https://www. nature. com/articles/d41586-
020-00288-x[2020-02-03].

[38] White House. Pioneering the future advanced computing ecosystem:a strategic plan. https://www.
house. gov/wp-content/uploads/2020/11/Future-Advanced-Computing-Ecosystem-Strategic-Plan-
Nov-2020. pdf[2020-11-18].

[39] European Union. State of the union:commission sets out new ambitious mission to lead on super-
computing. https://ec. europa. eu/newsroom/dae/document. cfm? doc_id=69379[2020-09-18].

[40] European Union. Digital Europe Programme: a proposed € 7. 5 billion of funding for 2021-2027. https://ec. europa. eu/digital-single-market/en/news/digital-europe-programme-proposed-eu75-billion-funding-2021-2027[2020-12-14].

Information Science and Technology

Tang Chuan ,Wang Lina ,Zhang Juan ,
Yangkuang Junyu ,Huang Ru

In 2020, science and technology powers continue to deepen the strategic layout in the field of information technology (IT), promote the innovation and development of IT, seek major breakthroughs and industrial applications of key technologies, inject new impetus into digital industrialization and industrial digital transformation, and seize the initiative in the development of digital economy. This paper focuses on the analysis of important research progress and national strategic plans in the leading edges of four key areas in IT, i. e. artificial intelligence, semiconductor, quantum information and high-performance computing, to reveal new strategic trends, technological trends, key challenges and opportunities in the future.

4.10　能源科技领域发展观察

陈　伟[1]　郭楷模[1]　岳　芳[1]
蔡国田[2]　汤　匀[1]　李岚春[1]

（1. 中国科学院武汉文献情报中心；2. 中国科学院广州能源研究所）

2020 年，全球新冠肺炎疫情深刻影响了能源供需格局，能源绿色低碳转型步伐加快。世界各国积极制定"绿色复苏"计划，争相提出碳中和目标，将可再生能源置于战略核心地位，加速推进新能源技术发展，创造新产业新业态，构建清洁低碳、安全高效、智慧融合的现代能源体系，推动经济可持续复苏，应对气候变化全球挑战。

一、重要研究进展

1. 碳基能源催化转化技术涌现新成果，天然气水合物海域试采获突破

（1）甲烷催化转化效率和选择性不断提升。美国布鲁克海文国家实验室的研究人员开发出新型氧化铈/氧化铜复合催化剂，在水分子的促进作用下将甲烷转化率提高 94%、甲醇选择性提高 4%，为甲烷高效制甲醇提供了新路径[1]。美国路易斯安那大学拉菲特分校的研究人员设计开发出基于铜基催化剂的双金属氧化物电催化剂，甲烷高效转化为丙醇的产率超过 56%，且可以稳定运行超过 40 h[2]。

（2）CO_2 电催化还原制高价值化学品获得新进展。美国卡内基梅隆大学和加拿大多伦多大学的研究人员合作将密度泛函理论计算和机器学习技术结合开发材料筛选系统，快速筛选的铜-铝合金电催化剂实现 CO_2 到乙烯高效催化还原，获得迄今最高的法拉第效率（80%）[3]。厦门大学的研究人员提出"氢助碳碳偶联"新机制，在氟修饰铜催化剂上达到 CO_2 电催化还原制乙烯和乙醇迄今最高的 C_{2+} 生成速率，实现了与传统 CO_2 加氢相当的 $C_{2\sim4}$ 单程收率和更显著的选择性[4]。

（3）中国率先实现水平井钻采深海天然气水合物。2020 年南海神狐海域试采成功，使中国成为全球首个采用水平井钻采技术试采海域天然气水合物的国家，创造了产气总量、日均产气量两项世界纪录，攻克了深海浅软地层水平井钻采核心技术，实

现了从"探索性试采"向"试验性试采"的重大跨越[5]。

2. 新一代太阳能高效转化利用技术产业化前景日趋明朗

(1) 钙钛矿太阳能电池技术稳步迈向商业化。美国北卡罗来纳大学的研究人员采用变幅电容分布(DLCP)技术首次通过实验手段揭示了金属有机卤化物中缺陷态密度的空间和能量分布,阐明了缺陷影响钙钛矿太阳能电池性能的作用机制[6]。澳大利亚新南威尔士大学的研究人员开发了新型封装工艺,助力钙钛矿太阳能电池首次通过国际电工委员会严苛的光伏稳定性标准测试,并首次使用了气相色谱-质谱分析法(GC-MS)清晰地揭示了有机/无机杂化钙钛矿太阳能电池的主要气相分解产物及其分解机制[7]。牛津光伏公司的研究人员制备的转换效率29.52%的晶硅-钙钛矿叠层太阳能电池,通过了德国弗劳恩霍夫太阳能研究所的权威认证,创造了晶硅-钙钛矿双结叠层电池转换效率新纪录,为转换效率突破30%提供了关键技术路径[8]。

(2) 人工光合系统研究取得新突破。中国科学院大连化学物理研究所与企业合作开发的千吨级液态太阳燃料合成示范项目成功运行,集成创新了液态太阳燃料合成全流程工艺装置,具有完全自主知识产权,整体技术处于国际领先水平[9]。德国马克斯·普朗克科学促进协会陆地微生物研究所的研究人员集成合成生物学与纳米微流控技术,研发人工叶绿体组装平台,实现了光驱动下CO_2的高效固定,展现出规模化应用潜力[10]。上海科技大学的研究人员制备新型气体扩散电极,突破三相界面扩散极限,电催化活性显著增强,实现了在低过电位下高活性、高选择性和高稳定性的CO_2催化还原,助力人工光合系统创造了20.1%的太阳能到燃料(CO_2还原制燃料)转换效率的世界纪录[11]。

3. 高效廉价催化剂推动氢能技术迈向低成本时代

(1) 高效廉价制氢催化剂及制备工艺取得突破。日本信州大学的研究人员制备基于铝掺杂钛酸锶复合催化剂,实现了紫外波段内量子产率接近100%,创造了全解水催化产氢析氧的量子产率纪录[12]。中国清华-伯克利深圳学院的研究人员开发了一种大通量制备二维二硫化钼(MoS_2)浆液电催化剂的方法,制备速率比已报道结果高出1~2个数量级,制备的催化剂具有良好的电解水产氢活性,在1 A/cm^2的大电流密度下所需过电压仅为412 mV,并呈现良好的稳定性,为利用低成本大储量矿石等自然资源大规模制备产氢催化剂开拓了新思路[13]。

(2) 燃料电池催化剂催化活性和稳定性不断提升。美国麻省理工学院的研究人员设计制备了一种过渡金属碳化物钛钨碳为核芯、原子级厚度金属铂为壳的核壳结构电催化剂，性能优于商用铂/碳催化剂性能，具备超万次循环稳定性，展现强劲的商业化潜力[14]。美国马里兰大学牵头的联合研究团队通过一步合成方法成功制备了分级多元素纳米材料，克服了单组分催化剂的单一选择性，相同条件下的氧还原反应质量催化活性是铂的 7 倍，氧析出反应质量催化活性是铱的 28 倍，为设计开发燃料电池高性能双功能催化剂提供了新的思路[15]。

4. 人工智能技术助推高性能电池研发效率，下一代储能电池新成果层出不穷

（1）机器学习技术加速高性能电池研发。美国斯坦福大学的研究人员利用机器学习技术开发闭环优化测试系统，将电池充电测试时间从近 2 年缩至 16 天，大幅提升了实验效率、缩减了实验周期，加速了新电池材料探索、发现和研发[16]。英国剑桥大学的研究人员整合电脉冲和机器学习技术开发高精度的电池健康状态和剩余寿命诊断系统，准确度超越商业化的传统方法，有助于改善电池性能和安全性[17]。Richpower公司的研发人员应用人工智能开发出一种"智能"电池管理系统，能够根据用户电池使用行为进行学习和优化，从而显著提高节能水平和效率；其人工智能算法可以预测电池未来故障，并为电池系统提供更稳定的操作和远程维护，实现对电池性能的优化[18]。

（2）全固态锂电池机制和性能研究双双取得突破。美国马里兰大学的研究人员借助原位中子深度分析表征手段定量测定了锂的沉积和剥离过程，首次揭示了与液态电池短路行为不同的石榴石型固态电池中"可逆短路"自恢复机制，阐明了石榴石型固态电池中锂枝晶的形成机制[19]。三星高等技术研究院的研究人员开发高性能硫银锗矿型固态电解质/银-碳复合负极固态电池体系，能量密度高达 942 Wh/kg，稳定循环超过 1000 余次，在电动汽车等高比能储能应用领域具有广阔应用前景[20]。日本京都大学的研究人员开发出一种全新的全固态氟离子电池原型，其理论能量密度是锂电池能量密度的 7 倍，有望让电动汽车实现一次充电续航 1000 公里的目标[21]。

5. 新型受控核聚变装置建造和试验取得新进展

英国"升级版兆安培球形托卡马克"装置首次获得等离子体，为 2040 年前建设一个紧凑型核聚变发电站奠定了关键技术基础[22]。中国核工业集团公司核工业西南物理研究院的研究人员自主设计建造的"中国环流器二号 M"（HL-2M）装置建成并实现首次放电，成为中国当前规模最大、参数最高的先进托卡马克装置，标志着我国自主掌握了大型先进托卡马克装置的设计、建造、运行技术[23]。

二、重大战略行动

1. 碳中和成为后疫情时代全球最受关注的议题，主要国家加快能源绿色低碳转型步伐

碳中和将成为未来技术和产业发展的全球性标准，甚至是贸易和投资的准入门槛，围绕这一新规则形成的国际秩序将重塑全球治理话语权。截至 2020 年底，已有 120 多个国家提出碳中和时间表，占到全球碳排放总量的 65％ 和全球经济体量的 70％[24]。欧盟提出《欧洲气候法案》[25]，拟正式确立《欧洲绿色协议》碳中和目标的法律地位，并早已开展低碳技术开发和各经济部门低碳产业布局。美国政府宣布重返《巴黎气候协定》，提出 2035 年电力零排放和 2050 年碳中和目标[26]。英国出台《能源白皮书：构建零碳未来》[27]，提出加速发展清洁低碳能源技术，构建排放交易体系，创建能源系统数字基础设施，打造低碳产业集群。日本经济产业省发布《绿色增长战略》[28]，确定到 2050 年实现碳中和目标，并提出了涵盖海上风电、氢能、核能在内的 14 个相关产业绿色增长实施计划和路线图。韩国政府通过《2050 碳中和促进战略》[29]，旨在加速发展可再生能源、氢能和电动汽车等低碳能源技术，发展环保产业，实现 2050 碳中和目标。

2. 化石能源利用加速向清洁化发展，油气行业加速绿色转型，煤炭资源高值转化成为布局重点

在低油价和去碳化大形势下，英国石油公司（BP Amoco）、荷兰皇家壳牌集团（Royal Dutch/Shell Group of Companies）、道达尔能源公司（Total Energies）等大型油气公司纷纷宣布 2050 年零碳排放战略目标，将低碳技术整合到石油生产中，并投资可再生能源、氢能、电池储能等技术以促进转型[30]。美国能源部发布的《化石能源路线图》[31]提出到 2030 年的发展目标及重点领域，通过升级现有煤电和开发变革性化石能源技术，实现化石能源资源的价值最大化和可持续绿色发展。同时，美国能源部还将投入 1.22 亿美元建立多个煤基高价值产品创新中心[32]，利用煤炭生产高价值产品并提取稀土元素和关键矿物，弱化煤炭能源的燃料属性，强化其化工原料和碳基材料的资源属性，为煤炭发展创造新的市场。

3. 氢能是未来清洁能源体系的关键组成，各国积极布局抢占氢能经济发展先机

欧盟发布《欧洲氢能战略》[33]，将氢能作为欧洲经济"绿色复苏"及构建 2050 年碳中和欧洲的重要支点，确定未来30年氢能发展三步走路线：①2020～2024年，在2024年前安装至少6 GW可再生能源电解槽，可再生能源制氢年产量达到100万t；②2024～2030年，在2030年前安装至少40 GW可再生能源电解槽，可再生能源制氢年产量达到1000万t；③2030～2050年，可再生能源制氢技术将逐渐成熟，其大规模部署将可以使所有脱碳难度系数高的工业领域用氢能代替。美国发布《氢能计划发展规划》[34]作为氢能研究、开发和示范总体框架，明确未来10年（2020～2030年）氢能发展核心技术领域、研发重点和主要技术经济指标。德国[35]、法国[36]均在2020年发布国家氢能战略，并成立国家氢能委员会，以推进氢能产业发展，谋求成为绿氢技术的全球领先者。

4. 新冠肺炎疫情凸显可再生能源优势，各国大力推进可再生能源发展

国际能源署的分析显示，受新冠肺炎疫情影响，2020年全球能源需求量下降约3.9%，但可再生能源成为唯一逆势增长（4.1%）的能源种类。预计到2025年，可再生能源将成为世界第一大电力来源。可再生能源产业在新冠肺炎疫情下表现出良好的韧性，各国政府加大对可再生能源发展的支持力度，以应对经济衰退，推动能源绿色转型[37]。欧盟委员会公布《海上可再生能源战略》[38]，提出到2050年需要投资近8000亿欧元用于发展海上风电、海洋能等，以将海上风电装机容量从2020年的12 GW提升到2050年的300 GW，并部署40 GW的海洋能及其他新兴技术（如浮动式海上风电和太阳能），助力欧盟实现碳中和目标。德国联邦政府通过《国家能源和气候计划》[39]，设定了2030年将可再生能源在电力结构中的占比提升到65%，在终端能源消费总量中的占比提到30%，以及将温室气体（较1990年）至少降低55%的目标。

5. 核能强国加速抢占未来核能发展制高点

（1）核电强国谋划布局强化国际市场竞争力。美国公布《核能战略愿景》[40]，启动"先进反应堆示范计划"，计划未来7年投入32亿美元，加速下一代先进核能技术研发和商业化进程，以增强美国技术优势并推动出口，提升美国核电竞争力。欧盟可持续核能技术平台发布新版《可持续核能战略研究议程》[41]，提出优先开展先进反应堆技术、核安全与燃料管理使能技术及交叉领域技术三大主题研究，推动欧洲先进核能技术不断向前发展，确保欧洲在民用核能领域保持技术和行业的领先地位。

（2）美国、欧盟、日本加速推进可控核聚变研究。美国能源部积极探索人工智能技术在核聚变研究方面的应用潜力[42]，旨在实现对海量实验数据的实时采集和快速高效分析，提升实验效率、缩短实验周期；开展聚变能新路线探索[43]，重点聚焦聚变能

新概念技术开发、新组件研发制备和新兴数字技术引入应用等三大主题研究内容，加速推进颠覆性技术研发。欧盟与日本实施核聚变研究战略合作协议第二阶段计划[44]，进一步密切合作推进现有聚变设施示范与优化运行，共同应对商业核聚变发电站建设面临的科学和工程挑战。

6. 美国、欧盟着力打造本土电池产业链，强化国内制造竞争力，抢占储能产业主动权

欧盟电池技术与创新平台发布《电池战略研究议程》[45]，从电池应用、电池制造与材料、原材料循环经济、欧盟电池竞争优势 4 个方面提出到 2030 年的研究主题及关键绩效指标，加速建立具有全球竞争力的欧盟电池产业。美国能源部制定《储能大挑战路线图》[46]，重点聚焦储能技术开发、储能制造和供应链、储能技术转化、政策与评估、劳动力开发五大主题领域的研究，并加速技术转化，增强美国国内制造竞争力，确保供应链安全，抢占全球储能市场的领导地位。

7. 多能融合综合能源系统成为各国新的战略竞争焦点

综合能源系统作为能源领域变革的重要方向，已成为当下世界各国研究和关注的焦点。欧盟在 2020 年相继发布《综合能源系统 2030 路线图》[47]和《2021～2024 年研发实施计划》[48]，提出未来 10 年拟投入 40 亿欧元开展综合能源系统研究和创新优先活动，确定了研发行动路线图，以实现 2050 年深度电气化、广泛数字化、完全碳中和的循环经济愿景。德国《国家氢能战略》强调实施"应用创新实验室"（Living Lab）创新资助形式，支持基于氢能的综合能源网络研究，构建未来融合高比例可再生能源的清洁能源系统。美国《核能战略愿景》明确核能-可再生能源多能融合系统发展目标，到 2027 年实现示范运行，到 2035 年实现广泛的商业化应用。日本推进在福岛试点建立融合可再生能源、氢能和智慧社区的"福岛系能源社会"，已建成全球最大可再生能源电力制氢示范厂[49]。

三、启示与建议

1. 研究制定支撑碳达峰碳中和目标的能源科技中长期路线图

加强碳达峰碳中和战略顶层设计，分析不同行业能源相关碳排放现状和机制，探讨能源相关减排技术潜力，研究制定支撑碳达峰碳中和战略目标的中长期技术发展路线图和行动方案，实施碳中和国家重大科技专项，加快零碳发电技术、大规模储能、

氢能、智慧能源网络、新能源交通、绿色化工、零能耗建筑等创新技术的研发与应用，创新碳市场政策机制（如覆盖各高排放行业的全国碳交易体系和碳金融市场建设），推动我国经济增长和碳排放量绝对脱钩，实现高质量发展。

2. 明确氢能定位统筹氢能产业健康有序发展

首先，研究明确氢能在我国能源体系中的定位、阶段性目标和发展路线图、发展原则和重点任务。其次，聚焦氢能核心技术及产业化能力突破，综合考虑资源禀赋、技术和产业基础、市场需求等多方面因素，因地制宜推动地方氢能产业示范应用工作。再则，建立氢能产业发展全国层面引导及协调机制，确定氢能产业相关部门分工，统筹和保障氢能产业健康有序发展。

3. 构建低碳化多能融合综合能源系统助力高质量发展

打造综合能源系统，整合热、电、冷、气、可再生能源等多能源品种，破除各能源种类之间的条块分割现状和多能耦合发展技术壁垒、市场壁垒和体制壁垒，实现多种能源系统之间相互补充和梯级利用、跨区域不同能源的联动，促进能源分系统逐步相互渗透、不断融合，实现能量流、物质流与信息流紧密融合，保障能源利用与生态文明同步协调发展。

致谢：中国科学院大连化学物理研究所蔡睿研究员、中国科学院广州能源研究所赵黛青研究员、中国科学院科技战略咨询研究院郭剑锋研究员审阅了本文并提出了宝贵的修改意见，特致谢忱。

参考文献

[1] Liu Z Y, Huang E W, Orozco I, et al. Water-promotedinterfacial pathways in methane oxidation to methanol on a CeO$_2$-Cu$_2$O catalyst. Science, 2020, doi: 10.1126/science. aba5005.

[2] Xu N N, Coco C A, Wang Y D, et al. Electro-conversion of methane to alcohols on "capsule-like" binary metal oxide catalysts. Applied Catalysis B: Environmental, 2020, doi: 10.1016/j. apcatb. 2020.119572

[3] Zhong M, Tran K, Min Y M, et al. Accelerated discovery of CO$_2$ electrocatalysts using active machine learning. Nature, 2020, 581: 178-183.

[4] Ma W C, Xie S J, Liu T T, et al. Electrocatalytic reduction of CO$_2$ to ethylene and ethanol through hydrogen-assisted C—C coupling over fluorine-modified copper. Nature Catalysis, 2020, doi: 10. 1038/s41929-020-0450-0.

[5] 中华人民共和国自然资源部. 三要素读懂我国天然气水合物第二轮试采. http://www.mnr.

gov. cn/dt/ywbb/202004/t20200401_2505126. html[2020-04-01].

[6] Ni Z Y, Bao C X, Liu Y, et al. Resolving spatial andenergetic distributions of trap states in metal halide perovskite solar cells. Science, 2020, 367: 1352-1358.

[7] Shi L, Bucknall M P, Young T L, et al. Gas chromatography-mass spectrometry analyses of encapsulated stable perovskitesolar cells. Science, 2020, doi: 10. 1126/science. aba2412.

[8] Oxford PV. Oxford PV hits new world record for solar cell. https://www. oxfordpv. com/news/ oxford-pv-hits-new-world-record-solar-cell[2020-12-21].

[9] 中国科学院大连化学物理研究所. 由大连化物所研发千吨级"液态太阳燃料合成示范项目"通过 科技成果鉴定. http://www. dicp. ac. cn/xwdt/mtcf/202010/t20201020_5719609. html[2020-10-19].

[10] Miller T E, Beneyton T, Schwander T, et al. Light-powered CO_2 fixation in a chloroplast mimic with natural and synthetic parts. Science, 2020, 368: 649-654.

[11] Xiao Y J, Qian Y, Chen A Q, et al. An artificial photosynthetic system with CO_2-reducing solar-to-fuel efficiency exceeding 20%. Journal of Materials Chemistry A, 2020, doi: 10. 1039/D0TA06714H.

[12] Takata T, Jiang J Z, Sakata Y, et al. Photocatalytic water splitting with a quantum efficiency of almost unity. Nature, 2020, doi: 10. 1038/s41586-020-2278-9.

[13] Zhang C, Luo Y T, Tan J Y, et al. High-throughput production of cheap mineral-based two-dimensional electrocatalysts for high-current-density hydrogen evolution. Nature Communications, 2020, doi: doi. org/10. 1038/s41467-020-17121-8.

[14] Göhl D, Garg A, Paciok P, et al. Engineering stable electrocatalysts by synergistic stabilization between carbide cores and Pt shells. Nature Materials, 2020, doi: 10. 1038/s41563-019-0555-5.

[15] Wu M L, Cui M J, Wu L P, et al. Hierarchical polyelemental nanoparticles as bifunctional catalysts for oxygen evolution and reduction reactions. Advanced Energy Materials, 2020, doi: 10. 1002/ aenm. 202001119.

[16] Attia P M, Grover A, Jin N, et al. Closed-loop optimization of fast-charging protocols for batteries with machine learning. Nature, 2020, 578: 397-402.

[17] Zhang Y W, Tang Q C, Zhang Y, et al. Identifying degradation patterns of lithium ion batteries from impedance spectroscopy using machine learning. Nature Communications, 2020, 578: 397-402.

[18] Globenewswire. EV battery tech closes acquisition to bring patented bms technology to North America, South America, Europe and Africa. https://www. globenewswire. com/news-release/2020/10/30/ 2118005/0/en/EV-Battery-Tech-Closes-Acquisition-to-Bring-Patented-BMS-Technology-to-North-America-South-America-Europe-and-Africa. html[2020-10-30].

[19] Ping W W, Wang C W, Lin Z W, et al. Reversible short-circuit behaviors in garnet-based solid-state batteries. Advanced Energy Materials, 2020, doi: 10. 1038/s41467-020-15235-7.

[20] Lee Y-G, Fujiki S, Jung C, et al. High-energy long-cycling all-solid-state lithium metal batteries

enabled by silver—carbon composite anodes. Nature Energy,2020,doi:10. 1038/s41560-020-0575-z.

［21］京都大学 . 新型電池 EV 1000キロに道 . http://www. uchimoto. jinkan. kyoto-u. ac. jp/wp-content/uploads/2020/08/20200810NKMTJM70136831. pdf［2020-08-10］.

［22］Department for Business,Energy & industrial strategy. All systems go for UK's £55M fusion energy experiment. https://www. gov. uk/government/news/all-systems-go-for-uks-55m-fusion-energy-experiment［2020-10-29］.

［23］核工业西南物理研究院 . 中国核聚变发展取得重大突破 新一代"人造太阳"装置中国环流器二号 M装置建成并实现首次放电 . https://www. swip. ac. cn/type/92. html? page=3［2020-12-06］.

［24］Energy & Climate Intelligence Unit. Net zero emissions race. https://eciu. net/netzerotracker［2021-04-06］.

［25］European Commission. European climate law. https://eur-lex. europa. eu/legal-content/EN/TXT/? qid=1588581905912&uri=CELEX:52020PC0080♯footnote10［2020-03-04］.

［26］The White House. FACT sheet:president biden takes executive actions to tackle the climate crisis at home and abroad,create jobs,and restore scientific integrity across federal government. https://www. whitehouse. gov/briefing-room/statements-releases/2021/01/27/fact-sheet-president-biden-takes-executive-actions-to-tackle-the-climate-crisis-at-home-and-abroad-create-jobs-and-restore-scientific-integrity-across-federal-government/［2021-01-27］.

［27］Department for Business,Energy&Industrial Strategy. Energy white paper:powering our net zero future. https://assets. publishing. service. gov. uk/government/uploads/system/uploads/attachment_data/file/945899/201216_BEIS_EWP_Command_Paper_Accessible. pdf［2020-12-14］.

［28］経済産業省 . 2050 年カーボンニュートラルに伴うグリーン成長戦略 . https://www. meti. go. jp/press/2020/12/20201225012/20201225012-2. pdf［2020-12-25］.

［29］대한민국 정책플러스 . 2050 탄소중립 추진전략 합동브리핑 . https://www. korea. kr/special/policyFocusView. do? newsId=156425378&pkgId=49500758［2020-12-07］.

［30］NS Energy. Which major oil companies have set net-zero emissions targets? https://www. nsenergybusiness. com/features/oil-companies-net-zero/♯［2020-12-16］.

［31］DOE. Fossil energy roadmap. https://www. energy. gov/sites/prod/files/2020/12/f81/EXEC-2018-003779% 20-% 20Signed% 20FE% 20ROADMAP_dated% 2009-22-20_0. pdf［2020-09-15］.

［32］DOE. DOE announces intent to provide ＄122M to establish coal products innovation centers. https:// www. energy. gov/articles/doe-announces-intent-provide-122m-establish-coal-products-innovation-centers［2020-06-26］.

［33］European Commission. EU hydrogen strategy. https://ec. europa. eu/commission/presscorner/detail/en/ip_20_1259［2020-07-08］.

［34］DOE. Energy department hydrogen program plan. https://www. hydrogen. energy. gov/pdfs/hydrogen-program-plan-2020. pdf［2020-11-12］.

［35］BMWi. The National Hydrogen Strategy. https://www. bmwi. de/Redaktion/EN/Publikationen/

Energie/the-national-hydrogen-strategy. pdf? __blob＝publicationFile&v＝6［2020-06-10］.

［36］ Ministère de l'Économie et des Finances. Stratégie nationale pour le développement de l'hydrogène décarboné en France. https：//www. ecologie. gouv. fr/sites/default/files/DP% 20-% 20Stratégie% 20nationale% 20pour% 20le% 20développement% 20de% 20l% 27hydrogène% 20décarboné% 20en% 20France. pdf［2020-09-08］.

［37］ IEA. Renewables 2020 analysis and forecast to 2025. https：//www. iea. org/reports/renewables-2020［2020-11-10］.

［38］ European Commission. An EU strategy to harness the potential of offshore renewable energy for a climate neutral future. https：//ec. europa. eu/energy/sites/ener/files/offshore_renewable_ energy_ strategy. pdf［2020-11-19］.

［39］ BMWi. Draft of the integrated national energy and climate plan. https：//www. bmwi. de/Redak-tion/EN/Downloads/E/draft-of-the-integrated-national-energy-and-climate-plan. pdf? _blob＝publication File&v＝4［2020-06-11］.

［40］ DOE. Office of nuclear energy：strategic vision. https：//www. energy. gov/sites/prod/files/2021/01/f82/DOE-NE% 20Strategic% 20Vision% 20-Web% 20-% 2001. 08. 2021. pdf［2021-01-08］.

［41］ SNE-TP. SNE-TP strategic research & innovation agenda. http：//www. snetp. eu/wp-content/uploads/2020/02/SNETP-Strategic-Research-Innovation-Agenda-SRIA-20-v5-aa. pdf［2020-02-04］.

［42］ DOE. Department of energy announces ＄30 million for new research on fusion energy. https：//www. energy. gov/articles/department-energy-announces-30-million-new-research-fusion-energy［2020-03-04］.

［43］ DOE. Department of energy announces ＄32 million for lower-cost fusion concepts. https：//www. ener-gy. gov/articles/department-energy-announces-32-million-lower-cost-fusion-concepts［2020-04-07］.

［44］ European Commission. Fusion energy collaboration between the EU and Japan：mastering the pow-er of the sun. https：//ec. europa. eu/info/news/fusion-energy-collaboration-between-eu-and-japan-mas-tering-power-sun-2020-mar-02_en［2020-03-02］.

［45］ European Technology and Innovation Platform on Batteries：Batteries Europe. Strategic research agenda for batteries. https：//ec. europa. eu/energy/sites/default/files/documents/batteries _ eu-rope_strategic_research_agenda_december_2020__1. pdf［2020-12-03］.

［46］ DOE. Energy storage grand challenge roadmap. https：//www. energy. gov/energy-storage-grand-challenge/downloads/energy-storage-grand-challenge-roadmap［2020-11-12］.

［47］ ETIP SNET. ETIP SNET R&I roadmap 2020-2030. https：//www. etip-snet. eu/wp-content/up-loads/2020/02/Roadmap-2020-2030_June-UPDT. pdf［2020-02-27］.

［48］ ETIP SNET. ETIP SNET R&I Implementation Plan 2021—2024. https：//www. etip-snet. eu/wp-content/uploads/2020/05/Implementation-Plan-2021-2024_WEB_Single-Page. pdf［2020-05-14］.

［49］ NEDO. The world's largest-class hydrogen production，fukushima hydrogen energy research field （FH2R） now is completed at Namie town in Fukushima. https：//www. nedo. go. jp/english/news/AA5en_100422. html［2020-03-07］.

Energy Science and Technology

Chen Wei ,Guo Kaimo ,Yue Fang ,Cai Guotian ,
Tang Yun ,Li Lanchun

The global novel coronavirus epidemic has profoundly affected the energy supply and demand pattern and accelerated the pace of green and low-carbon transformation. In response to promote sustainable economic recovery and the global challenge of climate change, countries around the world are actively formulating green recovery plans, vying to put forward carbon neutral goals, placing renewable energy at the core of their strategy, accelerating the development of new energy technologies, creating new industries and new formats, and building a clean, low-carbon, safe, efficient, and smart modern energy system. The major strategic plans for energy science and technology developed by major developed countries and regions, as well as the progress and important achievements of energy technology in 2020, are systematically sorted out and analyzed in this paper, which can help us accurately grasp the evolving technology directions. Finally, several constructive recommendations for the development of energy science and technology in China are proposed.

4.11　材料制造领域发展观察

万　勇　冯瑞华　黄　健　姜　山
（中国科学院武汉文献情报中心）

2020 年，材料制造领域的战略性地位更加重要，相关科技政策与布局持续加码。在信息技术等的驱动下，新材料研发与创新的发展不断加速；材料微观结构与宏观性能之间的基础理论取得突破，结合极限条件下制备加工技术的进步，推动新型高性能材料不断涌现，助力器件向高品质方向发展；材料的循环回收再利用引起各国重视。美国将新一代半导体材料及其制造视为科技与国家安全的重要基石，"制造业美国"新布局网络安全、生物工业制造等方向；欧盟持续聚焦原材料获取，"欧盟石墨烯旗舰计划"进入推广应用新阶段；韩国、日本等则着力发展人工智能半导体、智能制造，重视材料技术的基础作用。

一、重要研究进展

1. 信息技术助力材料设计创新，材料研发持续加速

材料科学与人工智能、计算技术等的交叉推动材料研发模式发生了变革。英国利物浦大学的研究人员将汽车装配线移动机器人改造为"人工智能化学家"。它每天可工作 21.5 h，在 8 天的时间里独立完成了 668 个实验，并研发出一种全新的化学催化剂[1]。美国国家标准与技术研究院开发出名为"材料探索和优化闭环自主系统"的人工智能算法，用来识别和开发对光子设备和生物启发计算机具有潜在应用的新化合物[2]。俄罗斯斯科尔科沃科学技术研究院开发出一种新算法，可在所有可能的化学元素组合中寻找具有所需特性的材料，并在超硬材料和磁性材料上进行了测试[3]。谷歌量子计算研究团队使用 10～12 个量子比特，采用变分量子本征解算器在空间三维的 3个方向上模拟了重氮化合物的异构化，这是迄今首次、最大规模的化学量子计算[4]。

2. 制备技术提升发展，推动新型材料不断涌现并刷新纪录

材料制备是材料科学发展的基础和技术源泉，先进制备技术能够促进并带动一系

列新材料的发展。美国纽约州立大学宾汉顿分校利用一种全新的混合制造工艺——集成了增材制造（又称三维打印）、真空铸造及从电子技术中衍生出的共形涂层技术，将由铋、铟和锡组成的菲尔德金属材料包裹在橡胶壳中，制备出全球首款液态金属晶格（点阵）结构[5]。北京高压科学研究中心与加拿大萨斯喀彻温大学利用激光加温金刚石对顶砧技术，在 150 万 atm 和 2200 K 的极端条件下，首次发现了一种具有褶皱蜂窝层状结构的聚合氮——褶层聚合氮[6]。美国罗切斯特大学在金刚石对顶砧中，用光化学方法合成出含碳的硫化氢体系，在 267 GPa 压力、15 ℃室温环境下观测到超导现象，首次跨越 0 ℃节点[7]。美国麻省理工学院研制出导电聚合物墨水，首次实现了导电聚合物微结构的高分辨、高通量、快速直接三维打印，为导电聚合物的加工制造提供了一种简单快速、成本低廉的技术[8]。

3. 材料结构与性能不断突破极限，积极赋能应用领域

一般地，金属的强度、延展性和韧性这 3 个性能指标是此消彼长的，无法同时获得。中国香港大学和美国劳伦斯伯克利国家实验室合作突破超高强钢的屈服强度-韧性组合极限，在高端钢材要求的高强度（约 2 GPa）、延展性（均匀延伸率 19 %）和韧性（$102 \text{ MPa} \cdot \text{m}^{1/2}$）3 个重要指标均达史无前例的水平[9]。自弛豫铁电单晶 PMN-PT 发现以来，其压电性能尚无新的突破，并且透光率低，无法满足压电器件多功能、高灵敏度的需求。西安交通大学研发出钐掺杂的 PMN-PT 单晶，通过交变电场进行极化，消除对光有散射作用的铁电畴壁，使产品兼具高压电性和高透光性，解决了长期以来二者难以共存的难题[10]。作为典型的一维纳米材料，碳纤维的超强特性一直是研究热点之一。美国莱斯大学的研究人员利用湿法纺丝的方法，制备出具有优异力学和电气性能的碳纳米管纤维：它的导电性约为铜的 80 %，首次突破 10 兆级阈值，达 10.9 MS/m；抗拉强度高达 4.2 GPa，高于杜邦芳纶纤维"凯夫拉"的 3.6 GPa，是迄今强度最大和导电性能最佳的碳纳米管纤维[11]。中国科学院金属研究所制备出一种不存在已知母体材料的全新的二维范德瓦耳斯层状材料 $MoSi_2N_4$，并获得了厘米级单层薄膜[12]。在此基础上，研究人员制备出的膜材料具有超快离子传输性能，在90 ℃和 98 % 相对湿度的传导率达 0.95 S/cm，是已报道的水相质子传输材料的性能最高值，在低温、低湿条件下仍可保持很高的质子传导率[13]。

4. 新材料不断革新器件形态与机制，高性能成为发展潮流

存储设备的尺寸减小，可为超高密度存储、神经形态计算系统、射频通信系统等铺平道路。美国得克萨斯大学奥斯汀分校利用二硫化钼纳米材料的孔洞实现高密度存储，研制出当前世界上最小的存储设备，存储密度是当前商用闪存设备的 100 倍[14]。

有机发光二极管（organic light emitting diode，OLED）的轻薄形态和耐用性往往不可兼得。英国圣安德鲁斯大学利用有机电致发光分子、金属氧化物和具有生物兼容性的聚合物保护层，研制出迄今最耐用、最轻、最薄的有机发光二极管，为未来电子设备的可穿戴设计提供了新的路径。并且，研究人员还利用其引导幼虫运动，进行神经科学研究[15,16]。半导体材料钙钛矿逐渐成为激光材料的热点之一，但存在长时间工作容易熄灭的缺陷。中国科学院长春应用化学研究所与日本九州大学联合开发出一种基于钙钛矿的新型半导体激光器，突破了以往仅能在低温下连续稳定工作的瓶颈，率先实现室温可连续激光输出[17]。

5. 材料回收日益受到重视，助力循环型社会发展

塑料废弃物会导致严重的全球环境问题，因此迫切需要找到低成本地有效处理塑料废弃物的解决方案。英国发布《可持续复合材料计划》，力求加快复合材料回收新技术的开发，解决当前存在的困难，找到更加低成本化、回收材料性能满足要求的解决方案；另一方面，将开发利用蔬菜废料、玉米、坚果壳和藻类等生物基材料，制成新型复合材料，实现可持续发展[18]。中国科学技术大学的研究人员模拟自然环境，基于光诱导碳—碳键裂解和偶联机制，在短短的 40 h 内就将聚乙烯降解为二氧化碳等产物，并进一步光还原为 C_2 燃料，为解决"白色污染"开辟了新的路径[19]。美国普渡大学采用超快微波辐射工艺技术，将废弃的聚对苯二甲酸乙二醇酯转变为对苯二甲酸二钠，并用作电池阳极材料，为可再生能源转化与存储提供了新的思路[20]。

二、重要战略规划

1. 美国：关注机器人、半导体制造等关键技术，新建"制造业美国"研究所

2020 年 9 月，美国计算机社区联盟发布第四版《机器人路线图：从互联网到机器人》，探讨了机器人在未来 5 年、10 年和 15 年促进经济发展的关键使能作用，尤其是在制造、医疗和服务行业。路线图将提出的 6 个方面的挑战映射到相关研究领域，突出了在新材料、集成传感、规划/控制方法等方面的新研究，以及多机器人协作、鲁棒计算机视觉识别、建模和系统级优化等的新研发内容[21]。9 月，美国半导体产业协会和波士顿咨询公司联合发布《半导体制造领域的政府激励措施与美国竞争力》报告，分析并预测了美国联邦政府的激励措施对美国半导体制造业的影响。报告指出，强大的激励措施将有望扭转美国几十年来芯片生产下降的趋势，美国国会正在考虑通

过立法来要求对半导体制造和研究进行大力投资[22]。10月，白宫安全委员会发布《关键与新兴技术国家战略》，先进工程材料、先进制造等位列20项技术清单。报告从推进国家安全创新基地建设、保护技术优势等维度介绍了需要采取的关键行动[23]。

"制造业美国"网络继续布局新方向。保护关键基础设施免受网络攻击是美国面临的最重要挑战之一。5月成立的网络安全制造业创新研究所聚焦节能制造的网络安全建设[24]；10月设立的生物工业制造研究所关注生物工业产品的国内供应链建设，为化学品、试剂、电子薄膜、聚合物、农产品等提供长期可靠的生物工业生产能力[25]。至此，研究所数量由2017年的14家升至16家。

2. 欧盟：聚焦原材料资源获取，推进石墨烯产业化应用

为指导欧洲工业实现气候中立和数字领军的双重转型，增强在全球产业竞赛中的战略自主性，欧盟委员会于2020年3月发布了《欧洲新工业战略》，提出建设"清洁氢联盟"、"低碳工业联盟"、"工业云及平台联盟"和"原材料联盟"等欧洲共同利益的重要项目[26]。"欧盟石墨烯旗舰计划"进入新阶段，专注推进产业化应用。欧盟委员会拟在2020～2023年资助1.5亿欧元，其中的1/3将专门用于将高技术成熟度的材料和技术推向市场[27]。石墨烯旗舰先锋项目在2020年扩大规模，进一步加速了产业化步伐，已用于汽车电池、水与空气过滤系统、飞机除冰等设备和产品[28]。9月，欧盟委员会发布《关键原材料韧性：绘制更高安全性和可持续性路线》和《欧盟战略技术和行业关键原材料前瞻研究》，提出了2020年关键原材料清单及未来行动计划，并分析了电池、风能、无人机和三维打印等行业的关键原材料需求与风险[29]。

3. 韩国：积极应对零部件出口限制，着力发展人工智能半导体和智能制造等产业

继2019年制定材料、零部件和装备的研发策略及创新措施之后，2020年7月，韩国出台了应对日本零部件出口限制的《材料、零部件、装备2.0战略》，大幅扩充相应供应链管理名单，增加了与美国、欧洲、中国等相关的供应链管理核心商品，范围也由先前的半导体、显示器、汽车、电子设备、机械、金属、基础化学、纺织等拓宽至生物、能源、机器人等领域。8月，韩国财政部在《基于数字的产业创新发展战略》中提出，通过"数字＋制造业"提高产业数据的利用率，增强主力产业的竞争力，提出了4项实施战略和9项具体推进任务[30]。10月，韩国政府发布《人工智能半导体产业发展战略》，计划到2030年人工智能半导体的全球市场占有率达20％，培育20家创新企业和3000名高级人才，把人工智能半导体培育为第二个动态随机存取存储器产业，实现人工智能半导体强国目标[31]。11月，韩国发布智能制造创新实施

计划，预计在 2025 年前，韩国中小企业和创业部将智能工厂的推进率从 22％ 提至 30％，数量增至 1000 家，从中选出 100 家作为 "灯塔"，并在工业园区建立数字集群，推动制造业与第一产业和第三产业的结合[32]。

4. 日本：重视材料技术的基础作用，重组供应链应对不确定的外部环境

为重新评估日本技术创新停滞的基本问题，日本经济产业省在 2020 年 5 月发布了《2020 工业技术展望报告》。报告提出了当前到 2050 年应集中资源的重要技术研发方向，其中包括作为所有领域基础的材料技术，作为实现 "社会 5.0" 而关注的 4 个技术领域之一[33]。5 月，日本经济产业省还发布了 2020 版《制造业白皮书》。白皮书指出，面对新型冠状病毒肺炎疫情蔓延、美中贸易争端和持续的地缘政治风险等各种不确定因素，日本制造业需采取重组供应链等新战略，日本企业需推进数字化转型、提升设计水平、强化人力资源等途径，以提高动态适应能力[34]。

三、发展启示与建议

1. 优先突破面临重大风险的关键材料

当前经济全球化面临种种困境，不少先进国家通过改变内部规则进而影响国际贸易规则。对此，我国也需做好应对。在材料领域，与国外先进水平相比，我国部分关键与战略材料及其重点环节仍存在较大差距，亟须提升自主研发水平和自主保障能力。需强化应用牵引、突破瓶颈，坚持问题导向，将面临重大风险的关键新材料作为攻克重点，加速突破关键核心材料，解决 "卡脖子" 技术的材料问题。

2. 加强领域的基础研究和应用研究

作为新一代高新技术的基础和先导，材料是新工业革命的物质保障，发达国家历来重视该领域的基础性研究工作。材料领域大且复杂，需要从经济社会发展和国家安全面临的实际问题中凝练出基础前沿及关键共性方向和目标，遴选出重点支持的材料方向，重视原始创新和颠覆性技术创新，抢占未来先进材料竞争的制高点，争取引领新一轮的科技革命和工业革命，建立国家材料技术创新中心，贯通从基础研究到工程放大的创新平台体系。

3. 优化稳定材料制造产业链和供应链

在当前全球新一轮产业转移的大背景下，以及受中美经贸摩擦、新冠肺炎疫情全

球蔓延的影响，增强产业链和供应链自主可控能力受到空前关注。关键矿物和稀土等原材料的供应链安全问题向来是美国和欧洲的重大关切，拟通过建立清单名录、资助相关研究、扩大供应商网络等途径降低其对第三国的依赖。在我国，新材料和先进制造同属战略性新兴产业。从产业链方面来看，我国在关键环节、核心零部件和主要设备等方面较薄弱，需要引导产业结构调整升级，建立科技创新体系，实现重要链条环节的自主可控。

4. 重视材料全生命周期绿色化

随着社会经济的发展和科学技术的进步，材料的循环使用日益受到重视。发达国家将新材料的发展与绿色发展相结合，重视新材料与资源、环境和能源的协调发展，大力推进从开发、生产、使用到循环回收的全生命周期的低消耗、低成本、少污染和综合利用等。我国可以通过产品设计改进、设备工艺改革等，从源头上节约资源能源、减少环境污染，打造更加高效、清洁、低碳的循环经济。

致谢：中国科学院金属研究所黄粮研究员、中国科学院长春应用化学研究所王鑫岩研究员、中国航空工业发展研究中心胡燕萍高级工程师、中国科学院宁波材料技术与工程研究所应华根研究员、中国科学院沈阳自动化所王楠副研究员对本文初稿进行了审阅并提出了宝贵的修改意见，在此表示感谢！

参考文献

[1] Burger B, Maffettone P M, Gusev V V, et al. A mobile robotic chemist. Nature, 2020, 583: 237-241.

[2] Kusne A G, Yu H, Wu C, et al. On-the-fly closed-loop materials discovery via Bayesian active learning. Nature Communications, 2020, 11: 5966.

[3] Allahyari Z, Oganov A R. Coevolutionary search for optimal materials in the space of all possible compounds. NPJ Computational Materials, 2020, 6: 55.

[4] Google AI Quantum and Collaborators, Arute F, Arya K, et al. Hartree-Fock on a superconducting qubit quantum computer. Science, 2020, 369(6507): 1084-1089.

[5] Binghamton University. Liquid metal research invokes "terminator" film—but much friendlier. https://www.binghamton.edu/news/story/2371/liquid-metal-research-invokes-images-of-terminator-film-but-much-friendlier[2020-04-13].

[6] Ji C, Adeleke A A, Yang L, et al. Nitrogen in black phosphorus structure. Science Advances, 2020, 6(23): eaba9206.

[7] Snider E, Dasenbrock-Gammon N, McBride R, et al. Room-temperature superconductivity in a carbonaceous sulfur hydride. Nature, 2020, 586: 373-377.

[8] Yuk H, Lu B, Lin S, et al. 3D printing of conducting polymers. Nature Communications, 2020, 11: 1604.

[9] Liu L, Yu Q, Wang Z, et al. Making ultrastrong steel tough by grain-boundary delamination. Science, 2020, 368(6497): 1347-1352.

[10] Qiu C, Wang B, Zhang N, et al. Transparent ferroelectric crystals with ultrahigh piezoelectricity. Nature, 2020, 577: 350-354.

[11] Taylor L W, Dewey O S, Headrick R J, et al. Improved properties, increased production, and the path to broad adoption of carbon nanotube fiber. Carbon, 2021, 171: 689-694.

[12] Hong Y, Liu Z, Wang L, et al. Chemical vapor deposition of layered two-dimensional $MoSi_2N_4$ materials. Science, 2020, 369(6504): 670-674.

[13] Qian X, Chen L, Yin L, et al. $CdPS_3$ nanosheets-based membrane with high proton conductivity enabled by Cd vacancies. Science, 2020, 370(6516): 596-600.

[14] Hus S M, Ge R, Chen P-A, et al. Observation of single-defect memristor in an MoS_2 atomic sheet. Nature Nanotechnology, 2021, 16: 58-62.

[15] Keum C, Murawski C, Archer E, et al. A substrateless, flexible, and water-resistant organic light-emitting diode. Nature Communications, 2020, 11: 6250.

[16] Murawski C, Pulver S R, Gather M C. Segment-specific optogenetic stimulation in *Drosophila melanogaster* with linear arrays of organic light-emitting diodes. Nature Communications, 2020, 11: 6248.

[17] Qin C, Sandanayaka A S D, Zhao C, et al. Stable room-temperature continuous-wave lasing in quasi-2D perovskite films. Nature, 2020, 585: 53-57.

[18] National Composites Centre. UK to lead the development of the next generation of sustainable composite materials. https://www. nccuk. com/news/uk-to-lead-the-development-of-the-next-generation-of-sustainable-composite-materials/[2020-07-03].

[19] Jiao X, Zheng K, Chen Q, et al. Photocatalytic conversion of waste plastics into C_2 fuels under simulated natural environment conditions. Angewandte Chemie International Edition, 2020, 59(36): 15497-15501.

[20] Ghosh S, Makeev M A, Qi Z, et al. Rapid upcycling of waste polyethylene terephthalate to energy storing disodium terephthalate flowers with DFT calculations. ACS Sustainable Chemistry & Engineering, 2020, 8(16): 6252-6262.

[21] Computing Community Consortium. Robotics roadmap for US robotics: from internet to robotics, 2020 edition. https://www. cccblog. org/2020/09/09/robotics-roadmap-for-us-robotics-from-internet-to-robotics-2020-edition/? utm_source=feedblitz&utm_medium=FeedBlitzRss&utm_campaign=cccblog[2020-09-09].

[22] Semiconductor Industry Association. Study finds federal incentives for domestic semiconductor manufacturing would strengthen America's chip production, economy, national security, supply

chains. https：//www. semiconductors. org/study-finds-federal-incentives-for-domestic-semiconductor-manufacturing-would-strengthen-americas-chip-production-economy-national-security-supply-chains/ ［2020-09-16］.

［23］White House. Statement from the press secretary regarding the national strategy for critical and emerging technologies. https：//trumpwhitehouse. archives. gov/briefings-statements/statement-press-secretary-regarding-national-strategy-critical-emerging-technologies/［2020-10-15］.

［24］DOE. Department of Energy Selects the University of Texas—San Antonio to lead cybersecurity manufacturing innovation institute. https：//www. energy. gov/articles/department-energy-selects-university-texas-san-antonio-lead-cybersecurity-manufacturing［2020-05-20］.

［25］DOD. DOD approves ＄87 million for newest bioindustrial manufacturing innovation institute. https：//www. defense. gov/Newsroom/Releases/Release/Article/2388087/dod-approves-87-million-for-newest-bioindustrial-manufacturing-innovation-insti/［2020-10-20］.

［26］European Commission. Making Europe's businesses future-ready：a new industrial strategy for a globally competitive, green and digital Europe. https：//ec. europa. eu/commission/presscorner/detail/en/ip_20_416［2020-03-10］.

［27］Graphene Flagship. European commission signs €150M grant funding the graphene flagship. https：//graphene-flagship. eu/news/Pages/European-Commission-Signs-%E2%82%AC150M-Grant-Funding-the-Graphene-Flagship. aspx［2020-02-25］.

［28］Graphene Flagship. The graphene flagship sails into Core 3. http：//graphene-flagship. eu/news/Pages/The-Graphene-Flagship-Sails-into-Core-3. aspx［2020-04-03］.

［29］European Commission. Commission announces actions to make Europe's raw materials supply more secure and sustainable. https：//ec. europa. eu/commission/presscorner/detail/en/IP_20_1542［2020-09-03］.

［30］Ministry of Economy and Finance. 디지털 기반 산업 혁신성장 전략 . https：//www. moef. go. kr/com/synap/synapView. do? atchFileId＝ATCH_000000000015090&fileSn＝2［2020-08-20］.

［31］Ministry of the Interior and Safety. 관계부처 합동 '인공지능(AI)반도체 산업 발전전략' 확정 . https：//www. gov. kr/portal/ntnadmNews/2298498［2020-10-12］.

［32］Korea IT News. South Korean government planning to supply 1,000 5G ＋ AI smart factories by 2025. https：//english. etnews. com/20201113200002［2020-11-13］.

［33］經濟産業省 .「産業技術ビジョン2020」を取りまとめました. https：//www. meti. go. jp/press/2020/05/20200529010/20200529010. html［2020-05-29］.

［34］METI. FY2019 measures to promote manufacturing technology（white paper on monodzukuri 2020）released. https：//www. meti. go. jp/english/press/2020/0529_002. html［2020-05-29］.

Materials and Manufacturing

Wan Yong , Feng Ruihua , Huang Jian , Jiang Shan

In 2020, the materials and manufacturing field showed the following trends: driven by machine learning, research and development of new materials continue to accelerate; breakthroughs in preparation and processing technology promote the emergence of new materials; research on material structure and performance break through the limits to help devices develop in the direction of high performance and low energy consumption; and materials recycle and reuse attract the attention of advanced countries. The United States pays attention to semiconductor manufacturing, and "Manufacturing USA" prepare new institute on Cybersecurity and Bioindustrial Manufacturing. The EU continues to focus on the supply of critical raw materials, and the Graphene Flagship entered a new phase of advancing industrial applications.

4.12　重大科技基础设施发展观察

李泽霞　李宜展　郭世杰　魏　韧　董　璐
（中国科学院文献情报中心）

2020年，新冠肺炎疫情席卷全球，对全球重大科技基础设施（简称重大设施）的发展产生一定影响。很多重大设施内部现场运维活动和外部访问活动大幅缩减，许多正在进行的设施建设与升级工作放缓，甚至完全停止。例如，激光干涉引力波天文台（Laser Interferometer Gravitational-wave Observatory，LIGO）在一季度关闭，欧洲核子研究中心暂停大型强子对撞机升级。尽管如此，全球重大设施快速发展的态势并未改变。这一判断来自以下几个方面的观察。2020年，国内外重大设施的建设和升级稳步推进；新技术的研发为重大设施的革新发展提供了动力；重大设施建立快速响应机制、提供先进技术方法助力新冠病毒研究及药物研发，应对人类的共同挑战，并发挥了重要作用；重大设施绩效评价工作更加系统。

一、领域重要进展

1. 重大科技基础设施建设稳步推进

虽然受到新冠肺炎疫情影响，世界各地在2020年仍稳步推进重大设施建设，主要表现在以下两个方面。

一方面，一批重大设施建设或升级项目在新冠肺炎疫情肆虐的艰难条件下顺利完成或取得重要进展。1月，被誉为"中国天眼"的500米口径球面射电望远镜（Five-hundred-meter Aperture Spherical radio Telescope，FAST）通过国家验收，正式开放运行。7月，历经7年的美国直线加速器相干光源（LINAC Coherent Light Source，LCLS）升级工作完成。升级设施LCLS-Ⅱ使用全新的加速器、低温超导射频加速技术及可精确控制X射线光束的波荡器，使性能得到大幅提升，X射线激光亮度提高了数亿倍[1]。8月，欧洲同步辐射光源（European Synchrotron Radiation Facility，ESRF）经过20个月的升级改造顺利出光，亮度提高了两个数量级，成为全球首个第四代高能同步辐射光源，将在未来相当长一段时间里保持同步辐射科学领域的领先地位[2]。8月，X射线自由电子激光试验装置通过工艺验收；11月，全球在建的最大地基光学望远镜——极大望远镜（Extremely Large Telescope，ELT）完成主镜的研制，标志着 ELT 建设的新阶段[3]。

另一方面，世界各地不失时机地启动了一批重要的建设项目，为稳步推进重大设施的未来发展奠定了基础。美国电子-离子对撞机（Electron Ion Collider，EIC）经过近 20 年的预研，于 2020 年 1 月获批启动建设，建造成本约在 16 亿~26 亿美元，计划于 2030 年前建成，以确保美国在核科学领域的全球领导地位[4]。4 月，英国研究与创新署资助 290 万英镑建设英国大气测量和观测设施（AMOF），计划于 2023 年建成，使英国具备世界一流的大气观测能力，以支持大气、气候和空气污染的最新研究[5]。10 月，美国先进加速器实验测试设施（Facility for Advanced Accelerator Experimental Tests，FACET）启动 FACET-Ⅱ升级改造项目，作为世界上唯一能够提供高能电子束和正电子束的加速器技术试验研究设施，将为下一代粒子对撞机和高功率光源提供机遇[6]。12 月，瑞士的同步辐射光源 Swiss Light Source 获得 9900 万瑞士法郎的预算支持，将在 2024 年前完成 SLS 2.0 的升级改造项目，以满足未来新兴领域的创新需求[7]。

2. 新技术推动重大设施的能力提升

发展新原理、新技术是重大设施发展的重要环节，在新冠肺炎疫情肆虐的 2020 年也取得了可喜的成就。2020 年 1 月，美国布鲁克海文国家实验室实现了世界上首个加速器粒子能量回收和再利用技术，通过粒子能量再利用系统来回收前一批粒子能量并加速下一批粒子，有望成为未来高性能直线加速器和自由电子激光的最节能技术，并为未来超亮粒子加速器铺平道路[8]。3 月，美国费米实验室研发出的新型铌锡超导磁体，产生创纪录的 11.4 T 磁场，远高出此前大型强子对撞机（large hadron collider，LHC）铌钛磁体产生的 7.5 T 磁场[9]。4 月，位于德国汉堡的欧洲 X 射线自由电子激光（European X-rays Free Electron Laser，EuXFEL）产生 25 keV 能量的强 X 射线脉冲，刷新了飞秒强 X 射线光子能量的世界纪录，为超短时间内探测原子尺度的材料内部动态过程提供了前所未有的科学机遇[10]。9 月，美国斯坦福直线加速器中心（Stanford Linear Accelerator Center，SLAC）国家加速器实验室发明了一种利用太赫兹辐射提高加速器粒子能量的新型加速器结构，在相同距离内加速可达到传统加速器能量的 10 倍，并可使加速器长度缩短为原来的 1/10[11]。

3. 继续推进重大设施绩效评价工作走向规范化和系统化

随着重大设施的快速发展，科学界和政府部门越来越认识到做好设施绩效评价和整体发展状况专题评估，对于保证重大设施良性发展极为重要。近几年，欧盟正在逐步推动设施绩效评价工作的规范化和系统化。2020 年，这一努力并没有因新冠肺炎疫情而停顿。6 月，欧盟"地平线 2020"资助的"研究基础设施影响评估路径"项目（RI-Path）发布《研究基础设施社会经济影响评估指南》[12]，在大范围调研的基础上，与设施资助者和决策者共同制定研究基础设施社会经济影响路径分析框架，阐述设施在促进科学发展、解决现实问题、塑造科学与社会结构方面的 13 条影响路径，给出跟踪每条路径和影响的常用指标，对深入理解影响产生的机制和过程做出有益探索。6 月，欧盟委员会发布高级专家组报告《研究基础设施对欧洲变革性科研的支撑》[13]，从全生命周期角度评估了 43 个研究基础设施在建设、运行和长期可持续性方面的进展。评估结果和标准作为 2021 年版路线图设施选取和支持的重要标准。

4. 支撑新冠病毒研究及药物研发应对人类共同挑战

尽管新冠肺炎疫情不同程度地影响了众多设施的开放利用，但也是重大设施及时有效地发挥支撑作用应对疫情挑战的机会。

（1）整合实验方法，建立快速响应机制积极支持新冠肺炎应对。新冠肺炎疫情暴发初期，多数重大设施启动新冠肺炎研究的专用快速访问机制[14,15]，积极支撑病原体检测和治疗方法研发，帮助理解病毒对人类的影响。5 月，欧洲基于加速器的光子源设施联盟（LEAPS）发布《LEAPS 抗击新冠肺炎疫情》研究报告[16]，向科研界介绍 LEAPS 的设施及其相关实验方法，分享疫情相关科研成果与资讯，分配专用束线时间为新冠病毒、疗法和疫苗研究开辟绿色通道。为抗击新冠肺炎疫情，美国能源部迅速建立实验设施资源快速访问门户[17]，汇总国家实验室下辖的同步辐射光源［先进光子源（Advanced Photon Source，APS）、先进光源（advanced light source，ALS）等］、自由电子激光（LCLS 等）、冷冻电镜、高性能计算等用户设施，开辟快速访问通道，通过滚动式的提案审查和快速调度给予优先快速访问机会[18-20]，集中公布疫情相关研究成果[21]，为科研人员和管理人员提供一站式资讯平台。

（2）支撑新冠肺炎结构解析及预防机制研究。1 月，中国科学家利用上海光源的生物大分子晶体学方法对新冠病毒的结构进行了解析，率先发布 2019-新冠病毒 3CL 水解酶（Mpro）的高分辨率晶体结构，为认识病毒及后续基于结构的药物研发提供了重要理论基础[22]。3 月，欧洲研究团队利用德国 DESY PETRA Ⅲ 光源，以 0.1 nm 分辨率解析 SARS-CoV-2 蛋白质三维结构，为筛选潜在药物提供知识基础[23]。5 月，国际团队利用 ALS 和冷冻电镜研究出可识别中和抗体的补充方法，可用于预防性治疗和暴露后治疗[24]。12 月，德国生物技术公司 BioNTech 公司和美国辉瑞公司利用 DESY PETRA Ⅲ 进行下一代 RNA 药物研究，提出改进 RNA 冠状病毒疫苗的方法[25]。

5. 依托重大设施实现多领域重大科技突破

（1）支撑天文和粒子物理的重大进展。2020年1月，《科学》报道平方公里阵列射电望远镜项目（Square Kilometre Array，SKA）团队利用帕克斯望远镜（Parkes Telescope）和莫龙洛天文台综合望远镜（Molonglo Observatory Synthesis Telescope，MOST）第一次在双星系统中探测到时空拖曳现象[26]，为广义相对论的成立提供证据。3月，英国华威大学领衔的国际研究小组利用威廉·赫歇尔望远镜（William Herschel Telescope，WHT）进行光谱检测，发现一颗大气中富含碳的合并白矮星[27]。这是首次通过大气成分确定白矮星起源。4月，欧洲南方天文台甚大望远镜（Very Large Telescope，VLT）首次观测到恒星在银河系中心环绕超大质量的黑洞运转，其运转轨道形似"玫瑰花结"，验证了爱因斯坦广义相对论的预测[28]。5月，FAST望远镜首次发现新快速射电暴[29]。7月，欧洲核子研究中心发现超稀有过程的首个证据——带电的钾离子变成带电的介子和两个中微子（即 $K^+ \rightarrow \pi^+ \nu\nu$），这可能导致新物理学并为其理论预测提供重要证据[30]。9月，天文学家发现迄今最大规模的黑洞碰撞事件，证实了"中等质量"黑洞的存在[31]。

（2）支撑材料科学的研究和发展。2020年2月，荷兰奈梅亨大学领衔的研究小组利用欧洲同步辐射光源，首次在温和条件下以原子精度观察氮化镓生长的初始阶段——半导体材料与液态镓界面的原子结构，为更好地理解和控制氮化镓生长、发展低缺陷生产工艺迈出第一步[32]。同月，科学家使用瑞士的同步辐射光源 Swiss Light Source（SLS）和新的断层扫描方法，首次以纳米级分辨率拍摄材料内部磁性过程的"3D电影"，相关见解将有助发展更加紧凑高效的磁性数据存储设备[33]。同月，美国橡树岭国家实验室利用中子源分析发现金属氢化物中的氢原子排列比模型预测的更紧密。这种特征可能是常温常压超导研究的关键[34]。5月，中国的科研人员利用稳态强磁场装置发现新型拓扑材料外尔半导体，成功地将外尔物理拓展到半导体体系[35]。10月，国际研究团队利用欧洲自由电子激光设施（EuXFEL）首次在磁性材料中观测到斯格明子（Skyrmion）的形成，揭示其微观过程及时间周期[36]。这些知识将有助于降低未来数据存储的能耗。

（3）支持新能源的探索和发展。2020年2月，美国斯坦福直线加速器中心国家实验室利用 LCLS 探索光敏剂分子铁卡宾原子级催化机制，为改进铁卡宾设计和提高催化效率提供知识[37]。同月，美国布鲁克海文国家实验室利用先进光子和散裂中子源（spallation neutron source，SNS）进行电池阴极材料高精度（灵敏度达0.1%）缺陷密度测量，提出改变原子排布的阴极材料优化策略，助力实现电池最佳性能[38]。

5 月，国际合作团队利用德国 BESSY Ⅱ 光源，首次详细地展示电解质中金属传导带的形成过程，为理解电解质从不导电转变为具有金属特性做出重要贡献，相关成果被选为《科学》封面文章[39]。

（4）支撑解决生命健康重大科学问题。1 月，德国科学家使用 DESY 的超亮 X 射线激光揭示布鲁氏锥虫的关键酶结构，为设计特异性阻断该酶并杀死寄生虫的药物提供蓝图，有助于研发治疗非洲锥虫病的新型药物[40]。6 月，日本理化学研究所发现太赫兹辐射可以破坏活细胞中的蛋白质，但不会杀死细胞，表明长期以来被认为不实用的太赫兹辐射有望用于癌症治疗[41]。9 月，瑞士保罗谢尔研究所质子治疗中心首次测试质子的超快、大剂量辐照。这项新的实验性 FLASH 技术可彻底改变癌症放疗方法，节省数周治疗时间[42]。9 月，国际研究团队使用欧洲同步辐射光源极亮光源（EBS）进行毫米级神经元连接（neuronal wiring）成像，并利用人工智能方法重建三维神经元，展现对厚脑组织样本前所未有的解析能力，建立精确的大脑图谱，揭示神经系统疾病机制[43]。

（5）支撑应对社会经济重大挑战。2020 年 1 月，英国的研究人员利用钻石光源研究铀的氧化态及其生化反应，更好地理解铀与环境的相互作用和迁移过程，有助于改进储存和处置放射性废物、清理涉核场所等的方式[44]。4 月，加拿大研究团队使用加拿大光源（Canadian Light Source，CLS）观察高锌饮食鸟类的羽毛中锌的含量和分布，实现在自然界中追踪金属的目标，为环境监测提供新路径[45]。11 月，美国劳伦斯伯克利国家实验室使用 X 射线自由电子激光研究将甲烷转化为甲醇的细菌——甲烷单加氧酶（sMMO），首次确定其结构，帮助设计该化学过程的高效工业催化剂，减少温室气体排放[46]。

二、重要战略计划与部署

1. 多国出台设施发展路线图，加强支持数据设施和中型设施

2020 年，多个国家的政府和国际性科学组织发布了新的设施发展路线图，表现出的一个显著特点是加强支持数据设施和中型设施。将数据设施、中型设施和大型设施统一规划的优越性值得重视。

2020 年 7 月，中国国家高能物理科学数据中心、国家空间科学数据中心、国家天文科学数据中心签订战略合作协议，联合推进科学数据资源建设、技术和应用的创新发展[47]。7 月，英国政府发布《英国研发路线图》[48]，提出要建立国际领先的国家数字研究基础设施，为研究领域带来数字化转型，从而发展英国的数字研究基础设施能

力（数据、超级计算机、软件和人才）。9 月，英国发布《国家数据战略》[49]，提出确保数据基础设施的安全性和弹性，转变政府对数据的使用以提高效率并改善公共服务，并支持创建一个可互操作的联合数据基础设施。10 月，美国国家科学基金会宣布将向 3 个中等规模的研究设施投入约 1.25 亿美元，以应对量子技术、海洋研究和新一代能源等领域的关键挑战[50]。这 3 个中型设施分别是强磁场束线装置（3269 万美元）、全球海洋生物地球化学阵列（5294 万美元）、用于分布式能源资源网络控制的并网测试基础设施（3946 万美元）。

2. 规划加速器长期发展，重视平台型设施对产业技术的带动作用

2020 年 3 月，韩国国家科学技术咨询委员会发布《大型加速器长期路线图与运营战略》，提出针对自由电子激光、质子加速器、重离子加速器和粒子加速器等大型加速器的 4 项中长期运营战略，以建立基础研究和产业发展的基础，创造世界级水平的优秀成果[51]。3 月，俄罗斯批准《2019～2027 年发展同步和中子加速器基础设施联邦科学技术计划》，总计拨款 1383 亿卢布（约合 133 亿元人民币），将在 2027 年建成四代同步辐射光源和脉冲型散裂中子源样机，并基于高通量中子束流反应堆 PIK 建成至少 25 个实验终端并投入使用[52]。7 月，英国科学与技术基础设施理事会（Science and Technology Facilities Council，STFC）发布 X 射线自由电子激光器（European X-ray Free Electron Laser，XFEL）的科学案例草案[53]，分析了未来几十年 XFEL 将带来的科学和技术机会，以及从 2030 年左右开始运行的具有新功能的英国 XFEL 设施将带来哪些影响。指出 XFEL 不仅可以回答诸多重大科学问题，而且对于发展具有潜在重要意义的新兴技术也至关重要。10 月，美国能源部宣布在 3 年内投资 1800 万美元资助"超强激光器设施网络"（LaserNetUS），旨在改善研究人员获取独特激光的途径，助力恢复美国在高强度激光研究的主导地位。

3. 制定粒子物理研究战略，支持专用型设施探索宇宙奥秘

2020 年 1 月，英国研究与创新署和美国能源部签署了一项协议，指出英国将为国际深层地下中微子实验（DUNE）及美国费米国家加速器实验室主持的相关项目提供价值 6500 万英镑的支持[54]。英国科研界作为 DUNE 的主要贡献者，英国 14 所大学和 2 个实验室为设施建设和实验开展提供了必要的专业和技术支持。此次投资是支持DUNE 项目的第一阶段，该项目将持续至 2026 年。6 月，欧洲核子研究中心理事会通过《2020 欧洲粒子物理战略》[55]，提出粒子物理近期和长期发展愿景，将正负电子对撞机（希格斯工厂）作为优先级最高的下一代对撞机。后续将考虑建设未来的强子对撞机（质心能量至少为 100 TeV），其能量范围比现有的大型强子对撞机高一个数

量级。该战略还提到，大规模数据密集型软件和计算基础结构是粒子物理研究计划的重要组成部分。该战略中提到的设施仍是发展战略设想，不同于路线图，今后的实际发展具有很大的不确定性。

三、启示与建议

1. 大力推广重大设施远程服务模式并加强相关技术研发

"远程操作"已有较长历史，随着网络技术、自动控制技术和人工智能的发展，远程操作已逐步应用于场景复杂、操作精度要求高的医学及科学实验等领域。近年来，许多重大设施都在努力升级和扩展设施支撑能力，特别是实验终端的远程操作能力，并在全球抗击疫情中发挥了独特的作用。远程操作不但对抗击疫情这样的特殊情况有重要意义，对于重大设施提高使用效率、降低使用费用、扩大开放共享等都具有普遍的重要意义，必然成为重大设施发展的一个重要趋势。因此建议建立不受流行性疾病影响的、高效且灵活的重大设施操作和访问原则、标准与实施细则；探索高质量、全方位的远程服务，包括从样品制备到多实验技术协同支撑；并为这些工作提供相应技术和专项资金的支持。

2. 加快加强数据基础设施建设

科技发达国家很早就开始关注科研数据问题。作为国家战略资源，我国在科学数据设施布局和建设方面远远落后于国际发达国家。近年来，数据基础设施成为发达国家规划布局的重点，发达国家不断强化对科学数据、模型和计算资源的战略部署。而我国目前大量生物研究的数据上传到美国国家生物技术信息中心（National Center for Biotechnology Information，NCBI）等国外生物数据中心，地理科学数据上传到4TU. ResearchData 等国外数据平台，存在因科研数据流失和数据垄断而受制于人的风险。因此建议加强加快我国科学数据设施的建设，更好地整合、管理和利用目前的设施。

3. 加快推进重大设施规范化、标准化、制度化绩效评价工作

科技发达地区（特别是欧盟）近年来非常重视对重大设施的绩效评价。规范化的评价将进一步促进重大设施的创新管理，加强重大设施规范化布局，提升其研究潜力和运行效能。随着近年来我国对重大设施布局和建设的加快，应同步加快推进重大设施规划化、标准化和制度化的绩效评价工作，加强对重大设施统筹规划、运行监测，

通过科学规范的管理方式来促进重大设施的快速健康高效发展。

　　致谢：中国科学院高能物理研究所阎永廉研究员、张闯研究员和中国科学院物理研究所金铎研究员审阅了全文并提出宝贵的修改意见和建议，谨致谢忱！

参考文献

[1] SLAC. SLAC's upgraded X-ray laser facility produces first light. https://www6. slac. stanford. edu/news/2020-07-17-slac-upgraded-x-ray-laser-facility-produces-first-light. aspx[2020-07-17].

[2] ESRF. Opening of ESRF-EBS, a new generation of synchrotron. https://www. esrf. fr/home/news/general/content-news/general/opening-of-esrf-ebs-a-new-generation-of-synchrotron. html [2020-08-25].

[3] ESO. Design achievements for actuators and sensors of the main mirror of ESO's ELT. https://www. eso. org/public/announcements/ann20032[2020-11-24].

[4] BNL. U. S. Department of Energy selects Brookhaven National Laboratory to host major new nuclear physics facility. https://www. bnl. gov/newsroom/news. php? a=116996[2020-01-09].

[5] STFC. STFC supporting new UK facility for atmospheric measurements. https://www. wired-gov. net/wg/news. nsf/articles/STFC+supporting+new+UK+facility+for+atmospheric+measurements+02042020154300[2020-04-02].

[6] SLAC. SLAC starts up new facility to revolutionize particle accelerators. https://www6. slac. stanford. edu/news/2020-10-12-slac-starts-new-facility-revolutionize-particle-accelerators. aspx[2020-10-12].

[7] PSI. PSI equips the Swiss Light Source SLS for the future. https://www. psi. ch/en/media/our-research/psi-equips-the-swiss-light-source-sls-for-the-future[2020-12-21].

[8] BNL. Transformative'green'accelerator achieves world's first 8-pass full energy recovery. https://www. bnl. gov/newsroom/news. php? a=116982[2020-01-21].

[9] FNAL. Three national laboratories achieve record magnetic field for accelerator focusing magnet. https://news. fnal. gov/2020/03/three-national-laboratories-achieve-record-magnetic-field-for-accelerator-focusing-magnet[2020-01-21].

[10] EuXFEL. European XFEL reaches world record photon energies. https://www. xfel. eu/news_and_events/news/index_eng. html? openDirectAnchor=1772&two_columns=0[2020-04-08].

[11] SLAC. SLAC invention could make particle accelerators 10 times smaller. https://www6. slac. stanford. edu/news/2020-09-23-slac-invention-could-make-particle-accelerators-10-times-smaller. aspx [2020-09-23].

[12] RIPath. Guide book for socio-economic impact assessment of research infrastructures. https://ri-paths-tool. eu/files/RI-PATHS_Guidebook. pdf[2020-06-24].

[13] European Commission. HLG report：supporting the transformative impact of RIS on European research. https://www. esfri. eu/latest-esfri-news/hlg-report-supporting-transformative-impact-ris-euro-

pean-research/[2020-06-25].

[14] ESRF. COVID-19 scientific research. https://www. esrf. fr/home/news/general/content-news/general/covid-19-scientific-research. html[2021-05-06].

[15] LEAPS. LEAPS facilities research on SARS-CoV-2 and rapid access. https://leaps-initiative. eu/leaps-facilities-research-on-sars-cov-2/[2020-05-28].

[16] LEAPS. Research at LEAPS facilities fighting COVID-19. https://leaps-initiative. eu/research-at-leaps-facilities-fighting-covid-19/[2020-05-14].

[17] DOE. Coronavirus: DOE response. https://www. energy. gov/covid/coronavirus-doe-response[2020-03-14].

[18] SLAC. LCLS call for proposals for coronavirus research. https://lcls. slac. stanford. edu/news/lcls-call-for-proposals-for-coronavirus-research[2020-05-20].

[19] SLAC. COVID-19 rapid access program. https://lcls. slac. stanford. edu/rapid-access-program/covid-19. [2020-12-05].

[20] BNL. Combining expertise across disciplines to address drug development, information processing, and more. https://www. bnl. gov/science/covid19. php[2021-08-19].

[21] APS. SARS-CoV-2 publications, preprints, highlights, and features about research at the advanced photon source. https://www. aps. anl. gov/Science/Publications/sars-cov-2[2021-08-19].

[22] 中国科学院 . 新型肺炎病毒 3CL 水解酶高分辨率晶体结构公布 . http://www. cas. cn/zt/sszt/kjgzbd/mtbd/202001/t20200129_4732988. shtml[2020-01-29].

[23] DESY. Scientists X-ray coronavirus proteins. https://www. desy. de/news/news_search/index_eng. html? openDirectAnchor=1799&two_columns=0[2020-03-24].

[24] Berkeley Lab. X-ray experiments zero in on COVID-19 antibodies. https://newscenter. lbl. gov/2020/05/19/xray-covid19-antibodies/[2020-05-19].

[25] DESY. DESY's X-ray source PETRA Ⅲ points possible ways to better RNA vaccines. https://www. desy. de/news/news_search/index_eng. html? openDirectAnchor=1977&two_columns=0[2020-12-01].

[26] SKAO. Astronomers detect distant space-time 'dragging' for the first time. https://www. skatelescope. org/news/astronomers-detect-distant-space-time-dragging-first-time/[2020-01-31].

[27] STFC. Two stars merged to form massive white dwarf. https://webarchive. nationalarchives. gov. uk/20200923012508/https://stfc. ukri. org/news/two-stars-merged-to-form-massive-white-dwarf/[2020-03-06].

[28] ESO. ESO telescope sees star dance around supermassive black hole, proves Einstein right. https://www. eso. org/public/news/eso2006/[2020-04-16].

[29] 中国科学院重大科技基础设施共享服务平台 . FAST 望远镜首次发现新快速射电暴 . https://lssf. cas. cn/lssf/500mkjwyj/xwdt/202005/t20200501_4556144. html[2020-05-01].

[30] STFC. NA62 experiment at CERN reports first evidence for ultra-rare process that could lead to new

physics. https://webarchive. nationalarchives. gov. uk/20200923014619/https://stfc. ukri. org/news/na62- experiment-at-cern-reports-ultra-rare-process-could-lead-to-new-physics/[2020-07-29].

[31] Nature. 'It's mindboggling!': astronomers detect most powerful black-hole collision yet. https://www. nature. com/articles/d41586-020-02524-w[2020-09-02].

[32] DESY. X-rays show birth of semiconductor for blue LEDs. https://www. desy. de/news/news_search/index_eng. html? openDirectAnchor=1788&two_columns=0[2020-02-24].

[33] PSI. Short film of a magnetic nano-vortex. https://www. psi. ch/en/media/our-research/short-film-of-a-magnetic-nano-vortex[2020-02-24].

[34] Oak Ridge National Laboratory. Closely spaced hydrogen atoms could facilitate superconductivity in ambient conditions. https://www. ornl. gov/news/closely-spaced-hydrogen-atoms-could-facili-tate-superconductivity-ambient-conditions[2020-02-03].

[35] 中国科学院重大科技基础设施共享服务平台. 稳态强磁场用户发现新型拓扑材料外尔半导体. https://lssf. cas. cn/lssf/wtqccsy/xwdt/202005/t20200522_4556780. html[2020-05-22].

[36] EuXFEL. Controlling tiny magnetic swirls in the sea of spins. https://www. xfel. eu/news_and_events/news/index_eng. html? openDirectAnchor=1825&two_columns=0[2020-10-07].

[37] SLAC. How iron carbenes store energy from sunlight— and why they aren't better at it. https://www6. slac. stanford. edu/news/2020-02-06-how-iron-carbenes-store-energy-sunlight-and-why-they-ar-ent-better-it. aspx[2020-02-06].

[38] Brookhaven National Laboratory. Cathode 'defects' improve battery performance. https://www. bnl. gov/newsroom/news. php? a=116984[2020-02-05].

[39] HZB. BESSY Ⅱ:experiment shows for the first time in detail how electrolytes become metallic. https://www. helmholtz-berlin. de/pubbin/news_ seite? nid = 21345;sprache = en;seitenid = 1 [2020-06-05].

[40] DESY. Research team finds possible new approach for sleeping sickness drugs. https://www. desy. de/news/news_search/index_eng. html? openDirectAnchor=1773&two_columns=0[2020-01-30].

[41] SciTechDaily. Terahertz radiation can disrupt proteins in living cells—contradicting conventional belief. https://scitechdaily. com/terahertz-radiation-can-disrupt-proteins-in-living-cells-contradicting-conventional-belief/[2020-06-02].

[42] PSI. New technique for ultrafast tumour therapy. https://www. admin. ch/gov/en/start/docu-mentation/media-releases. msg-id-80542. html[2020-09-28].

[43] ESRF. Neural cartography. http://www. esrf. eu/home/news/general/content-news/general/neu-ral-cartography. html[2020-09-15].

[44] STFC. X-rays used to better illuminate how uranium moves in deep underground environments. https://stfc. ukri. org/news/x-rays-used-to-better-illuminate-how-uranium-moves-in-deep-under-ground-environments/[2020-01-06].

[45] CLS. Synchrotron technique promising for tracing metals in nature. https://www. lightsource. ca/

news/details/mapping_metals_in_feathers[2020-04-21].

[46] Berkeley Lab. How to reduce greenhouse gas? Tips from a methane-eating microbe. https://news-center. lbl. gov/2020/11/13/methane-eating-microbe-enzyme/[2020-11-13].

[47] 国家高能物理科学数据中心. 国家高能物理科学数据中心、国家空间科学数据中心、国家天文科学数据中心签订战略合作协议. https://www. nhepsdc. cn/news/detail/81[2020-07-20].

[48] UK Government. UK_research_and_development_roadmap. https://assets. publishing. service. gov. uk/government/uploads/system/uploads/attachment_data/file/896799/UK_Research_and_Development_Roadmap. pdf/[2020-07-01].

[49] DCMS. National data strategy. https://www. gov. uk/government/publications/uk-national-data-strategy/national-data-strategy/[2020-09-09].

[50] NSF. NSF enables groundbreaking science with $125 million in mid-scale infrastructure investment. https://www. nsf. gov/news/special_reports/announcements/102920. jsp/[2020-10-29].

[51] 대형가속기장기로드맵및운영전략. 대형가속기 장기로드맵 및 운영전략(안) . https://www. msit. go. kr/web/msip-Contents/contentsView. do? cateId=_policycom2&artId=2776307[2020-03-25].

[52] Правительство Российской Федерации. Об утверждении Федеральной научно-технической программы развития синхротронных и нейтронных исследований и исследовательской инфраструктуры на 2019-2027 годы. https://docs. cntd. ru/document/564524137? marker = 6560IO[2020-04-01].

[53] STFC. UK XFEL science case. https://stfc. ukri. org/files/uk-xfel-science-case/[2020-07-02].

[54] Kurt Riesselmann. UK invests £65 million in international science projects hosted by Fermilab. https://news. fnal. gov/2020/01/uk-invests-65-million-in-international-science-projects-hosted-by-fermilab/[2020-01-22].

[55] CERN. 2020 update of the european strategy for particle physics. https://home. cern/sites/home. web. cern. ch/files/2020-06/2020% 20Update% 20European% 20Strategy. pdf/[2020-06-19].

Major Research Infrastructure

Li Zexia , Li Yizhan , Guo Shijie , Wei Ren , Dong Lu

In 2020, COVID-19 swept the world and had a certain impact on the development of major research infrastructure globally. Many major facilities' internal on-site operation and maintenance activities and external visits have been greatly reduced, and many constructions and upgrades of facilities have slowed down or even stopped completely. For example, the Laser Interferometer Gravitational-wave

Observatory(LIGO)closed in the first quarter, and CERN suspended the upgrade of the Large Hadron Collider. Nevertheless, the trend of rapid development of major research infrastructure has not changed. This judgment comes from the following observations. In 2020, the construction and upgrading of major research infrastructure domestic and abroad are steadily advancing; the research and development of new technologies provide impetus for the innovative development of major research infrastructure; rapid response mechanisms for major research infrastructure was established to respond to the common humanity challenges, major research infrastructure provided advanced technology methods to assist COVID-19 research and drug development, and played an important role; the performance evaluation of major facilities is also more systematic.

4.13　世界主要国家和组织科技创新战略与规划发展观察

葛春雷　李　宏　刘　栋　张秋菊　王建芳

（中国科学院科技战略咨询研究院）

2020 年伊始，新冠肺炎疫情席卷全球，不仅冲击了各国的经济和社会发展，也对科技创新活动产生深远影响。在新一轮科技革命和产业变革深入推进并叠加疫情的背景下，世界主要国家坚持将科技创新作为提振经济、增长信心、加速经济社会转型升级的根本动力，冀望通过出台综合性科技创新规划，全方面规划科研愿景，增强科技竞争力；在数字化、绿色发展和卫生健康等领域进行战略性部署，力图推动国家向数字化、绿色化转型和健康发展。

一、综合性科技创新规划

1. 全面规划科研愿景，着力保持科技优势地位

新一轮科技革命和产业变革加速演进，以科技为核心的国际力量对比深刻调整。美国、英国等发达国家通过在科学研究、科技人才、基础设施、伙伴关系等重要领域提出全方面科研愿景，力争在国际格局调整中保持领先。

2020 年 6 月，美国国家科学理事会（National Science Board，NSB）发布《2030愿景》规划报告[1]，在科学与工程研究、科技人才、基础设施建设和伙伴关系 4 个关键要素领域提出倡议，旨在继续保持美国的全球领导者地位：①科学与工程研究。确保对人工智能、量子信息系统等攸关美国竞争力的关键领域的近期与长期资助，广泛投资基础研究，加快从基础研究发现到创新的转化。②科技人才。集聚世界上最有才华的科学与工程人才，建立包容且多样的人才库，确保强劲的国内和国际人才供给通道。③基础设施建设。采用更具战略性的方法开发国内研发基础设施，支持本国更多区域建设中型研究基础设施，扩大创新的地理分布范围，纠正空间和机构间创新资源不均衡的现象。④伙伴关系。加强国内合作，增进联邦政府与州政府、慈善机构的合作，建立官产学研紧密合作关系，扩展战略性双边和多边国际伙伴关系。

7月，英国发布《英国研究与开发路线图》，旨在通过加大基础设施投入、吸引全球人才等措施提升本国科研实力，巩固世界领先科学大国的地位[2]，主要举措包括：①优先投资基础设施。建立国际领先的数字研究基础设施。②制定人才战略。激励和支持卓越科技人才和团队，增加对处于职业早期阶段的科研人员的支持。③扩展全球合作。为与全球的研究创新合作伙伴关系创造机会。④加大对基础科学和关键领域的资助。保障变革性新技术的研发位居全球前列。⑤推动科学研究的转化和创新。释放未来可以推动英国增长和繁荣的行业和技术创新[3]。

8月，韩国发布《科学技术未来战略2045》[4]，提出面向2045年的科技发展长期目标和主要政策方向，强调以应对挑战和转型为核心，大力发展能够提高民众生活质量和经济发展质量、为人类社会做出贡献的科技，构建安全健康、丰饶便利、公平诚信、对人类有所贡献的社会。该战略文件从科技发展参与主体、科技发展空间（区域与全球）、科技发展政策环境三大维度确立了8大政策方向。

10月，美国国务院发布《关键与新兴技术国家战略》[5]，强调发展关键与新兴技术，并以推进美国国家安全创新基地（National Security Innovation Base，NSIB）建设为支柱，保持美国的全球领导力。美国国家安全创新基地被定义为包括学术界、国家实验室和私营机构在内的美国知识、能力和人员网络，将想法转化为创新，将发现转化为成功的商业产品和公司，并保护和改善美国人的生活方式。

2. 以科技创新推动经济复苏，重点发力数字化与绿色转型

科技创新为应对疫情、恢复经济社会发展提供强大动力。世界主要国家以数字化和绿色转型为抓手来复苏经济。

5月，欧盟委员会发布预算总额为7500亿欧元的"欧盟下一代"疫后复苏计划[6]。通过该计划，欧盟将在未来着力打造绿色、数字化和更具韧性的欧洲。短期措施主要包括：支持成员国的经济复苏，拉动私营部门投资，应对卫生危机。长期战略包含三大政策方向：以绿色、低碳、环保的理念促进经济"绿色复苏"，全面推动数字经济发展，强调公平的竞争环境。政策措施包括：加强研究、创新与外部行动计划，支持应对未来危机的关键计划，支持全球合作，等等。

9月，法国公布"法国复兴"经济复苏计划[7]，投资1000亿欧元，围绕生态转型和数字化等方面，使法国经济在2022年恢复至新冠肺炎疫情暴发前的水平。在生态方面，法国投入300亿欧元用于推进生态转型，重点支持建筑节能改造、工业脱碳、绿色交通、绿色铁路、绿色氢能、生物多样性、农业转型、核能等，致力于使法国在2050年实现碳中和。在数字化方面，法国将在巴黎建设7.3万 m^2 的数字健康科技园，依托海量健康数据和尖端设施，打造拥有培训-研究-创新-转化完整价值链的世界

一流园区，使法国成为数字健康研究的领军者[8]。

二、数字化领域的战略与规划

随着数字技术和数字经济在新冠肺炎疫情中发挥越来越重要的作用，世界主要国家对数字化发展的重视程度不断提升，从国家层面提出具体的发展战略，旨在把握未来发展新机遇，利用数字化转型带动整个社会的转型升级。

1. 利用数据促进数字经济和创新发展

新兴数字技术促进了人们对数据资源的价值发掘。数据逐步成为企业、产业乃至国家的关键创新要素和战略性资源。数据利用在全球竞争和数字社会发展中的价值日益凸显，受到世界主要国家的广泛关注。

2月，欧盟委员会发布《欧洲数据战略》[9]，提出欧盟未来5年发展数据经济所需的政策措施和投资策略，通过建立跨部门的治理框架、加强数据基础设施投资、提升个体数据权利和技能、打造公共欧洲数据空间等措施，致力于构建"单一数据市场"，将欧洲打造成全球数据赋能社会的典范和领导者。

8月，韩国财政部发布《基于数字的产业创新发展战略》[10]。通过制定"数字＋制造业"创新发展战略，将重点放在制造业这一韩国优势产业，针对产业特殊性制定大数据政策，以提高制造业中产业数据的利用率，通过数字创新促进韩国成为世界4大产业强国之一。

9月，英国数字、文化、媒体和体育部（Department for Digital，Culture，Media & Sport，DCMS）发布《国家数据战略》[11]，旨在推动数据在政府、企业和社会中的使用，进而推动创新。优先任务包括：释放数据在经济领域的价值，确立促增长、可信赖的数据体制，转变政府对数据的使用以提高效率并改善公共服务，确保数据基础设施的安全性和弹性，倡导数据的跨境流动。

10月，美国国防部发布部门《数据战略》[12]，提出国防部应加快向"以数据为中心的组织"过渡，快速地大规模利用数据提高作战优势和效率，以支持美国国防战略和数字现代化战略的实现。

11月，美国白宫发布《开拓未来的先进计算生态系统战略计划》[13]，将"打造创新型、可信赖、经过验证、可用且可持续的软件和数据生态系统"确立为四项战略目标之一，构想了一个面向未来的先进计算生态系统，为美国继续保持在科学工程、经济竞争和国家安全方面的领先优势奠定基础。

2. 塑造以数字解决方案为动力的数字化社会

随着新一轮科技革命和产业变革的加速推进,加快数字化应用发展、加速数字社会建设成为国际社会的重要议题[14]。

2月,欧盟委员会发布战略文件《塑造欧洲数字未来》[15],提出未来5年欧盟数字化变革的理念和行动,致力于塑造涵盖网络安全、关键基础设施及数字化教育等全方位的数字化未来。主要目标包括:①开发以人为本的技术。为所有欧洲的数字能力投资;保护人们免受网络威胁。②发展公平且有竞争力的数字经济。形成充满活力的创新社会,使快速成长的初创企业及中小企业能够获得融资并发展壮大;确保欧盟的相关规则适合数字经济目标;在确保保护个人信息和敏感数据的同时,增加对高质量数据的访问。③通过数字化塑造开放、民主和可持续的社会。

12月,加拿大发布《重启、复苏和重新构想加拿大人的繁荣:构建数字化、可持续和创新性经济的宏伟增长计划》,提出创造包容性增长轨道,鼓励产业数字化转型,投资数字和物理的战略性基础设施,实现数字和数据驱动的经济[16]。

三、绿色发展领域的战略与规划

新冠肺炎疫情大流行促使人们深刻反思人与自然的关系,世界主要经济体积极倡导疫后"绿色复苏",旨在促进工业和经济绿色、低碳和可持续发展,推动全球经济绿色转型。

1. 打造生态经济模式,促进产业绿色发展

生态经济是实现绿色发展的必然要求,产业发展生态化是发展生态经济的关键和核心。世界主要国家积极培育绿色经济,旨在把生态优势转化为经济发展优势。

1月,德国通过《国家生态经济战略》[17],整合以往的生态经济政策,为未来几年德国进一步发展生态经济确立新的战略框架。该战略遵循两个指导方针:①将生态知识和先进技术作为未来可持续和"气候中和"经济体系的支柱。②通过生物资源实现可持续经济和循环经济。其核心目标是在德国建立可持续、可循环和创新引领的经济。

11月,英国发布《绿色工业革命十点计划》,以期在2050年之前实现温室气体净零排放目标。十点计划包括:海上风能,氢能,核能,电动汽车,绿色公共交通、骑行与步行,"净零航空"和绿色航海,绿色建筑,碳捕集、利用与封存,自然保护,绿色金融与创新。为此英国政府将投入210亿英镑[18]。

2. 依靠科技创新支撑绿色发展

绿色技术研究与创新助力经济发展的绿色转型，依靠科技创新破解绿色发展中的难题。

12月，丹麦发布《国家投资绿色研究、技术和创新战略》[19]，将从加大绿色研究资助、建立绿色研究与创新伙伴关系、加强对话与国际合作、加强对绿色研究的监测与评估、发挥国家气候研究中心的支持和协同作用等方面，加速开发绿色解决新方案和技术，减少温室气体排放，增加绿色工作岗位，强化丹麦绿色产业的全球领先地位。

四、公共卫生领域的战略与规划

新冠肺炎疫情大流行给世界各国的公共卫生危机治理带来了巨大挑战。国际社会更加重视公共卫生领域的科技发展，通过加大研究投入，提高传染病防控与治疗的科学化水平，加强科学研究在公共卫生危机应对中的关键作用。

5月，欧盟委员会为确保"欧盟下一代"疫后复苏计划的有效实施，设立预算为94亿欧元的新的独立卫生计划，主要用于疾病预防、危机防范、重要药物和设备的采购及改善长期健康。计划提出在欧盟研发与创新框架计划"地平线欧洲"中加强对健康相关研究与创新活动的支持，加强对科学驱动的解决方案的投资，以应对冠状病毒大流行、临床试验、创新性保护措施、病毒学和疫苗研发、疾病治疗和诊断及将研究结果转化为公共卫生政策措施等挑战[20]。

7月，日本发布《统合创新战略2020》，提出要提高公共卫生危机的应对能力，快速高效开展诊断治疗技术、疫苗、医学器械的研发工作，重视对控制疫情必不可少的行动经济学等社会科学，灵活运用数字技术快速及时地发布信息，防止疫情扩散[21]。

9月，俄罗斯出台《2035年传染病免疫预防发展战略》[22]。该战略主要针对白喉、麻疹、风疹、乙型病毒性肝炎和季节性流感等传染病的免疫预防，目标是通过提供俄罗斯生产的免疫生物制剂，持续发展传染病免疫预防，从而做到预防、控制传播、消灭传染病和其他疾病。

11月，俄罗斯发布《2025年俄罗斯联邦卫生发展战略实施行动计划》[23]。计划分为6个主要任务：提高医疗服务的可用性和质量；疾病预防；开发、推广和应用新的医疗技术和药物；预防危险传染病的传播；完善公共卫生领域的监管体系；保障生物安全。

五、启示与建议

1. 增强科技创新原动力、支撑力与引领力

新冠肺炎疫情大流行加剧了国际科技竞争,国际社会着力在科技创新领域寻求应对公共卫生危机、复苏和提振经济、加速经济社会数字化转型、促进绿色低碳技术发展以缓解气候变化等复杂问题的解决方案。我国应贯彻科技强国战略,坚持"四个面向",在实践中进一步提升科技创新的核心地位,增强科技创新的原动力、支撑力和引领力,依靠科技创新应对重大风险挑战和国际竞争新格局,赢得未来发展的主动权。

2. 高度重视促进生命健康、数字技术和生态环保技术领域的科技创新

新冠肺炎疫情大流行促使国际社会更加重视生命健康、数字经济和绿色发展。这些领域成为全球科技创新的关键领域,面临前所未有的发展机遇。建议我国将保障人民生命健康、发展数字经济、建设数字社会、促进绿色发展等放在优先发展的战略地位,贯彻"面向人民生命健康"国家战略导向,强化科学技术在重大疾病防治中的关键作用;着力推动数字技术的研发和创新应用,全面推进数字中国建设;加快实施绿色可持续发展战略,以科技创新赋能绿色发展。

致谢:中国科学院武汉文献情报中心刘清研究员、中国科学院成都文献情报中心张志强研究员审阅了全文并提出宝贵的修改意见和建议,谨致谢忱!

参考文献

[1] NSB. Vision 2030. https://www.nsf.gov/nsb/publications/2020/nsb202015.pdf[2020-06-30].

[2] 中国国际科技交流中心. 英国政府发布国家研发路线图. http://kczg.ciccst.org.cn/index.php?m=content&c=index&a=show&catid=111&id=282[2022-02-18].

[3] UK Department for Business,Energy & Industrial Strategy. UK Research and Development Road-map. https://assets.publishing.service.gov.uk/government/uploads/system/uploads/attachment_data/file/896799/UK_Research_and_Development_Roadmap.pdf[2021-07-31].

[4] 陈奕彤. 韩国科学技术未来战略 2045. https://www.sohu.com/a/441335073_468720[2021-11-02].

[5] 黄婧. 美国《关键与新兴技术国家战略》明确 20 项关键技术. http://www.istis.sh.cn/list/list.aspx?id=12788[2021-10-27].

［6］ 王灏晨．"欧盟下一代"经济刺激计划与中欧经贸合作发展．https：//xueshu. baidu. com/usercent-er/paper/show？ paperid＝1j5x04k0pc3m00v0kx510pj0bd422352&site＝xueshu_se［2021-10-26］．

［7］ 李宏策．支撑当下　规划未来——2020 年世界科技发展回顾科技政策．http：//baijiahao. baidu. com/s？id＝1687963674288286461&wfr＝spider&for＝pc［2021-11-10］．

［8］ MESRI. PariSanté Campus：faire de la France un leader mondial de la santé. numérique. https：// www. enseignementsup-recherche. gouv. fr/cid155744/parisante-campus-faire-de-la-france-un-lead-er-mondial-de-la-sante-numerique. html［2020-12-28］．

［9］ 朱开鑫．欧盟距离单一数据市场有多远？《欧洲数据战略》解读．https：//www. sohu. com/a/ 376338480_455313［2021-12-02］．

［10］ 기획재정부．디지털기반산업혁신성장전략http：//www. moef. go. kr/com/synap/synapView. do?atchFileId ＝ATCH_000000000015090&fileSn＝2［2020-09-01］．

［11］ UK Department for Digital，Culture，Media & Sport. National Data Strategy. https：//www. gov. uk/government/publications/uk-national-data-strategy/national-data-strategy［2020-12-20］．

［12］ 张涛．美发布最新《国防部数据战略》全文详解．https：//xw. qq. com/cmsid/20201014A0BC7800 ［2022-02-18］．

［13］ 徐婧．美白宫发布《开拓未来的先机计算生态系统战略计划》．http：//qikan. cqvip. com/Qikan/ Article/ReadIndex？ id＝7104114737&info＝6nDThmnTIJvbqZ4zqnwITY0edlzPka% 2fsM6b4% 2ffVngt4A9jHqN9g87g% 3d% 3d［2022-01-11］．

［14］ 新华财经．余晓晖：加快数字化发展，推进数字中国建设．https：//baijiahao. baidu. com/s？ id＝ 1697176851873148221&wfr＝spider&for＝pc［2021-11-30］．

［15］ European Commission. Shaping Europe's digital future：commission presents strategies for data and artificial intelligence. https：//ec. europa. eu/commission/presscorner/detail/en/ip _ 20 _ 273 ［2020-03-02］．

［16］ Canada's Industry Strategy Council. Restart，recover and reimagine posperity for all Canadians：an ambitious growth plan for building a digital，sustainable and innovative economy. https：//www. ic. gc. ca/eic/site/062. nsf/vwapj/00118a_en. pdf/ $ file/00118a_en. pdf［2021-02-30］．

［17］ BMBF. Nationale Bioökonomiestrategie für eine nachhaltige，kreislauforientierte und starke Wirtschaft. https：//www. bmbf. de/de/nationale-biooekonomiestrategie-fuer-eine-nachhaltige-kreislauforientierte-und-starke-10654. html［2020-02-20］．

［18］ 张翼燕．英国"绿色工业革命"十点计划——主要国家实现碳中和的应对举措．https：//www. sohu. com/a/461165513_468720［2021-12-02］．

［19］ Ministry of Higher Education and Science. Green solutions of the future—strategy for investments in green research，technology，and innovation. https：//ufm. dk/en/publications/2020/green-solutions-of-the-future-strategy-for-investments-in-green-research-technology-and-innovation［2020-10-20］．

［20］ European Commission. The EU budget powering the recovery plan for Europe. https：//eur-lex. europa. eu/legal-content/EN/TXT/？ uri＝COM：2020：442：FIN［2020-06-02］．

[21] 日本内閣府. 統合イノベーション戦略. https：//www8. cao. go. jp/cstp/tougosenryaku/index. html[2020-08-20].

[22] Правительство Российской Федерации. Правительство утвердило Стратегию развития иммунопрофилактики на следующие 15 лет. http：//government. ru/docs/40490/[2020-09-30].

[23] Правительство Российской Федерации. Михаил Мишустин утвердил план мероприятий для реализации Стратегии развития здравоохранения до 2025 года. http：//government. ru/docs/40996[2020-12-30].

Strategy and Planning of S&T Innovation

Ge Chunlei，Li Hong，Liu Dong，Zhang Qiuju，Wang Jianfang

In 2020，the global COVID-19 pandemic not only impacted the economic and social development of all countries，but also has had a profound influence on their scientific and technological innovation activities. Whether it was for fighting the pandemic or for restoring the economy，the major countries in the world insisted on scientific and technological innovation as an important way to promote the rapid recovery of their economy and society from the impact of the COVID-19 pandemic. Through the introduction of comprehensive plans for long-term scientific and technological innovation，they have created a scientific research vision that covers all aspects and enhanced the scientific and technological competitiveness in the post-pandemic era. At the same time，they have deployed in the areas such as digitalization，green development and health，and promoted the country's transition to digitalization，green environment and healthy publics.

4.14　世界主要国家和组织科技创新体制机制变革发展观察

惠仲阳　李　宏　张秋菊　王建芳

（中国科学院科技战略咨询研究院）

2020 年，各国普遍面临防控新冠肺炎疫情、复苏经济的挑战。因此，在科技组织管理方面，世界主要国家持续推进科学研究组织管理体制机制改革，针对新冠肺炎疫情防控及经济社会发展面临的突出问题，专门成立新冠肺炎疫情应对决策科技咨询机构，设立针对新冠肺炎疫情防控和复工复产的资助计划；在创新体系方面，通过构建创新平台机构、集群式创新基地等方式，着力打造知识循环流动的创新体系、优势互补的创新生态；在法律法规方面，通过立法或修法弥补有关科技伦理、法律法规盲区的漏洞，促进研发活动规范化、激发创新主体研发新兴技术的积极性。

一、组织管理体制机制

1. 宏观决策与管理体系

（1）成立新冠肺炎疫情决策咨询与协调机构，提高新冠肺炎疫情防控的科学性和专业性。针对新冠肺炎疫情大流行，主要国家成立了由不同级别（政府最高领导人或相关部长）负责的新冠肺炎疫情应对领导小组，协调各级政府、大学和科研机构开展新冠肺炎疫情防控工作。随着新冠肺炎疫情在全球大流行，各国相继成立了专门应对新冠肺炎疫情的决策咨询机构，就科研攻关、出入境管控、产业扶持等具体事项向政府部门提供咨询建议。

1 月，日本成立了由首相任负责人的新型冠状病毒传染病应对本部[1]，作为日本应对新型冠状病毒的最高决策和协调机构。2 月，在新型冠状病毒传染病应对本部下成立专家咨询会议[2]，国立传染病研究所所长为负责人，成员包括日本医师协会、大学、研究机构的知名专家，针对新冠肺炎疫情研判和应对等问题为国家决策提供科学建议。3 月，法国先后成立了两个高级别政府咨询机构[3]，新冠肺炎科学委员会负责向总统和总理就新冠肺炎疫情防控措施提出建议；新冠肺炎研究与顾问委员会

（CARE）负责在诊断测试、新疗法临床试验、疫苗研发、数字化与人工智能应用于治疗新冠肺炎等方面为卫生部长和教研部长提供快速响应的科学建议[4]。同月，德国联邦教研部支持成立了全国性大学医院研究网络，其中设立了由联邦政府人员和其他学术网络成员组成的指导小组，用于领导和协调政府与大学、医院之间的工作及联系。4月，英国商业、能源与产业战略部（BEIS）、卫生与社会保障部（HSC）和研究与创新署（UKRI）共同成立了领导和推动新冠病毒疫苗快速开发与生产的工作组。

（2）完善组织管理机构，促进科研效率提升和关键技术研发。新一轮科技革命和产业变革加速演进、全球经济新旧动能转换，对政府管理服务科研活动效能提出更高要求。主要国家通过新建或整合政府管理机构，提升科研效率、推动关键核心技术研发。

英国政府科学办公室《通过科学重建英国的远大目标——政府科技能力评估》报告[5]，建议完善政府的科学工作体系：由英国政府指导全国的科学活动；各部委应每年公布其科学研究领域的相关资助、政策和管理制度，以鼓励科技合作和委托研发；在重要的跨部门管理的科学研究领域建立共同治理模式，以改进资助方式。

5月，美国国会提出《无尽前沿法案》①，除了明确 10 个优先发展领域以外，还建议加强在美国国家科学基金会下设技术学部。技术学部的项目官员有权采用美国国防部高级研究计划局的管理方法，增大使用短期聘任外部专家的权力。此外，新的技术学部的地位将更高，将接受美国国会任命的外部专家委员会的建议。外部专家委员会将定期更新优先技术清单。

2. 科研资助体系

（1）以应对新冠肺炎疫情流行问题为导向，助力经济社会发展。2020 年，各国均面临新冠肺炎疫情防控和维持经济社会运转所带来的挑战。主要国家一方面直接针对新冠肺炎治疗、新冠病毒疫苗开发等研究工作开展资助，另一方面围绕维持和恢复经济社会发展设立专项资助计划。

2月，法国建立新发传染病研究资助机构，整合原有的新发传染病应急研究与行动联合体（REACTing）和法国艾滋病与病毒性肝炎研究署（ANRS）两个机构，发挥资助与协调全国相关研究两大职能，以更好地应对新冠肺炎大流行和加强对新发传染病的研究。5月，欧盟发布了"下一代欧盟"（Next Generation EU）计划——7500亿欧元的疫后复苏计划[6]，帮助恢复新冠肺炎疫情对经济和社会造成的直接破坏，并酝酿提出 2021～2027 年的总值 1.1 万亿欧元的长期预算。该预算将推动社会经济复

---

① 2020 年 6 月被纳入《2021 年美国创新和竞争法案》的总体框架，并得到美国参议院的批准。

苏、修复和振兴单一市场，保证公平的竞争环境，并支持紧急投资，尤其是在绿色和数字化转型方面的投资，以科技为支撑助推经济复苏。

（2）完善资助体系，提高资助效率和执行力。针对当前科学技术迭代快、科研投入成本高等特点，主要国家根据本国国情不断完善资助体系，提高资助活动的效率和执行力。

6月，欧洲创新理事会（EIC）顾问委员会发布报告《欧洲创新理事会：影响导向的愿景与路线图》，提出"改善基金管理方式，提升效率和执行力"的要求，如针对深度技术、高影响高风险研究，关注早期、种子投资阶段等；提升资助流程的简洁性，避免企业花费时间和费用来准备完整的申请；注重欧洲创新理事会资助的灵活性，使项目能够重新定位、调整或中止。

7月，日本政府提出拟成立总额高达10万亿日元的"大学资助基金"[7]，用于弥补大学经费不足、建设国际顶尖研究型大学，完善日本的资助体系。其中，日本政府出资0.5万亿日元、财政融资4万亿日元，加之市场运作，最终形成10万亿日元的资金规模，由日本科学技术振兴机构（Japan Science and Technology Agency，JST）管理。

3. 科研评价体系

（1）不断完善评估体系和评估方法。针对日益专业和复杂的科研评估任务，以韩国为代表，确立分工明确、突出重点的评估体系，采用专业、高效的评估方法。

8月，韩国科学技术信息通信部制定了《第4次国家研发成果评价基本计划（2021—2025）》[8]。通过推进自主、负责任的评价，实现提升研究成果创造能力的目标。计划拟推进：①强化研究执行主体的评价自主性，以及通过公开评价提升责任性。②提高国家战略性，建立一致性评价体系。③尊重研发成果价值的多样性，提高其经济社会贡献度。④通过使用和积累信息，强化基于数据的定性评价等4个基本方向，以及推进提高评价自主性与责任性、加强"政策-投入-评价"三者间反馈、提升以效果为中心的成果评价、夯实成果评价的基础保障等4个重点项目。

（2）提高评估方法的科学性和公正性。针对科研评估过程中的问题，以英国为代表，提出通过改革不断提高科研评估工作的科学性和公正性。

10月，英国商业、能源与产业战略部负责科学与创新事务的部长阿曼达·索洛维指出[9]，科研评估是一种强有力地改变科技界行为的激励杠杆；通过将评估与资助挂钩，推动开放创新，使科研成果产生更大的影响力。目前英国科研评估体系面临的挑战有：评估体系可能会对科研工作本身的完整性产生负面影响，导致方向错误和重复的科研工作日益增多；对科技期刊的过度依赖，以至发表和引用似乎已经成为工作目的；科研评估工作程序日益复杂和无效。未来，英国的科研评估改革将继续深入，以建立一个有效的科研评估体系，从而激励科研人员从事创造性和冒险性的科研工作、推动公共资金的高效使用。

二、科技创新体系

1. 建立主体明确、知识循环流动的创新体系

（1）加强创新主体建设，明确责任义务和发展方向。为了促进大学、科研机构等知识创新主体的发展，主要国家通过优化治理守则、择优建设创新中心等方式，明确各方的责任和义务，推进重点领域发展。

3月，日本政府、文部科学省、国立大学协会联合发布《国立大学法人治理守则》，进一步明确国立大学的使命，大学负责人①及管理层、相关协会的责任义务。

8月，美国国家科学基金会宣布在全美创建新的人工智能研究所，旨在加速研究并培养未来的劳动力[10]。

俄罗斯持续推进世界一流科学中心和科教中心建设。10月公布了10家世界一流科学中心名单[11]，每个科学中心由多所高校和科研机构参与建设；12月又批准了第二批5家科教中心名单。这些中心将整合高校和科研机构，加强与实体经济部门合作，培养新技术专业人才。

（2）构建创新平台机构，促进交流协作和开放创新。各国政府建立各类创新平台，促进不同创新主体针对新冠肺炎疫情等社会挑战、前沿研究领域的交流协作和开放创新。

4月，澳大利亚成立了"快速研究信息论坛"（RRIF）[12]，由国家首席科学家发起建立、澳大利亚科学院管理，既迅速汇集了多学科研究的专门知识，向政府提供咨询，又促进政府首席科学家和其他部门科学顾问的协作，凸显研究和创新在推动经济社会进步方面的价值。

同月，西班牙高等科研理事会（Consejo Superior de Investigaciones Científicas，CSIC）启动名为"全球健康"的跨学科研究平台（PTI）[13]，集中西班牙150多个生物技术、纳米技术、统计学、人工智能等领域在内的跨学科研究小组，特别针对新冠病毒所带来的挑战进行集中攻关，寻求解决办法。

---

① 国立大学负责人即为"国立大学法人代表"；国立大学校长有时由该负责人担任，有时由该负责人委托"校长遴选委员会"聘请他人担任。

2. 构建优势互补、资源集聚的研发创新生态

（1）政府牵头，为跨部门、跨领域科研合作创造条件。各国政府积极推进，通过签订合作条约、认证优质企业等形式，推动不同创新主体开展跨部门、跨领域的合作。

3 月，日本文部科学省发布了首批 8 家"研究支撑服务和合作伙伴"入选企业名单[14]，旨在借助优秀企业的力量优化本国的创新生态。获得该称号的企业一方面可以获得与文部科学省的优先合作权，向科研人员提供优质的产品和服务；另一方面可以获得文部科学省的合作支持，进一步提升企业在科技界的被认可度。

11 月，西班牙科学与创新部与西班牙国内重要的大学、科研机构、企业和协会等共同签署《科技创新条约》，旨在推动公立与私立机构在科技创新方面的可持续性协调合作，在资源、组织和人才奖励方面做出承诺。

（2）集聚资源，开展高水平的产学研合作。各国政府通过打造科技园区、创新基地等形式，将人才、资源、技术等集聚在一起，推动高水平的产学研合作。

6 月，日本设立了"共创环境支援计划"（COI-NEXT），规划新型冠状病毒肺炎疫情大流行背景下未来社会应有的景象，将大学、科研机构、企业等各方力量聚集起来，资助一批符合国家战略、发挥大学或地区优势的产学官共创基地。每个基地最多资助 3.2 亿日元/年，资助期限长达 10 年。

12 月，法国宣布在巴黎建设数字健康科技园[15]，依托海量健康数据和尖端设施开展产学研协同创新，使法国成为数字健康研究的领军者。该科技园由法国国家健康与医学研究院、国家信息与自动化研究所、法国健康数据中心等 5 大公共机构共同建设，旨在打造拥有培训-研究-创新-转化完整价值链的世界一流园区。

三、法律法规体系

1. 完善科研环境，促进研发活动规范化

针对科研工作的职业吸引力下降、诚信伦理等方面出现的问题，主要国家立法细化对科研活动的支持和管理办法，以改善科研环境，规范研发活动。

5 月，韩国国会审议通过了规范各部门开展国家研发工作的《国家研发创新法案》[16]，于 2021 年 1 月 1 日生效。法案涵盖了提高研究人员的自律性、营造创新环境等国家研发创新的核心原则和内容，旨在将政府各部门不同的研发管理规定体系化，减轻研究人员的行政负担，营造专注的研发环境。

12月，法国颁布《2021—2030年研究规划法》[17]（LPR），将在未来10年通过增加科研投入与改善科研环境的方法推动法国科研可持续发展，保持法国的世界科技强国地位。主要内容包括大幅提高科研投入，增强科研工作的吸引力，完善科研机构的组织形式，简化对实验室与研究人员的管理等。

2. 扫清法律障碍，提高研发新兴技术的积极性

针对信息通信等新兴技术研发过程中可能面临的法律问题，主要国家以法律手段明确各主体的法律责任，扫除法律的灰色地带，激发创新主体的研发积极性。

5月，巴西政府宣布新的《信息和通信技术法》生效[18]。新的法律通过改革激励模式，促进对信息和通信技术公司研究、开发和创新活动的投资。

6月，日本政府公布了《促进特定信息通信技术应用系统研发与应用法》[19]。由于信息通信系统的研发、应用可能涉及网络安全、个人隐私等问题，该法律明确了政府、从业者等主体的权利和义务关系，制定了研发、应用计划的认定制度，有助于促进信息通信技术的快速发展与应用。

四、启示与建议

针对新冠肺炎疫情反复的不确定性，未来一段时间，各国仍然会将新冠肺炎疫情防控、经济复苏作为一项优先事务，尤其是发挥科学技术的支撑作用。各国都需要"面向人民生命健康"，快速、彻底地控制新冠肺炎疫情的流行和反复，快速促进和恢复经济社会发展，完善科技创新管理体制机制和法律法规，以科技的力量为民众的生命健康保驾护航。

1. 设立专门的决策咨询部门，为政府的疫情研判、疫苗研发等工作提供专业的咨询意见

未来出现新冠肺炎疫情局部暴发或零星散发的概率较高，需要处理好新冠肺炎疫情防控和经济复苏之间的关系。通过设立高层次决策咨询部门，听取领域专家的建议，为政府科学防控、精准管理提供智力支持。

2. 加快建设类型多样、学科交叉的创新平台，促进知识和人才流动

在新一轮科技革命和产业变革孕育兴起的背景下，需要加快建设一批新型创新平台，推动重点尖端前沿技术研发，攻克产业界面临的共性技术问题，构建能够发挥主体积极性、集聚优势资源的创新环境，促进领域间、机构间的知识和人才流动。

3. 加强立法工作，解决新技术研发和应用过程中可能出现的法律伦理问题

人工智能等新兴技术和产品可能造成个人隐私信息泄露、主体责任不明等法律伦理问题，必须加强立法工作，健全伦理规范和法律体系，既激发创新主体的研发积极性，又确保新兴技术安全可靠、服务人类福祉。

致谢：中国科学院武汉文献情报中心刘清研究员、中国科学院成都文献情报中心张志强研究员审阅了全文并提出宝贵的修改意见和建议，谨致谢忱！

参考文献

[1] 首相官邸. 新型コロナウイルス感染症対策本部. https：//www. kantei. go. jp/jp/singi/novel_coronavirus/taisaku_honbu. html[2020-02-03].

[2] 首相官邸. 新型コロナウイルス感染症対策本部第9回（令和2年2月14日開催）. https：//www. kantei. go. jp/jp/singi/novel_coronavirus/th_siryou/sidai_r020214. pdf[2020-03-01].

[3] Élysée. Médecins, chercheurs et scientifiques mobilisés contre le COVID-19. https：//www. elysee. fr/emmanuel-macron/2020/03/24/medecins-chercheurs-et-scientifiques-mobilises-contre-le-covid-19 [2020-03-30].

[4] MESRI. Le Comité analyse, recherche et expertise (CARE) Covid-19. https：//www. enseignementsup-recherche. gouv. fr/cid151204/le-comite-analyse-recherche-et-expertise-care-covid-19. html[2020-05-15].

[5] Government Office for Science. A review of government science capability. https：//assets. publishing. service. gov. uk/government/uploads/system/uploads/attachment_data/file/844502/a_review_of_government_science_capability_2019. pdf[2020-10-15].

[6] European Commission. The EU budget powering the recovery plan for Europe. https：//eur-lex. europa. eu/legal-content/EN/TXT/? uri＝COM：2020：442：FIN[2020-10-20].

[7] 総合科学技術・イノベーション会議. 世界と伍する研究大学の実現に向けた大学ファンドの資金運用の基本的な考え方（案）. https：//www8. cao. go. jp/cstp/tyousakai/sekai/7kai/siryo2. pdf[2020-10-21].

[8] 국가과학기술자문회의. [제12회] 국가과학기술자문회의 제12회 심의회의 결과. https：//www. pacst. go. kr/jsp/post/postCouncilView. jsp? post_id＝1773&board_id＝11&etc_cd1＝COUN01＃this [2020-09-05].

[9] BEIS. Science minister on "the research landscape". https：//www. gov. uk/government/speeches/science-minister-on-the-research-landscape[2020-10-28].

[10] NSF. NSF advances artificial intelligence research with new nationwide institutes. https：//www.

nsf. gov/news/special_reports/announcements/082620. js[2020-10-16].

[11] Правительство Российской Федерации. Утверждён список получателей грантов среди научных центров мирового уровня. http://government. ru/docs/all/130521/[2020-10-30].

[12] Australia's Chief Scientist. Rapid research information forum(RRIF). https://www. chiefscientist. gov. au/RRIF/[2020-10-20].

[13] Consejo Superior de Investigaciones Científicas. Interdisciplinary thematic platforms (PTIs). https://www. csic. es/en/research/;interdisciplinary-thematic-platform[2021-01-01].

[14] 文部科学省. 令和元年度「研究支援サービス?パートナーシップ認定制度」認定サービスの決定について. https://www. mext. go. jp/a_menu/kagaku/kihon/1422215_ 00003. htm[2020-04-05].

[15] MESRI. PariSanté campus ;faire de la France un leader mondial de la santé. numérique. https://www. enseignementsup-recherche. gouv. fr/cid155744/parisante-campus-faire-de-la-france-un-leader-mondial-de-la-sante-numerique. html[2020-12-28].

[16] 과학기술정보통신부 . 연구자가연구에만전념할수있는제도적기반이마련되었습니다 .https://www. msit. go. kr/web/msipContents/contentsView. do? cateId=_policycom2&artId=2884356[2020-06-05].

[17] MESRI. Loi de programmation de la recherche 2021-2030. https://www. enseignementsup-recherche. gouv. fr/pid39124/www. enseignementsup-recherche. gouv. fr/pid39124/www. enseignementsup-recherche. gouv. fr/pid39124/loi-de-programmation-de-la-recherche-2021-2030. html[2021-09-12].

[18] MCTI. MCTIC recebe crédito extraordinário de R $ 352 milhões para combate à COVID-19. https://antigo. mctic. gov. br/mctic/opencms/salaImprensa/noticias/arquivos/2020/05/MCTIC_recebe_credito_extraordinario_ de _ R _ 352 _ milhoes _ para _ combate _ a _ Covid19. html? searchRef = MCTIC% 20recebe% 20credito% 20extraordinario&tipoBusca=expressaoExata[2021-11-03].

[19] e-Gov 法令検索. 特定高度情報通信技術活用システムの開発供給及び導入の促進に関する法律、https://elaws. e-gov. go. jp/document? lawid=502AC0000000037[2020-11-12].

Institutional and Mechanism Reform of S&T Innovation

Xi Zhongyang, Li Hong, Zhang Qiuju, Wang Jianfan

In 2020, major countries and organizations in the world were faced with the challenges posed by COVID-19's prevention and control and post epidemic economic recovery. Therefore, in terms of organization and management, these countries and organizations continued to promote the reform of the scientific research organization and management system; in view of the prominent problems faced by economic and social development such as epidemic prevention and control，they established special decision-making consultation agencies for epidemic response and funding plans. In terms of innovation system，through the construction of innovation platform institutions and cluster innovation bases，they strove to create an innovation system with knowledge circulation and an innovation ecosystem with complementary advantages. In terms of laws and regulations，through legislation to make up for ethics and legal blind spots，these countries and organizations promoted the standardization of research and development activities，and stimulated the enthusiasm of innovative entities to develop new technologies.

4.15 世界主要国家和组织未来产业
创新发展观察

陈晓怡 刘 澌 贾晓琪 王建芳 李 宏

（中国科学院科技战略咨询研究院）

2020 年，新冠肺炎疫情全球大流行带来的经济社会危机对世界各国的发展造成极大冲击，公共卫生与人口健康、产业链与供应链安全、数字经济与数字主权等问题凸显，通过技术和产业变革引领经济社会转型发展成为共识并取得重大进展。为此，世界主要国家和组织加快了对未来产业的布局，一是出台战略计划，布局、部署应对未来挑战、引领未来发展的重点领域；二是大力投资创新，力图使技术创新和未来产业发展成为国家经济复苏和竞争力提升的持续动力。

一、未来产业部署重点

1. 制定综合性产业战略，布局引领产业变革的关键技术

面对新冠肺炎疫情、气候变化、能源危机、数字主权等全球性重大挑战，2020 年世界主要国家和组织积极出台综合性产业战略，识别并布局引领新技术和新产业变革的关键技术。

2 月，法国政府发布《使法国成为突破性技术经济体》报告[1]，建议国家支持 22 个新兴产业，优先支持氢能、量子技术、网络安全、精准农业与农机设备、可持续食品、生物控制、数字健康、生物疗法与创新疗法生物产品、去碳工业、新一代可持续复合材料等十大产业；关注数字教育、可回收材料、可持续燃料等 12 个新兴市场。

3 月，欧盟委员会发布《欧洲新产业战略》[2]，提出对欧洲工业未来发展具有重要战略意义的关键使能技术——机器人技术、微电子技术、高性能计算和数据云基础设施、区块链、量子技术、光子学、工业生物技术、生物医学、纳米技术、制药、先进材料和技术。

5 月，日本经济产业省发布《产业技术愿景 2020》[3]，展望 2025～2050 年产业技术发展的五大趋势，提出应优先发展支撑超智能社会的物联网、数字技术等关键技

术；重点发展生物技术、材料技术、环境能源技术。具体包括：支持物联网的机器人技术、传感器技术、网真和远程操控技术、脑机接口技术、机器翻译技术；后工业时代的下一代超级计算机技术，如新型存储技术、量子计算技术；等等。

2. 支持氢能与绿色交通技术创新，推动绿色低碳能源产业发展

面对日益严峻的能源安全和气候变化压力，2020 年，各国在"去碳化"的道路上纷纷选择绿色、高效的氢能作为清洁能源部署重点，并推动以氢能和电能为动力的绿色交通产业等发展。

欧盟、美国接连发布氢能战略，欧洲国家共同选择推动绿氢技术研发和产业发展。7 月，欧盟委员会发布《欧洲氢能战略》，为欧洲未来 30 年清洁能源（特别是氢能）的发展指明方向，将通过降低可再生能源成本并加速发展相关技术扩大可再生能源制氢在难以"去碳化"领域的大规模应用。德国《国家氢能战略》计划投入 70 亿欧元用于资助绿氢研发和创新，使氢能成为工业原材料，开发氢能运输、存储和配送基础设施，建立绿氢国际市场。法国《国家无碳氢能战略》计划到 2030 年投入 70 亿欧元用于发展绿氢，通过发展电解制氢行业、开发氢能交通、支持绿氢技术研究创新和人才培养提升氢能产业竞争力，并促进工业脱碳。挪威《国家氢能战略》将促进氢能技术商业化，投入主攻氢能技术的 ENERGIX 项目，并支持氢能在交通和工业领域的应用。《芬兰氢能路线图》展望未来 10 年低碳氢生产、绿色化学物质和燃料领域的氢利用，以及氢存储、运输和最终用途等发展方向。10 月，俄罗斯发布《2024 年前俄罗斯联邦氢能发展行动计划》[4]，发展高生产效率的出口导向型氢能。11 月，美国能源部发布《氢能计划发展规划》，提出至 2030 年氢能发展的技术和经济指标，明确以制氢、输运氢、储氢、氢转化、终端应用为关键技术领域重点，实现氢能转化相关技术跨领域的广泛应用。

一些国家还重视配套的绿色交通技术及产业发展。9 月，法国在经济复苏计划中投入 85 亿欧元用于支持清洁能源汽车、铁路现代化改造等绿色交通产业。10 月韩国科学技术信息通信部发布《未来汽车推广与市场领先战略》[5]提出至 2025 年完成电动汽车 113 万辆、氢能汽车 20 万辆的国内市场普及目标；至 2030 年，1000 家汽车零部件企业向未来汽车企业转型。12 月，丹麦公布《国家投资绿色研究、技术和创新战略》[6]，将加速开发交通、工业、农业等绿色解决新方案和技术的推出，建立绿色未来基金会。

3. 发展数字信息技术，推进产业数字化转型与数字产业化发展

2020 年，欧洲国家以数字化引导产业与社会转型；在数字信息领域的前沿技术布

局上，美国、欧盟、俄罗斯、韩国进一步加强在人工智能、半导体与电子等关联产业上的投入。

（1）推动产业与社会的数字化转型。2月，欧盟委员会发布战略文件《塑造欧洲数字未来》[7]，拟建立以数字解决方案为动力的欧洲社会。战略涵盖从网络安全到关键基础设施、数字教育到技能的广泛内容。6月，俄罗斯发布《俄罗斯2035年制造业综合发展战略》[8]，加速数字技术在航空、造船、电子、医疗、汽车、农机、石化等优先工业领域的应用。9月，法国经济复苏计划投入18.85亿欧元支持中小企业和政府公共服务的数字化升级。

（2）持续推进人工智能产业化发展。2月，欧盟提出人工智能具体行动，为人工智能产业营造更好的政策环境，鼓励公立机构与私营部门加强合作，调动价值链各环节的资源和积极性。6月，德国提出人工智能投入从原计划的30亿欧元增至2025年的50亿欧元，提升"人工智能欧洲制造"的竞争力。

（3）支持先进半导体技术和电子产业发展。1月，俄罗斯发布《俄罗斯2030年电子工业发展战略》[9]，拟通过优化生产能力、更新设备、开发新的技术和方向、掌握突破性电子工业技术等，创建有竞争力的电子工业。6月，美国国会两党议员提出《为生产半导体创造有力激励措施》（Creating Helpful Incentives to Produce Semiconductors，CHIPS）法案[10]，提议为半导体研发计划投资150亿美元，支持先进半导体研究和原型设计、半导体标准制定、建立新的先进制造研究院支持半导体基础研究。10月，韩国科学技术信息通信部等部门共同制定《人工智能半导体产业发展战略》[11]，推进确保掌握领先型创新技术与人才、激活创新发展型产业生态系统等战略。

二、支持未来产业及企业创新的政策措施

1. 面向未来产业需要，建设产学研深度协同机构和网络

各国通过新建研发机构，建设贯穿创新链条的网络组织和科创园区，积极探索支持未来产业发展的新型研发生产组织模式。

（1）建立储备未来产业基础的量子前沿研究中心。8月，美国能源部计划在5年内投资6.25亿美元建立5个量子科学与工程类研究中心。每个中心由一个或多个跨科学和工程学科及跨机构的协作研究团队组成。9月，德国巴伐利亚州提出在未来2年建立量子技术园区，由大学外科研机构联盟与慕尼黑相关大学及产业界合作成立"慕尼黑量子科学技术中心"，覆盖量子技术从基础到应用的所有方面。

（2）建设面向产业应用和升级的人工智能研究所和创新网络。8月，美国国家科

学基金会等提出建设气象与气候学、分子发现与合成、下一代食品系统、未来农业、管理学、机器学习等 7 个领域的人工智能研究所[12]。10 月，巴西科技与创新部（MCTI）联合巴西研究和工业创新公司的 17 个研究中心成立了全国最大的人工智能创新网络[13]，提供机器学习、物联网、大数据等不同领域的技术解决方案，重点在汽车和农业产业进行人工智能开发和应用。

（3）设立支持绿色低碳产业发展的能源技术研发中心和集群。3 月，英国商务部和交通部宣布资助 3000 万英镑，建立 4 个"推动电力革命中心"（DER），研发绿色电机，并为虚拟产品开发、数字制造和先进组装技术研发提供场所，推动世界领先的电机测试和制造。7 月，德国联邦教研部投资 1 亿欧元用于资助 4 个新的电池研究能力集群，聚焦电池从生产、高效使用、回收到质量保证的所有问题。

（4）成立产学研协同创新的未来医学科创园区。9 月，德国巴伐利亚自由州决定建立"马丁斯里德生命科学园区"，开展跨学科研究，并为生物技术初创企业和疾病诊疗技术研发提供支持。12 月，法国宣布在巴黎建设致力于未来医学研究的数字健康科技园，开展基于大学、科研机构、医院和健康数据机构的产学研协同创新。

（5）建立引领先进数字产业发展的半导体先进制造研究所。6 月，美国 CHIPS 法案提出，未来 5 年每年将投资 3000 万美元用于支持美国国家标准技术研究院建立新的半导体先进制造研究院。重点是：开发先进的测试、封装和装配能力；支持半导体设备维护自动化研究；开展半导体人才的教育和技能培训，从而支持美国芯片产业发展。

2. 大力扶持企业技术创新，提供解决方案与多元资金投入

世界主要国家和组织通过提供解决方案，支持技术扩散，出台税收、补贴、投资等优惠政策，给予企业全方位的关注和支持。

（1）支持企业开展突破性技术创新与应用。2 月，英国科学技术委员会（Council for Science and Technology，CST）发布"促进技术扩散提升生产力"建议[14]：建立面向企业的"国家生产力中心"，帮助企业提高对新技术的熟练掌握程度；由英国商务部和研究与创新署、产业战略委员会及地方企业合作伙伴合作审查创新企业资助计划，创建具有全面影响的各地区技术扩散系统。6 月，欧洲创新理事会（EIC）顾问委员会发布《欧洲创新理事会：以影响为导向的愿景与路线图》[15]，持续支持欧盟创新企业和初创企业，提出促进深度技术研究与市场机会间的联系等。

（2）拓展多元研发投资形式。2 月，英国科学技术委员会发布新的研发投资建议[16]对政府提出以下要求：为企业提供投资方向，确定拟发展的新行业或技术能力，识别新出现的科技和产业机遇方向；完善融资机制以提高企业参与度，鼓励养老基金投资新兴高科技企业，支持政府与企业以投资组合资助创新活动。此外，德国执政联

盟的"未来一揽子计划"和法国"经济复苏计划"都以政府公共投资、企业投资、国有银行投资、免税等形式支持数字技术和健康技术等未来产业关键技术研发。

（3）实施税收优惠政策。6月发布的美国CHIPS法案提议为半导体制造公司提供税收优惠和补贴组合政策：从2021年起为合格的半导体设备或制造设施的支出提供税收抵免；建立100亿美元的信托基金，为投资国内半导体制造设施的企业提供补助金；建立多边微电子安全基金。9月，法国"经济复苏计划"为创新型中小企业提供200亿欧元的减税。12月，巴西总统签署《物联网法》，提出在未来5年免除对机器通信系统的安装和运行税等[17]。

3. 加强战略性产业自主性，保障产业链供应链安全

欧洲因新冠肺炎大流行而引发产业链供应链危机，所以积极出台举措以加强战略性原材料和产业链的安全。

9月，欧盟委员会发布《关键原材料弹性：寻求更安全和更可持续供应的路径》[18]，提出2020年关键原材料清单和未来行动计划，旨在建立欧盟工业生态系统的弹性价值链；减少对主要关键原材料的依赖；加强欧盟内部原材料的采购；实现第三国采购来源多样化。报告提出欧盟应在燃料电池、风能、稀土、机器人原材料、三维打印等方面加强自主研发和扩大材料供应来源。法国"经济复苏计划"要求投入10亿欧元用于支持战略性产业回迁与本土产业链构建，把健康（药品和医疗器械）、必要工业材料（化学品、金属）、电子、农业食品和5G列为五大战略性产业。

三、启示与建议

1. 各国未来产业布局聚焦数字、健康、低碳等方向

面对数字主权、新冠肺炎疫情、气候变化、能源危机等全球共同挑战，各国在数字领域重点布局人工智能、大数据、量子技术等，在健康领域聚焦生物技术、数字医疗等，在绿色低碳领域聚焦氢能、绿色交通等，尝试依托颠覆性技术的突破引领新的产业变革，解决当前与未来的发展问题。

2. 数字技术与绿色技术成为政府引导企业向未来产业转型升级的重要路径

欧盟及其成员国通过战略引导、资金投入和政策帮扶大力支持企业的数字化和绿色转型，以新技术与传统产业的融合促进传统产业数字化升级，提升整个国家乃至整个欧洲的工业竞争力，以实现数字主权和碳中和目标。

3. 技术主权与产业自主因疫情大流行被提升到新的高度

新冠肺炎大流行带来的产业链和供应链"掉链"问题使欧美国家意识到产业自主的重要性，加大"制造业回流"力度，同时提升本土生产能力，强调保障原材料、生物医药等战略性产业和技术的韧性。

4. 以新型研发和生产组织方式的转变支持未来产业的发展

未来产业在技术与技术、技术与产业之间的深度融合，需要依靠前沿研究机构和企业等创新主体的充分互动。各国积极探索产学研合作的跨机构研究所、研发网络/集群等新型研发模式，拓宽协同创新的深度和广度。

致谢：中国科学院武汉文献情报中心刘清研究员、中国科学院成都文献情报中心张志强研究员审阅了全文并提出宝贵的修改意见和建议，谨致谢忱！

参考文献

[1] Ministre de l'Économie et des Finances. Remise du rapport《 Faire de la France une économie de rupture technologique 》. https://www. economie. gouv. fr/remise-rapport-faire-france-economie-rupture-technologique#[2020-02-26].

[2] European Commission. A new industrial strategy for Europe. https://ec. europa. eu/info/sites/info/files/communication-eu-industrial-strategy-march-2020_en. pdf[2020-03-29].

[3] 経済産業省.「産業技術ビジョン2020」を取りまとめました. https://www. meti. go. jp/press/2020/05/20200529010/20200529010. html[2020-06-20].

[4] Правительство Российской Федерации. Михаил Мишустин утвердил план мероприятий по развитию водородной энергетики. http://government. ru/docs/40703/[2020-10-31].

[5] 과학기술정보통신부.관계부처합동'인공지능 (AI)반도체 산업 발전전략'확정 . https://www. msit. go. kr/web/msipContents/contentsView. do?cateId=_policycom2&artId=3158372[2020-11-01].

[6] Minstry of higher education and science. Green solutions of the future—strategy for investments in green research,technology, and innovation. https://ufm. dk/en/publications/2020/green-solutions-of-the-future-strategy-for-investments-in-green-research-technology-and-innovation[2020-12-30].

[7] European Commission. Shaping Europe's digital future:commission presents strategies for data and Artificial Intelligencehttps://ec. europa. eu/commission/presscorner/detail/en/ip_20_273[2020-02-28].

[8] Правительство Российской Федерации. Михаил мишустин утвердил стратегию развития обрабатывающей промышленности. http://government. ru/docs/39844/[2020-06-29].

[9] Правительство Российской Федерации. Утверждена Стратегия развития электронной

промышленности Российской Федерации на период до 2030 года. http://government.ru/docs/38795/[2020-01-30].

[10] Congress of the United States. Creating helpful incentives for producing semiconductors(CHIPS) for America act. https://www.aip.org/sites/default/files/aipcorp/images/fyi/pdf/CHIPS_for_America_Act.pdf[2020-06-30].

[11] 과학기술정보통신부.관계부처합동'인공지능 (AI) 반도체 산업 발전전략'확정. https://www.msit.go.kr/web/msipContents/contentsView.do?cateId=_policycom2&artId=3137757[2020-11-01].

[12] NSF. NSF advances artificial intelligence research with new nationwide institutes. https://www.nsf.gov/news/special_reports/announcements/082620.jsp[2020-09-30].

[13] MCTI. MCTI e Embrapii lançam a maior rede de Inovação em Inteligência Artificial do país. https://www.gov.br/mcti/pt-br/acompanhe-o-mcti/noticias/2020/10/mcti-e-embrapii-lancam-a-maior-rede-de-inovacao-em-inteligencia-artificial-do-pais[2020-11-25].

[14] CST. Diffusion of technology for productivity: CST letter. https://assets.publishing.service.gov.uk/government/uploads/system/uploads/attachment_data/file/868334/CST_Technologies_for_productivity_letter.pdf[2019-09-28].

[15] European Commission. The European innovation council—a vision and roadmap for impact. https://ec.europa.eu/research/eic/pdf/ec_rtd_eic-vision-roadmap-impact.pdf[2020-07-02].

[16] CST. Investing in UK research and development: CST letter. https://www.gov.uk/government/publications/investing-in-research-and-development[2020-03-28].

[17] MCTI. Lei da Internet das Coisas é sancionada pelo presidente da República. https://www.gov.br/mcti/pt-br/acompanhe-o-mcti/noticias/2020/12/lei-da-internet-das-coisas-e-sancionada-pelo-presidente-da-republica[2021-01-17].

[18] European commission. Critical raw materials resilience: charting a path towards greater security and sustainability. https://eur-lex.europa.eu/legal-content/EN/TXT/PDF/?uri=CELEX:52020DC0474&from=EN[2020-12-10].

Future Industrial Innovation

Chen Xiaoyi, Liu Si, Jia Xiaoqi, Wang Jianfang, Li Hong

In 2020, COVID-19 pandemic and the economic and social crisis it brings have had and will continue to have a great impact on the development of countries around the world. The ensuing problems of public health, industrial chain security, digital sovereignty and others were highlighted. Leading economic and social transformation with the help of technological and industrial changes has become the consensus of major economies. The major countries and international organizations have accelerated the future industries overall arrangement. Firstly, they released strategic plan to develop key areas to meet future challenges and lead future development; Secondly, they make great efforts to invest in innovation, and strive to make technological innovation and future industrial development a sustainable driving force for economic recovery and competitiveness improvement.

4.16　世界主要国家和组织国际科技合作及人才交流发展观察

王文君　叶　京　张秋菊

（中国科学院科技战略咨询研究院）

2018年中美贸易争端引发科技战略博弈，2020年新冠肺炎疫情大流行进一步加剧了国际创新合作的不稳定性、不确定性，中国、美国等经济体间的科技竞争更加激烈复杂，给国际科技合作及人才交流带来重大变化。但随着公共卫生、气候变化等全球性挑战问题的出现和加剧，各国携手合作的共识在加强且行动更加积极，希望通过国际科技合作来共同应对挑战。这造成当前国际科技交流呈现"合作中有竞争、竞争中有合作"的复杂态势。

一、强化国际科技合作顶层设计

大国之间的战略竞争特别是科技竞争重现，相关国家及组织调整国际科技合作战略，意图加强关键技术保护及争夺技术标准的国际主导权。各国在国际科技合作的顶层设计上更加注重自身发展需求，旨在通过巩固双边或多边合作"朋友圈"提升本国科技实力及影响力。

1. 维护自身发展和战略利益，制定国际科技合作战略

相关国家以提升本国科技实力为目标，更加突出考虑本国的战略利益，开展战略性国际科技合作顶层设计。

4月，俄罗斯科学与高等教育部发布的《2020年科学与高等教育部目标和任务公开声明》[1]中指出，国际科技合作在科学国际化背景下能够捍卫俄罗斯在科学领域的认同感和国家利益，俄罗斯将通过互利的国际合作提高本国的科学工作成效。

7月，英国政府发布《国家研发路线图》[2]，其中在国际科技合作部分明确提出通过战略性双边合作及对多边研究与创新组织和网络的领导，加强和发展与外国政府及国际资助者的合作，在强大的国际合作与有效保护英国知识产权、敏感领域研究、个

人信息、学术价值之间取得平衡等。

12月，美国国防部发布《国际科技合作战略》[3]，旨在确保美国在作战和科技转化等方面的主导地位。战略报告指出，国际科技合作需要基于"强调美方利益"，并提出加强技术保护、注重合作效益、提高执行效率等原则。在合作考虑因素中，为在合作中充分获得美国利益，明确了美国科技需求及合作现状、国外科技资源、区域合作驱动因素、国外科技合作意愿共 4 个方面因素。

2. 巩固本国国际"朋友圈"，加强合作伙伴关系建设

通过与合作对象制定合作战略蓝图、签署合作协议、召开对话会议、资助合作项目等方式，主要国家及组织不断巩固国际科技合作伙伴或联盟关系。

2月，美国国务院与德国联邦教研部续签《美国政府与德国政府关于科学和技术的协议》[4]，确定未来两国将加深访问、共享和管理人工智能基础设施及数据，量子技术，先进的能源科学和技术（如聚变材料、人造光合作用和先进的电池技术），公共卫生安全合作研究，生物经济等新兴技术领域的合作。8月，美国与澳大利亚科学与前沿技术对话联合委员会召开网络会议[5]，双方就加强人工智能、量子信息科学与海洋探测合作交换了意见，并讨论了确保国际研究诚信的办法。

10月，欧盟与拉美和加勒比国家共同体（Comunidad de Estados Latinoamerica-nos y Caribeños，CELAC）联合启动《欧盟-拉共体研究与创新联合倡议》（JIRI）[6]，并制定《2021—2023 年拉共体-欧盟科技合作战略蓝图》，主要围绕研究人员流动性、基础研究设施合作、医疗、环境可持续发展、数字转型等方面的合作进行规划。12月，欧盟在"泛非计划"（Pan-Afirica Programme）中向非洲 4 个科技领域新项目提供资助[7]，具体包括粮食安全和生态系统监测 2500 万欧元、地质调查和自然资源可持续管理 800 万欧元、非洲-欧洲数字发展中心 800 万欧元、空中导航系统现代化 450 万欧元。

3. 强调技术合作监管，积极参与或主导国际标准制定

为保护本国利益，在国际合作中，以美国为代表的发达国家强调对技术合作进行"风险管理"，实质上强化了对本国技术的保护和控制。此外，为强化技术互操作性及科技转化的主导地位，发达国家还积极部署和参与乃至主导技术的国际标准制定。

在技术合作监管方面，美国政府于 10 月发布的《关键与新兴技术国家战略》中明确提出加强国际合作中的"技术风险管理"。针对合作国提出：将在技术开发的早期阶段要求进行安全设计，并与盟国和合作伙伴一起采取类似行动；让盟国和合作伙伴参与制定他们自己的流程，类似于美国外国投资委员会执行的流程。针对竞争国提

出：确保竞争对手不使用非法手段获取美国知识产权、研发资料或科技成果；通过促进学术机构、实验室和行业的研究安全来保障研发活动的完整性，同时平衡外国研究人员的贡献；评估全球各国的科技政策、能力和趋势，以及如何影响或破坏美国的战略和计划。

在技术标准制定方面，3月，美国特朗普政府发布《美国保护5G安全国家战略》[8]，提出通过布拉格5G安全会议等机制参与制定国际5G安全原则，与外国合作伙伴和盟国进行双边与多边合作。

7月，欧盟委员会发布的《欧洲氢能战略》[9]也提出：①在多边论坛上促进国际标准的制定，加强欧盟在国际论坛上有关氢的技术标准、法规和定义的领导地位。②通过国际标准化机构和联合国全球技术法规机构（联合国欧洲经济委员会、国际海事组织）扩大国际合作，如加强协调制定氢动力汽车的法规等。

二、加强重点领域国际科技合作

在新科技变革叠加新冠肺炎疫情大流行的形势下，公共卫生与健康、数字与信息技术、能源与气候环境等领域的重大全球性挑战对科技合作的紧迫性与导向性提出了新的要求，主要国家及组织通过合作计划、联盟等形式在重点领域开展了相关合作。

1. 加强卫生健康科技合作，携手共抗疫情

新冠肺炎大流行成为需要世界各国共同努力克服的全球重大卫生安全挑战。应对疫情危机，各国政府及国际组织积极发挥相关作用，采取了发起"全球疫苗计划"、建设卫生领域国际合作研究中心等措施。

为应对疫情，各地加强疫苗合作及公平分配，让新冠病毒疫苗成为人人可得的全球公共产品。4月，流行病防范创新联盟（The Coalition for Epidemic Preparedness Innovations，CEPI）、全球疫苗免疫联盟（The Global Alliance for Vaccines and Immunisation，GAVI）和世界卫生组织联合发起"新冠肺炎疫苗实施计划"（CO-VAX）[10]。截至10月，全球已有包括中国在内的184个国家和经济体加入其中。

4月，韩国组建新冠肺炎治疗药物和疫苗开发"泛政府支援小组"，提出将加强国际互助合作：①提前获取人体样本、病原体等传染病研究资源等，建立本地研究中心并进行事前研究。优先推进与韩国病例相关度高的亚洲地区的研究中心建设，后续增加欧洲、非洲等地区的研究中心建设。②加强国际社会交流。通过与海外主要国家、国际机构等进行定期会议，建立紧密的新冠肺炎信息国际合作体系。

2. 联合开展绿色低碳能源技术研发，推动可持续发展

气候变化、能源危机等问题给人类的生存和发展带来日益严峻的挑战，相关国家通过签署、制定相关协议和计划，开展绿色低碳领域的研发合作，其中欧盟及其成员国尤为积极。

在气候变化方面，9月，丹麦政府启动《全球气候行动战略：绿色、可持续发展的世界》[11]。该战略明确将强化丹麦与中国、日本、韩国、印度尼西亚、墨西哥的原有绿色战略伙伴关系，并建立与印度、南非的新绿色战略伙伴关系，为欧盟和世界贸易组织（World Trade Organization，WTO）贸易政策的绿色化而努力，在绿色转型和创新研究中发挥领导作用，积极推动国际气候谈判。

在能源绿色转型方面，7月，欧盟委员会发布《欧洲氢能战略》[12]。该战略中重新设计了欧洲与邻国和地区及其国际、区域和双边合作者的能源伙伴关系，旨在促进供应多样化及帮助设计稳定和安全的供应链。10月，俄罗斯政府发布《2024年前俄罗斯氢能发展行动计划》[13]。该行动计划中涉及了国际合作的具体计划：拟定与氢能生产国和消费国（德国、日本、丹麦、意大利、澳大利亚、荷兰、韩国等）进行双边合作的提案；制定有关发展氢能国际合作的建议；为俄罗斯参加氢能领域的多边合作、相关国际机构的活动制定建议；提出相关建议，在国外形成俄罗斯作为无碳排放环保型氢能供应商的声誉；提出在国际市场上推广俄罗斯氢产品和氢能技术的建议。

3. 应对疫情防控需求，推动数字领域国际合作加速发展

新冠肺炎的全球大流行，促进了远程医疗、在线教育、共享平台、协同办公、跨境电商等技术服务的快速发展和广泛应用，促进各国加强在数字领域的国际合作。

2月，奥地利、保加利亚、丹麦和罗马尼亚4国正式加入"欧盟量子通信基础设施计划"（QCI）[14]，代表已有24个欧盟成员国加入QCI计划，未来10年将共同研发和部署欧盟的量子通信基础设施。欧盟通过该计划将量子技术和系统整合到现有通信基础设施，利用量子基础设施以超级安全的方式传输、存储信息和数据，并实现欧盟通信资产全连接。6月，加拿大、法国、德国、澳大利亚、美国、日本、韩国、欧盟等15个创始成员正式成立全球首个"人工智能全球合作伙伴组织"（GPAI）[15]。该组织重点关注合理使用人工智能、数据管理、人工智能对未来就业的影响、创新和商业化等4个领域发展。9月，英国发布的《国家数据战略》[16]中强调将倡导数据的跨境流动。脱欧后，英国将推广英国做法并开展国际合作，确保数据不受国界和碎片化监管制度的不当限制。

三、国际科技人才争夺加剧而国际交流受阻

高科技的竞争，本质上是高端人才的竞争。一方面，顶尖科技人才成为竞相争夺的焦点，各国加大对国际人才的吸引力度，以应对新冠肺炎疫情暴发后经济下滑和未来科技创新发展的人才需求；另一方面，新冠肺炎进一步破坏了国家间互信，核心技术垄断性与科技国家化趋势日渐增强，使得科技人才的合作交流在一定程度上受到阻碍。

1. 主要国家争夺全球顶尖科技人才

近年来，掌握尖端科技的创新型科技人才成为国际人才争夺战的焦点。2020 年，世界发达国家发布了一系列人才吸引新措施和计划。

1 月，英国政府宣布开始实施一项新的快速签证计划[17]，将不限制到英国的人才数量，以吸引世界顶尖科学家、研究人员和数学家，促进英国的科学发展，显示出英国政府支持吸引顶尖人才的决心。新体系将提供一个由英国研究与创新署管理的全新快速签证通道，使科学家个人能够快速进入签证申请阶段。为配合"全球人才"签证计划，英国政府将在未来 5 年为来英国的全球顶级人才的实验和数学研究提供高达 3 亿英镑的支持资金。

12 月，法国颁布《2021—2030 研究规划法》（LPR）[18]。在吸引国际人才部分，该规划法提出的短期目标为以奖金形式吸引外籍研究人员，长期目标为提升法国机构在国际上的排名。

2. 主要国家创造有利于人才交流合作的科研环境

为复苏受新冠肺炎疫情影响的社会经济，主要国家及组织通过各类战略及行动计划，为人才交流与合作创造具有吸引力的科研环境，以在吸引人才后留住人才。

6 月，韩国国家科学技术咨询会议审议通过了由科学技术信息通信部制定的《迈向 2030 年人才强国——科学技术人才政策中长期创新方向》[19]，提出将激活吸引各类人才的环境。具体包括：增加回国吸引力、吸引海外青年人才、扩大全球研究网络、构建高层次科技人才支持平台、完备高层次科技人才支持的制度基础等。

9 月，欧盟委员会发布《为科研和创新的新欧洲研究区》（ERA）战略文件[20]，提出以人才驱动为基础的行动计划。欧盟力图通过加强成员国间的合作，促进研究人员的流动和知识与技术的自由流通，确保每个人都可以从研究及其成果中受益，特别强调从协调的方法转向国家政策之间的更深层次融合。具体为，强调通过人才流动计

划、培训等方式，提供支持研究人员职业生涯发展的政策措施组合，使欧洲对人才更具有吸引力。

3. 美国等遏制与竞争对象国的人才交流

新冠肺炎疫情暴发后，各国临时关闭边境，暂停签发签证和工作许可，全球人才流动因此几乎陷入停滞状态。自 2020 年 6 月起，美国政府宣布取消与中国军方有关联的留学生及研究人员的签证[21]。西北工业大学、哈尔滨工程大学、哈尔滨工业大学、北京航空航天大学、北京理工大学、南京理工大学和南京航空航天大学等 7 所高校，以及船舶工业、兵器工业、航空航天、核工业等国防工业系统高校的中国学生和访问学者，均受到这一政策带来的签证限制。特别是，美国政府不断加大力度并全方位地审查在其科研机构（如美国国立卫生研究院）和研究型大学工作的华裔科学家与中国的科技合作行为，甚至以间谍、非法合作等名义起诉了一些科学家，减少中国科学、技术、工程与数学（STEM）专业留学生的留学签证等。美国采取类似"科技脱钩"举动，对中国高层次人才交流与中美科技合作造成了极为不利的影响。

四、启示与建议

面对国际竞争新格局及新冠肺炎疫情的新挑战，世界主要国家和组织依据自身发展的特点及需要，强化战略性国际科技合作顶层设计，加强技术监管并推动标准国际化，争夺国际人才并限制交流。合作抗疫和经济复苏的迫切需求成为全球科技创新合作的新的连接点，各国在公共卫生与健康、能源与气候环境、信息技术等领域不断拓展新合作。基于前述态势，我们提出以下建议。

1. 加强国际科技合作政策监测和风险预警

当前竞争与合作并存的复杂环境下，急须加强有关国家科技合作政策的监测评估研究，强化对国际科技合作风险的管控和预警。密切跟踪、监测世界主要国家和地区国际科技合作的政策变化，尤其关注美国等国专门针对与我国科技合作的限制政策，建立国际科技合作风险预警预判机制，敏锐把握相关政策动态，加强我国相关部门之间的协作与互动。

2. 结合"一带一路"建设，探索国际科技合作新路径

中国"一带一路"倡议成为引领新型全球化、推动世界经济包容和强劲增长的有效路径之一，未来应积极关注国际科技发展动态，尤其是人类健康、气候变化、信息

技术等全球性发展议题的科技发展趋势与我国比较优势，探索新的科技合作升级发展路径；推动生物多样性保护、"数字丝路"等大科学计划，完善多边合作机制，加强与国际标准化机构和联合国全球技术法规机构间的合作，积极参与国际标准制定，贡献全球科技治理的中国方案。

3. 调整合作方式，转危为机，抢占新一轮科技人才制高点

中美战略竞争、新冠肺炎大流行等破坏了国家间互信，核心技术垄断性与科技国家化趋势日渐增强，国际科技人才合作与交流在一定程度上受到阻碍，特别是美国政府限制中美两大科技国家之间的科技合作。与此同时，随着中国实施创新驱动发展战略、推进科技强国建设，中国的科技快速发展及科技投入大幅增长，对全球科技人才的吸引力不断增强。我国应抓紧制定具有针对性的人才政策，大力吸引在美国等西方科技强国工作的华裔科学家回国工作。同时，完善高端人才移民政策，积极搭建官方、半官方、民间等各类国际科技人才合作交流平台，推进"走出去"与"请进来"同步并行，以多样化渠道吸引人才，冲破西方国家对我国的科技封锁和打压，抢占新一轮科技人才制高点。

致谢：中国科学院武汉文献情报中心刘清研究员、中国科学院成都文献情报中心张志强研究员对本文进行了审阅，并提出了宝贵的修改建议，特致谢忱。

参考文献

[1] Министерство науки и высшего образования Российской Федерации. Публичная декларация целей и задач министерства науки и высшего образования на 2020 год. https：//minobrnauki. gov. ru/ru/documents/card/?id_4＝1161&cat＝/ru/documents/docs/[2021-01-01].

[2] Department for Business，Energy and Industrial Strategy. UK_research_and_development_roadmap. https：//assets. publishing. service. gov. uk/government/uploads/system/uploads/attachment_data/file/896799/UK_Research_and_Development_Roadmap. pdf[2021-01-01].

[3] U. S. Department of Defense. DOD international science and technology engagement strategy. https：//www. cto. mil/dod-ists/[2021-01-01].

[4] United States Government. Joint statement on German-U. S. science and technology cooperation. https：//www. state. gov/joint-statement-on-german-u-s-science-and-technology-cooperation/[2021-01-01].

[5] United States Government. U. S. -Australia joint commission meeting on science and frontier technologies dialogue. https：//www. state. gov/u-s-australia-joint-commission-meeting-on-science-and-frontier-technologies-dialogue/[2021-01-01].

［6］ Gobierno de Mexico. Iniciativa conjunta CELAC-UE sobre investigación e innovación(JIRI)，octava reunión de altas autoridades sobre ciencia y tecnología，30 de octubre de 2020. https：//www. gob. mx/sre/prensa/iniciativa-conjunta-celac-ue-sobre-investigacion-e-innovacion-jiri-octava-reunion-de-altas-autoridades-sobre-ciencia-y-tecnologia? idiom＝es［2021-01-01］.

［7］ European Commission. EU mobilises 82. 5 million to strengthen cooperation with Africa in digital and space technology. https：//ec. europa. eu/international-partnerships/news/eu-mobilises-eu825-million-strengthen-cooperation-africa-digital-and-space-technology_en［2021-01-01］.

［8］ United States Government. National strategy to secure 5G of the United States of America. https：//www. whitehouse. gov/wp-content/uploads/2020/03/National-Strategy-5G-Final. pdf［2021-01-01］.

［9］ European Commission. EU hydrogen strategy. https：//ec. europa. eu/commission/presscorner/detail/en/ip_20_1259［2021-01-01］.

［10］ The Global Alliance for Vaccines and Immunisation. COVAX explained. https：//www. gavi. org/vaccineswork/covax-explained［2021-01-01］.

［11］ Ministry of Foreign Affairs of Denmark. Global climate action strategy. https：//um. dk/en/foreign-policy/new-climate-action-strategy/［2021-01-01］.

［12］ European Commission. EU hydrogen strategy. https：//ec. europa. eu/commission/presscorner/detail/en/ip_20_1259［2021-01-01］.

［13］ Правительство Российской Федерации. Михаил мишустин утвердил план мероприятий по развитию водородной энергетики. http：//government. ru/docs/40703/［2021-01-01］.

［14］ European Commission. Austria，Bulgaria，Denmark and Romania join initiative to explore quantum communication for Europe. https：//digital-strategy. ec. europa. eu/en/news/austria-bulgaria-denmark-and-romania-join-initiative-explore-quantum-communication-europe［2021-01-01］.

［15］ European Commission. About GPAI. https：//gpai. ai/［2021-01-01］.

［16］ Department for Digital，Culture，Media and Sport. National data strategy. https：//www. gov. uk/government/publications/uk-national-data-strategy/national-data-strategy［2021-01-01］.

［17］ Department for Business，Energy & Industrial Strategy. Boost for UK science with unlimited visa offer to world's brightest and best. https：//www. gov. uk/government/news/boost-for-uk-science-with-unlimited-visa-offer-to-worlds-brightest-and-best［2021-01-01］.

［18］ Ministère de l'Enseignement Supèrieur，de la Rechercher et de l'Innovation. MESR. Présentation du projet de loi de programmation de la recherche en Conseil des ministres. https：//www. enseignementsup- recherche. gouv. fr/cid153284/presentation-du-projet-de-loi-de-programmation-de-la-recherche-en-conseil-des-ministres. htm l［2021-01-01］.

［19］ 연구자가연구에만전념할수있는제도적기반이마련되었습니다 . https：//www. pacst. go. kr/jsp/post/postCouncilView. jsp? post_id＝1684&board_id＝11&etc_cd1＝COUN01♯this［2021-01-01］.

［20］ European Commission. A new era for research and innovation. https：//eur-lex. europa. eu/legal-content/EN/TXT/? uri＝COM：2020：628：FIN［2021-01-01］.

[21] United States Government. Remarks by president Trump on actions against China. https://www. whitehouse. gov/briefings-statements/remarks-president-trump-actions-china/[2021-01-01].

International S&T Cooperation and Talent Exchanges

Wang Wenjun, Ye Jing, Zhang Qiuju

In 2020, confronting the COVID-19 pandemic, the world was facing unprecedented changes marked by uncertainties and unknowns. The competition among major countries and organizations intensified. At the same time, from the COVID-19 pandemic to climate change, facing the emergence and intensification of global challenges, the pace of international cooperation has not stopped. At present, international cooperation and competition are intertwined, showing a complex situation of "competition in cooperation and cooperation in competition".

第五章

重要科学奖项巡礼

Introduction to Important Scientific Awards

5.1　黑洞形成的理论和银河系中心黑洞的观测

——2020 年诺贝尔物理学奖评述

吴学兵　傅煜铭

（北京大学天文学系，科维理天文与天体物理研究所）

2020 年的诺贝尔物理学奖授予了三位科学家（图 1），分别为：

罗杰·彭罗斯（Roger Penrose），1931 年生于英国科尔切斯特。1957 年获英国剑桥大学博士学位。英国牛津大学教授。

赖因哈德·根策尔（Reinhard Genzel），1952 年生于德国巴特洪堡。1978 年获德国波恩大学博士学位。德国马克斯-普朗克地外物理研究所所长，美国加州大学伯克利分校教授。

安德烈娅·盖兹（Andrea Ghez），1965 年生于美国纽约。1992 年获美国加州理工学院博士学位。美国加州大学洛杉矶分校教授。

罗杰·彭罗斯　　　　　　赖因哈德·根策尔　　　　　　安德烈娅·盖兹

图 1　2020 年诺贝尔物理学奖获得者

英国科学家罗杰·彭罗斯因证明黑洞形成是爱因斯坦广义相对论的直接预言而获

得一半奖金；德国科学家赖因哈德·根策尔和美国科学家安德烈娅·盖兹因在银河系中心发现超大质量致密天体而获得另一半奖金。目前，超大质量黑洞是对银河系中心的超大质量致密天体的唯一解释，因此可以说 2020 年的诺贝尔物理学奖授予了黑洞形成的理论和银河系中心黑洞的观测研究。

一、彭罗斯：黑洞的形成不可避免

罗杰·彭罗斯是一位杰出的数学家、数学物理学家、科学哲学家，他对物理学最重要的贡献就是奇点定理。

1915 年，阿尔伯特·爱因斯坦（Albert Einstein）发表了著名的广义相对论，指出引力会影响光线的运动。很快，卡尔·史瓦西（Karl Schwarzschild）对广义相对论场方程的研究发现，如果将球对称的物体压缩到一个临界半径（即史瓦西半径）以内，外界观测者将无法看到该半径内的情况；更奇怪的是，这个坍缩物体的中心有一个密度无限大的点——奇点（singularity），广义相对论并不能描述这个点的时空几何性质。奇点的存在成为了广义相对论的一大挑战，许多物理学家猜测某种物理机制会阻止恒星或者其他天体坍缩到史瓦西半径以内，从而避免奇点形成；也有人猜测奇点只不过是由高度对称的假设导致的。1939 年，物理学家罗伯特·奥本海默（J. Robert Oppenheimer）首次计算了大质量恒星剧烈坍缩的情况[1]，发现当大质量恒星耗尽燃料时，它们首先以超新星的形式爆炸，然后坍缩成致密的恒星遗迹，连光都无法逃出这种致密遗迹的引力束缚①。

第二次世界大战后的天文学发现促使人们进一步考虑引力坍缩的结果。20 世纪 50 年代，天文学家使用射电望远镜发现了一些具有致密形态和强烈射电辐射的天体——类星射电源（后来统称为"类星体"）。1963 年，马尔滕·施密特（Maarten Schmidt）首次证认了类星体 3C 273 的光学发射线，发现它具有较高的谱线红移，即该类星体位于银河系外的宇宙深处[2]。观测证据表明类星体具有较小的尺度和极大的能量，类星体的产能机制只能通过数百万倍太阳质量的黑洞吸积过程来解释。可是，黑洞这种天体真的可以自然形成吗？

这个问题一直令罗杰·彭罗斯感到困惑。1964 年秋天，担任伦敦伯克贝克学院数学教授的彭罗斯在和同事散步时，突然有了一丝灵感。他很快付诸实践，使用整体微分几何的数学工具，从理论上证明了广义相对论中，即使坍缩物质不是球对称的，也可以形成奇点。彭罗斯指出，当物质收缩到事件视界（史瓦西半径）以内时，一定会

---

① 现在一般认为坍缩成为黑洞的恒星应至少具有 25 倍太阳质量。

产生所谓的"俘获面"（trapped surface），该表面处的所有光线都会因为强引力而向内汇聚，这表明引力坍缩过程无法中止，奇点（singularity）必然形成（图2）[3]。彭罗斯的奇点定理表明黑洞可以通过引力坍缩自然形成。随后斯蒂芬·霍金（Stephen W. Hawking）进一步发展了奇点定理，指出宇宙大爆炸可能始于一个奇点。彭罗斯与霍金关于奇点的一系列研究结果被合称为彭罗斯-霍金奇点定理。

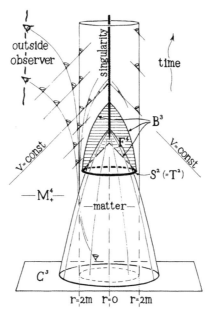

图 2　球对称情形下引力坍缩形成奇点的时空图[3]（一个空间维度被压缩以方便展示）
注：本图为彭罗斯论文原图

彭罗斯的奇点定理是 20 世纪六七十年代广义相对论"黄金时代"的璀璨明珠。1967 年，约翰·惠勒（John A. Wheeler）采用和推广了"黑洞"一词，用于指代引力坍缩到视界以内的天体。黑洞迅速成为公众热衷了解、天文学家竞相搜寻的"明星"天体，一个黑洞研究的新时代随之到来。彭罗斯的奇点定理实际上证明了黑洞形成是广义相对论的直接预言，奠定了黑洞形成最重要的科学理论基础。

二、根策尔和盖兹对银河系中心黑洞的观测研究

如何寻找黑洞？这是天文学家十分感兴趣的问题。黑洞本身并不发光，但这并不意味着黑洞没有可观测的效应。理论研究表明，黑洞通过吸积周围的气体物质可以产生很强的辐射。在气体物质形成的吸积盘结构中，距离黑洞越近，物质温度越高，产

生的辐射能量也越高，中心垂直于吸积盘方向还可以产生强烈的喷流。最早的一批类星体就是因为具有强烈的射电辐射（主要来自喷流）而被发现的。随着越来越多的类星体（以及其他活动星系）被发现，天文学家意识到所有的活动星系中心都具有超大质量黑洞（至少百万倍太阳质量）。类星体在近邻宇宙中反而不如在更遥远（也更早期）的宇宙中常见，表明近邻正常星系可能历史上都曾经历过活跃阶段，正常星系中心可能也有超大质量黑洞！那我们所在的银河系（一个漩涡状的正常星系）中心是否也存在超大质量黑洞呢？

1931 年，射电天文学之父卡尔·央斯基（Karl G. Jansky）探测到了来自人马座附近的银河系中心的射电辐射。1974 年，天文学家布鲁斯·巴里克（Bruce Balick）和罗伯特·布朗（Robert L. Brown）使用美国国家射电天文台的射电干涉仪在银河系中心发现了一个致密的射电源人马座 A*[4]，之后的研究指出人马座 A* 很可能就是银河系中心的黑洞。但要让这个结论板上钉钉，天文学家还需要更多确凿的证据。

由于银河系中心视线方向上有大量的恒星、气体和尘埃，要从可见光波段看到银心的景象十分困难。天文学中用"消光"来描述被观测的天体发出的电磁辐射被气体和尘埃吸收和散射的过程，电磁波的波长越大，受到的消光影响越小，因此对于人马座 A* 的观测研究主要集中在红外、毫米波和射电波段。

20 世纪 70 年代起，加州大学伯克利分校的查尔斯·汤斯（Charles H. Townes）领导团队率先使用红外成像和光谱观测研究了人马座 A* 附近星际气体云的运动，估计出人马座 A* 的质量为 200 万～400 万倍太阳质量，但无法区分人马座 A* 究竟是单一的黑洞，还是包含众多恒星的星团[5]。赖因哈德·根策尔在 1981 年加入汤斯的团队，随后在加州大学伯克利分校任教。从 90 年代开始，洲际的毫米波甚长基线干涉测量（very long baseline interferometry，VLBI）揭示出人马座 A* 的角直径只有 0.02～0.05″，且人马座 A* 几乎没有运动，这与附近恒星的高速运动形成了强烈的对比。种种迹象显示，银河系中心的许多恒星都在围绕人马座 A* 运动。

20 世纪 90 年代中期，已经成为德国马克斯-普朗克地外物理研究所所长的赖因哈德·根策尔及其团队使用欧洲南方天文台位于智利的 3.5 m 新技术望远镜（New Technology Telescope，NTT），对人马座 A* 附近 0.1″内的恒星运动进行了观测[6]。几乎与此同时，加州大学洛杉矶分校的安德烈娅·盖兹领导的银河系中心研究团队使用夏威夷 10 m 凯克望远镜（Keck Telescope，包括两台望远镜）开展了相似的观测[7]。新技术望远镜和凯克望远镜都使用了主动光学技术，望远镜的镜片形状可通过计算机控制进行调节，消除了风、热膨胀和机械应力变形产生的影响，获得优质的成像。它们还使用了自适应光学技术（图 3），通过观察自然导星或激光导星的亮度变化，测量大气湍流造成的影响，并使用辅助镜面进行实时大气改正。

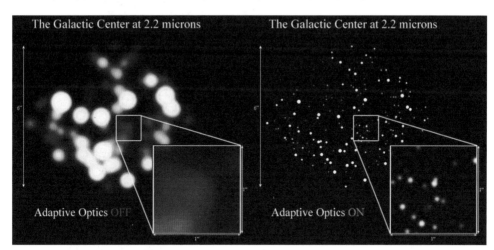

图 3　使用自适应光学前（左）后（右）凯克望远镜拍摄的 2.2 μm 波段
银河系中心图像，自适应光学可以校正地球大气湍流造成的模糊现象

图片来源：加州大学洛杉矶分校安德烈娅·盖兹研究团队网站：https://www.astro.ucla.edu/~ghezgroup/gc/images.html

正是得益于大口径望远镜、高灵敏度的红外探测器和自适应光学等技术，根策尔和盖兹的团队分别独立发现人马座 A* 附近的恒星围绕其做开普勒运动，同时还测定出人马座 A* 具有大约 400 万倍太阳质量[8,9]。两个团队积累了超过 20 年的观测数据，对人马座 A* 附近恒星的运动轨道的分析证明这些恒星都围绕同一个单一的点质量运动，且该点质量的位置在毫角秒量级上与人马座 A* 射电源重合。2017 年，根策尔和盖兹的团队测定出人马座 A* 几乎全部质量集中于与致密射电源重合、小于 17 个光时的区域内，该区域的大小仅仅为 400 万倍太阳质量黑洞的视界尺度的几倍（图 4）。这一观测证据直接排除了中子星集团、恒星质量黑洞、褐矮星、恒星遗迹等可能性，人马座 A* 只可能是一个具有 400 万倍太阳质量的黑洞！

根策尔和盖兹的宏伟计划还不止于发现银河系中心黑洞本身。广义相对论预言任何大质量物体周围的时空都是弯曲的，物体越致密，附近时空的曲率越大。黑洞周围恒星发出的光也必须穿过弯曲的时空、失去部分能量才能到达地球，结果是恒星的光出现与多普勒效应无关的引力红移。在恒星到达轨道的近星点（离黑洞最近处）时，恒星光谱中引力红移的效应最明显，此时对恒星光谱的观测十分有助于检验广义相对论。2018 年，轨道周期为 16 年的恒星 S0-2 到达近星点，根策尔领导的团队使用位于智利的甚大望远镜（Very Large Telescope，VLT，包括 4 台 8m 望远镜）上的

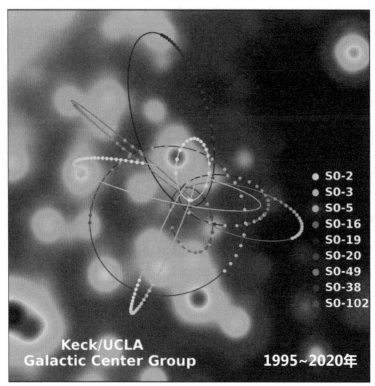

图4 银河系中心 1.0×1.0″内的恒星轨道，背景是 2015 年拍摄的衍射极限图像的中心部分

这些恒星的年平均位置被绘制成彩色的点，随着时间的推移，点的颜色饱和度逐渐增加。曲线是拟合的最优轨道解

图片来源：加州大学洛杉矶分校安德烈娅·盖兹研究团队网站：https://www.astro.ucla.edu/~ghez-group/gc/images.html

GRAVITY 干涉仪器，盖兹领导的团队利用凯克望远镜、双子座望远镜和昴星团望远镜分别对这颗恒星进行了观测[10,11]。两个团队都以相当高的精度探测到了广义相对论所预言的引力红移。不过，两个团队测量到的近星点到黑洞的距离却存在数百光年的偏差，这对未来的测量精度提出了更高的要求。

三、总结与展望

黑洞是如今广为人知的宇宙中最神秘的天体，而帮助我们理解黑洞最重要的理论和观测研究几乎都是在过去 60 年内完成的。彭罗斯、根策尔和盖兹作为这些重要研究的杰出贡献者，获得诺贝尔奖实至名归。对于黑洞的重要观测研究还包括对于双黑洞系统的引力波探测（2017 年诺贝尔物理学奖）、对于恒星质量黑洞（如黑洞 X 射线双星）的观测、事件视界望远镜（Event Horizon Telescope，EHT）对 M87 星系中心超大质量黑洞进行成像等。所有这些观测突破都是通过挑战技术极限而实现的，凯克望远镜、甚大望远镜干涉仪、激光干涉引力波天文台等大型天文装置和尖端仪器对于获得革命性的观测证据起到了决定作用。可以期待，未来对于黑洞内部的物理性质、黑洞周围的极端环境、宇宙中各类黑洞的形成与演化等方面的研究一定还会为物理学带来新的启示。

参考文献

[1] Oppenheimer J R, Snyder H. On continued gravitational contraction. Physical Review, 1939, 56(5): 455-459.

[2] Schmidt M. 3C 273: A star-like object with large red-shift. Nature, 1963, 197(4872): 1040.

[3] Penrose R. Gravitational collapse and space-time singularities. Physical Review Letters, 1965, 14(3): 57-59.

[4] Balick B, Brown R L. Intense sub-arcsecond structure in the galactic center. The Astrophysical Journal, 1974, 194: 265-270.

[5] Wollman E R, Geballe T R, Lacy J H, et al. Ne II 12.8 micron emission from the galactic center. II. The Astrophysical Journal Letters, 1977, 218: L103-L107.

[6] Eckart A, Genzel R. Observations of stellar proper motions near the Galactic Centre. Nature, 1996, 383(6599): 415-417.

[7] Ghez A M, Klein B L, Morris M, et al. High proper-motion stars in the vicinity of Sagittarius A*: Evidence for a supermassive black hole at the center of our Galaxy. The Astrophysical Journal, 1998, 509(2): 678-686.

[8] Ghez A M, Salim S, Weinberg N N, et al. Measuring distance and properties of the Milky Way's central supermassive black hole with Stellar Orbits. The Astrophysical Journal, 2008, 689(2): 1044-1062.

[9] Genzel R, Eisenhauer F, Gillessen S. The Galactic Center massive black hole and nuclear star cluster. Reviews of Modern Physics, 2010, 82(4): 3121-3195.

[10] Gravity Collaboration, Abuter R, Amorim A, et al. Detection of the gravitational redshift in the orbit of the star S2 near the Galactic Centre massive black hole. Astronomy and Astrophysics, 2018, 615: L10-L25.

[11] Do T, Hees A, Ghez A, et al. Relativistic redshift of the star S0-2 orbiting the Galactic Center supermassive black hole. Science, 2019, 365(6454): 664-668.

Theory of Black Hole Formation and Observations of a Supermassive Black Hole in the Galactic Center
—Commentary on the 2020 Nobel Prize in Physics

Wu Xuebing, Fu Yuming

The 2020 Nobel Prize in Physics was awarded to Roger Penrose for his contribution to the black hole formation theory, and to Reinhard Genzel and Andrea Ghez for their discovery of a supermassive compact object(actually a black hole)in the Galactic center. This article reviews their seminal works on black hole formation theory and observations. Penrose invented ingenious mathematical methods to explore Einstein's general theory of relativity, and showed that the theory leads to the formation of black holes. Genzel and Ghez each lead a group of astronomers who have mapped the orbits of the brightest stars that are closest to the center of Milky Way, and found something that is both invisible and heavy (four million solar masses)there. According to the current theory of gravity, there is only one candidate—a supermassive black hole.

5.2　魔剪——CRISPR-Cas 极简史

——2020 年诺贝尔化学奖评述

程田林[1]　仇子龙[2]

（1. 复旦大学脑转化研究院；2. 中国科学院脑智
卓越创新中心，神经科学国家重点实验室）

2020 年法国科学家埃玛纽埃勒·沙尔庞捷（Emmanuelle Charpentier）与美国科学家珍妮弗·道德纳（Jennifer Doudna）因为发明了 CRISPR-Cas 基因编辑技术而共同获得诺贝尔化学奖（图 1）。从 2012 年两位科学家的里程碑论文发表到获奖只有短短的八年时间，这可谓生物医学与化学领域史上最快的诺贝尔奖之一。究竟 CRISPR/Cas 技术有何神奇之处可以如此快速地斩获诺奖？其实在 CRISPR-Cas 技术被发明之前，已有好几个基因编辑技术达到了可实用的阶段，如锌指酶（zinc finger nucleases）和 TALEN 系统。为何 CRISPR-Cas 技术能够杀出重围，成为科学研究中的利器和有望拯救千万身患遗传疾病的病患的"神药"的呢？本文尝试从 CRISPR-Cas 的科学发现、最新进展及生物制药的应用展望几个方面加以简述。

埃玛纽埃勒·沙尔庞捷　　　珍妮弗·道德纳

图 1　2020 年诺贝尔化学奖获得者

一、CRISPR-Cas：细菌对抗病毒的免疫系统

CRISPR 的全称是成簇的规律间隔的短回文重复序列。CRISPR 重复序列在 1987 年由日本研究者石野良纯（Yoshizumi Ishino）等最先报道[1]。但是直到十余年后，科学家才逐渐明白 CRISPR 序列存在的意义。首先，研究者发现 CRISPR 序列可以转录成 RNA，而且序列附近常有 Cas 蛋白编码基因与之共存[2,3]。2005 年，一系列研究报道提示 CRISPR 序列可能与细菌的免疫作用有关，进化生物学家尤金·库宁（Eugene Koonin）整合相关信息提出了一个很有开创性的假说：CRISPR-Cas 系统是细菌对抗病毒的免疫系统[4]。这个假说被一个有趣的实验验证了——微生物学家鲁道夫·巴兰古（Rodolphe Barrangou）带领团队发现，对 CRISPR 这个重复序列进行编程，可以让细菌针对性地对某些噬菌体产生免疫作用[5,6]。这一发现很快得到了工业应用，生产酸奶的厂家在 CRISPR-Cas 系统被完全发现之前就用此方法来研发更耐噬菌体感染的产酸奶菌。

而 CRISPR 序列究竟是如何帮助细菌抵抗病毒感染的？沙尔庞捷与道德纳最后携手彻底解开了 CRISPR-Cas 系统的层层面纱[7-9]。两位科学家在 2012 年发表的里程碑式工作揭示了 CRISPR-Cas 系统的核心组分——与引导 RNA 序列共同作用的 Cas9 核酸酶，完整了解释了此系统的分子机制。CRISPR-Cas 系统的机制研究是 CRISPR-Cas 被改造应用于基因组编辑的基础，这也是沙尔庞捷与道德纳共同获得诺奖的核心贡献。

CRISPR-Cas 系统的作用机制被厘清之后，其作为基因编辑工具的潜力便展露无遗。2013 年 1 月，张锋、乔治·丘奇（George Church）与道德纳实验室分别发表文章，将 CRISPR-Cas9 系统首次用于真核生物细胞和人类细胞系中，实现了在真核生物基因组中的高效率靶向基因敲除[10-12]。具体应用时间线参见图 2。

二、CRISPR-Cas：系统优化与局限

1. CRISPR-Cas 系统的脱靶效应及潜在应用风险

当 CRISPR-Cas 系统被改造用于在真核细胞中进行基因编辑后，研究者很快发现了这个系统具有一个很大的缺陷，就是由于引导 RNA 的靶向不精确，或者由于 Cas9 核酸酶的活性过高，很容易在真核生物的基因组中产生意外的切割，被称为脱靶效应。尽管以前在干扰小 RNA（small interfering RNA，siRNA）的研究中也发现过类

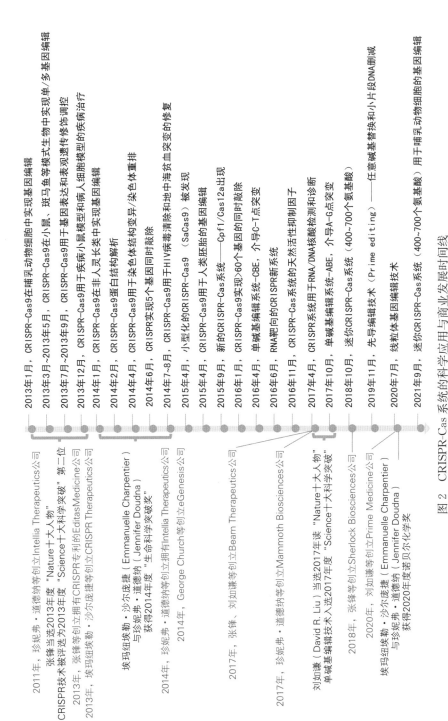

图 2 CRISPR-Cas 系统的科学应用与商业发展时间线

2011年，珍妮弗·道德纳等创立Intellia Therapeutics公司

张锋当选2013年度"Nature十大人物"
CRISPR技术被评选为2013年度"Science十大科学突破"第二位
2013年，张锋等创立拥有CRISPR专利的EditasMedicine公司
2013年，埃玛纽埃勒·沙尔庞捷等创立CRISPR Therapeutics公司

埃玛纽埃勒·沙尔庞捷（Emmanuelle Charpentier）
与珍妮弗·道德纳（Jennifer Doudna）
获得2014年度"生命科学突破奖"

2014年，George Church等创立eGenesis公司

2017年，张锋、刘如谦等创立Beam Therapeutics公司

2017年，珍妮弗·道德纳等创立Mammoth Biosciences公司

刘如谦（David R. Liu）当选2017年读"Nature十大人物"
单碱基编辑技术入选2017年度"Science十大科学突破"

2018年，刘如谦等创立Prime Medicine公司

2020年，刘如谦等创立Sherlock Biosciences公司
埃玛纽埃勒·沙尔庞捷（Emmanuelle Charpentier）
与珍妮弗·道德纳（Jennifer Doudna）
获得2020年度诺贝尔化学奖

2013年1月，CRISPR-Cas9在哺乳动物细胞中实现基因编辑
2013年3月~2013年5月，CRISPR-Cas9在小鼠、斑马鱼等模式生物中实现单/多基因编辑
2013年7月~2013年9月，CRISPR-Cas9用于基因表达和表观遗传修饰调控
2013年12月，CRISPR-Cas9用于疾病小鼠病模型和病人细胞模型的疾病治疗
2014年1月，CRISPR-Cas9在非人灵长类中实现基因编辑
2014年2月，CRISPR-Cas9蛋白结构解析
2014年4月，CRISPR-Cas9用于染色体结构变异/染色体重排
2014年6月，CRISPR实现5个基因同时敲除
2014年7~8月，CRISPR-Cas9用于HIV病毒清除和地中海贫血突变的修复
2015年4月，小型化的CRISPR-Cas9（SaCas9）被发现
2015年4月，CRISPR-Cas9用于人类胚胎的基因编辑
2015年9月，新的CRISPR-Cas系统——Cpf1/Cas12a出现
2016年1月，CRISPR-Cas9实现>60个基因的同时敲除
2016年4月，单碱基编辑系统-CBE，介导C-T点突变
2016年6月，RNA靶向的CRISPR新系统
2016年11月，CRISPR-Cas系统的天然活性抑制因子
2017年4月，CRISPR系统用于RNA/DNA核酸检测和诊断
2017年10月，单碱基编辑系统-ABE，介导A-G点突变
2018年10月，迷你CRISPR-Cas系统（400~700个氨基酸）
2019年11月，先导编辑技术（Prime editing）——任意碱基替换和小片段DNA删减
2020年7月，线粒体基因编辑技术
2021年9月，迷你CRISPR-Cas系统（400~700个氨基酸）用于哺乳动物细胞的基因编辑

似的脱靶效应，但 CRISPR-Cas 的脱靶效应却有可能破坏基因组，对生物的遗传物质产生永久影响，因此潜在后果比较严重。

在此之外，研究者还发现在真核生物细胞中，由于 CRISPR-Cas 在基因编辑的过程中产生了 DNA 的双链断裂，因此还有可能导致一系列意外激活真核细胞损伤修复的机制，有可能导致被基因编辑的细胞出现不稳定性。由于 CRISPR-Cas 毕竟是细菌中的"免疫系统"，因此被应用在真核细胞中出现这一些意外反应也不足为怪。因此，自从 CRISPR-Cas 技术诞生之日起，如何减少脱靶，尽可能地降低潜在风险就成了此基因编辑技术走向临床应用的最重要挑战。

2. 新型 Cas9 的筛选

最初的 Cas9 核酸酶编码基因取自化脓性链球菌（*Streptococcus pyogenes*），该基因的编码序列接近 4 kb，超过了腺相关病毒（adeno-associated virus，AAV）载体的装载上限，因而限制了其在动物体内的应用以及潜在的临床药物开发。为此研究者努力地寻找小型化的 Cas9 基因，并最终在金黄色葡萄球菌（*Staphylococcus aureus*）中发现了编码序列长度为 3 kb 的 Cas9 基因，其被称作 SaCas9[13]。此外，研究者还发现了另一种具有不同特性的 CRISPR-Cpf1/Cas12 系统，找到了更小型的 CRISPR-Cas12f 系统，其蛋白编码序列长度在 1.2～2 kb，这极大地改善了 CRISPR 体积过大的问题[14-17]。

3. 单碱基编辑系统

与传统技术相比，CRISPR-Cas 系统能实现高效的靶向基因敲除，但其通过同源重组实现精准编辑的效率仍有很大不足。2016 年以来，哈佛大学刘如谦（David R. Liu）实验室以 CRISPR-Cas9 系统为基础，通过将不同的脱氨基酶与 Cas9 切口酶（nCas9）或失活型 Cas9（dCas9）融合，构建出两种单碱基编辑系统——胞嘧啶碱基编辑器（cytosine base editors，CBEs）和腺嘌呤碱基编辑器（adenine base editors，ABEs）[18,19]。CBEs 和 ABEs 能精准高效地介导基因组特定位点的碱基对 C·G—T·A 和 A·T—G·C 的高效转换。与经典的 CRISPR-Cas 系统相比，CBEs 和 ABEs 能高效实现精准基因编辑。目前已知的人类致病遗传变异中，约 58% 都属于点突变，这其中 C·G—T·A 和 A·T—G·C 变异占所有致病点突变的 2/3 以上[20]。因此，单碱基编辑工具 CBEs 和 ABEs 在疾病模型构建和人类遗传病的基因治疗中极具应用潜力[21]。不过对于点突变之外的其他变异类型，以及 C·G—T·A 和 A·T—G·C 点突变之外的其他点突变，现有的单碱基编辑系统难以应用。2019 年，刘如谦实验室将逆转录酶 MMLV 与 Cas9 切口酶（nCas9）融合，构建出新的精准基因编辑策略——

引导编辑（prime editing，PE），这一策略理论上可以对任意的碱基点突变和小片段的插入缺失突变进行修复，这为精准基因编辑提供了新的工具[22]。不过进行引导编辑需要的一系列基因体积过大且编辑效率有限，仍然有很大的改进空间。

4.RNA 编辑系统

除了经典的定向编辑 DNA 的 CRISPR-Cas9/Cas12 系统之外，研究者还发现了一大类特异性识别单链 RNA（single-stranded RNA，ssRNA）的 CRISPR-Cas13 系统，并在此基础上将其改造应用于 RNA 的检测和标记，以及 RNA 的表达调控、表观遗传修饰、RNA 编辑和定位、RNA 编程等领域[23]。CRISPR-Cas13 系统在以核酸检测为基础的疾病诊断中也极具应用潜力，在新型冠状病毒感染、寨卡病毒感染、登革热等多种疾病的诊断中有重要的应用前景[23]。此外，研究者还将 RNA 靶向的脱氨基酶与 CRISPR-Cas13 系统融合，获得了可定向编辑 RNA 的碱基编辑器[24,25]。

三、CRISPR-Cas：基因编辑技术临床应用进展

CRISPR-Cas 系统为基础的基因编辑技术自诞生以来，其临床应用前景便备受关注。随着技术的进步和成熟，CRISPR 基因编辑技术已经被众多研究者用于临床疾病治疗，相关的临床研究结果也陆续发布。目前，有代表性的研究成果如下：

1. 血红蛋白病（β-地中海贫血和镰状细胞病）

2018 年，沙尔庞捷教授参与创立的 CRISPR Therapeutics 公司宣布开展了第一项基因编辑的临床研究，研究者分离 β-地中海贫血和镰状细胞病患者的造血干细胞，通过电穿孔将 Cas9-sgRNA（small guide RNA）核糖核蛋白复合物（RNP）递送至造血干细胞以破坏 BCL11A 基因增强子位点（该位点被破坏可增强 β 球蛋白表达）。在清除患者自身的造血干细胞后，将编辑后的造血干细胞输回患者体内。两名患者（一名 β-地中海贫血患者、一名镰状细胞病患者）在接受治疗后，体内的 β 球蛋白表达明显增加，且在接受治疗的一年内，升高的 β 球蛋白表达水平一直维持，首次验证了基因编辑治疗疾病的可行性[26,27]。值得一提的是，国内研究者参与创立的邦耀生物科技公司从哈佛大学获得 BCL11A 基因增强子位点基因编辑靶点的专利授权，在国内多家医院同期进行了重症地中海贫血的基因编辑临床试验，也获得了非常好的临床效果。这两个基因编辑的临床试验成功都说明了 CRISPR-Cas 作为基因治疗方法的巨大潜力。

2. 癌症治疗

2020年，宾夕法尼亚大学研究者卡尔·朱恩（Carl H. June）发表报道，在三名难治性癌症患者（两名患有晚期难治性骨髓瘤，一名患有转移性肉瘤）中开展CRISPR-Cas技术的临床研究：研究者分离了患者的T细胞，随后通过电穿孔将Cas9-sgRNA核糖核蛋白复合物（RNP）递送至患者T细胞中以破坏 *TRAC*、*TRBC* 和 *PDCD1* 三种基因以增强T细胞对肿瘤的杀伤力。研究显示，经改造的T细胞在患者体内可持续存留9个月[28]。中国华西医院的卢铀教授在难治性非小细胞肺癌患者中开展了类似的实验。以上研究证实了CRISPR-Cas9基因编辑技术在难治性癌症患者中具有安全性和可行性[29]。

3. 甲状腺素转运蛋白淀粉样变性

由于CRISPR-Cas系统的潜在脱靶效应和切割基因组的未知风险，如果用于疾病治疗，AAV载体荷载CRISPR-Cas系统会导致体内长期表达，因此对基因组造成了一定的潜在风险。基于此考虑，运用脂质纳米颗粒（lipid nanoparticle，LNP）荷载CRISPR-Cas系统进行短期表达，完成基因编辑后系统即降解，提高了基因编辑技术临床应用的安全性。这种方法被应用于一种罕见的遗传病——甲状腺素转运蛋白淀粉样变性病（transthyretin amyloid polyneuropathy，ATTR），ATTR患者因错误折叠的甲状腺素转运蛋白（transthyretin，TTR）在神经和心肌等组织中累积而致病。道德纳教授参与创立的Intellia公司研发人员在6名患者中开展了临床试验，利用脂质纳米颗粒将sgRNA和Cas9的mRNA通过静脉注射递送至患者体内以靶向肝脏中高表达的TTR编码基因。结果显示，该策略能通过对肝脏中的突变TTR基因进行敲除，而有效降低患者血液中TTR的水平且患者无明显不良反应[30]。以上临床研究结果表明，CRISPR-Cas9技术在临床应用上具有安全性和可行性。

作为CRISPR-Cas技术第一次被成功用于在体的疾病治疗，此研究被国际著名学术期刊《科学》评为2021年"十大科学突破"之一。

4. 单碱基编辑技术临床应用的前景

单碱基编辑技术已经广泛用于多种遗传性疾病小鼠模型的治疗研究。2021年，研究者将该技术用于非人灵长类模型中，实现了前蛋白转换酶枯草溶菌素9（proprotein convertase subtilisin/kexin type 9，PCSK9）基因位点的精准编辑，最终有效降低血液中PCSK9和低密度脂蛋白的水平。非人灵长类中的研究策略如下：研究者利用脂纳米颗粒将sgRNA和Cas9的mRNA通过静脉注射递送至非人灵长类体内，结果显示该策略能精准编辑 PCSK9 基因位点并降低血液中 PCSK9 和低密度脂蛋白的水平[31,32]。这一结果为单碱基编辑技术对一系列代谢疾病的临床治疗奠定了基础。

四、CRISPR-Cas 基因编辑技术应用展望

随着 CRISPR-Cas 系统的横空出世，此技术已经成为生命科学研究中对基因进行操作的利器。研究者在许多领域中实现了 CRISPR-Cas 系统对动植物和微生物的高效基因编辑，极大地推动了生命科学研究的进程。

对于这个基因组"神刀"，今后最重要、也是最具挑战性的任务就是如何能安全有效地将 CRISPR-Cas 技术应用在人类遗传疾病的治疗中。随着递送手段的不断改进和脱靶效应与潜在风险的不断降低，相信 CRISPR-Cas 技术很快会在地中海贫血与ATTR 疾病中的成功应用后继续大放光彩。

参考文献

［1］Ishino Y，Shinagawa H，Makino K，et al. Nucleotide sequence of the IAP gene，responsible for alkaline phosphatase isozyme conversion in *Escherichia coli*，and identification of the gene product. Journal of Bacteriology，1987，169：5429-5433.

［2］Mojica F J，Díez-Villaseñor C，Soria E，et al.，Biological significance of a family of regularly spaced repeats in the genomes of Archaea，Bacteria and mitochondria. Molecular Microbiology，2000，36：244-246.

［3］Makarova K S，Aravind L，Grishin N V，et al. A DNA repair system specific for thermophilic Archaea and Bacteria predicted by genomic context analysis. Nucleic Acids Research，2002，30：482-496.

［4］Makarova K S，Grishin N V，Shabalina S A，et al. A putative RNA-interference-based immune system in prokaryotes：Computational analysis of the predicted enzymatic machinery，functional analogies with eukaryotic RNAi，and hypothetical mechanisms of action. Biology Direct，2006，1：7.

［5］Barrangou R，Fremaux C，Deveau H，et al. CRISPR provides acquired resistance against viruses in prokaryotes. Science，2007，315：1709-1712.

［6］Garneau J E，Dupuis M E，Villion M，et al. The CRISPR/Cas bacterial immune system cleaves bacteriophage and plasmid DNA. Nature，2010，468：67-71.

［7］Jinek M，Chylinski K，Fonfara I，et al. A programmable dual-RNA-guided DNA endonuclease in adaptive bacterial immunity. Science，2012，337：816-821.

［8］Gasiunas G，Barrangou R，Horvath P，et al. Cas9-crRNA ribonucleoprotein complex mediates specific DNA cleavage for adaptive immunity in bacteria. Proceedings of the National Academy of

Sciences of the United States of America,2012,109:E2579-E2586.

[9] Deltcheva E,Chylinski K,Sharma C M,et al. CRISPR RNA maturation by trans-encoded small RNA and host factor RNaseIII. Nature,2011,471:602-607.

[10] Cong L,Ran F,Cox D,et al. Multiplex genome engineering using CRISPR/Cas systems. Science, 2013,339:819-823.

[11] Jinek M,East A,Cheng A,et al. RNA-programmed genome editing in human cells. eLife,2013,2: e00471.

[12] Mali P ,Yang L,Esvelt K M,et al. RNA-guided human genome engineering via Cas9. Science, 2013,339:823-826.

[13] Ran F A, Cong L, Yan W X, et al. *In vivo* genome editing using *Staphylococcus aureus* Cas9. Nature,2015,520:186-191.

[14] Karvelis T,Bigelyte G,Young J K,et al. PAM recognition by miniature CRISPR-Cas12f nucleases triggers programmable double-stranded DNA target cleavage. Nucleic Acids Research,2020,48: 5016-5023.

[15] Kim D. ,Lee J M,Moon S B,et al. Efficient CRISPR editing with a hypercompact Cas12f1 and engineered guide RNAs delivered by adeno-associated virus. Nature Biotechnology,2022,40:94-102.

[16] Wu Z, Zhang Y, Yu H, et al. Programmed genome editing by a miniature CRISPR-Cas12f nuclease. Nature Chemical Biology,2021,17:1132-1138.

[17] Xu X,Chemparathy A,Zeng L,et al. Engineered miniature CRISPR-Cas system for mammalian genome regulation and editing. Molecular Cell,2021,81(20):4333-4345. e4.

[18] Komor A C,Kim Y B,Packer M S et al. Programmable editing of a target base in genomic DNA without double-stranded DNA cleavage. Nature,2016,533(7603):420-424.

[19] Gaudelli N M,Komor A C,Rees H A,et al. Programmable base editing of A * T to G * C in genomic DNA without DNA cleavage. Nature,2017,551(7681):464-471.

[20] Landrum M J, Lee J M, Mark B, et al. ClinVar: Public archive of interpretations of clinically relevant variants. Nucleic Acids Research,2016,44(D1):D862-868.

[21] Rees H A,Liu DR. Base editing:precision chemistry on the genome and transcriptome of living cells. Nature Reviews Genetics,2018,19:770-788.

[22] Anzalone A V,Randolph P B,Davis J R,et al. Search-and-replace genome editing without double-strand breaks or donor DNA. Nature,2019,576:149-157.

[23] Smargon A A,Shi Y J,Yeo G W. RNA-targeting CRISPR systems from metagenomic discovery to transcriptomic engineering. Nature Cell Biology,2020,22:143-150.

[24] Cox D B T, Gootenberg J S, Abudayyeh O O, et al. RNA editing with CRISPR-Cas13. Science,

2017,358:1019-1027.

［25］Abudayyeh O O,Gootenberg J S,Franklin B,et al. A cytosine deaminase for programmable single-base RNA editing. Science,2019,365:382-386.

［26］Zeng J,Wu Y,Ren C,et al. Therapeutic base editing of human hematopoietic stem cells. Nature Medicine,2020,26:535-541.

［27］Frangoul H,Altshuler D,Cappellini M D,et al. ,CRISPR-Cas9 gene editing for sickle cell disease and β-thalassemia. New England Journal of Medicine,2020,384(3):252-260.

［28］Stadtmauer E A,Fraietta J A,Davis M M,et al. ,CRISPR-engineered T cells in patients with refractory cancer. Science,2020,367(6481):976.

［29］Lu Y,Xue J,Deng T,et al. ,Safety and feasibility of CRISPR-edited T cells in patients with refractory non-small-cell lung cancer. Nature Medicine,2020,26(5):732-740.

［30］Gillmore J D,Gane E,Taubel J,et al. CRISPR-Cas9 *in vivo* gene editing for transthyretin amyloidosis. New England Journal of Medicine,2021,385(6):493-502.

［31］Musunuru K,Chadwick A C,Mizoguchi T,et al. ,*In vivo* CRISPR base editing of PCSK9 durably lowers cholesterol in primates. Nature,2021,593(7859):429-434.

［32］Rothgangl T,Dennis M K,Lin P,et al. *In vivo* adenine base editing of PCSK9 in macaques reduces LDL cholesterol levels. Nature Biotechnology,2021,39:949-957.

The Magic Scissors—The Brief History of CRISPR-Cas
——Commentary on the 2020 Nobel Prize in Chemistry

Cheng Tianlin,Qiu Zilong

The 2020 Nobel Prize in Chemistry was awarded to two scientists Emmanuelle Charpentier and Jennifer Doudna,for the development of CRISPR-Cas gene editing systems. Only eight years have passed from landmark researches demonstrating functions and mechanisms of CRISPR-Cas systems in 2012 to Nobel Prize announcement in 2020. Why CRISPR-Cas systems are so compelling and appealing in basic and clinical applications as compared to other similar technologies such as zinc finger nucleases (ZFNs) and TALENs? Here is a brief summary to demonstrate the discovery,latest advances and outlook in biopharmaceutical applications of CRISPR-Cas systems.

5.3 丙型肝炎病毒的发现及未来展望

——2020 年诺贝尔生理学或医学奖评述

童一民 钟 劲

（中国科学院上海巴斯德研究所）

在全球范围内约有 7100 万名丙型肝炎病毒（hepatitis C virus，HCV）感染者。HCV 感染可引发急性或慢性肝炎，即丙型肝炎（简称丙肝），部分慢性丙肝可进一步发展为肝硬化、甚至引发肝癌。因此，HCV 对人类健康造成重大威胁。2020 年 10 月 5 日，诺贝尔奖委员会将 2020 年度诺贝尔生理学或医学奖授予了美国国立卫生研究院的哈维·詹姆斯·阿尔特（Harvey J. Alter）、加拿大阿尔伯塔大学的迈克尔·霍顿（Michael Houghton）和美国洛克菲勒大学的查尔斯·赖斯（Charles M. Rice）（图 1），以表彰这三位科学家在"发现丙型肝炎病毒"方面所做出的贡献。HCV 的发现、随后细胞模型的建立，以及后续高效的抗病毒药物的成功研发，是人类与病毒斗争历程中的辉煌篇章。但是，彻底清除丙肝不仅需要好的抗病毒药物，同时也需要有预防性疫苗。丙肝预防性疫苗的开发是完成世界卫生组织制定的 2030 年消灭 HCV 宏伟目标的最后的一块关键性的"拼图"。因此，还需要更精准地投入足量的资源到相应的研究领域中。

哈维·詹姆斯·阿尔特　　　　迈克尔·霍顿　　　　查尔斯·赖斯

图 1　2020 年诺贝尔生理学或医学奖获得者

一、丙型肝炎病毒的发现之旅

早在 1947 年，英国肝脏病专家麦卡勒姆根据流行病学研究数据分析推测认为：至少存在两类肝炎，一类经由粪-口途径传播，称为甲型肝炎（简称甲肝），由甲型肝炎病毒（hepatitis A virus，HAV）感染引起；另一类通过血液传播，称为乙型肝炎（简称乙肝），由乙型肝炎病毒（hepatitis B virus，HBV）感染引起。美国国立卫生研究院的阿尔特注意到，尽管排除了 HAV 和 HBV 感染的因素，不少患者在输血后依然罹患肝炎。将这些 HAV 及 HBV 阴性的肝炎患者血清接种给黑猩猩，可导致黑猩猩出现肝炎症状。这些结果表明，除了甲肝和乙肝之外，还存在一种"非甲非乙型肝炎"（non-A，non-B hepatitis，NANBH）[1]。然而阿尔特按照经典病毒学方法，通过特异性抗体或分离纯化病毒颗粒的手段未能成功鉴定出这种可导致 NANBH 的新病原体。

1982 年，病毒学家霍顿加入凯龙（Chiron）公司，主持开展了鉴定 NANBH 病原体的研究。他与其同事一起，纯化了 NANBH 患者样本中的 RNA，并反转录成为 DNA，然后通过限制性内切酶将 DNA 消化成短片段，并连接到噬菌体载体中，建立了携带有重组 DNA 片段的噬菌体文库。随后他们采用 NANBH 患者来源的抗血清来筛选表达 NANBH 病原体 cDNA 片段的重组噬菌体，最终在 1989 年克隆出新病毒的近全长基因组序列，将该病毒命名为"hepatitis C virus"[2,3]。随着 HCV 基因组序列成功解析，HCV 检测筛查方法也得以顺利建立，极大地提高了输血及血制品的安全性，使数百万人避免被 HCV 感染。

霍顿团队克隆的 HCV 基因组序列，并不能产生具有感染力的病毒。当时在华盛顿大学工作的赖斯研究团队发现这个 HCV 基因组序列并不完整，在病毒基因组 3′ 端的多聚尿嘧啶序列下游还有一段病毒序列。他们克隆了病毒 3′ 最末端的约 98 个核苷酸的高度保守的非编码序列，并且证实了带有该末端序列的 HCV 全长序列在黑猩猩里具有感染性，从而补全了 HCV 病毒发现历程中的最后一环[4,5]。

这三位诺奖获得者的 HCV 发现研究之路，如同一场接力赛。阿尔特发现了有别于甲肝或乙肝的新型肝炎疾病；霍顿鉴定并克隆了导致该新型肝炎病原体的近全长基因组序列，建立了该病毒的检测方法；赖斯补全了病毒基因组全长序列，并通过实验证明了 HCV 是导致丙肝的病原体。HCV 的鉴定发现和克隆是 HCV 研究史上的首个里程碑式胜利，促成了后来 HCV 细胞培养模型的相继建立，为 HCV 研究以及抗病毒药物的研发提供了重要的模型工具[6]。

二、丙肝防治现状及挑战

在 HCV 基础研究的支撑下，靶向病毒蛋白抑制病毒复制的小分子化合物药物研发得到了迅速发展。这些药物统一命名为抗 HCV 的直接抗病毒药物（direct-acting antiviral agents，DAA），包括非结构蛋白 NS3/4A 蛋白酶抑制剂、NS5A 抑制剂和 NS5B 聚合酶抑制剂等。这些 DAA 的联合使用能够清除 95％ 的患者体内的病毒，提升了针对 HCV 的治疗效果[7]。由于这些高效抗病毒药物的问世，世界卫生组织制定了 2030 年消灭丙肝的宏伟目标。然而，由于病毒的高度变异性，DAA 治疗导致病毒基因组出现耐药突变的可能性几乎是无法避免的。根据以往的经验，消灭一种传染病离不开有效的预防性疫苗，目前还没有任何一种病毒能单纯依靠抗病毒药物得以彻底清除。预防性疫苗具备抗病毒药物无法替代的优势，这是因为虽然 DAA 治愈率高，但被治愈的病人仍然能会被 HCV 再次感染，尤其吸毒等高风险人群的复发率更是高达 10.67％[8]。

由于 HCV 高度的变异性、病毒基因组的多样性，以及病毒蛋白较低的免疫原性，丙肝疫苗研发遇到很多的困难。2019 年，一项基于腺病毒载体的丙肝疫苗在大规模人体实验中没有展示出预期的保护性效果[9]；另一种基于重组 HCV 包膜蛋白的候选疫苗早期在黑猩猩模型中进行初步的测试，目前还在临床开发中。动物感染模型是疫苗研发的关键工具。HCV 的天然宿主只有人类和黑猩猩，而之前唯一可用的黑猩猩动物模型也在 10 年前就被禁止应用于实验动物研究，导致目前所有的丙肝候选疫苗均无法在体内进行保护性评价研究，阻碍了丙肝疫苗研发的进展。有科学家尝试建立新型 HCV 小动物感染模型，如人源化小鼠模型等，但这些替代模型都还存在着各自的不足之处，目前尚不能替代黑猩猩模型用于丙肝疫苗的保护性评价工作[10]。

2021 年，一些 HCV 研究者提出了建立可控的人感染模型（controlled human infection model，CHIM）来进行丙肝疫苗保护性测试的设想[11]。具体操作为：对入组参与临床试验的受试者进行丙肝疫苗或疫苗安慰剂对照接种，然后用 HCV 在人体内开展攻毒实验，通过比较疫苗及安慰剂接种组的感染率来评价候选疫苗的保护效果。基于目前 DAA 对 HCV 的高水平的清除率，受试者如果被 HCV 感染，能及时通过 DAA 治疗来清除其体内的病毒，这使得体内攻毒保护实验处于"可控"的状态。毫无疑问，推进丙肝疫苗研发的紧迫性与缺乏可用于在体内开展疫苗评价的动物模型的这一现状的矛盾，直接催生了 CHIM 设想的提出及其可行性的讨论。但是我们认为，虽然目前 DAA 抗病毒治疗具有接近 100％ 的治愈率，使得通过 CHIM 开展临床试验具有较高的可控性，但直接在人体开展攻毒保护试验存在较大的伦理问题，HCV

不同于新型冠状病毒这类可导致大规模快速传染的病原体，通过CHIM进行临床试验的申请很难通过监管机构的批准。如果受试者本身肝脏存在其他未诊断的潜在问题时，HCV感染可能会增加其发生爆发性肝炎的风险以及罹患肝癌的风险。此外，用于开展体内保护性评价试验所用的HCV毒株来源有待进一步明确。体外细胞培养制备的HCV虽然能满足大规模生产和质控标准化的要求，但这些病毒含有的一些体外细胞培养的适应性突变往往会降低其在人体内感染能力，不利于疫苗保护效果的客观评价。总体而言，CHIM的设想为丙肝疫苗保护性测试提供了一条新的思路，但距离真正利用CHIM来重启新一轮的疫苗研发还有很长的路要走。

三、总结和展望

HCV的发现历程以及随之建立的病毒感染细胞模型工具，包括后续抗病毒药物的成功研发，是人类与病毒斗争历程中的辉煌篇章。直接抗病毒药物的出现，显著地降低了HCV对人类健康的威胁，同时也为丙肝预防性疫苗的研发争取到了宝贵的时间窗口。丙肝疫苗既是消灭HCV宏伟目标的最后的，也是最关键的一块拼图。而疫苗成功研发迫切需要建立一个可用于在体内开展疫苗保护性评价的动物感染模型。因此，现阶段社会各界还需要更精准地投入足量资源到相应的研究领域中。

参考文献

[1] Tabor E,Drucker J A,Hoofnagle J H,et al. Transmission of non-A,non-B hepatitis from man to chimpanzee. Lancet,1978,311:463-466.

[2] Choo Q L,Kuo G,Weiner A J,et al. Isolation of a cDNA clone derived from a blood-borne non-A, non-B viral hepatitis genome. Science,1989,244:359-362

[3] Kuo G,Choo Q L,Alter H J,et al. An assay for circulating antibodies to a major etiologic virus of human non-A,non-B hepatitis. Science,1989,244:362-364.

[4] Kolykhalov A A,Feinstone S M,Rice C M. Identification of a highly conserved sequence element at the 3' terminus of hepatitis C virus genome RNA. Journal of Virology,1996,70(6):3363-3371.

[5] Kolykhalov A A,Agapov E V,Blight K J,et al. Transmission of hepatitis C by intrahepatic inoculation with transcribed RNA. Science,1997,277:570-574.

[6]Lohmann V,Körner F,Koch J-O,et al. Replication of subgenomic hepatitis C virus RNAs in a hepatoma cell line. Science,1999,285:110-113.

[7] Sofia M J,Bao D,Chang W,et al. Discovery of a β-d-2'-Deoxy-2'-α-fluoro-2'-β-C- methyluridine nucleotide prodrug (PSI-7977) for the treatment of hepatitis C virus. Journal of Medicinal Chemistry, 2010,53:7202-7218.

[8] Li D, Huang Z, Zhong J. Hepatitis C virus vaccine development: old challenges and new opportunities. National Science Review, 2015, 2: 285-295.

[9] Cox A L, Page K, Melia M, et al. LB10. A randomized, double-blind, placebo-controlled efficacy trial of a vaccine to prevent chronic hepatitis C virus infection in an at-risk population. Open Forum Infectious Diseases, 2019, 6(Supplement_2): S997.

[10] Berggren K A, Suzuki S, Ploss A. Animal models used in Hepatitis C Virus research. International Journal Of Molecular Sciences, 2020, 21(11): 3869.

[11] Liang T J, Feld J J, Cox A L, et al. Controlled human infection model—Fast track to HCV vaccine? The New England Journal of Medicine, 2021, 385: 1235-1240.

The Discovery of Hepatitis C Virus and Its Future Prospects

——Commentary on the 2020 Nobel Prize in Physiology or Medicine

Tong Yimin, Zhong Jin

A total of 71 million people in the world are estimated to be infected with hepatitis C virus (HCV), and some of them are subject to developing liver cirrhosis or liver cancer. The 2020 Nobel Prize in Physiology or Medicine was awarded to Harvey J. Alter, Michael Houghton and Charles M. Rice for their outstanding contributions in "the discovery of hepatitis C virus". The subsequent establishment of HCV replicon that led to the development of direct-acting antivirals stands as a milestone accomplishment in the human's long history of fight against viruses. While we bask in the glory of this triumph, we need to be reminded that the fight against HCV is far from over. A prophylactic HCV vaccine is the last and most important piece of the jigsaw puzzles in completing the HCV elimination by 2030, an ambitious goal set up by the World Health Organization. This rings a bell requiring more investments in projects related to HCV vaccine development.

5.4　2020 年沃尔夫数学奖获奖者简介①

陈秀雄[1]　Oscar García-Prada[2]　Francisco Presas[2]

（1. 中国科学技术大学几何与物理研究中心；
2. 西班牙国家研究委员会数学科学研究所）

2020 年 1 月 13 日，帝国理工学院数学教授、纽约州立大学石溪分校西蒙斯几何与物理中心成员西蒙·唐纳森（Simon Donaldson）和斯坦福大学教授雅科夫·叶利阿什贝格（Yakov Eliashberg）教授凭借多年来为微分几何与拓扑领域所作出的杰出贡献，共同摘得本年度沃尔夫数学奖桂冠。

一、西蒙·唐纳森

1. 简历

西蒙·唐纳森（图 1），1957 年 8 月 20 日出生于英国剑桥。他于 1983 年在牛津大学获得博士学位，师从迈克尔·阿蒂亚（Michael Francis Atiyah）和奈杰尔·希钦（Nigel J. Hitchin）。他先后担任过牛津大学万灵学院青年研究员（1983～1985 年），普林斯顿高等研究院访问学者（1983～1984 年），牛津大学沃利斯数学教授（1985～1991 年），斯坦福大学讲座教授（1997～1998 年）。自 1998 年起，他在伦敦帝国理工学院担任教授（2001～2002 年任皇家学会研究教授），并于 2014 年成为纽约州立大学石溪分校西蒙斯几何与物理中心的永久成员。

1983 年西蒙·唐纳森首次应邀在国际数学家大会（International Congress of Mathematicians，ICM）上作报告，并在之后的 1986 年、1998 年及 2018 年国际数学家大会上应邀作 1 小时大会报告。

西蒙·唐纳森获得的奖项无数，包括菲尔兹奖（Fields Medal，1986 年）、瑞典皇家科学院的克拉福德奖（The Crafoord Prize，1994 年，与丘成桐共获）、邵逸夫数学

---

①　本文原文为英文，由上海科技大学数学科学研究所蔡明亮、李龙、姚成建，加拿大蒙特利尔大学数学研究中心张俊翻译成中文。

科学奖（2009 年，与克利福德·陶布斯共获）、数学突破奖（Breakthrough Prize in Mathematics，2015 年）、奥斯瓦尔·德-维布伦（Oswald Veblen）几何奖（2019 年，与陈秀雄、孙崧共获）、沃尔夫数学奖（Wolf Prize in Mathematics，2020 年，与亚科夫·伊利埃伯格共获）。他是英国皇家学会会员、美国国家科学院外籍院士、法国科学院外籍院士和瑞典皇家科学院外籍院士。

西蒙·唐纳森和诺拉·唐纳森于 1986 年结婚，婚后育有两儿一女。

图 1　西蒙·唐纳森
2017 年摄于西班牙国家研究委员会数学科学所（ICMAT-CSIC）

2. 成就

西蒙·唐纳森的研究工作基本可归类于以下领域：①全纯向量丛上的微分几何；②规范场论在四维流形拓扑中的应用；③辛流形与近复几何；④高维规范场论与凯勒几何。

1980 年，唐纳森博士学习伊始，导师奈杰尔·希钦给了他一个关于紧致凯勒流形上全纯向量丛的杨-米尔斯联络问题。此前，奈杰尔·希钦和小林昭七曾分别独立地猜测全纯向量丛上厄米特-杨-米尔斯联络的存在性应该与代数几何的稳定性条件密切相关。厄米特-杨-米尔斯方程是一个非线性偏微分方程组，在实四维时与瞬子方程一致，在实二维时与射影平坦条件一致。由纳拉辛汉（Narasimhan）和塞沙德里（Seshadri）证明并在之后被迈克尔·阿蒂亚与拉乌尔·博特（Raoul Bott）用杨-米尔斯理论重新表述的一个关于黎曼曲面的定理为希钦-小林昭七猜想提供了重要证据。唐纳森从丘成桐关于卡拉比猜想的工作以及 Eells-Sampson 1964 年关于调和映射的工作中得到启发，试图以分析的方法入手将所要的联络转化成一个非线性热方程解的极限。沿着这个思路，他在 1983 年给出了那拉西姆汉-塞沙德里定理的一个新证明，并在几年后证明了代数曲面上的希钦-小林昭七猜想[1]。在此之后，凯伦·乌伦贝克（Karen Uhlenbeck）和丘成桐在一般的紧致凯勒流形上证明了上述猜想，唐纳森则在任意维数的代数流形上给出了一个不同的证明[2]。

唐纳森采纳的辛动量映射的观点（阿蒂亚和博特于 1984 年首先提出）以及他所引进的分析工具启发了众多涉及各种类型希格斯场的规范场论方面的工作，如希格斯丛、涡旋等。唐纳森研究 Eells-Sampson 1964 年那篇文章的另一个收获则是关于扭曲调和映射的存在性定理。唐纳森将这个定理与希钦的希格斯丛理论以及 Simpson 和 Corlette 的推广相结合，给出了非阿贝尔霍奇对应。这个对应为研究曲面群特征簇提供了一条代数几何途径，并促使了高阶泰希米勒理论的诞生[3]。

唐纳森在 1981 年秋季意识到杨-米尔斯理论中的瞬子方程可以用来研究四维流形的拓扑问题，他于是暂时放下了希钦-小林昭七猜想的研究。1982 年，作为一名二年级研究生，唐纳森做出一个震惊整个数学界的证明[4]。他的结果结合迈克尔·弗里德曼（Michael Friedman）关于四维庞加莱猜想的工作便可推出存在一个与标准的四维欧氏空间拓扑同胚但不微分同胚的微分流形。诚然，弗里德曼和唐纳森在 1986 年荣获菲尔兹奖的工作主要是关于封闭四维流形的，它们表明微分流形在微分意义与拓扑意义上有着惊人的差异性。

唐纳森成就的惊人之处在于他对理论物理的借鉴，即利用杨-米尔斯方程来研究几何学。这种洞见在那个时代对几何学家来说是完全陌生的。之前与此最相似的例子当是调和微分形式的线性霍奇理论。实际上，霍奇（William Vallance Douglas Hodge）也是受到了麦克斯韦电磁场理论的启发，而瞬子恰是调和微分形式的一种自然的非线性推广。杨-米尔斯方程涉及的分析则要困难很多，幸运的是陶布斯（Taubes，Clifford Henry）以及乌伦贝克 1982 年的工作为唐纳森提供了所需要的分析基础。这条研究途径激发了唐纳森和其他数学家很多重要的工作，特别是四维闭流形的唐纳森不变量的引进。唐纳森-乌伦贝克-丘定理建立了稳定全纯向量丛与厄米特-杨-米尔斯联络之间关系，而正是这种关系使得一些复代数曲面上唐纳森不变量得以被计算出来。Kronheimer-Mrowka 1994 年给出了唐纳森不变量的基本结构公式，但紧跟着这一重要工作而出现的 Seiberg-Witten 方程从根本上改变了四维规范场论的研究焦点。和唐纳森理论形成对比的是，这个新的理论本质上依赖于与一个新的 S 对偶原则相关的量子场论的方法。Witten 和其他人很快证明利用这个新的方程组能以更简单的方法来获得旧理论的几乎所有的结果，同时还可以证明一些新的结果。

1994 年唐纳森提出了一个引人注目的研究纲领，该纲领试图把凯勒几何中的一些经典结果拓展到辛几何领域。他证明，如果一个紧致辛流形的辛结构所在的上同调类是整的，那么这个上同调类的任何一个充分大的正整数倍的庞加莱对偶可以用余二维的辛子流形表示[5]。在凯勒流形上，经典的 Bertini 定理告诉我们这个子流形可以被取成复子流形。唐纳森的证明利用到了辛流形上的预量子线丛的存在性，并在选取了一个相容的近复结构后证明了一个在扰动辛结构意义下的量化的萨德定理。他的理论使人们可以对辛流形找到类似于凯勒流形的小平嵌入的嵌入，并使得人们可以将射影代数簇上"莱夫谢茨线性束"（Lefschetz pencil）的经典理论扩展到辛几何中。这或许是这个理论最有影响的成果，已经在辛拓扑学中得到了广泛的应用，特别是于同调镜像对称猜想的形成和证明中，因为莱夫谢茨线性束可以用来编译辛拓扑结构的信息。

1998 年，唐纳森和他当时的学生也是后来的合作者托马斯（Richard Thomas，现为帝国理工学院教授）开启了一项研究高维规范场论的重要计划。该计划旨在把三维和四维流形的基本想法拓展到更高维数，尤其是六、七与八维。这些流形上相应的有卡拉比-丘结构，G2 结构或者 Spin（7）-结构。与低维情形类似，这项计划也是基于一些理论物理的思想，特别是弦论。他与托马斯最初的工作导致了现在被称为 Donaldson-Thomas 不变量的概念。唐纳森近年来有相当一部分的学生在这个领域做了博士论文，促使了该研究计划成为当前非常活跃的研究课题。

唐纳森在凯勒几何学上的贡献同样是划时代的。在其开创性的论文[6]中，唐纳森提出了一个关于凯勒几何的宏伟纲领，重新激发了对卡拉比（Eugenio Calabi）在 20 世纪 50 年代提出的常标量曲率存在性问题的研究兴趣。唐纳森的这个纲领仍是受到了辛动量映射的启发，这时凯勒度量所形成的空间在形式上是一个非紧型的对称空间，而标量曲率函数则是与某个固定的辛形式相容的所有复结构组成的空间上的动量映射。这样，寻找卡拉比的常标量曲率度量就转化成了寻找这个无限维空间上动量映射的零点。这个绝妙的视角自然是得益于杨-米尔斯方程的研究。由此，唐纳森提出了一系列凯勒几何学中的问题，进而在过去 20 年里带来了许多令人兴奋的进展，其集大成者则是陈-唐纳森-孙给出的丘成桐关于凯勒-爱因斯坦度量的稳定性猜想的证明。

卡拉比 1958 年的一篇论文[7]给出了 Monge-Ampere 方程的基本估计。这个估计在丘成桐 1976 年给出的卡拉比猜想的证明中发挥了关键的作用。这里的证明是针对第一陈类小于或等于零的凯勒流形。第一陈类大于零时情况则更加微妙，此时卡拉比本来的猜想是有反例的。在 20 世纪 80 年代，丘成桐提出了一个猜想，即法诺流形（即第一陈类大于零的凯勒流形）上凯勒-爱因斯坦度量的存在性与切丛上某种有待确定的代数稳定性联系了起来。唐纳森纲领的第一个基本结果是他的一个猜想的证明，即凯勒势空间是一个度量空间[8]。唐纳森 2001 年证明了当自同构群是平凡的时候，常标量曲率度量是唯一的，并在几何不变理论中建立了常标量曲率度量的存在性与代数稳定性之间的第一个一般联系。在 2002 年的另一篇论文中，唐纳森用代数几何的语言重新阐述了田刚对 K-稳定性的解析描述，并将其扩展到一般极化流形上。这个工作引出了凯勒几何学中的一个核心猜想，即 Yau-Tian-Donaldson 猜想，此猜想把在极化流形上的常标量曲率凯勒度量的存在性与 K-稳定性联系了起来。2005 年，唐纳森证明了 Yau-Tian-Donaldson 猜想中必要条件的一个弱形式。2002～2009 年，唐纳森发表的三篇文章解决了在凯勒环对称曲面上的 Yau-Tian-Donaldson 猜想。2009～2012 年，陈秀雄-唐纳森在任意维数下证明了正标量曲率凯勒-爱因斯坦度量在局部 L2-能量足够小时的正则性，接着在 2012 年，唐纳森-孙崧证明了凯勒-爱因斯坦度量在 Gromov-Hausdorff 极限下的代数性，这对法诺流形模空间理论的研究起到了重要作用。与此同时，在 Yau-Tian-Donaldson 猜想的驱动下，众多学者发展了凯勒几何中的代数几何理论和复分析理论。最后，在 2013 年，陈秀雄-唐纳森-孙崧首次给出了丘成桐稳定性猜想的证明[9]。在 2015 年，唐纳森-孙崧研究了凯勒-爱因斯坦度量的奇性理论，并在其代数几何与黎曼几何之间建立了联系，这引出了 K-稳定性的局部理论。

二、雅科夫·叶利阿什贝格

1. 简历

雅科夫·叶利阿什贝格（图 2）1946 年 12 月 11 日生于苏联列宁格勒（现为圣彼得堡）。他于 1972 年在列宁格勒大学（Leningrad University）获得博士学位，师从 Vladimir Rokhlin。攻读博士学位期间，他开始与长他四岁的米哈伊尔·格罗莫夫（Mikhail Gromov）讨论学术问题，这对他后来的学术生涯产生了深远的影响。他曾于 1972～1979 年在位于俄罗斯联邦科米共和国的瑟克特夫卡尔国立大学（Syktyvkar State University）任职，后于 1980～1987 年负责一家计算机软件公司，并在业余时间从事数学工作。1988 年，他移居美国，一年后受聘为斯坦福大学数学教授，至今已有 30 多年。他的主要研究兴趣是辛拓扑（symplectic topology），并且他与格罗莫夫一起被认为是辛拓扑学的创始人。

叶利阿什贝格是 1986 年和 1998 年国际数学家大会的特邀报告人，并于 2006 年在国际数学家大会上做了一小时报告。

叶利阿什贝格获得的奖项包括奥斯瓦尔德-维布伦几何奖（2001 年）、海因茨·霍普夫奖［Heinz Hopf Prize，2013 年，与赫尔穆特·霍弗（Helmut Hofer）共获］、克拉福德奖（2016 年）和沃尔夫奖（2020 年，与西蒙·唐纳森共获）。叶利阿什贝格是美国国家科学院院士。

叶利阿什贝格和艾达结婚 40 余年。

图 2　雅科夫·叶利阿什贝格
2020 年摄于斯坦福大学

2. 成就

雅科夫·叶利阿什贝格的数学生涯就是一部辛拓扑的诞生和发展史。受格罗莫夫的影响，他一生中的大部分时间都在试图揭示辛拓扑的本质，成就斐然。虽然众多人参与了辛拓扑的发展，但毋庸置疑的是叶利阿什贝格的启迪和思想是辛拓扑的灵魂。

辛拓扑与叶利阿什贝格的人生是完全交织在一起的，想要了解叶利阿什贝格，我们先从辛拓扑的发展史谈起。若从基本的出发点来看待辛几何，首要的问题便是辛结构背后是否存在真正的几何学，格罗莫夫的博士论文第一次尝试回答这个问题。格罗莫夫的看法是整个辛拓扑只是微分拓扑的一个分支。换句话说，辛几何中的分类问题通常要求映射或张量满足某种特定的偏微分方程或更一般的 PDR（偏微分关系），即满足某一个不等式。如果任何形式上的解都可以同伦到一个真正意义的解，那我们就说该关系满足 h-原理。这里我们想指出的是，在由映射所对应的射流空间（jet space）内，的确存在只是广义射流的解，即不记入导数所诱导的偏微分关系［通常称作和乐条件（holonomy condition）］。格罗莫夫得到了一系列结果来推行一个观念：如果有一个形式解，就会有一个实质解。叶利阿什贝格在 1971 年左右开始和格罗莫夫交流。

一方面，他对格罗莫夫的 h-原理开始形成了自己的看法，h-原理可以在皱纹嵌入理论（wrinkled embedding theory）[10-12]中得到体现；另一方面，叶利阿什贝格觉得格罗莫夫关于辛拓扑的看法可能是不正确的。随后格罗莫夫改变了他的观点，并立即提出了他那著名的定理：格罗莫夫二择一（Gromov's alternative）定理。

格罗莫夫二择一定理是说一个辛流形的辛同胚群（group of symplectomorphisms）在保持体积不变的微分同胚群内要么是 C^0-闭的，要么它的 C^0-闭包是整个群。叶利阿什贝格在 1979～1980 年开始尝试证明辛同胚群实际上只能是 C^0-闭的。这一旦得到证实，辛拓扑学便由此发端：因为可以容易地得到诸多推论来说明辛拓扑实际上不同于微分拓扑。然而，叶利阿什贝格并没完全写下他的证明，原因是他被研究院开除了，而为了照顾家庭，他一直在找临时的工作，从而游离于学术圈之外，直到他最终进入一家会计软件公司。

叶利阿什贝格并不是唯一研究该问题的人。实际上，贝内坎（D. Bennequin）于 1983 年发表了一篇文章[13]，指出在三维切触流形上可以得到非常相似的结论。不久之后，康利（C. C. Conley）和策恩德（E. Zehnder）发表了一系列文章。作为推论，他们也得到了格罗莫夫 的二择一定理中的刚性结论[14]。当叶利阿什贝格注意到这些发展后，他设法完成了自己的证明并寄给了自 1974 年以来一直待在法国的格罗莫夫。遗憾的是，叶利阿什贝格当时的证明有一个小错误，他想方设法将正确的证明寄出，但在当时的政治环境下，此事谈何容易，他便放弃了。一直到 1986 年，他才得以在当年的国际数学家大会论文集上发表了他的证明[15]。

即使没有任何资助或帮助，叶利阿什贝格也一直坚持数学研究。比如，他给苏联最好的期刊之一提交了一篇论文，其中证明了阿诺德猜想（Arnold's conjecture）在任何黎曼曲面上都成立。因审稿人在审稿期间与叶利阿什贝格讨论了文中的证明，并透露了自己为审稿人的身份，期刊不能发表该文章。但因文中的证明是正确的，又没充分的理由退稿，所以期刊编辑部要求他主动撤稿，他拒绝了这个要求。40 年后，这篇论文仍然在该期刊编辑部存着。

在 20 世纪 80 年代后期，叶利阿什贝格移居美国。他参加了几次学术会议，并获得了加州大学伯克利分校一个一年的博士后职位。那一年中，他公布了一系列结果：Stein 流形的拓扑表征[16]、三维过扭曲（overtwisted）切触结构的分类[17]，还有几个使用伪全纯曲线（pseudo-holomorphic curves）获得的刚性结果。他是伪全纯曲线技术的世界级专家，然而这些却是他担任工程师时，利用空余时间学习数学所掌握的。1989 年，他获得了斯坦福大学的正教授职位。很明显，一个关于辛拓扑的新世界正在显现，而斯坦福大学正是它的中心。人们之前认为，一旦格罗莫夫二择一定理被证明，就意味着一切都结束了。然而，实际情况却完全相反：一方面，对格罗莫夫二择一定理的多种证明给出了辛拓扑中不同的刚性例子，并且所使用的技术非常适用于解决辛拓扑领域中的各种问题；另一方面，格罗莫夫的论文提供了一系列反方向的结果。所以，辛拓扑比微分拓扑更具有刚性，但是这样的刚性又有多强呢？

我们来看一看这个理论是如何在具体问题中进行演变的：三维切触结构的分类的问题。正如我们之前简要提到的，贝内坎已经得到了一定的刚性[13]，实际上他能够证明在三维欧式空间中至少存在两种不同的切触结构。之后叶利阿什贝格可以对任何三维切触流形上的某一类切触结构进行分类，即过扭曲的切触结构。一个切触结构若不是过扭曲的，我们就称它为紧的（tight）。因此，接下来我们只需要对紧的切触结构进行分类即可。实际上，叶利阿什贝格概述了如何在三维中对紧的接触结构进行分类，并引入了一系列几何的方法[17]。该论文具有深远的影响，为三维接触结构的分类铺平了道路。三十多年后，又一个惊喜出现了[18,19]，这次叶利阿什贝格同样也参与其中。这两篇文章合在一起证明了，即使是紧的接触结构也只是在 π_0 上是刚性的，即一个固定光滑的三维切触流形上的非同痕的切触结构的数量是一个完全几何的、不可计算的量，并且需要更精细的不变量去理解。这再次表明，在一个固定的流形上，切触结构的模空间的每个连通分量的拓扑都满足 h-原理。所以，整个问题都解决了，并且在这个具体的问题中，柔性（flexibility）和刚性之间的边界可以被精确的刻画出来：过扭曲的结构是柔性的；模空间的每个连通分量虽然满足柔性，但是紧的结构在 π_0 上是刚性的。

换句话说，这正是叶利阿什贝格数学方法的核心思想。在任何辛拓扑问题中，总有两个选择：通过 h-原理的工具来尝试证明这是一个柔性问题；如果这不起作用，再更改为刚性的工具（如生成函数，伪全纯曲线等），并证明这是一个刚性问题。

刚性方面的主要工具是 Floer 理论（全纯曲线理论）。在 20 世纪 90 年代，叶利阿什贝格开始思考能让这一理论发挥作用的最广泛的情形。为此，他开始寻找这些不变量背后的代数结构。在 1998 年的国际数学家大会上，叶利阿什贝格公布了第一个与之相关的结果，这着实让人大吃一惊，因为他正在建立一个纯代数的理论，这与他以往做数学的风格完全不同。

叶利阿什贝格曾不断声明："虽然我不喜欢这些做超流形、超代数和超微积分的超人，但是我不得不说我已经改变了这个观点，现在我正在成为他们中的一员。"在亚历山大·吉文特尔（A. Givental）和 H. Hofer 的帮助下，他建立了一个适用性非常广泛的理论，该理论包含很多非常具体的例子：完整的 Gromov-Witten 不变量、Floer 同调、不同的切触同调等。即使是 Chas-Sulliva 引入的在流形环空间（loop space）的同伦上的复杂操作也可以看作是该理论的一种极限情况。因为该理论具有场论的形式，因而取名为辛场论。事实上，存在一个使用亏格为零的全纯曲线定义的经典场论和一个通过形变全纯曲线的亏格而定义的量子化（quantization）。如今，仍然有几个神秘的特征（即使过了 20 年）亟待被理解：

（1）从代数角度验证和发展的场论形式无法通过几何的方式来理解；

（2）令人惊讶的是粘连公式（gluing formula）［即如何粘连两个辛配边（symplectic cobordisms）］可以给出可积系统的层次结构（hierarchy）。即使是最简单的粘连也可以给出如 Toda 层次结构和 KdV 层次结构。

在叶利阿什贝格提出所谓的"唯有全纯曲线这一原则"时，他引入了辛拓扑中最具争议的且尚未解决的问题。具体陈述是"如果辛几何中的一个具体问题在辛场论中没有任何阻碍（obstruction），那么它应该是一个柔性的问题"。证明它的方法很简单：只需检查所有例子中它是否为真。当然这并不是一个证明，但随着被检验的例子越来越多，你可以开始掂量叶利阿什贝格的猜想是否正确。这是叶利阿什贝格过去 14 年工作的一个缩影，而且他已经从研究辛拓扑的刚性转向了研究辛拓扑的柔性。这里我们列出叶利阿什贝格在过去几年中引入的几个柔性例子：柔性勒让德类的定义（在 E. Murphy 的论文[20]中完成）、柔性 Stein 流形的定义、过扭曲切触结构在高维的定义，以及对这些例子的所对应的 h-原理的完整证明。此外，一个清晰的收敛过程指出，基于相似的原因，所有柔性的研究对象都是柔性的。

参考文献

［1］Donaldson S K. Anti-self-dual Yang – Mills connections on a complex algebraic surface and stable vector bundles. Proceedings of the London Mathematical Society,1985,3:1-26.

［2］Donaldson S K. Infinite determinants,stable bundles and curvature. Mathematical Journal,1987,54:231-247.

［3］Hitchin N J. The self-duality equations on a Riemann surface,Proceedings of the London Mathematical Society,1987,55:59-126.

［4］Donaldson S K. Self-dual-connections and the topology of smooth 4-manifolds,Bulletin of the American Mathematical Society,1983,8:81-83.

［5］Donaldson S K. Symplectic submanifolds and almost complex geometry. Journal of Differential Geometry,1996,44:666-705.

［6］Donaldson S K. Symmetric spaces,Kahler geometry and Hamiltonian dynamics. American Mathematical Society Translations,Series2,1999,196:13-33.

［7］Calabi E. Improper affine hyperspheres of convex type and a generalization of a theorem by K. Jorgens. Michigan Mathematical Journal,1958,5(2):105-126.

［8］Chen X X. Space of Kahler metrics. Journal of Differential Geometry,2000,56:189-234.

［9］Chen X X. Donaldson S,Sun S. Kahler-Einstein metrics on Fano manifolds I,II and III. Journal of the American Mathematical Society,2015,28:183-278.

[10] Eliashberg Y, Mishachev N. Wrinkling of smooth mappings and its applications. Inventiones Mathematicae,1997,130(2):345-369.

[11] Eliashberg Y, Mishachev N. Wrinkling of smooth mappings-II Wrinkling of embeddings and K. Igusa's theorem. Topology,2000,39 (4):711-732.

[12] Eliashberg Y, Mishachev N. Wrinkling of smooth mappings-III Foliations of codimension greater than one. Topological Methods in Nonlinear Analysis,1998,11 (2):321-350.

[13] Bennequin D. Entrelacements et' equations de Pfaff. Third Schnepfenried geometry conference, Vol. 1 (Schnepfenried,1982),87161,Ast'erisque,107-108,Soc. Math. France,Paris,1983.

[14] Conley C C, Zehnder E. The Birkhoff-Lewis fixed point theorem and a conjecture of V. I. Arnold. Inventiones Mathematicae,1983,73 (1):33-49.

[15] Eliashberg Y. Combinatorial methods in symplectic geometry. Proceedings of the International Congress of Mathematicians, Vol. 1, 2 (Berkeley, Calif. , 1986), 531 – 539, Amer. Math. Soc. , Providence,RI,1987.

[16] Eliashberg Y. Topological characterization of Stein manifolds of dimension>2. International Journal of Mathematics,1990,1 (1):29-46.

[17] Eliashberg Y. Classification of overtwisted contact structures on 3-manifolds. Inventiones Mathematicae, 1989,98 (3):623-637.

[18] Eliashberg Y, Mishachev N. The space of tight contact structures on R^3 is contractible. arXiv: 2108. 09452

[19] Ferna'ndez E,Martínez-Aguinaga J,Presas F. The homotopy type of the contactomorphism groups of tight contact 3-manifolds,part I. arXiv:2012. 14948.

[20] Murphy E. Loose Legendrian embeddings in high dimensional contact manifolds. arXiv:1201. 2245.

Introduction to the 2020 Wolf Prize for Mathematics

ChenXiuxiong,Oscar Garc'ia-Prada ,Francisco Presas

Simon Donaldson has been a leader in geometry and topology in the last thirty five years. He has drawn inspirations from physics to solve problems in pure mathematics. His use of moduli spaces of solutions of physical equations has had a profound impact in many branches of mathematics and physics. His revolutionary invariants of four dimensional manifolds have changed our geometrical understanding of space and time,and his transformative insights inKähler geometry have revealed the deep connection between stability in algebraic geometry and existence of interesting metrics in complex geometry. Yakov Eliashberg has helped lay the foundations of symplectic and contact topology, one of the most striking and enduring developments in mathematical research in the last forty years. The Eliashberg-Gromov symplectic rigidity theorem is one of the cornerstones of symplectic topology. The Eliashberg dichotomy about contact structures has transformed the landscape of the contact topology. Eliashberg has discovered a number of astonishing appearances of homotopy principle in symplectic and contact topology, which has completely changed our view of symplectic world.

5.5　2020年度维特勒森奖获奖者简介

魏艳红

（中国科学院西北生态环境资源研究院）

维特勒森奖（Vetlesen Prize）是地球科学领域的世界性重要奖项，于1959年由维特勒森基金设立，是以挪威人G. U. 维特勒森（G. Unger Vetlesen）的名字命名的。该奖项向世界各国开放，旨在表彰在地球科学领域对地球、地球演化史以及地球与宇宙的关系等方面取得重大突破的科学家。维特勒森奖被誉为地球科学领域的"诺贝尔奖"，由哥伦比亚大学拉蒙特-多尔蒂地球观测站管理，维特勒森基金会颁发奖金，该奖项每三年颁发一次，获奖者可获得25万美元奖金和一枚金质奖章，还被邀请在哥伦比亚大学做维特勒森演讲[1]。

2020年度的维特勒森奖授予了法国地球物理学家阿尼·卡泽纳夫（Anny Cazenave），一位在记录现代海平面上升及其原因方面做出了杰出贡献的科学家（图1）。在过去20多年里，卡泽纳夫率先使用卫星数据来绘制海平面上升图，同时分析了冰盖、陆地和淡水体的相关变化，并将许多以前有关海平面上升和气候变化的松散线索联系了起来。

阿尼·卡泽纳夫

图1　2020年维特勒森奖获得者

　　阿尼·卡泽纳夫是法国图卢兹大学地球物理学与空间海洋学研究实验室的高级科学家，也是国际空间科学研究所（总部设在瑞士伯尔尼）的地球科学主任。她还担任"世界气候研究计划"（World Climate Research Program，WCRP）科学委员会的官员，也是"未来地球"（Future Earth）计划咨询委员会的成员。作为绘制现代海平面上升图的先驱，阿尼·卡泽纳夫的研究主要是关于空间技术在地球科学中的应用，主要分为3个主题：①卫星大地测量及其在大地测量问题中的应用；②海洋地球物理学，主要是卫星测高；③海平面变化及其与气候的关系，是政府间气候变化专门委员会2007年和2014年综合报告中海平面部分的主要作者。

　　自19世纪80年代以来，全球平均海平面上升了21～24 cm，而且上升速度还在加快，目前每年上升约3.6 mm，是20世纪大部分时间的2.5倍。直到20世纪90年代，海平面的高度主要还是由停泊在沿海港口的潮汐仪记录的，开阔的海洋海平面数据在很大程度上是一个"黑洞"。其后，欧美航天机构发射了一系列新的卫星雷达高度计，能够或多或少地实时监测世界各地的海平面和陆地高度。卡泽纳夫及其团队迅速开发出数据分析方法，并将其与其他信息分析方法结合起来。到21世纪初，他们已经确定全球海平面每年上升约3 mm，有些地方的海平面上升幅度是这个数字的三倍。通过分析1993～2008年海平面变化趋势的空间格局，可以看出西太平洋地区海平面上升的速度是全球平均水平的五倍，而其他地区（如东太平洋），海平面在过去15年里一直在下降[2]。这种差异是由区域因素造成的，包括风和波浪模式，以及土地本身的沉降或反弹。

　　通过将极地冰盖的高度测量值，以及与质量相关的冰盖重力场变化的最新卫星数据和全球海水的盐度、温度等数据整合到一张图中，卡泽纳夫及其团队发现，近几十年来导致海平面上升的一半原因是由于海水的膨胀，另一半原因是由于冰川和冰盖的融化。但随着气候变暖的加剧，在过去10年或12年里海平面上升的原因已经发生变化：2/3来自冰川融化，还有一小部分是由于河流、湖泊、土壤和含水层中的水向海洋转移。这似乎是由于多种因素的综合作用，包括人类对地表和地下水的过度抽取，以及气温上升导致的蒸发加剧。

参考文献

[1] Vetlesen Prize. 2020 Vetlesen Prize Laureate. https://lamont. columbia. edu/about/vetlesen-prize.

[2] Cazenave A，Lombard A，Llovel W. Present-day sea level rise：A synthesis. Comptes Rendus Geoscience，2008，340(11)：761-770.

Introduction to the 2020 Vetlesen Prize Laureate

Wei Yanhong

The 2020 Vetlesen Prize for achievement in the Earth sciences was awarded to French geophysicist Anny Cazenave, for her pioneering work in charting modern sea-level rise and linking it with climate change.

5.6　2020 年度泰勒环境成就奖获奖者简介

廖　琴

（中国科学院西北生态环境资源研究院）

2020 年度的泰勒环境成就奖（Tyler Prize for Environmental Achievement）授予保护生物学家格蕾琴·戴利（Gretchen C. Daily）和环境经济学家帕万·苏克戴夫（Pavan Sukhdev）（图 1）。这两位科学家都是阐明和量化自然环境经济价值的先驱，开创性地以严谨的科学和经济术语来评估自然资本，使人们认识到自然在支持人类福祉方面的重要作用[1]。

格蕾琴·戴利　　　　　　　　帕万·苏克戴夫

图 1　2020 年泰勒环境成就奖获得者[2]

一、戴利获奖工作简介

戴利对保护生物学的格局演变产生了巨大影响，被评为"科学界最重要的 50 位女性"之一。戴利目前是美国斯坦福大学生物系教授、伍兹环境研究所高级研究员、保护生物学中心主任，是美国国家科学院院士、美国艺术与科学院院士、美国哲学院院士，同时也是瑞典皇家科学院贝耶尔国际生态经济研究所（Beijer Institute for

Ecological Economics）和大自然保护协会（The Nature Conservancy）的董事会成员。

戴利被誉为乡村生物地理学领域的先驱。她将乡村生物地理学描述为阐明乡村人口、物种和生态系统命运的一个新概念框架，并指出地球上越来越多的陆地表面的生态系统质量受到人类强烈影响。戴利的研究试图确定"哪些物种最重要和最值得保护"以及"决定物种在特定生态系统中相对重要性的科学依据是什么"。戴利主持编著的《自然的服务：社会对自然生态系统的依赖》（*Nature's Services：Societal Dependence on Natural Ecosystems*，1997 年出版）一书，被誉为"过去 30 年来出版的最具影响力的环境书籍之一"，在全球掀起了生态系统服务功能研究的热潮。

2005 年，戴利与大自然保护协会、明尼苏达大学和世界自然基金会合作建立了"自然资本项目"组织，旨在通过激励更大和更有针对性的自然资本投资，改善自然和人类福祉。他们共同开发了一套名为"InVEST"的免费开源软件模型，用于绘制和评估维持人类生活的自然商品和服务。本质而言，该软件是政府、非营利组织、国际贷款机构和企业平衡环境与经济目标的工具。

戴利与乔治娜·梅斯（Georgina Mace）共同获得了 2018 年西班牙对外银行（BBVA）基金会生态和保护生物学领域的"知识前沿奖"（Frontiers of Knowledge Award）。她获得的其他重要奖项还包括 2017 年 Asahi Glass 基金会"蓝色星球奖"（Blue Planet Prize）、2012 年沃尔沃环境奖（Volvo Environment Prize）和 2010 年生物多样性绿色奖（MIDORI Prize for Biodiversity）。

二、苏克戴夫获奖工作简介

苏克戴夫出生于印度，是著名的环境经济学家，在理解自然价值方面做出了非常了不起的工作。苏克戴夫担任世界自然基金会（World Wide Fund For Nature，WWF）主席，联合国环境规划署（United Nations Environment Programme，UNEP）亲善大使[3]，以及生态系统与生物多样性经济学（TEEB）咨询委员会、斯德哥尔摩复原力中心（Stockholm Resilience Centre）和剑桥保护倡议组织（Cambridge Conservation Initiative）的董事会成员。

作为职业银行家，苏克戴夫在澳大利亚和新西兰银行集团工作了 11 年，在德意志银行工作了 14 年，以参与印度货币、利率和衍生品市场的发展而闻名。凭借数十年在全球领先的经济学职位经验，苏克戴夫领导了几个著名的项目，并在更广泛的经济理论中牢固地建立了绿色经济学体系。

2008 年，应德国联邦环境、自然保护和核安全部（Federal Ministry for the Environment，Nature Conservation and Nuclear Safety）部长的邀请，苏克戴夫主持

了具有里程碑意义的"生态系统和生物多样性经济学"（TEEB）研究。由于揭示了自然服务损失带来的经济影响，以及将生物多样性和生态系统的经济学与伦理、公平和减少贫困联系起来，TEEB 的第一份报告在全球范围内受到欢迎。TEEB 的最终报告汇集了来自各大洲的 550 多名作者和审稿人，使人们前所未有地关注生态系统服务的经济和社会重要性，以及生态系统服务损失造成的真实代价。随后，苏克戴夫领导了联合国环境规划署发起的"绿色经济倡议"，以证明绿色经济不是增长的负担，而是增长的新引擎、新就业的来源以及减少贫困的手段。苏克戴夫在 2011 年创立了"可持续未来全球倡议"（GIST）咨询集团[4]，帮助企业和政府实施 TEEB、绿色经济倡议以及其撰写的《企业 2020》（*Corporation 2020*，2012 年出版）一书中提出的许多策略和方法。

　　苏克戴夫曾获 2016 年 Asahi Glass 基金会"蓝色星球奖"、2015 年德国复兴信贷银行（KfW）基金会 Bernhard Grzimek 生物多样性奖和 2013 年哥德堡可持续发展奖（Gothenburg Award for Sustainable Development）。

参考文献

［1］Tyler Prize. 2020 Laureates. https：//tylerprize. org/laureates/［2021-11-29］.

［2］Stockholm Resilience Center. Gretchen Daily and Pavan Suhkdev win 2020 Tyler Prize. https://www. stockholmresilience. org/news--events/general-news/2020-01-27-gretchen-daily-and-pavan-suhkdev-win-2020-tyler-prize. html［2021-11-29］.

［3］United Nations. 'Green economy'pioneer Pavan Sukhdev wins 2020 Tyler Prize for Environmental Achievement. https：//news. un. org/en/story/2020/01/1056082［2021-11-30］.

［4］University College Oxford. Tyler Prize for Environmental Achievement. https：//www. univ. ox. ac. uk/news/tyler-prize-for-environmental-achievement/［2021-11-30］.

Introduction to the 2020 Tyler Laureates

Liao Qin

　　The 2020 Tyler Prize for Environmental Achievement was awarded to Conservation Biologist Gretchen C. Daily，and Environmental Economist Pavan Sukhdev. They were both pioneers in illuminating and quantifying the economic value of the natural environment. Each has trailblazed the valuing of natural capital in rigorous scientific and economic terms，and recognized nature's vital role in supporting human wellbeing.

5.7 2020年度图灵奖获奖者简介

唐 川

（中国科学院成都文献情报中心）

2021年3月31日，国际计算机协会（Association for Computing Machinery，ACM）宣布阿尔弗雷德·瓦伊诺·阿霍（Alfred Vaino Aho）和杰弗里·大卫·厄尔曼（Jeffrey David Ullman）获得2020年图灵奖（Turing Award）（图1），以表彰他们对编程语言实现领域基础算法和理论研究所做的贡献[1]。除此以外，这两位教授还将该领域的诸多研究成果编纂成教材，深刻影响了数代计算机科学家。

阿尔弗雷德·瓦伊诺·阿霍　　　　　杰弗里·大卫·厄尔曼

图1　2020年图灵奖获得者

阿尔弗雷德·瓦伊诺·阿霍1941年出生于加拿大，本科毕业于多伦多大学，在普林斯顿大学获得电气工程硕士和计算机科学博士学位，现为美国哥伦比亚大学劳伦斯·古斯曼名誉教授。他曾在美国贝尔实验室工作了30多年，担任计算科学研究中心副主席一职，之后于1995年入职哥伦比亚大学计算机科学系。他曾获得IEEE约翰冯·诺伊曼奖章（John von Neumann Medal）和 NEC C&C 基金会 C&C 奖，是美国国家工程院（US National Academy of Engineering）、美国艺术与科学院（American Academy of Arts and Sciences）和加拿大皇家学会（Royal Society of Canada）的院士[2]。同时他也是 ACM、IEEE、贝尔实验室和美国科学促进会的会士（Fellow）。

　　杰弗里·大卫·厄尔曼 1942 年出生，是斯坦福大学名誉教授，也是计算机科学在线学习平台 Gradiance Corporation 的首席执行官。他本科毕业于哥伦比亚大学，在普林斯顿大学获得计算机科学博士学位。1966～1969 年，在贝尔实验室担任技术人员，1969～1979 年在普林斯顿大学任教，1979 年入职斯坦福大学。他所获得的荣誉包括 IEEE 约翰·冯·诺依曼奖章、NEC C&C 基金会 C&C 奖、高德纳奖（Donald E. Knuth Prize）、国际计算机协会 Karl V. Karlstrom 杰出教育家奖和 SIGMOD Codd 创新奖[3]。厄尔曼是美国国家工程院、美国国家科学院、美国艺术与科学院的成员，也是国际计算机协会的会员。

　　阿霍和厄尔曼奠定了关于算法、形式语言、编译器和数据库的基本概念，阐明了不同领域的研究是如何紧密联系的，引入了关键的核心技术概念，包括不可或缺的特定算法等概念。如今，无论在手机、汽车，还是网络公司内部的大型服务器上运行的程序，都是人类用高级编程语言编写出来后再编译成低级代码来执行。尽管无数研究人员和行业翘楚都做出了贡献，但阿霍和厄尔曼的成果具有特别的影响力，他们帮助我们理解了算法的理论基础，还为编译器和编程语言设计方面的研究和实践指明了方向，这种将高级编程语言翻译成计算机能识别的低级语言的技术很大程度都归功于他们。

　　阿霍和厄尔曼从 1967 年到 1969 年都在美国贝尔实验室工作，合作开发了用于分析和翻译编程语言的高效算法。1969 年，厄尔曼入职斯坦福大学，阿霍则继续在贝尔实验室工作了 30 年，之后入职哥伦比亚大学。两人一直保持着密切的合作，共同撰写书籍和论文，并为算法、编程语言、编译器和软件系统引入了新技术，奠定了编程语言理论、编程语言实现、算法设计与算法分析的基础。他们共合著了九本计算机领域的热门书籍（包括初版和更新版本），其中有两本最为著名，其一为《计算机算法设计与分析》（1974），其二为《编译器设计原则》（1977）[1]。

　　国际计算机协会主席加布里埃尔·科蒂斯（Gabriele Kotsis）[1] 称赞阿霍和厄尔曼自 20 世纪 70 年代初以来，一直是计算机领域的思想领袖，他们的工作指引着一代又一代的程序员和研究人员。谷歌人工智能高级研究员兼高级副总裁杰夫·迪恩（Jeff Dean）[1] 也称赞他们的教材已成为学生、研究人员和从业人员培训的黄金准则。

参考文献

[1] A. M. Turing Award. A. M. Turing Award Honors Innovators Who Shaped The Foundations of Programming Language Compilers And Algorithms. https://amturing. acm. org/[2021-12-30].

[2] Alfred V Aho. Author's Profile. https://dl. acm. org/profile/81100024612 [2021-12-30].

[3] Jeffrey D. Ullman. Author's Profile. https://dl. acm. org/profile/81100314798 [2021-12-30].

Introduction to the 2020 Turing Award Winners

Tang Chuan

The 2020 Turing Award Winners for Computer Achievement was Professor Alfred Aho of Columbia University, and Professor Jeffrey Ullman of Stanford University. They have all made outstanding contributions to the basic algorithms and theoretical research in the field of programming language implementation, and gathered these results into their extremely influential works, which have influenced generations of computer scientists.

5.8　2020 年国家最高科学技术奖及自然科学奖简介

叶　京

（中国科学院科技战略咨询研究院）

2021 年 11 月 3 日，中共中央、国务院在北京大会堂隆重举行 2020 年度国家科学技术奖励大会。中共中央总书记、国家主席、中央军委主席习近平向获得 2020 年度国家最高科学技术奖的中国航空工业集团有限公司顾诵芬院士和清华大学王大中院士（图 1）颁发奖章、证书[①]。

顾诵芬　　　　　　　　　　王大中

图 1　2020 年度国家最高科学技术奖获奖人

---

①　新华社. 中共中央国务院隆重举行国家科技奖励大会 习近平出席大会并为最高奖获得者等颁奖. http://www.gov.cn/xinwen/2020-01/10/content_5468098.htm.

一、2020 年度国家最高科学技术奖获奖人概况

2020 年度国家最高科学技术奖获奖人概况如下[1]。

1. 顾诵芬

顾诵芬，男，1930 年 2 月出生，江苏苏州人，1951 年毕业于上海交通大学航空工程系。现任中国航空工业集团有限公司科技委高级顾问、中国航空研究院名誉院长。1991 年当选中国科学院学部委员①（院士），1994 年当选中国工程院院士。

顾诵芬是我国著名飞机设计大师、飞机空气动力设计奠基人。他建立了新中国飞机空气动力学设计体系，开创了我国自主研制歼击机的先河，持续开展航空战略研究，为我国航空科技事业作出了重大贡献。

2. 王大中

王大中，男，1935 年 2 月出生，河北昌黎人。1958 年毕业于清华大学工程物理系。曾任清华大学核能技术研究所所长、清华大学校长。1993 年当选中国科学院院士。

王大中是国际著名核能科学家，致力于发展具有固有安全特性的先进核能系统。他带领产学研联合团队实现了我国高温气冷堆技术从跟跑、并跑到领跑的整体发展过程，为我国在先进核能领域逐步走向世界前沿奠定了重要技术基础。

二、2020 年度国家自然科学奖概况

2020 年度的国家自然科学奖共授予 46 个项目。具体获奖项目及其完成人情况如表 1 所示[2]。

表 1 2020 年度国家自然科学奖获奖项目

序号	编号	项目名称	主要完成人	提名者
一等奖（2 项）				
1	Z-103-1-01	纳米限域催化	包信和（中国科学院大连化学物理研究所） 潘秀莲（中国科学院大连化学物理研究所） 傅　强（中国科学院大连化学物理研究所） 邓德会（中国科学院大连化学物理研究所）	中国科学院

---

① 1993 年 10 月后，中国科学院学部委员改称中国科学院院士。

续表

序号	编号	项目名称	主要完成人	提名者
2	Z-103-1-02	有序介孔高分子和碳材料的创制和应用	赵东元（复旦大学） 李　伟（复旦大学） 邓勇辉（复旦大学） 张　凡（复旦大学）	上海市
二等奖（44 项）				
1	Z-101-2-01	p进霍奇理论及其应用	刘若川（北京大学）	张继平
2	Z-101-2-02	不可压流体方程组的非线性内蕴结构	雷　震（复旦大学） 周　忆（复旦大学）	张平文　张平 黄飞敏
3	Z-101-2-03	同余数问题与 L-函数的算术	田　野（中国科学院数学与系统科学研究院）	席南华
4	Z-101-2-04	波动方程反问题的数学理论与计算方法	包　刚（浙江大学）	江　松
5	Z-102-2-01	基于超冷费米气体的量子调控	张　靖（山西大学） 王鹏军（山西大学） 黄良辉（山西大学） 孟增明（山西大学）	山西省
6	Z-102-2-02	基于量子信息技术研究量子物理基本问题	李传锋（中国科学技术大学） 许金时（中国科学技术大学） 黄运锋（中国科学技术大学） 柳必恒（中国科学技术大学）	中国科学院
7	Z-102-2-03	基于高精度脉泽天体测量的银河系旋臂结构研究	徐　烨（中国科学院紫金山天文台青海观测站） 郑兴武（南京大学） 张　波（中国科学院上海天文台） 李晶晶（中国科学院紫金山天文台青海观测站） 吴元伟（中国科学院紫金山天文台青海观测站）	青海省
8	Z-103-2-01	活细胞化学反应工具的开发与应用	陈　鹏（北京大学） 赵　劲（南京大学） 昌增益（北京大学） 李　劼（北京大学） 林世贤（北京大学）	教育部

续表

序号	编号	项目名称	主要完成人	提名者
9	Z-103-2-02	碳链与金属的螯合化学	夏海平（厦门大学）	福建省
			张　弘（厦门大学）	
			朱　军（厦门大学）	
			朱从青（厦门大学）	
			王铜道（厦门大学）	
10	Z-103-2-03	荧光探针性能调控与生物成像应用基础研究	张晓兵（湖南大学）	湖南省
			谭蔚泓（湖南大学）	
			赵子龙（湖南大学）	
			陈　卓（湖南大学）	
11	Z-103-2-04	手性金属-有机多孔固体的设计构筑及性能研究	崔　勇（上海交通大学）	教育部
			刘　燕（上海交通大学）	
			袁国赞（上海交通大学）	
			宣为民（上海交通大学）	
			刘泰峰（上海交通大学）	
12	Z-103-2-05	单壁碳纳米管的可控催化合成	李　彦（北京大学）	高　松 任咏华 卜显和
			杨　烽（北京大学）	
			杨　娟（北京大学）	
			褚海斌（北京大学）	
			金　钟（北京大学）	
13	Z-104-2-01	峨眉山大火成岩省与地幔柱研究	徐义刚（中国科学院广州地球化学研究所）	广东省
			何　斌（中国科学院广州地球化学研究所）	
			王　焰（香港大学）	
			肖　龙（中国科学院广州地球化学研究所）	
			钟玉婷（中国科学院广州地球化学研究所）	
14	Z-104-2-02	黄土高原生态系统过程与服务	傅伯杰（中国科学院生态环境研究中心）	陈发虎 崔　鹏 于贵瑞
			陈利顶（中国科学院生态环境研究中心）	
			吕一河（中国科学院生态环境研究中心）	
			冯晓明（中国科学院生态环境研究中心）	
			王　帅（中国科学院生态环境研究中心）	
15	Z-104-2-03	华南陆块中生代陆内成矿作用	胡瑞忠（中国科学院地球化学研究所）	侯增谦 杨春和 周忠和
			毛景文（中国地质科学院矿产资源研究所）	
			苏文超（中国科学院地球化学研究所）	
			王岳军（中国科学院广州地球化学研究所）	
			袁顺达（中国地质科学院矿产资源研究所）	

序号	编号	项目名称	主要完成人	提名者
16	Z-104-2-04	二万年以来东亚古气候变化与农耕文化发展	吕厚远（中国科学院地质与地球物理研究所） 肖举乐（中国科学院地质与地球物理研究所） 杨晓燕（中国科学院地理科学与资源研究所） 张健平（中国科学院地质与地球物理研究所） 吴乃琴（中国科学院地质与地球物理研究所）	中国科学院
17	Z-104-2-05	寒武纪特异保存化石与节肢动物早期演化	张喜光（云南大学） 侯先光（云南大学） 马晓娅（云南大学） 丛培允（云南大学） 刘煜（云南大学）	云南省
18	Z-105-2-01	水稻高产与氮肥高效利用协同调控的分子基础	傅向东（中国科学院遗传与发育生物学研究所） 黄先忠（中国科学院遗传与发育生物学研究所） 王少奎（中国科学院遗传与发育生物学研究所） 刘倩（中国科学院遗传与发育生物学研究所）	李振声
19	Z-105-2-02	早期胚胎发育与体细胞重编程的表观调控机制研究	高绍荣（同济大学） 高亚威（同济大学） 张勇（同济大学） 陈嘉瑜（同济大学） 鞠振宇（杭州师范大学）	季维智 裴钢 魏辅文
20	Z-105-2-03	水稻驯化的分子机理研究	孙传清（中国农业大学） 谭禄宾（中国农业大学） 朱作峰（中国农业大学） 谢道昕（清华大学） 付永彩（中国农业大学）	刘耀光 武维华 陈温福
21	Z-105-2-04	成年哺乳动物雌性生殖干细胞的发现及其发育调控机制	吴际（上海交通大学） 邹康（上海交通大学） 孙斐（中国科学技术大学） 赵小东（上海交通大学） 刘以训（中国科学院动物研究所）	上海市

续表

序号	编号	项目名称	主要完成人	提名者
22	Z-106-2-01	造血干细胞调控机制与再生策略	程　涛（中国医学科学院血液病医院） 刘　兵（中国人民解放军总医院第五医学中心） 王前飞（中国科学院北京基因组研究所） 竺晓凡（中国医学科学院血液病医院） 程　辉（中国医学科学院血液病医院）	周　琪 陈赛娟 裴端卿
23	Z-106-2-02	麻风危害发生的免疫遗传学机制	张福仁（山东第一医科大学附属皮肤病医院） 张学军（安徽医科大学第一附属医院） 刘　红（山东第一医科大学附属皮肤病医院） 王真真（山东第一医科大学附属皮肤病医院） 孙勇虎（山东第一医科大学附属皮肤病医院）	沈　岩 张　学 沈洪兵
24	Z-106-2-03	非酒精性脂肪性肝病及相关肝癌自然史、发病机制、诊断和防治研究	于　君（香港中文大学） 黄炜燊（香港中文大学） 陈力元（香港中文大学） 张　翔（香港中文大学） 沈祖尧（香港中文大学）	香港特别行政区
25	Z-106-2-04	新型纳米载药系统克服肿瘤化疗耐药的应用基础研究	李亚平（中国科学院上海药物研究所） 于海军（中国科学院上海药物研究所） 尹　琦（中国科学院上海药物研究所） 张志文（中国科学院上海药物研究所） 张鹏程（中国科学院上海药物研究所）	中国科学院
26	Z-107-2-01	面向多义性对象的新型机器学习理论与方法	周志华（南京大学） 耿　新（东南大学） 高　尉（南京大学） 张道强（南京大学） 王　魏（南京大学）	教育部
27	Z-107-2-02	真实感图形的实时计算理论与方法	周　昆（浙江大学） 邵天甲（清华大学） 潘志庚（杭州师范大学） 彭群生（浙江大学） 石教英（浙江大学）	浙江省

续表

序号	编号	项目名称	主要完成人	提名者
28	Z-107-2-03	面向多租户资源竞争的云计算基础理论与核心方法	金　海（华中科技大学） 吴　松（华中科技大学） 刘海坤（华中科技大学） 刘方明（华中科技大学） 廖小飞（华中科技大学）	教育部
29	Z-107-2-04	非线性切换系统的分析与控制	孙希明（大连理工大学） 赵　军（东北大学） 王　伟（大连理工大学） 马瑞诚（东北大学）	黄　捷 柴天佑 何　友
30	Z-107-2-05	特种光电器件的超快激光微纳制备基础研究	孙洪波（吉林大学） 陈岐岱（吉林大学） 张永来（吉林大学） 王海宇（吉林大学） 夏　虹（吉林大学）	教育部
31	Z-107-2-06	深度学习处理器体系结构新范式	陈云霁（中国科学院计算技术研究所） 陈天石（中国科学院计算技术研究所） 杜子东（中国科学院计算技术研究所） 孙凝晖（中国科学院计算技术研究所） 郭　崎（中国科学院计算技术研究所）	中国科学院
32	Z-107-2-07	视觉运动模式学习与理解的理论与方法	胡卫明（中国科学院自动化研究所） 刘成林（中国科学院自动化研究所） 李　兵（中国科学院自动化研究所） 张笑钦（中国科学院自动化研究所） 王　恒（中国科学院自动化研究所）	陈熙霖 邓中翰 高新波
33	Z-107-2-08	分布式动态系统的自学习优化协同控制理论与方法	张化光（东北大学） 罗艳红（东北大学） 孙秋野（东北大学） 刘振伟（东北大学） 王占山（东北大学）	教育部
34	Z-108-2-01	面心立方材料弹塑性力学行为及原子层次机理研究	韩晓东（北京工业大学） 张　泽（浙江大学） 王立华（北京工业大学） 张跃飞（北京工业大学） 郑　坤（北京工业大学）	北京市

续表

序号	编号	项目名称	主要完成人	提名者
35	Z-108-2-02	光催化材料的能带与微观结构调控	刘　岗（中国科学院金属研究所） 成会明（中国科学院金属研究所） 杨勇强（中国科学院金属研究所） 牛　萍（中国科学院金属研究所） 康宇阳（中国科学院金属研究所）	中国科学院
36	Z-108-2-03	限域反应构建晶态能量转换材料及调控机制	李春忠（华东理工大学） 江　浩（华东理工大学） 杨化桂（华东理工大学） 朱以华（华东理工大学） 沈建华（华东理工大学）	上海市
37	Z-108-2-04	基于结构基元的新电磁材料和新效应的发现	陈小龙（中国科学院物理研究所） 郭建刚（中国科学院物理研究所） 王　刚（中国科学院物理研究所） 钱　天（中国科学院物理研究所） 金士锋（中国科学院物理研究所）	吴以成 陈仙辉 陈延峰
38	Z-108-2-05	秉承自然生物精细构型的遗态材料	张　荻（上海交通大学） 范同祥（上海交通大学） 顾佳俊（上海交通大学） 周　涵（上海交通大学） 张　旺（上海交通大学）	上海市
39	Z-109-2-01	河流动力学及江河工程泥沙调控新机制	方红卫（清华大学） 何国建（清华大学） 王光谦（清华大学） 吴保生（清华大学） 黄　磊（清华大学）	教育部
40	Z-109-2-02	状态相关非饱和土本构关系及应用	吴宏伟（香港科技大学） 李焯芬（香港大学） 戴福初（中国科学院地理科学与资源研究所） 赵仲辉（香港科技大学） 周　超（香港科技大学）	香港特别行政区
41	Z-109-2-03	内燃机复合循环理论与方法	舒歌群（天津大学） 田　华（天津大学） 卫海桥（天津大学） 梁兴雨（天津大学）	金东寒 樊建人 齐　飞

续表

序号	编号	项目名称	主要完成人	提名者
42	Z-109-2-04	耗散最小化多场协同对流传热强化理论和方法	刘　伟（华中科技大学） 刘志春（华中科技大学） 明廷臻（华中科技大学） 张晓屿（华中科技大学） 郭　剑（华中科技大学）	姜培学 陈学东 王如竹
43	Z-110-2-01	考虑非均匀结构效应的金属材料剪切带	戴兰宏（中国科学院力学研究所） 白以龙（中国科学院力学研究所） 蒋敏强（中国科学院力学研究所） 刘龙飞（中国科学院力学研究所） 陈　艳（中国科学院力学研究所）	张统一 汪卫华 赵亚溥
44	Z-110-2-02	具有界面效应的复合材料细观力学研究	段慧玲（北京大学） 王建祥（北京大学） 黄筑平（北京大学）	教育部

注：按照现行国家科学技术奖学科分类代码，101 代表数学学科组、102 代表物理与天文学学科组、103 代表化学学科组、104 代表地球科学学科组、105 代表生物学学科组、106 代表基础医学学科组、107 代表信息科学学科组、108 代表材料科学学科组、109 代表工程技术科学学科组、110 代表力学学科组。

参考文献

[1] 2020 年度国家科学技术奖励大会 . 2020 年度国家最高科学技术奖获奖人 . http：//www. most. gov. cn/ztzl/gjkxjsjldh/jldh2020/[2021-12-21].

[2] 2020 年度国家科学技术奖励大会 . 2020 年度国家自然科学奖获奖项目目录 . http：//www. most. gov. cn/ztzl/gjkxjsjldh/jldh2020/jlgb/202110/t20211029_177639. html[2021-12-21].

Introduction to the 2020 State Preeminent Science and Technology Award and the State Natural Science Award

Ye Jing

The 2020 National Top Science and Technology Award of China was awarded to two distinguished academicians，Pros. Gu Songfen and Prof. Wang Dazhong，for their outstanding achievements in their respective fields. Prof. Gu is one of the famous aircraft design masters and founders of aircraft aerodynamic design. As an internationally renowned nuclear scientist, Prof. Wang is dedicated to the development of advanced nuclear energy systems with inherent safety features. The 2020 National Natural Science Awards of China has been conferred on 46 projects.

5.9　2020 年未来科学大奖获奖者简介

叶　京

（中国科学院科技战略咨询研究院）

2020 年 9 月 6 日，未来科学大奖科学委员会在北京公布 2020 年获奖名单[1]。张亭栋、王振义因其发现三氧化二砷（ATO，俗称砒霜）和全反式维甲酸（all-*trans*-retinoicacid，ATRA）对急性早幼粒细胞白血病的治疗作用，摘得"生命科学奖"；卢柯因其开创性的发现和利用纳米孪晶结构及梯度纳米结构以实现铜金属的高强度、高韧性和高导电性，获得"物质科学奖"；彭实戈因其在倒向随机微分方程理论，非线性 Feynman-Kac 公式和非线性数学期望理论中的开创性贡献，荣膺"数学与计算机科学奖"。

一、未来科学大奖生命科学奖

未来科学大奖生命科学奖获奖人哈尔滨医科大学第一附属医院教授张亭栋、上海交通大学教授王振义（图 1）发现了三氧化二砷和全反式维甲酸对急性早幼粒细胞白血病的治疗作用。

张亭栋　　　　　　　　　　　王振义

图 1　2020 年度未来科学大奖生命科学奖获奖人

癌症仍然是人类健康的一个主要威胁。在人类探索癌症治疗的过程中，张亭栋和王振义对治愈急性早幼粒细胞白血病（acute promyelocytic leukemia，APL）做出了决定性的贡献。APL曾经是最凶险和致命的白血病之一，张亭栋和王振义的工作使APL治愈率达到90%。几千年来，三氧化二砷（ATO，俗称砒霜）曾被试用于多种不同的疾病，但其疗效一直没有得到可靠的、可重复的和公认的结论。20世纪70年代，张亭栋及其同事的研究首次明确ATO可以治疗APL。20世纪80年代，王振义和同事们首次在病人体内证明全反式维甲酸（ATRA）对APL有显著的治疗作用。张亭栋和王振义的工作在国际上得到了验证和推广，使ATO和ATRA成为当今全球治疗APL的标准药物，拯救了众多患者的生命。

二、未来科学大奖物质科学奖

未来科学大奖物质科学奖获奖人中国科学院院士、中国科学院金属研究所研究员、沈阳材料科学国家研究中心主任卢柯（图2）开创性的发现和利用纳米孪晶结构及梯度纳米结构以实现铜金属的高强度、高韧性和高导电性。

卢柯

图2　2020年度未来科学大奖物质科学奖获奖人

卢柯及其研究团队发现了两种新型纳米结构可以提高铜金属材料的强度，而不损失其良好的塑性和导电性，在金属材料强化原理上取得了重大突破。

提高金属材料的强度一直是材料物理领域中最核心的科学问题之一。通常材料的强化均通过引入各种缺陷以阻碍位错运动来实现，但材料强度提高的同时会丧失塑性和导电性，这导致了材料领域著名的长期未能解决的材料强度与塑性（或导电性）的倒置关系。如何克服这个矛盾，成为国际材料领域几十年以来一个重大科学难题。

卢柯团队发现，在金属铜中引入高密度纳米孪晶界面，可使纯铜的强度提高一个数量级，同时保持良好的拉伸塑性和很高的电导率（与高纯无氧铜相当），获得了超高强度高导电性纳米孪晶铜。这个发现突破了强度-导电性倒置关系并开拓了纳米金属材料一个新的研究方向。纳米孪晶强化原理已经在多种金属、合金、化合物、半导体、陶瓷和金刚石中得到验证和应用，成为具有普适性的材料强化原理。

卢柯团队还发现了金属的梯度纳米结构及其独特的强化机制。梯度纳米结构可有效抑制应变集中，实现应变非局域化，其拉伸塑性优于普通粗晶结构。具有梯度纳米结构的纯铜样品其强度较普通粗晶铜高一倍，同时拉伸塑性不变，也突破了传统强化机制的强度-塑性倒置关系，被应用在工业界并取得显著经济效益。

三、未来科学大奖数学与计算机科学奖

未来科学大奖数学与计算机科学奖获奖人中国科学院院士、山东大学教授彭实戈（图3）在倒向随机微分方程理论、非线性 Feynman-Kac 公式和非线性数学期望理论中做出了奠基性和开创性贡献。

彭实戈

图3　2020年度未来科学大奖数学与计算机科学奖获奖人

彭实戈和 Pardoux 合作于1990年发表的文章被认为是倒向随机微分方程理论（BSDE）的奠基性工作。这项工作开创了一个重要的研究领域，其中既有深刻的数学

理论，又有在数学金融中的重要应用。彭实戈在这个领域一直持续工作，做出了一系列重要贡献。

彭实戈于 1992 年创建了非线性 Feynman-Kac 公式，从而对一大类二阶非线性微分方程给出了 BSDE 表示。

彭实戈发展了非线性数学期望的理论，这与传统的线性数学期望有本质上的不同，但相似的数学理论仍能够建立。其对风险的定义和定量有重大应用。

参考文献

[1] 未来科学大奖官网 . 2020 未来科学大奖获奖名单公布：张亭栋、王振义、卢柯、彭实戈获奖 . http：//www. futureprize. org/cn/nav/detail/898. html [2021-12-21].

Introduction to the 2020 Future Science Prize Laureates

Ye Jing

The four winners of 2020 were announced by Future Science Prize Committee in Beijing on September 6，2020. The Prize in life sciences is awarded to Prof. Tingdong Zhang and Prof. Zhenyi Wang for their discovery of therapeutic effects of arsenic trioxide and all-trans retinoic acid on acute promyelocytic leukemia. The Prize in physical sciences is awarded to Prof. Ke Lu for his discovery by using nano-twin structure and gradient nanostructure to achieve high strength，high toughness and high conductivity of copper metal. The Prize in mathematics and computer sciences is awarded to Prof. Shige Peng for his seminal contributions to backward stochastic differential equation theory，nonlinear Feynman-Kac formula and nonlinear mathematical expectation theory.

第六章

中国科学发展概况

A Brief of Science Development in China

6.1　持之以恒加强基础研究夯实科技自立自强根基

李　哲　崔春宇　李春景　李　华

（科技部基础研究司）

　　基础研究是整个科学体系的源头，是所有技术问题的总机关。强大的基础研究是科技自立自强的前提和根基。党中央、国务院高度重视基础研究，习近平总书记在两院院士大会、中国科协十大、科学家座谈会等重要会议上多次强调基础研究的重要性。科技部坚决贯彻落实习近平总书记重要指示批示精神和党中央、国务院重大决策部署，真抓实干、开拓进取，不断推动基础研究发展迈上新台阶。

一、推动基础研究各项重大任务落实落地

1. 加强顶层设计与统筹谋划

　　2018 年，国务院印发《关于全面加强基础科学研究的若干意见》（国发［2018］4号）。科技部会同相关部门制定《加强从 0 到 1 基础研究工作方案》《新形势下加强基础研究若干重点举措》等政策文件，围绕优化总体布局、强化原创导向、激发创新主体活力、营造良好发展环境、完善支持机制等方面，对新时代加强我国基础研究做出全面部署。科技部、教育部、中国科学院、国家自然科学基金委四部门联合印发《加强数学科学研究工作方案》，提出稳定支持基础数学，加强应用数学和数学的应用研究，已布局 13 个国家应用数学中心。新时期加强基础研究的政策体系初步形成。

　　科技部落实中央决策部署，会同相关部门研究制定《基础研究十年规划（2021—2030 年）》（简称《十年规划》），着眼科技自立自强目标，描绘未来 10 年我国基础研究发展蓝图；编制"十四五"国家基础研究专项规划，明确发展思路和目标，凝练未来 5 年基础研究发展重点和主要任务；面向未来 15 年科学前沿发展趋势和国家重大需求，在 2021～2035 年国家中长期科技发展规划中加强基础研究系统布局。中长期科技发展规划、《十年规划》、"十四五"国家基础研究专项规划等紧密衔接，形成远中近相结合的基础研究战略布局。

2. 强化基础研究系统部署

面向世界科学前沿和国家重大战略需求,在关系国计民生和长远发展的领域强化基础研究和应用基础研究系统部署。"十三五"期间,在国家重点研发计划基础前沿领域部署干细胞及转化研究、量子调控与量子信息、纳米科技、蛋白质机器与生命过程调控、全球变化及应对、大科学装置前沿研究、合成生物学、发育编程及其代谢调节、变革性技术关键科学问题、引力波探测等 10 个重点专项,组织实施国家磁约束核聚变能发展研究专项和国家质量基础的共性技术研究与应用重点专项,在全链条一体化重点专项中加强基础研究和应用基础研究任务部署。

面向 2030 年国家长远发展和战略必争领域,率先启动实施量子通信与量子计算机、脑科学与类脑研究等科技创新 2030—重大项目。其中,脑科学围绕脑与认知、脑机智能和脑的健康三个核心问题完成首批项目立项,量子项目主要布局量子通信、量子计算机、量子精密测量三项任务。

3. 推动国家科技创新基地建设发展

推动国家重点实验室体系重组取得重大进展。国家重点实验室是国家战略科技力量的重要组成部分,在新形势下,为提升国家重点实验室的原始创新能力,更好引领支撑经济社会高质量发展,按照党中央、国务院决策部署,科技部制定并印发《重组国家重点实验室体系方案》,开展重组国家重点实验室体系工作。

优化调整国家科技资源共享服务平台。根据科技资源类型,科技部对原有国家科技基础条件平台进行优化调整,形成 20 个国家科学数据中心、30 个国家生物种质与实验材料资源库。同时,为支撑疫情科研攻关,批准建设国家人类疾病动物模型资源库。

推动国家野外科学观测研究站的建设发展,将原有 105 个国家野外科学观测研究站优化调整为 98 个。综合考虑国家战略需求和学科发展需要,聚焦重点方向和空白区域,遴选新建国家野外科学观测研究站 69 个,国家野外科学观测研究站体系布局进一步优化。

4. 促进科技资源开放共享

为推动落实《关于国家重大科研基础设施和大型科研仪器向社会开放的意见》,科技部先后发布《国家重大科研基础设施和大型科研仪器开放共享管理办法》《国家科技资源共享服务平台管理办法》,建设重大科研基础设施与大型科研仪器国家网络管理平台,开展开放共享评价考核和后补助工作,推动科研设施与仪器向社会开放。

为落实《科学数据管理办法》，科技部印发《科技计划项目科学数据汇交工作方案（试行）》，推进科学数据向国家科技资源共享服务平台汇交，向各类创新主体开放。

5. 营造有利于基础研究发展的创新环境

科技部持续深化评价制度改革，推动出台《关于完善科技成果评价机制的指导意见》，坚持"破四唯"和"立新标"并举，建立以质量、绩效和贡献为核心的评价导向，坚持科学分类、多维度评价，尊重科技创新规律。持续推进减轻科研人员负担专项行动，通过减表、解决"报销繁"、检查"瘦身"、精简"牌子"、精简"帽子"、"四唯"清理、信息共享等一系列具体行动，建立服务于人的创造性活动的科研管理机制，取得积极成效。持续推进科研经费管理改革，贯彻落实《关于进一步完善中央财政科研项目资金管理等政策的若干意见》，扩大科研项目经费管理自主权，完善科研项目经费拨付机制，加大科研人员激励制度，减轻科研人员的事务性负担，进一步激发科研人员的创造性和创新活力。

二、新时期基础研究整体水平显著提升

经过长期努力，我国基础研究整体水平显著提高，国际影响力日益提升，支撑引领经济社会发展的作用不断增强，已进入从量的积累向质的飞跃、从点的突破向系统能力提升的关键阶段。

1. 基础研究经费投入持续增加

我国基础研究经费投入从 2015 年的 716.12 亿元增加到 2021 年的 1696 亿元，增加超过一倍。基础研究经费占全社会研发经费比重从 2015 年的 5.05％增长到 2021 年的 6.09％，且 2019～2021 年连续 3 年超过 6％。

2. 学科发展呈现良好态势

数学、物理学、化学、天文学、地球科学、生命科学等基础学科得到更多倾斜支持，农业、能源、资源环境、海洋、信息、制造、材料、工程、医学等应用学科建设稳步推进，量子、认知、纳米、空间等新兴学科得到高度重视并快速发展，多学科交叉融合和跨学科研究日益活跃。基础研究产出水平和质量进一步提升。国际科技论文总量连续 11 年稳居世界第 2 位，总被引次数由 2015 年的第 4 位升至 2017 年的第 2 位并保持至今。高被引论文、国际热点论文、最具影响力期刊论文数量排名均为世界第 2 位。2020 年，我国科学家在《细胞》（*Cell*）、《自然》（*Nature*）、《科学》（*Science*）

三大顶尖学术期刊上发表的论文数量为516篇，连续4年排名世界第4位。2021年，我国材料科学、化学、计算机科学和工程技术4个领域的论文的被引用次数排名世界第1位，农业科学、生物与生物化学、环境与生态学等10个领域的论文的被引用次数排名世界第2位。

3. 基础研究人才队伍不断壮大

我国从事基础研究的全时人员总量从2015年的25.32万人年增加到2020年的42.68万人年。全球高被引科学家数量从2015年的168人次增加到2021年的1057人次，连续3年位居世界第2位。中国科学家获得诺贝尔生理学或医学奖、克利夫兰奖、维加奖等一系列国际重要科技奖项。在纳米限域催化、聚集诱导发光、水稻高产优质性状形成的分子机制等领域涌现出一批由顶尖科学家领衔的优秀创新团队。中青年人才已成为基础研究中坚力量，青年科技工作者开始在基础研究领域挑大梁，成为科技创新队伍中最具活力的生力军。

4. 国家重点实验室建设取得重要进展

改革开放以来，科技部依托高校、科研院所、企业等共建设运行国家重点实验室500多家，覆盖医药、材料、信息、农业、能源、制造、地球科学等15个学科领域，成为推动学科发展、孕育重大原始创新、解决国家经济社会发展重大科技问题的重要支撑。"十三五"期间，国家重点实验室获得了全部国家自然科学奖一等奖、通用项目国家科学技术进步奖特等奖，并获得其他各类国家级奖励500多项；聚集我国48%的中国科学院院士和30%的中国工程院院士，培养大批高水平领军人才；在第四轮全国学科评估中，A+学科国家重点实验室占比近半。经过30多年的发展，国家重点实验室已成为科技界的"金字招牌"，代表了我国基础研究、应用基础研究和前沿技术研究的最高水平。

5. 国家重大科技基础设施建设稳步拓展

国家重大科技基础设施的规模持续增长，覆盖领域不断扩大，布局更加合理，涵盖物理学、地球科学、生物学、材料科学和力学等20多个一级学科，涉及粒子物理与核物理、天文、同步辐射、海洋、能源和国家安全等众多领域。"十三五"期间，500米口径球面射电望远镜、稳态强磁场、散裂中子源、高海拔宇宙线观测站等一批"国之重器"建成并投入运行。目前国家重大科技基础设施规划布局总量达57个，其中31个已建成运行，基本覆盖重点学科领域和事关科技长远发展的关键领域，在支撑高水平科学研究、促进经济社会发展中发挥了引领作用。

6. 科技资源开放共享成效显著

建成科研设施与仪器国家网络管理平台并上线运行，截至 2021 年底，全国 4000 家单位的 10.8 万台（套）大型科研仪器和 91 个重大科研基础设施纳入国家网络管理平台统一对外开放；各部门 300 多个中央单位和 20 多个省（自治区、直辖市）建设了仪器在线服务平台，与国家平台实现互联互通，形成跨部门、跨地区、多层次的网络服务体系；建立科技资源开放共享市场化的网络服务平台，形成互联网＋创新服务＋共享经济的新模式，重大科研基础设施和大型科研仪器开放共享服务体系逐步健全。

7. 基础研究国际影响力日益提升

我国科学家继续深度参与国际热核聚变实验堆（International Thermonuclear Experimental Reactor，ITER）计划、平方公里射电阵（Square Kilometre Array，SKA）、地球观测组织（Group on Earth Observations，GEO）等国际大科学计划和大科学工程，支持科学界发起全脑介观神经联接图谱、深时数字地球（Deep-time Digital Earth，DDE）等国际大科学计划。国家科技计划对外开放的力度加大，更多外籍科研人员承担国家重大科研任务，实现中央财政科研经费过境香港、澳门使用。国际联合研究广泛深入开展。在 2020 年《科学引文索引》（Science Citation Index，SCI）收录的中国科学家发表的论文中，国际合著论文数量为 14.45 万篇，较 2019 年增长 11.1％，占中国科学家发表论文总数的 26.2％。一批中国科学家在国际学术组织和学术期刊担任重要职务。

8. 重大创新成果竞相涌现

中国科学家在若干重点基础前沿方向取得一批具有国际影响力的原创成果。首次观测到三维量子霍尔效应、非常规新型手性费米子；发射国际上首颗量子科学实验卫星"墨子号"并率先实现星地间千公里级量子纠缠和密钥分发及隐形传态，76 个光子的量子计算原型机"九章"和 62 比特、66 比特可编程超导量子计算原型机"祖冲之号""祖冲之二号"成功问世；首次实现原子级石墨烯可控折叠，提出纳米限域催化新概念；研制世界首款异构融合类脑计算芯片"天机芯"，首次实现人工合成淀粉，灵长类动物早期胚胎发育机制取得新突破等。

对经济社会发展的支撑引领作用不断增强。在信息、材料、能源、制造等领域取得高性能碳基互补金属氧化物半导体器件（complementary metal oxide semiconductor，CMOS）集成电路、共格纳米析出强化的超高强钢、先进重型燃气轮机制造等重要成果，为传统产业转型升级和战略性新兴产业培育提供科学支撑。围绕农业、健康、环境等领域，在水稻功能基因组筛查、埃博拉病毒及新型冠状病毒等病毒致病分子机制和传播机制、大气 $PM_{2.5}$ 污染特征和来源成因等方面取得重大突破，为国家可持续发展和民生改善做出重要贡献。

三、全面加强基础研究的几点考虑

当前，新一轮科技革命和产业变革蓬勃兴起，科研范式发生深刻变革，学科交叉融合日益深入，基础前沿领域孕育重大突破，科学成果加速转化应用，持续催生颠覆性创新，这为我国实现跨越赶超提供了重要机遇。进入新发展阶段，我国经济转型升级步伐加快，国际环境发生深刻复杂变化，我国关键核心技术受制于人的问题更加凸显，国家发展和安全各领域都迫切需要基础研究这个创新源头提供有力支撑。

新形势下，科技部将以贯彻落实《基础研究十年规划（2021—2030 年）》为重点，抓好基础研究工作，基本思路是：以习近平新时代中国特色社会主义思想为指导，全面贯彻党的十九大和十九届历次全会精神，完整、准确、全面贯彻新发展理念，着眼构建新发展格局，坚持"四个面向"，以服务国家战略和引领高质量发展为主线，以重大原始创新和关键核心技术突破为主攻方向，以深化体制机制改革为根本动力，持之以恒加强基础研究，提高原始创新能力，为加快实现高水平科技自立自强、建设科技强国奠定坚实基础。

1. 突出需求导向，加强基础研究前瞻部署

面向"十四五"，强化国家战略需求导向，聚焦基础前沿领域，围绕世界科学前沿重点方向，启动实施物态调控、合成生物学、纳米前沿、催化科学、大科学装置前沿研究等一批重点专项；在全链条一体化设计重点专项中，面向国家重大需求和国民经济主战场，在农业、健康、制造、材料、信息、资源环境、能源环保、社会治理与公共安全等领域加强应用基础研究任务部署，解决制约国家发展和安全的关键难题，为创新驱动发展提供源头供给。

2. 稳步推进国家重点实验室重组

落实《重组国家重点实验室体系方案》，突出国家战略导向和重大需求，按照试点先行、分批推进的原则，通过充实、调整、整合、撤销、新建等方式对现有国家重点实验室进行优化整合，创新管理体制机制，强化实体化建设，完善考核评价和优胜劣汰机制，建立国家重点实验室承担国家重大科技任务机制。通过重组，推动形成基础研究、应用基础研究和前沿技术融通发展的实验室新体系。坚持系统观念，做好与

国家实验室的衔接，打造以国家实验室为核心、以国家重点实验室为支撑的中国特色国家实验室体系，打造国家使命驱动的战略科技力量。重组后，国家重点实验室统一更名为"全国重点实验室"。

3. 持续加强科研基础条件平台建设

建设一批国家科学数据中心，完善科学数据管理和开放共享机制。优化国家生物种质与实验材料资源库（馆）建设布局，持续开展收集保藏、整理加工和共享服务。开展科技基础资源调查，持续获取和整编自然本底数据、种质、标本等重要科技基础资源。优化国家野外科学观测研究站布局，制定完善科学观测规范与标准，持续提升野外科学观测研究站观测、研究和示范能力。开展基础科研条件与重大科学仪器设备研发攻关，以关键核心部件国产化为突破口，支持高端科学仪器工程化研制与应用开发；加强高端科研试剂、实验动物、科学数据软件工具等条件手段自主研发。构建国家科研论文和科技信息高端交流平台，服务科技创新。

4. 完善支持基础研究发展的体制机制

（1）完善多元化投入机制。中央财政持续加大基础研究投入，建立健全竞争和稳定相协调的支持机制。建立中央和地方政府共同出资、共同组织基础研究项目的协同机制，带动地方政府支持基础研究。对企业投入基础研究实行税收优惠，鼓励社会力量以捐赠和建立基金等方式多渠道投入，扩大基础研究资金来源。

（2）完善项目形成和组织管理机制。充分发挥政府作为重大创新活动组织者作用，组织科技界、产业界、战略研究机构等共同研究提出需求，构建从国家安全和经济社会发展实践中凝练科学问题的机制。对于自由探索类项目，尊重科学家的学术灵感，鼓励科学家将国家需求和个人兴趣相结合。对于原创性、非共识项目，采取特殊方式开辟快速立项通道。创新组织实施机制，实行"揭榜挂帅"、"赛马"制、首席科学家负责制等新型管理方式。在国家重点研发计划中全面推行青年科学家项目。

（3）深化评价制度改革。落实《关于完善科技成果评价机制的指导意见》要求，开展基础研究评价改革试点，强化质量、绩效、贡献为核心的评价导向，坚持"破四唯"和"立新标"并举，推行代表作评价制度，开展长周期评价，探索实行国际同行评价。完善自由探索型和任务导向型科技项目分类评价制度，建立非共识科技项目评价机制。探索构建符合基础研究规律和人才成长规律的评价体系。

5. 推动企业加强基础研究

强化企业创新主体地位，提升企业开展基础研究的动力和能力。支持科技领军企业面向长远发展前瞻布局基础研究，深度参与国家科技计划论证实施，牵头承担国家基础研究任务。支持企业牵头整合集聚创新资源，形成跨领域、大协作、高强度创新基地，促进产学研深度融合。

6. 加强基础研究国际合作

实施开放包容、互惠共享的国际合作战略，积极融入全球创新网络。深化政府间的基础研究合作，积极构建双多边合作机制。加大国家科技计划和国家重点实验室等创新基地对外开放力度，支持高水平外籍科学家牵头或参与实施国家科技计划项目。结合国家自然科学基金改革，设立面向全球的科学研究基金。积极牵头组织实施国际大科学计划和大科学工程。深化科研人员国际交流，支持和推荐我国科学家到国际学术组织交流和任职，适时发起成立国际科技组织，提升中国科技的国际话语权和影响力。

Persevering in Strengthening Basic Research, Consolidating the Foundation of Sci-tech Self-reliance and Self-strengthening

Li Zhe, Cui Chunyu, Li Chunjing, Li Hua

The Ministry of Science and Technology（MOST）firmly implemented the spirit of the important address and instructions given by President Xi Jinping and important decisions and deployments of the Chinese Party Central Committee and the State Council. Efforts have been made to improve top-down design and overall planning and strengthen basic research and systematic deployment, constantly promoting the construction and development of the country's sci-tech innovation bases, advocating the openness and sharing of sci-tech resources, fostering a favorable innovation environment of basic research and persisting in pushing China's development of basic research to a new stage. Through our long-term efforts, the overall level of China's basic research has obviously enhanced with its international influence increasingly growing. It plays an ever-growing role in supporting and leading the development of China's economic society, and it is entering the key stage from the "quantitative accumulation" to the "qualitative leap" and from point breakthrough to systematic capability improvement.

6.2　2020 年度国家自然科学基金项目申请与资助情况综述

赵英弘　郑知敏　郝红全　高阵雨　张韶阳

李志兰　车成卫　王　岩　王长锐

（国家自然科学基金委员会计划局）

2020 年，国家自然科学基金委员会（以下简称自然科学基金委）以习近平新时代中国特色社会主义思想为指导，全面贯彻中共十九大和十九届二中、三中、四中、五中全会精神和习近平总书记关于科技创新和基础研究的重要论述，按照《国务院关于全面加强基础科学研究的若干意见》《关于深化项目评审、人才评价、机构评估改革的意见》《国务院关于优化科研管理提升科研绩效若干措施的通知》《关于进一步弘扬科学家精神加强作风和学风建设的意见》等文件的要求和部署，以 "构建理念先进、制度规范、公正高效的新时代科学基金治理体系" 为改革目标[1]，聚焦明确资助导向、完善评审机制、优化学科布局三大核心改革任务，扎实推进国家自然科学基金深化改革。同时，深入贯彻习近平总书记关于新冠肺炎疫情防控有关重要讲话和指示批示精神，全面落实党中央国务院关于打赢疫情防控阻击战工作部署要求，认真研究应对措施，妥善部署相关工作，按计划完成了全年各类项目的申请、受理、评审和资助工作。

一、项目申请与受理情况

2020 年，国家自然科学基金项目的申请量继续大幅增加，全年共接收各类项目申请 281 170 项，较 2019 年增加 12.19 %[2]。其中，在项目申请集中接收期间共接收 2 369 个依托单位提交的 15 类项目申请 269 671 项，同比增加 28 960 项[3]，增幅为 12.03 %。经初步审查，全年共受理项目申请 278 659 项；不予受理 2511 项，占接收项目申请总数的 0.89 %，较 2019 年（1.72 %）[2]有大幅降低。在不予受理的项目申请中，"未按要求提供证明材料、推荐信、导师同意函、知情同意函、伦理委员会证明等""不属于项目指南资助范畴""申请人不具备该类项目的申请资格"是主要的不予受理原因。根据《国家自然科学基金条例》（以下简称《条例》）要求，共受理复审申请 218 项。经审查，维持原不予受理决定的有 211 项，占全部不予受理项目的 8.40%；原不予受理决定有误、重新送审的有 7 项，占全部不予受理项目的 0.28%。

2020 年，自然科学基金委积极采取措施应对新冠肺炎疫情影响，解除科研人员后顾之忧，为打赢疫情防控的人民战争、总体战、阻击战做出贡献[3]。具体为：①配合疫情防控工作需要，将项目申请集中接收、结题材料接收等关键管理环节整体推迟一个月以上。②在项目申请集中接收阶段全面实行无纸化申请。2020 年，受新冠肺炎疫情影响，集中接收的 269 671 份项目申请全部纳入无纸化申请试点范围。这样一方面减少了申请人一方大量纸质资源的消耗；另一方面也降低了依托单位一方盖章、邮寄和审核等的管理成本。③面向疫情防控一线科研人员定向开放项目申请。对于因抗击疫情延误申请的科研人员，将申请截止日推迟至 2020 年 5 月 20 日。④充分发挥信息系统辅助功能，简化填写信息与申请材料要求。2020 年，通过信息系统对研究起止日期、合作研究单位信息、特定申请人群所需附件材料等进行提示和前端控制；将同行专家推荐信、导师同意函、境外人员知情同意书等附件材料，以标准模板形式嵌入信息系统，供申请人下载填写。上述举措切实减轻了申请人和依托单位管理人员的负担，为其提供了更好的服务。⑤简化初审管理。通过进一步挖掘和提升信息系统辅助功能，一方面对申请人填写信息实现了一些前端控制；另一方面，对于申请书初步审查（即形式审查）时需要人工审查的事项，自然科学基金委按照能简则简的原则，尽量以判断"有"或"无"作为标准。2020 年项目申请的不予受理率仅为 0.89%，较 2019 年的 1.72% 有大幅降低。

2020 年，自然科学基金委按照中共中央办公厅、国务院办公厅《关于进一步弘扬科学家精神加强作风和学风建设的意见》指示，贯彻落实"科研人员同期主持和主要参与的国家科技计划（专项、基金等）项目（课题）数原则上不得超过 2 项"的要求，对科学基金项目的限项申请规定进行了相应调整，将高级职称人员同时负责和参与的项目总数由 3 项调整为 2 项。此外，以优化申请代码设置为切入点，启动学科布局调整工作，选择工程与材料科学部、信息科学部为试点部门，重新梳理一级和二级申请代码，不再设置三级申请代码。

二、项目评审情况

2020 年，科学基金项目评审工作总体呈现以下特点。

1. 扩大分类评审试点范围

2020 年，全部面上项目和重点项目试点开展基于四类科学问题属性的分类评审工作，实施分类评审项目的数量占接收项目总数的 41.53 %。为使申请人、评审专家更准确地理解新时期科学基金的资助导向，自然科学基金委加强宣传引导，发布了四类科学问题属性典型案例库，更新了项目分类申请与评审宣讲视频；在面上项目和重点项目评审会议上，以不同形式向评审专家宣讲了四类科学问题属性的具体内涵和评审要点。经过多种渠道宣传，申请人对四类科学问题属性的理解得到逐步修正，选择科学问题属性的比例趋向理性，在撰写申请书时注意基于所选科学问题属性思考和整合自身的研究内容；评审专家更加关心、重视按照四类科学问题属性进行评审，更有针对性地遴选创新性项目；新时期科学基金资助导向逐步深入人心。

2. 稳步推进"负责任、讲信誉、计贡献"评审机制试点工作

2020 年，自然科学基金委从 8 个科学部中选择 10 个学科（每个试点学科至少选择一个项目类型），在通讯评审环节开展"负责任、讲信誉、计贡献"（Responsibility，Credibility，Contribution，RCC）评审机制试点工作。为帮助科学界更好地理解 RCC 评审机制，自然科学基金委制定了通讯评审专家行为规范；从全自然科学基金委 RCC 评审机制通用试点方案中提炼要点，形成面向评审专家的多种形式的宣传培训材料。RCC 评审机制试点工作得到评审专家的大力支持。总体上，通讯评审专家的反馈意见快于以往，评审意见过于简单的现象得以减少，申请人对评审专家意见的评价总体上较满意。

3. 实施原创探索计划，引导和激励原始创新

2020 年，自然科学基金委启动实施原创探索计划，设置专门资助渠道，探索支持具有颠覆性、非共识、高风险等特征的原创思想的新机制。自然科学基金委将"自下而上"的自由申请与"自上而下"的顶层设计相结合，引入专家推荐、预申请、双盲评审、评审意见反馈和答复等创新的申请与评审方式，项目评审以研究思想的原创性和预期成果的引领性作为评价重点，并力求尊重科学规律、推动学术民主、宽容失败，鼓励科研人员挑战高风险性研究，为项目的甄别和遴选提供自由、宽松的学术氛围，探索有别于现行模式的项目遴选方式和实施机制。

4. 进一步完善评价机制

自然科学基金委注重弘扬科学精神，营造优良创新环境，把避免"唯论文、唯职称、唯学历、唯奖项"（以下简称"四唯"）倾向的要求落到实处。在通讯评审和会议评审环节，自然科学基金委向评审专家强调注意克服"四唯"倾向，以创新能力、质量、贡献、绩效为项目评价导向。进一步落实代表作评价制度，在个人简历中强调其

他代表性研究成果中不得再罗列论文和专著,特别是取消了国家杰出青年科学基金项目、优秀青年科学基金项目和创新研究群体项目申请书中的"论文收录与被引用情况统计表",引导科技界更加关注标志性成果本身的质量、贡献和影响。

5. 加强评审工作规范化管理

2020年,自然科学基金委进一步明确严格选择评审专家、严格执行回避和保密要求、强化实施维护公正性承诺制度、重视评审专家培训、严明会议答辩评审现场纪律等若干要求,并发布《关于加强作风和学风建设营造风清气正评审环境的公开信》,着力营造公平公正的评审环境。

6. 针对疫情,推行网络会议评审方式

针对因疫情防控要求而无法正常组织现场会议评审的情况,自然科学基金委及时调整会议评审方式。提出"线下、线上相结合""网络会议平台+信息系统支持"的替代方式,确保会议评审工作安全、规范、有序进行。评审专家普遍反映采用这种线上答辩的方式能够有效避免干扰评审工作行为,进一步提升了评审公正性。

三、项目资助情况

经过评审和审批程序,2020年共批准资助项目45 656项,直接费用2 830 251.27万元。

1. 稳定"自下而上"自由探索项目经费占比,保证源头创新活力

2020年,自然科学基金委继续稳定对面上项目、青年科学基金项目和地区科学基金项目的资助力度,保持自由探索类项目经费占比,支持科研人员在科学基金资助范围内进行自由探索。促进各学科均衡、协调和可持续发展,资助面上项目19 357项,直接费用1 112 994万元。培养青年科学技术人员独立主持科研项目、进行创新研究的能力,资助青年科学基金项目18 276项,直接费用435 608万元。培养和扶植特定地区的科学技术人员,为区域创新体系建设与经济、社会发展服务,资助地区科学基金项目3 177项,直接费用110 738万元。上述3类项目直接费用合计1 659 340万元,占2020年批准资助项目总直接费用的58.63%,与2019年基本持平。

2. 瞄准世界科学前沿和国家重大需求,推动重点领域取得突破

基础研究是提出和解决科学问题的研究活动。自然科学基金委按照"四个面向"

的要求，着力引导科研人员面向世界科学前沿和国家重大需求凝练核心科学问题，开展创新性研究。2020 年，自然科学基金委资助重点项目 737 项，直接费用 216 527 万元，平均资助强度为 293.80 万元/项；资助重大项目 45 项，直接费用 79 139.38 万元，平均资助强度为 1 758.65 万元/项；新启动 5 个重大研究计划，分别为"第二代量子体系的构筑和操控""极端条件电磁能装备科学基础""未来工业互联网基础理论与关键技术""组织器官再生修复的信息解码及有序调控"和"冠状病毒-宿主免疫互作的全景动态机制与干预策略"，正在实施的 33 个重大研究计划共资助项目 460 项，直接费用 87 399.96 万元；加强对仪器研制相关基础研究的支持力度，打通自由申请和部门推荐两类仪器项目的计划额度，资助国家重大科研仪器研制项目 88 项，直接费用 94 494.78 万元，较 2019 年增加 16 154.02 万元，增幅为 20.62%。上述 4 类项目直接费用合计 477 561.12 万元，占 2020 年批准资助项目总直接费用的 16.87%。

3. 快速启动专项资助，积极应对疫情影响

2020 年，为有效应对新冠肺炎疫情，自然科学基金委深入贯彻习近平总书记关于新冠肺炎疫情防控的重要指示精神，及时响应，统筹部署，采取了一系列资助措施。一是根据"立足解决科学问题，兼顾应急与长远，避免重复资助"的原则，先后以专项项目、重大项目、重大研究计划等形式，安排经费 1.15 亿元，组织科研人员开展基础研究领域攻关，为疫情防控提供科技支撑。二是组织开展公共卫生应急管理政策研究，并启动"新冠肺炎疫情防控等公共卫生事件的应对、治理及影响"专项项目。三是加强交叉与协同，针对防疫物资重复利用、环境污染治理、气溶胶传播病毒等疫情防控需求中的科学问题及时组织开展研究。四是推动国际合作共同应对挑战，推动国际科学理事会发声支持我国的抗疫工作，与德国研究联合会、韩国国家研究基金会、瑞典研究理事会、土耳其科技研究理事会、东欧三国（捷克、奥地利、匈牙利）、金砖五国、比尔及梅琳达·盖茨基金会等联合资助新型冠状病毒相关研究。另外，自然科学基金委设置专门渠道，面向全社会征集新冠病毒相关研究创新思路建议 1691 条，为下一步部署相关研究的资助工作奠定了重要基础，也为完善重大类型项目立项机制进行先期探索。

4. 推进人才资助体系升级，助力优秀人才成长

2020 年，自然科学基金委推进人才资助体系升级，优化资助政策，打造科学有效、功能完整的基础研究人才资助体系，为建设世界科技强国提供有效支撑。

持续优化资金管理，进一步提升资助效能。在 2019 年 60 家依托单位试点提高智力密集型和纯理论基础研究项目间接费用比例的基础上，自然科学基金委于 2020 年

进一步调整了人才类项目的经费资助结构，所有依托单位获批的青年科学基金项目、优秀青年科学基金项目和创新研究群体项目均采用新的经费资助结构，在总经费固定的前提下，进一步提高间接费用比例。2020 年，自然科学基金委在《2020 年度国家自然科学基金项目指南》中明确，国家杰出青年科学基金项目的经费不再分为直接费用和间接费用，无需编制项目预算。实行承诺制，项目负责人在规定范围内自主使用经费，承诺项目经费全部用于与本项目研究工作相关的支出，不截留、挪用、侵占，不用于与科学研究无关的支出等。

落实加强人才计划统筹衔接的工作要求，自然科学基金委参与制定并积极落实加强国家杰出青年科学基金项目和优秀青年科学基金项目与其他科技人才计划的统筹衔接要求，优化资源配置，避免重复支持。

自然科学基金委于 2020 年修订印发优秀青年科学基金项目和国家杰出青年科学基金项目管理办法，取消非华裔外籍的申请限制，扩大对外开放的程度，吸引更多外籍优秀学者来华开展科学研究工作，促进我国科技水平的提高。

2020 年，自然科学基金委资助优秀青年科学基金项目 600 项，直接费用 72 000 万元；资助优秀青年科学基金项目（港澳地区）25 项，直接费用 3000 万元；资助国家杰出青年科学基金项目 298 项，资助经费 116 920 万元；资助创新研究群体项目 37 项，直接费用 36 010 万元；资助基础科学中心项目 13 项，直接费用 77 000 万元。

5. 完善多元投入机制，激发协同创新活力

2020 年，自然科学基金委深入实施联合基金资助模式，引导与整合政府、行业、企业及个人等社会资源投入基础研究，吸引和集聚全国优势科研力量，关注国家战略需求及区域与产业发展需求，解决地方、行业和企业发展中的关键和共性问题中的重大科学问题，提升地区行业企业自主创新能力、促进区域创新体系建设。

稳步拓展联合基金合作范围，持续深入推进联合基金实施机制改革。2020 年，福建、山西、河南、甘肃等省份加入区域创新发展联合基金；中国广核集团有限公司加入企业创新发展联合基金；水利部及中国气象局等行业部门与自然科学基金委设立联合基金。截至 2020 年 12 月，20 个省（自治区、直辖市）加入区域创新发展联合基金，协议期内联合资助方将投入 62.90 亿元；5 家企业加入企业创新发展联合基金，协议期内联合资助方将投入 10.94 亿元；与 6 个行业部门设立联合基金，协议期内联合资助方将投入 15.90 亿元。新时期联合基金模式下共吸引委外资金 89.74 亿元，多元投入机制初步建成。

包含正在实施的其他联合基金在内，2020 年 20 个联合基金共接收项目申请 5806 项，资助联合基金项目 1084 项，较 2019 年增加 159 项，增幅为 17.19%，直接费用

238 750.40 万元，较 2019 年增加 53 660.40 万元，增幅为 28.99%。

6. 实施原创探索计划，提升原始创新能力

2020 年，自然科学基金委先后发布了《2020 年度国家自然科学基金原创探索计划项目申请指南》及"肿瘤研究新范式探索""面向复杂对象的人工智能理论基础研究""深时地球科学知识图谱与知识演化研究""管理与经济科学新理论方法和新范式"等 4 个指南引导类原创探索计划项目指南。2020 年，自然科学基金委收到预申请 1766 项，其中指南引导类 1000 项，专家推荐类 766 项。61 项指南引导类及 142 项专家推荐类原创探索计划项目通过预申请审查。经过评审和审批程序，批准资助原创探索计划项目 53 项，其中指南引导类 18 项，专家推荐类 35 项，涵盖数学物理科学部、化学科学部、生命科学部、地球科学部、工程与材料科学部、信息科学部、医学科学部 7 个科学部，直接费用 11 234.18 万元。

四、2021 年工作展望

2021 年是我国"十四五"规划开局之年，并迎来中国共产党成立一百周年。自然科学基金委全面贯彻党中央、国务院关于科技创新和基础研究的最新决策部署，深入落实科技创新要"面向世界科技前沿、面向经济主战场、面向国家重大需求、面向人民生命健康"的重要指示，继续深化科学基金改革，推动基础研究高质量发展，不断提升我国源头创新能力。

1. 加强顶层设计，优化资助布局

推动构建"基础科学、技术科学、生命和医学、交叉融合"4 个板块，重塑资助布局，强化治理能力；全面实施新的申请代码，统筹部署交叉领域研究，优化科学基金学科布局，积极应对科研范式变革。

2. 激励原始创新，培育原创成果

深入实施原创探索计划，鼓励支持高风险、颠覆性、非共识的原创思想，培育从无到有的引领性原创成果，培育未来重大科学突破。

3. 深入推进人才资助体系升级

实行更加积极开放的人才资助政策，加强对创新人才和团队的支持力度，优化人才资助体系，夯实人才队伍基础。

4. 继续优化管理，提高治理能力和管理水平

进一步扩大基于四类科学问题属性的分类评审范围，引导科研人员准确凝练科学问题，提升项目申请和评审质量；全面实施无纸化申请，简化申请管理要求；扩大"包干制"试点范围，推进经费管理改革；扩大联合基金规模，探索社会个人捐赠基础研究机制，强化多元投入、促进协同创新；完善依托单位准入和退出机制，构建信誉评价体系，提升服务效能。

5. 扎实营造良好学术生态

持续开展"负责任、讲信誉、计贡献"评审机制试点工作，提升项目评审质量、构建良好评审环境；深入实施科学基金学风建设行动计划，引导四方主体开展负责任的科研、评审和管理活动，提倡良好科学文化、营造有利于创新的科研环境。

参考文献

[1] 李静海. 深化科学基金改革推动基础研究高质量发展. 中国科学基金,2020,34(5):529-532.

[2] 郝红全,郑知敏,李志兰,等. 2019年度国家自然科学基金项目申请、评审与资助工作综述. 中国科学基金,2020,34(1):46-49.

[3] 郝红全,郑知敏,严博,等. 2020年度国家自然科学基金项目申请集中接收与受理情况. 中国科学基金,2020,34(5):615-620.

Proposal Application and Funding of NSFC in 2020：an Overview

Zhao Yinghong，*Zheng Zhimin*，*Hao Hongquan*，
Gao Zhenyu，*Zhang Shaoyang*，
Li Zhilan，*Che Chengwei*，*Wang Yan*，*Wang Changrui*

This article gives a summary of proposal applications，peer review and funding of National Natural Science Fund in 2020. In 2020，the total amount of direct cost is about 28.30 billion Yuan，and funding statistics for various kinds of projects are listed.

6.3　中国正深刻改变世界科学版图
——基于科学结构图谱的分析

王小梅　李国鹏　陈　挺

（中国科学院科技战略咨询研究院）

　　为分析中国于"十三五"时期在基础科学领域取得的成就，本文基于 2010～2015 年和 2014～2019 年两个时段（简称两个时段）的科学结构图谱[1]，对比分析了"十二五"时期和"十三五"时期中国基础科学在科技界普遍关注的热点或潜在热点前沿领域中的发展态势。分析发现，在"十三五"时期，中国的基础科学领域发展迅速，已经深刻改变了世界的科学版图。同时，通过对比分析中国与美国、德国、英国、日本、法国等科技发达国家在全球科学热点前沿领域中的产出规模、学科布局、研究合作及优势领域等，找到我国与科技发达国家的差距，以期为我国科技优先领域确定、战略重点选择和科技政策导向提供参考。

一、科学结构图谱的研究方法与数据

　　科学结构图谱是通过可视化技术展现高度抽象的科学，特别是自然科学基础研究的宏观结构的工具，可以揭示科学热点前沿领域间的关联关系与发展进程。中国科学院科技战略咨询研究院科学结构研究组（简称科学结构研究组）自 2007 年开始开展相关研究，每两年绘制一期科学结构图谱[1-5]，周期性监测科学研究结构及其演变规律。科学结构图谱的原理是通过科睿唯安（Clarivate Analytics）的基本科学指标库（Essential Science Indicators，ESI）中高被引论文的同被引聚类形成研究领域（research area），揭示了在科学研究实践中自然、客观和动态形成（尤其是通过交叉融汇形成）的世界普遍关注的热点研究前沿领域。这些领域超越了传统的学科分类，客观地反映了科学家相互引证所表征的科研的某种共性。

　　为支持更大量的数据分析，使聚类和可视化算法更加准确、细致，从 2018 年开始，科学结构研究组全面升级了算法，具体研究方法参见《科学结构图谱 2021》[1]。图 1 为两个时段的科学结构图，两个时段的科学结构布局总体保持稳定，直观地反映了当前的科学结构及科学研究活动情况。

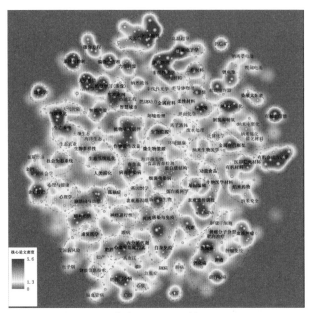

（a）2010～2015年

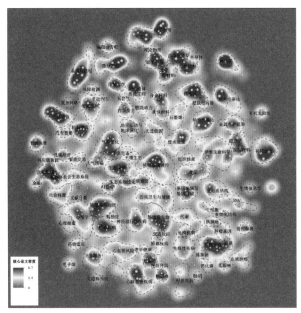

（b）2014～2019年

图1 两个时段的科学结构图

论文数量越多，密度越大，颜色越偏暖（红）；反之，论文数量越少，密度越小，颜色越偏冷（蓝）。图中圈出了一些研究领域的群组区域，标识主体研究内容是为了掌握科学结构的研究内容科学结构图谱分析中使用的高被引论文被称为"核心论文",在后续研究中引用这些核心论文的论文被称为"施引论文"。2014～2019 年的核心论文被聚类成 1333 个研究领域。在此基础上,我们按照研究领域之间的相似性或其共性概念,从共同研究解决的科学问题的角度将研究领域归类为若干个研究领域群。位于图 1(b)顶部的是天文学与粒子物理学,其右下方是凝聚态物理学与光学,主体包括"量子物理"、"自旋电子学"、"非线性光学"(图中简称光学)、"半导体物理"(图中简称半导体)、"超材料"、"二维材料"等研究领域群,其中的"二维材料"也属于纳米科技;物理学的右下方以化学与材料科学为主,且大多属于纳米科技,包括"锂电池""纳米电催化""纳米光催化""纳米生命科学"等研究领域群,化学中还有"有机合成方法学"(图中简称有机合成)、"超分子自组装材料"(图中简称超分子)、"医用发光材料"(图中简称医用材料)、"金属有机框架"等研究领域群;数学、计算机科学与工程学位于图的左上方,包括"系统与控制""无线通信""机器学习""智能电网"等研究领域群,工程科学的覆盖范围较广,与地球科学、化学等有交叉;图的中心偏左部分是地球科学与生态环境,包括"地壳运动与地球演化""气候变化""生态保护""废水处理"等研究领域群;植物学与动物学、生物学位于图的中心位置,主体包括"植物基因调控""基因编辑与基因合成""蛋白质结构""干细胞""RNA""公共卫生与健康""食品营养与健康"(图中简称食品)等研究领域群,医学位于生物学的下方,也是整个图的下方,包括"肿瘤免疫"、"心脑血管疾病"、"神经与精神疾病"(图中简称精神疾病)、"神经退行性疾病"、"糖尿病"、"免疫性疾病"、"肠道微生物"等研究领域群;左偏下为社会科学等。

科学结构图谱的高被引论文和研究领域取自科睿唯安的 ESI 数据库,时间跨度是 6 年。表 1 显示了两个时段的科学结构图谱的数据情况。

表 1　两个时段的科学结构图谱的数据情况

项目	数据情况	
时间范围	2010～2015 年	2014～2019 年
检索时间	2016 年 3 月	2020 年 3 月
研究领域数量/个	969	1 333
核心论文数量/篇	44 495	50 767
施引论文数量/篇	1 801 996	2 504 043

二、中国的总体科研态势

中国在热点前沿领域中快速崛起,与科技发达国家的差距快速缩小。对比分析两个时段的科学结构图谱后发现，中国核心论文的总量快速增加，核心论文在所有研究领域的覆盖面及覆盖强度呈明显增大和增强的趋势。这表明，中国的科研结构正在逐步完善，虽然各学科间仍然存在不均衡现象，但是其均衡性正在逐步改善。

1. 中国的科研活跃度迅猛提升

中国在世界热点前沿领域中的科研发展迅猛，从两个时段的科学结构图谱可以发现，中国的核心论文数量世界份额稳居世界第二位，从 11.5％ 上升到 21.0％，增幅达 82.6％，且自 2018 年起，中国的核心论文数量世界份额已经超过美国的核心论文数量世界份额（表 2、图 2）。在 2014～2019 年这个时段的科学结构图谱中，中国的核心论文数量世界份额已大幅超越德国、英国、日本、法国 4 国的核心论文数量世界份额之和，而同期科技发达国家整体的核心论文数量世界份额普遍呈下降的趋势，美国、德国、日本、法国的降幅接近或超过 20％。

表 2   中国及科技发达国家在两个时段的核心论文世界份额及世界排名

时段	项目	国家					
		美国	中国	德国	英国	日本	法国
2010～2015 年	核心论文数量世界份额/％	37.1	11.5	6.3	7.4	2.6	3.4
	核心论文数量世界份额排名/位	1	2	4	3	9	5
2014～2019 年	核心论文数量世界份额/％	29.7	21.0	4.9	6.9	2.0	2.6
	核心论文数量世界份额排名/位	1	2	4	3	10	8
核心论文数量世界份额增长率/％		−19.9	82.6	−22.2	−6.8	−23.1	−23.5

注：本文采用分数计数法（按每篇论文中每个国家或机构的作者数量占全部作者数量的比例计数，一篇论文的计数为 1）计算国家的核心论文数量。

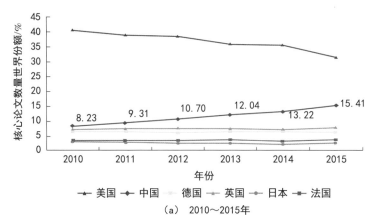

（a）2010～2015年

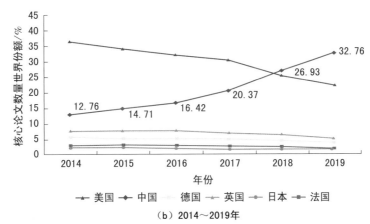

（b）2014～2019年

图2 中国及科技发达国家在两个时段的核心论文数量世界份额的年度变化趋势

施引论文反映了在这些热点前沿领域的持续研究成果。中国的施引论文数量世界份额也排名世界第二位，并超过核心论文数量世界份额，且有较强的上升后劲（表3）。美国、德国、英国、日本、法国5国整体表现为下降趋势。

表3 中国及科技发达国家在两个时段的施引论文数量世界份额及世界排名

时段	项目	国家					
		美国	中国	德国	英国	日本	法国
2010～2015年	施引论文数量世界份额/%	27.1	14.8	5.8	5.6	4.0	3.6
	施引论文数量世界排名/位	1	2	3	4	5	6
2014～2019年	施引论文数量世界份额/%	22.7	22.2	4.9	5.1	3.3	2.9
	施引论文数量世界排名/位	1	2	4	3	5	8
施引论文数量世界份额增长率/%		−16.2	50.0	−15.5	−8.9	−17.5	−19.4

2. 中国的科研结构正在逐步完善，各学科的研究水平仍然严重不均衡

从中国在所有研究领域的覆盖率、覆盖强度看，中国的科研结构布局正在逐步完善，学科结构布局逐步稳定，但在各学科的研究水平均衡性方面还有提升空间。

（1）中国的研究领域覆盖率显著增加。中国的核心论文在所有研究领域的覆盖面及覆盖强度呈明显增大和增强趋势，研究领域的覆盖率从世界排名第五位升到第三

位，覆盖率从 66.7% 升到 78.6%，与英国（80.3%）接近，较美国（94.8%）还有一定的差距。新增研究领域的覆盖率也明显增加，从 48.3% 升到 66.7%。同时，科技发达国家的研究领域覆盖率均出现小幅下降，其中法国、德国的降幅超过 5%（表4）。

表4　中国及科技发达国家科研覆盖率情况

时段	项目	国家					
		美国	中国	德国	英国	日本	法国
2010~2015 年	发文研究领域数（共 1084 个）	1051	723	857	901	562	746
	全研究领域数占比/%	97.0	66.7	79.1	83.1	51.9	68.8
	发文新增研究领域（共 87 个）	71	42	40	47	18	27
	新增研究领域数占比/%	81.6	48.3	46.0	54.0	20.7	31.0
2014~2019 年	发文研究领域数（共 1333 个）	1263	1048	955	1070	639	826
	全研究领域数占比/%	94.8	78.6	71.6	80.3	47.9	62.0
	发文新增研究领域（共 180 个）	147	120	65	97	34	51
	新增研究领域数占比/%	81.7	66.7	36.1	53.9	18.9	28.3

（2）中国的优势研究领域不断加强，学科结构布局逐步稳定。每个研究领域中叠加不同国家核心论文数量世界份额构成了各国的核心论文分布图，对比分析中国和科技发达国家的核心论文数量世界份额分布图（图3、图4），中国核心论文在研究领域的覆盖面积及密度大大增加，并且形成了若干密度极高的深红色区域。这代表，通过 5 年的深耕，中国在这些研究领域占据了主导地位。

从 2010~2015 年到 2014~2019 年，中国科研的优势研究领域正在逐步稳定，主要集中位于图谱的上半部分，涵盖纳米科技、系统与控制、无线通信、机器学习、岩土安全、废水处理等研究领域。而在位于图谱下部的医学、生物学、社会科学，以及天文学与粒子物理学、地球科学与生态环境等学科的研究领域群中研究领域所占份额相对较小。美国则相反，与中国具有明显的互补关系。

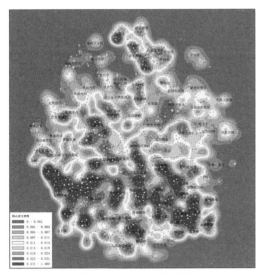
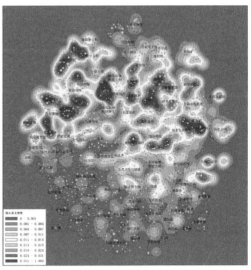

（a）2014～2019年美国　　　　　　　　　（b）2014～2019年中国

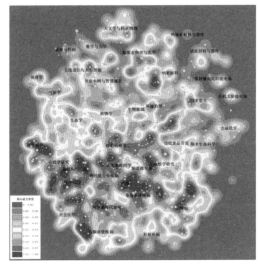
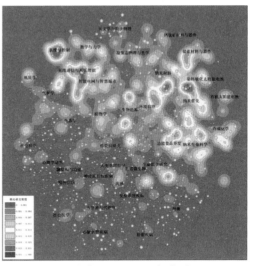

（c）2010～2015年美国　　　　　　　　　（d）2010～2015年中国

图3　两个时段中国、美国的核心论文数量世界份额分布图

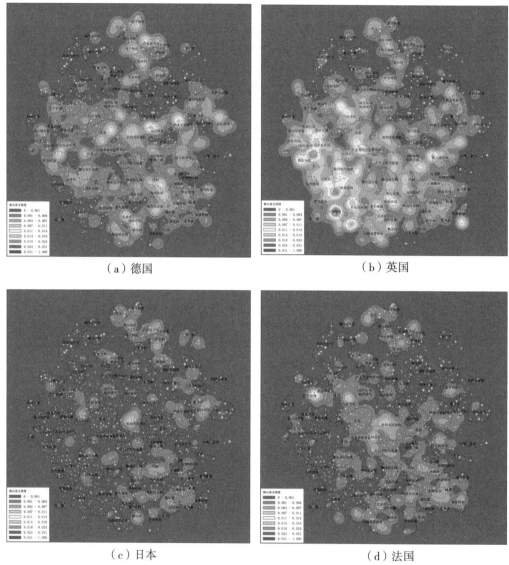

（a）德国　　　　　　　　　　　　　　（b）英国

（c）日本　　　　　　　　　　　　　　（d）法国

图 4　2014～2019 年德国、英国、日本、法国的核心论文数量世界份额分布图

从两个时段的科学结构图谱可以看出，中国各个学科的研究水平仍然很不均衡，尽管在相对较弱的医学和生物学发展较快，但在相对较强的工程学、计算机科学、化学与材料科学等学科的进步更快。相比之下，代表性科技发达国家在各研究领域的研究相对较均衡。

3. 中国的国际合作覆盖面及覆盖强度明显增大和增强

基于科学结构图谱的国际合著率对比（图 5）显示出，世界范围内的国际合作整体呈上升趋势，合作力度越来越强。

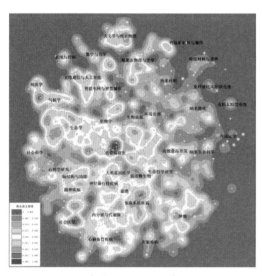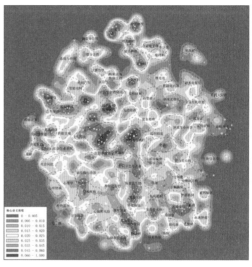

（a）2010～2015年　　　　　　　　　　（b）2014～2019年

图 5　基于科学结构图谱的国际合著率对比

中国在各个研究领域的国际合著覆盖面及覆盖强度呈现明显的增加和增强趋势（表5），而且在国际合作中的引领度逐年增强，国际合著论文通讯作者的占比大幅增加。2014～2019 年，中国合著论文的通讯作者占比达到 63.7％，跃居美国、中国、德国、法国、英国、日本 6 国之首；美国、德国、英国、日本合著论文通讯作者占比呈现增加的趋势，其中中国合著论文通讯作者占比的增幅远高于美国、德国、法国、英国、日本，达到 32.0％（表6）。在美国、中国、德国、法国、英国、日本 6 国中，中国论文的国际合著率（均值）最低，但完全依赖国际合作（国际合著率等于100％）和缺乏国际合作（国际合著率等于 0）的研究领域占比均下降了，说明中国参与国际合作研究活动的范围已经加大，对国际合著的依赖性逐步降低，科研实力越来越强。

表5　中国及科技发达国家的国际合著率情况

项目		国家					
		美国	中国	德国	英国	日本	法国
国际合著率 (均值)/%	2010～2015年	52.3	50.1	79.7	78.8	66.8	85.1
	2014～2019年	62.0	52.1	85.8	83.7	78.8	89.0
	变化率	18.5	3.9	7.6	6.2	17.9	4.6
国际合著率等 于100%的研究 领域占比/%	2010～2015年	7.0	39.4	44.9	38.3	51.1	56.3
	2014～2019年	17.0	36.0	57.7	48.1	65.3	64.9
国际合著率等 于0的研究领 域占比/%	2010～2015年	4.2	10.4	6.7	6.0	16.9	6.6
	2014～2019年	3.6	7.7	4.2	5.2	8.0	5.1

表6　中国及科技发达国家的合著论文中通讯作者占比情况

项目	国家					
	美国	中国	德国	英国	日本	法国
2010～2015年占比情况/%	50.5	48.3	29.2	31.8	21.3	25.8
2014～2019年占比情况/%	51.6	63.7	31.0	33.5	21.6	22.8
变化率/%	2.2	32.0	6.4	5.2	1.4	−11.7

4. 中国在对技术创新有影响的研究领域中的表现

论文被专利引用,表明该项科学研究对技术创新有一定的影响,本文分析中国及科技发达国家在对技术创新有影响的研究领域(以下简称创新研究领域)的科研表现。在2014～2019年这个时段的科学结构中,被专利引用的核心论文分布在864个研究领域中。因部分研究领域中被专利引用的核心论文数量占比很低,笔者选择了至少含有10篇被专利引用核心论文的研究领域进行统计,这样的研究领域共有267个。从国家分布(表7)来看,各国在创新研究领域的覆盖率差别不明显,美国略高于其他5个国家。

表7　2014～2019年中国及科技发达国家覆盖创新研究领域统计

项目	国家					
	美国	中国	英国	德国	日本	法国
各国覆盖创新研究领域数量/个	264	214	215	216	200	169
各国被专利引用核心论文数量世界 份额/%	57.8	20.4	15.9	14.6	5.6	9.2

注:研究领域内至少有10篇核心论文被专利引用。

　　但从被专利引用核心论文数量世界份额来看，美国的优势明显，为 58％，远远高于其他 5 个国家。中国被专利引用核心论文数量世界份额为 20％，排名第二，与中国全部核心论文数量世界份额的 21％（表 2）接近。美国等科技发达国家被专利引用核心论文数量世界份额却远高于全部核心论文数量世界份额。

　　图 6 展示了 2014～2019 年美国、中国、英国、德国和日本 5 个国家论文被专利引用的研究领域科研活跃度。从图 6 可以看出，美国和中国论文被专利引用的研究领域数量远超英国、德国和日本，且覆盖范围的差异更加明显。美国占主导的研究领域主要集中在生物学与医学研究领域群，少部分在材料与器件研究领域群，且优势较明显。英国、德国在医学研究领域群占有一定的份额，中国在大多数医学和生物学研究领域群中鲜有被专利引用，只在"蛋白质结构""RNA"和"干细胞"等几个研究方向较少的研究领域内占有一定份额。

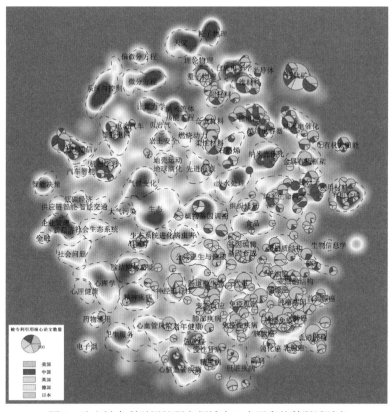

图 6　论文被专利引用的研究领域中 5 个国家的科研活跃度

饼图表示份额，饼的大小和被专利引用论文的数量成正比，用颜色区别不同的国家；图中仅展示了中国、美国、英国、德国和日本 5 个国家的情况在对技术创新有影响的研究领域中，中国的表现与中国总体的科研态势类似，仍然在化学与材料科学、数学、计算机科学与工程学、废水处理中比较有优势，但中国的优势在全领域中没有那么明显。并且，在中国表现突出的研究领域中，美国被专利引用论文数量份额仍然占比较大，在一些研究领域甚至超过中国，如"机器学习""锂电池"等。

三、中国各学科领域的科研态势

通过对比中国与科技发达国家的科研结构差异，可以分析中国的科研工作特点，进而描述中国的科研结构布局。

表8统计了两个时段中国和科技发达国家在学科领域中的通讯作者论文数量份额。因为学科交叉融合的范围越来越广，难以做到严格意义上的学科划分，我们计算了每个研究领域群中核心论文所属学科的占比，把该研究领域群归为占比排位第一的学科。

表8　两个时段中国和科技发达国家在学科领域中的通讯作者论文数量份额　单位:%

学科领域	2010～2015 年						2014～2019 年					
	美国	中国	德国	英国	日本	法国	美国	中国	德国	英国	日本	法国
天文学与粒子物理学	39.3	2.1	10.4	7.9	3.8	5.9	37.7	4.4	10.1	8.6	4.3	6.9
凝聚态物理学与光学	38.3	13.0	9.3	4.6	4.8	3.8	36.1	21.2	8.7	5.2	4.2	3.4
化学与材料科学	29.0	29.7	6.1	3.9	3.5	2.0	24.1	47.8	5.6	3.3	2.6	1.3
数学、计算机科学与工程学	19.1	27.6	3.5	2.8	1.2	1.9	9.6	51.7	2.1	3.1	1.0	1.5
地球科学与生态环境	39.9	8.6	7.1	9.2	4.1	4.0	27.5	21.4	6.1	8.0	1.9	3.0
生物学	36.8	11.2	6.3	7.5	2.2	4.1	33.4	20.4	5.8	7.2	1.9	3.6
医学	49.3	1.5	5.9	10.3	1.3	3.9	48.8	3.5	6.1	9.9	1.3	3.3
社会科学	49.3	1.5	5.9	10.3	1.3	3.9	35.7	9.9	3.7	15.3	0.2	1.9

注：红色字体为通讯作者论文数量份额上升幅度高的学科，蓝色字体为通讯作者论文数量份额下降幅度大的学科。社会科学不是本文的分析重点，后文未做展开论述。

相较2010～2015 年，中国在2014～2019 年的通讯作者论文数量份额在各个研究领域群都有很大的提升，接近或超过倍增。在化学与材料科学、数学、计算机科学与工程学中的通讯作者论文数量份额远远超过美国的通讯作者论文数量份额。天文学与粒子物理学、医学两个学科中与美国、德国、英国的差距还较大。以下对比分析中国与科技发达国家在各个研究领域群的科研结构。

1. 天文学与粒子物理学

在天文学与粒子物理学研究领域群，中国的通讯作者论文数量份额超过日本，在

6 个国家中排名第五位（表 9）。

表 9　中国及科技发达国家在天文学与粒子物理学研究领域群的通讯作者论文数量份额

单位:%

研究领域群	研究领域数	美国	中国	德国	英国	日本	法国
天文学	15	44.0	2.4	9.8	10.2	5.2	5.8
粒子物理	5	19.8	8.2	12.2	4.8	2.9	9.5
理论物理	4	56.4	5.3	4.5	10.5	2.3	5.3

注：粉色为无；蓝色为 0％；绿色为（0，1％）；黄色为［1％，3％）；橙色为［3％，7％）；紫色为［7％，12％）；红色为［12％，100％］。以下同。

在"天文学"研究领域，中国的通讯作者论文数量份额在 6 个国家中是最少的，与美国 44％ 的通讯作者论文数量份额有较大差距，但比 2010～2015 年的 1.9％ 通讯作者论文数量份额有所提升。"粒子物理学"研究领域，中国的通讯作者论文数量份额在 6 个国家中排名第四位，低于美国、德国和法国，相较 2010～2015 年的 2.5％ 通讯作者论文数量份额，进步较快；2010～2015 年，"理论物理"研究领域相关研究论文在科学结构图谱中未形成明显的研究领域群。

在天文学与粒子物理学和光学研究领域群，德国在各个研究领域的发展水平比较均衡，其中在"粒子物理"研究领域的优势比较明显，英国在"天文学"研究领域表现较好。

2. 凝聚态物理学与光学

在凝聚态物理学与光学，中国的通讯作者论文数量份额在 6 个国家中排名第二位（表 10），较上一期增长 8.2％，仅次于美国的 36.1％，远超英国、德国、法国和日本。

表 10　中国及科技发达国家在凝聚态物理学和光学研究领域群的通讯作者论文数量份额

单位:%

研究领域群	研究领域数	美国	中国	德国	英国	日本	法国
量子物理	9	29.9	10.8	12.6	10.0	1.9	3.3
自旋电子学	8	40.0	17.9	9.3	4.3	9.3	5.2
二维材料	6	42.9	26.9	4.6	2.5	3.4	2.3
半导体物理	5	34.0	29.2	8.3	3.5	7.6	2.1
非线性光学	8	36.1	22.6	12.3	2.7	1.8	3.6
超材料	11	30.6	30.6	3.8	5.9	1.6	2.5

中国在"超材料"、"半导体物理"和"非线性光学"三个群组的通讯作者论文数量份额上升快速，几乎与美国平起平坐。中国在量子物理群组中通讯作者论文数量份额排名第三。相比 2010～2015 年，中国在凝聚态物理学和光学全部的研究领域群中通讯作者论文数量份额都有明显提升，尤其是量子力学、自旋电子学、超材料和非线性光学。

德国在量子物理、非线性光学研究领域群表现突出，英国在量子物理研究领域群表现比较好。

3. 化学与材料科学

中国的化学与材料科学领域研究领域群排名第一（表 11），研究领域群达到 47.8％，较上一期（2010～2015 年）增长 18.1％。通讯作者论文数量份额大大超过美国的 24.1％，远超英国、德国、法国和日本。

表 11　中国及科技发达国家在化学与材料科学研究领域群的通讯作者论文数量份额

单位：%

研究领域群	研究领域数	美国	中国	德国	英国	日本	法国
电储能材料与器件	18	24.3	57.0	4.4	2.0	3.3	1.1
钙钛矿材料与器件	4	29.2	34.3	2.3	9.0	2.5	0.2
陶瓷材料	3	20.0	51.0	2.0	7.0	0.0	0.0
界面化学	3	8.0	70.5	2.3	2.3	0.0	4.5
纳米药物	8	28.7	46.2	4.1	0.5	1.1	2.5
合金材料	6	32.6	32.6	8.4	6.7	0.6	1.7
柔性材料与器件	6	51.1	25.5	2.6	1.5	0.7	1.1
有机太阳能电池	21	25.4	61.9	8.1	2.5	1.5	0.5
纳米碳材料应用	3	27.5	68.2	0.9	3.8	0.0	0.9
金属有机框架	8	32.2	35.9	5.1	6.3	2.7	4.1
有机合成方法学	9	22.9	35.0	14.9	4.8	4.0	2.1
超分子自组装材料	16	15.7	38.5	7.7	9.8	11.9	2.4
医用发光材料	5	5.6	68.0	0.8	1.6	6.4	0.4
纳米材料制备	5	41.2	35.3	5.9	0.0	2.9	0.0
纳米生物医学	3	21.4	47.9	4.6	2.2	1.7	1.7
纳米安全	13	21.8	13.6	2.7	1.8	3.6	1.8

除了柔性材料与器件、纳米安全和纳米材料制备 3 个研究领域群外，其他 13 个群组中国通讯作者论文数量份额均排在 6 个国家中的第一位，其中陶瓷材料、界面化

学、纳米药物、有机太阳能电池、纳米碳材料应用、医用发光材料和纳米生物医学 7
个研究领域群的通讯作者论文数量份额接近或超过 50％。

化学与材料科学一直是我国表现较好的领域，相比 2010～2015 年，中国在该领
域大部分研究领域群中通讯作者论文数量份额仍然都有提升，尤其是 2014～2019 年 3
个中国排名不是第一位的研究领域群增长尤其明显，柔性材料与器件从 20.6％增长到
25.5％，纳米安全从 1.8％增长到 13.6％，纳米材料制备从 26.4％增长到 35.3％。

4. 数学、计算机科学与工程学

中国的工程/数学/计算机领域通讯作者论文数量份额排名第一（表 12），通讯作
者论文数量份额达到 51.7％，较上一时段（2010～2015 年）增长 24.1％，远超其他
5 个国家。

表 12　中国及科技发达国家在数学、计算机科学与工程学研究领域群的通讯作者论文数量份额

单位:%

研究领域群	研究领域数	美国	中国	德国	英国	日本	法国
微分/偏微分方程	20	7.9	48.2	2.3	2.8	0.8	1.5
计算力学	6	11.4	20.1	4.7	2.0	2.5	3.5
系统与控制	21	1.4	81.1	1.2	1.4	0.2	0.8
无线通信	15	12.4	55.1	2.2	2.5	1.8	0.9
人工智能	33	12.5	60.0	2.4	2.9	0.8	2.1
能源互联网	17	14.5	32.0	2.0	6.6	0.5	0.8
热能工程	13	6.7	25.1	0.9	3.0	0.7	1.9
先进能源	14	21.8	17.8	4.0	7.6	2.4	2.1

除了先进能源，其他 7 个研究领域群中国通讯作者论文数量份额均排在 6 个国家
中的第一位，其中微分/偏微分方程、系统与控制、无线通信、人工智能 4 个研究领
域群的通讯作者论文数量份额接近或超过 50％。

工程/数学/计算机是我国在 2014～2019 年这个时段表现最好的学科。相较
2010～2015 年这个时段，中国在微分/偏微分方程、无线通信和人工智能 3 个研究领
域群中的通讯作者论文数量份额排名提升显著。在 2010～2015 年时段，微分/偏微分
方程通讯作者论文数量份额为 21％，无线通信通讯作者论文数量份额为 20.6％，人
工智能通讯作者论文数量份额为 37.9％。

5. 地球科学与生态环境

在地球科学与生态环境领域，中国的通讯作者论文数量份额在 6 个国家中排名第

二位（表13），通讯作者论文数量份额达到21.4%，仅次于美国。较上一时段（2010～2015年）增长显著，增长12.8%。

表13　中国及科技发达国家在地球科学与生态环境研究领域群的通讯作者论文数量份额

单位：%

研究领域群	研究领域数	美国	中国	德国	英国	日本	法国
岩土开采	19	8.2	71.2	0.6	1.8	0.8	1.0
地壳运动与地球演化	11	29.9	26.5	7.0	8.6	3.2	3.7
气候变化	17	42.0	9.2	6.9	15.4	1.0	4.4
大气污染	7	41.7	33.5	5.5	4.7	0.4	0.0
生态系统保护	15	30.7	1.0	7.5	9.8	0.3	3.5
水文生态	6	31.7	16.7	3.3	10.0	0.0	6.7
土壤生态学	5	26.5	21.2	10.6	6.6	0.0	4.0
污染治理	20	6.9	42.3	3.6	1.7	4.4	1.4
环境与健康	5	33.3	25.4	7.9	0.0	0.0	1.6

在岩土开采和污染治理研究领域群，中国通讯作者论文数量份额在6个国家中排名第一位。在地壳运动与地球演化、大气污染、水文生态、土壤生态学、环境与健康5个研究领域群中，中国通讯作者论文数量份额在6个国家中排名第二，仅次于美国。在气候变化、生态系统保护两个研究领域群，中国通讯作者论文数量份额较小，分别为9.2%和1%，位于第三位和第五位。

相较2010～2015年，在2014～2019年这个时段，中国地球科学与生态环境领域中通讯作者论文数量份额有显著提升。在2010～2015年这个时段，气候变化研究领域群的中国通讯作者论文数量份额为3.5%、大气污染研究领域群的中国通讯作者论文数量份额为1.5%、污染治理研究领域群的中国通讯作者论文数量份额为23.4%，水文生态研究领域群的中国通讯作者论文数量份额仅为0.5%。

英国在气候变化研究领域群表现突出，美国、德国、英国除在中国非常占优的岩土开采、污染治理研究领域群中通讯作者论文数量份额很小外，总体上各研究领域群间比较均衡。

6. 生物学

在生物学研究领域，中国通讯作者论文数量份额在6个国家中排名第二位（表14），达到20.4%，较上一时段（2010～2015年）增长显著，增长了14.1%。中国通讯作者论文数量份额仅次于美国的33.4%，远超英国、德国、法国和日本。

表 14　中国及科技发达国家在生物学研究领域群的通讯作者论文数量份额

单位:%

研究领域群	研究领域数	美国	中国	德国	英国	日本	法国
生态系统进化	8	42.3	4.5	5.5	17.5	0.3	3.8
植物基因调控	39	26.2	21.5	8.4	9.0	4.9	6.1
动植物病虫害	9	25.3	15.8	5.0	6.9	0.0	5.3
食品营养与健康	23	6.5	30.8	0.5	0.4	0.1	2.1
基因编辑与基因合成	19	56.6	7.6	6.9	8.2	3.6	2.0
蛋白质结构	10	53.1	6.5	12.6	11.9	0.4	1.9
RNA	7	27.4	52.1	5.3	2.2	1.3	0.7
干细胞功能研究	6	58.2	7.2	4.6	5.7	2.7	2.3
生物信息学方法	3	27.4	76.9	0.9	0.0	0.0	0.0
公共卫生与健康	21	35.5	12.4	3.6	8.5	0.1	5.7

在食品营养与健康、RNA、生物信息学方法 3 个研究领域群中,中国的通讯作者论文数量份额在 6 个国家中排名第一位。在植物基因调控、动植物病虫害、干细胞功能研究和公共卫生与健康 4 个研究领域群中,中国的通讯作者论文数量份额在 6 个国家中排名第二位,仅次于美国。在生态系统进化、基因编辑与基因合成、蛋白质结构三个研究领域群中,中国的通讯作者论文数量份额较小,分别排在 6 个国家中的第 3 位和第 4 位。

相较 2010～2015 年,在 2014～2019 年这个时段,中国在生物学的全部研究领域群中通讯作者论文数量份额都有显著提升。在 2010～2015 年这个时段,生态系统进化研究领域群的中国通讯作者论文数量份额为 2.5%、食品营养与健康研究领域群的中国通讯作者论文数量份额为 13%、生命组学相关领域为 8.4%。

德国在蛋白质结构研究领域群、英国在生态系统进化研究领域群表现突出。在中国占优的食品营养与健康研究领域群中,美国、德国、英国通讯作者论文数量份额较小。

7. 医学

在医学研究领域群,中国的通讯作者论文数量份额在 6 个国家中排名第 4 位(表 15),为 3.5%,但较上一时段(2010～2015 年)增长了 2%,通讯作者论文数量份额低于美国的 48.8%、英国的 9.9% 和德国的 6.1%,高于法国的 1.9% 和日本的 0.2%。

在大多数研究领域群中,中国的通讯作者论文数量份额较小,只有新兴的人工智能医疗和肠道微生物两个研究领域群在 6 个国家中排名第 2 位。通讯作者论文数量份额超过 5% 的研究领域群有 2 个,分别为代谢与免疫、肝脏疾病研究领域群。相比其他学科，中国在医学的研究领域群中通讯作者论文数量份额提升不明显。

在医学领域，美国具有绝对的优势。英国在卫生服务保健、脑科学、神经疾病和心理学等研究领域群表现突出。

表 15　中国及科技发达国家在医学研究领域群的通讯作者论文数量份额

单位：%

研究领域群	研究领域数	美国	中国	德国	英国	日本	法国
心理学	39	51.5	1.8	3.6	12.5	0.1	0.7
脑结构与功能	20	53.6	2.5	6.6	13.1	1.5	2.7
神经退行性疾病	12	53.4	2.7	6.4	10.0	2.1	2.1
精神疾病	19	42.1	1.6	3.9	19.3	1.4	0.7
人工智能医疗	9	42.9	15.3	5.2	7.0	3.8	2.8
肠道微生物	19	45.3	9.6	4.3	5.8	1.7	3.3
代谢与免疫	32	48.7	5.4	10.0	7.6	1.9	2.3
免疫性疾病	19	53.1	1.4	7.3	9.2	1.6	5.8
肿瘤	75	54.1	3.4	6.3	7.1	2.9	4.6
心脑血管疾病	34	42.1	1.8	6.1	11.7	0.9	3.0
肝脏疾病	8	38.7	6.4	7.5	8.5	3.9	6.2
糖尿病	6	43.9	0.3	12.0	12.3	1.3	2.3
危重症疾病	8	34.6	0.6	6.6	6.6	2.2	6.3
卫生服务保健	27	38.7	1.6	1.8	20.7	0.0	1.5

四、中国的优势与弱势研究领域

从科学结构图谱中我们可以看出，中国的优势研究领域主要集中在纳米科技、数学、计算机科学与工程科学，以及环境治理等方面。本节根据通讯作者的论文数量排名，分析中国具体的优势研究群组。同时，通过统计缺失核心论文产出的研究领域，可以分析中国的弱势研究领域群。

通过分析发现，中国在核心论文发表数量快速增加的同时，已经出现了一些以中国为主导的研究领域群，如岩土开采、系统与控制、纳米催化、人工智能、无线通信、废水处理等，已经开始深刻地影响着世界的科学版图。

1. 中国的优势研究领域群

如果在一个研究领域群中的绝大部分研究领域中，中国通讯作者论文数量排名第一，我们认为该研究领域群是中国具有优势的研究领域群。从表16中可以看出，中国的优势研究领域群主要集中在工程、数学、计算机科学、材料和化学等。中国通讯作者论文数量在 6 个国家中排名第一位的研究领域在研究领域群中的占比超过 80％（标红）的共有 7 个，分别是岩土开采、系统与控制、纳米催化、人工智能、电储能材料与器件、纳米生物医学、无线通信。此外，微分/偏微分方程、能源互联网两个研究领域群中通讯作者论文数量排名第一位的研究领域在研究领域群中的占比超过 60％。

表 16　中国优势研究领域群

岩土开采	系统与控制	纳米催化	人工智能	电储能材料与器件
12（12）	16（17）	8（9）	25（29）	15（18）
纳米生物医学	无线通信	微分/偏微分方程	能源互联网	金属有机框架
10（12）	12（15）	10（15）	10（15）	5（9）
食品营养与健康	废水处理	有机合成方法学	医用发光材料	合金材料
12（23）	7（16）	7（16）	5（5）	3（6）

注：括号外的数字为中国论文通讯作者排名第一位的研究领域数，括号内的数字为研究领域群中的研究领域数。

2. 中国的弱势研究领域群

在 2014～2019 年科学结构图谱中，没有中国通讯作者论文的研究领域数量为 494 个，占全部研究领域数量的 37％。表 17 统计了 2014～2019 年科学结构图谱中研究领域群内没有中国通讯作者发文的研究领域数量。从中可以看出，中国的弱势研究领域群主要在医学和社会科学。没有中国通讯作者发文的研究领域在研究领域群中占比超过 70％（标红）的研究领域群有 6 个：肿瘤、心脑血管疾病、卫生服务保健、危重症疾病、社会问题研究、社会生态系统。此外，在脑结构与功能、神经退行性疾病、心理学研究、生态系统保护 4 个研究领域群中，没有中国通讯作者发文的研究领域的占比超过 60％（黑体）。

表 17　中国弱势研究领域群

肿瘤	心脑血管疾病	卫生服务保健	危重症疾病	社会问题研究
72（75）	31（33）	23（27）	6（8）	15（20）
社会生态系统	脑结构与功能	神经退行性疾病	心理学研究	生态系统保护
8（11）	16（24）	8（12）	20（31）	9（15）
精神疾病	代谢与免疫	天文学	肠道微生物	气候变化
9（17）	14（30）	7（15）	7（19）	5（17）

注：括号外的数字为没有中国通讯作者发文的研究领域数，括号内的数字为研究领域群中的研究领域数。

五、结　语

"十二五"时期和"十三五"时期科学结构图谱的分析表明,中国的基础研究在两个五年中取得了显著进步。在总体科研态势上,中国的核心论文数量份额大幅提升,核心论文数量份额一直稳居世界第二位,且自2018年起超过了美国。在2014~2019年这个时段,中国的核心论文数量份额大幅超越德国、英国、日本、法国4个国家的世界数量份额之和,而同期传统科技强国的整体核心论文数量份额普遍呈下降趋势;在研究领域覆盖率、国际合著中的引领度等方面,我国都在迅速进步,逐步缩小与美国、英国、德国等国的差距。但在对技术创新有影响的研究领域中,中国论文的覆盖率和所占数量份额都没有在科学结构图谱中全部研究领域中的优势明显。在各学科的科研态势中,中国的发展也比较快速,在学科结构布局逐步稳定的同时,已经出现了一些由中国主导的研究领域群,并正在影响着世界的科学版图。

但同时我们也可以看出,与科技发达国家相比,中国的研究水平在各个学科间仍然很不均衡,有494个研究领域尚没有中国通讯作者发文;在以中国为主的新增优势研究领域群中,科技发达国家总体参与得较少,这可能与中国在发展过程中面临的特有问题相关,也可能与中国对发表国际论文的激励措施有关。中国与其他科技发达国家在科学研究结构上的差异,可能隐含了科研布局、创新重点和创新能力上的差异,后续还需要持续深入的研究。

中国高质量发展面临新起点,除了规模增长外,更需要提升研究的质量,重视基础科学中的实质性科学研究,促进学科交叉融合,大力推动理论创新。

参考文献

[1] 王小梅,李国鹏,陈挺. 科学结构图谱2021. http://sciencemap. casaid. cn/dist/[2022-01-27].

[2] 王小梅,韩涛,李国鹏,等. 科学结构图谱2017. 北京:科学出版社,2017.

[3] 王小梅,韩涛,王俊,等. 科学结构地图2015. 北京:科学出版社,2015.

[4] 潘教峰,张晓林,王小梅,等. 科学结构地图2012. 北京:科学出版社,2013.

[5] 潘教峰,张晓林,王小梅,等. 科学结构地图2009. 北京:科学出版社,2010.

China is Changing the Landscape of World Science
—Based on the analysis of Science Structure Map

Wang Xiaomei ,Li Guopeng ,Chen Ting

In order to understand China's achievements in basic science during the "Thirteenth Five-Year Plan" period, this report compared and analyzed the development of Chinese basic science in the Twelfth and Thirteenth Five-Year Plan periods in terms of research hotspots or potential hot frontiers based on the science structure map. The study found that the rapid development of basic science in China has profoundly changed the scientific landscape of the world. At the same time, by comparing and analyzing the scale of output, disciplinary layout, research cooperation and areas of strength in hot frontiers in Global Science between China and developed countries to identify the gaps with developed countries in science and technology, and provide a guide for the identification of priority areas, the choice of strategic priorities, and the orientation of science and technology policies in China.

第七章

中国科学发展建议

Suggestions on Science
Development in China

7.1 关于加强科普工作制度化建设的问题与建议

中国科学院学部咨询课题组①

2016 年，习近平总书记在"科技三会"上指出："科技创新、科学普及是实现创新发展的两翼，要把科学普及放在与科技创新同等重要的位置。没有全民科学素质普遍提高，就难以建立起宏大的高素质创新大军，难以实现科技成果快速转化。"[1] 新中国成立以来，尽管我国科学技术普及（以下简称"科普"）工作取得了长足进展，但是科普法治建设滞后于科普事业的发展需求、科研与科普工作结合不够及应急科普机制不健全等问题仍然突出。面对新时代建设现代化强国的使命与任务，亟须国家进一步加强科普工作的制度化建设，促进我国科普事业快速发展，为实现"到新中国成立 100 年时建成富强民主文明和谐美丽的社会主义现代化强国"奋斗目标奠定坚实基础。

一、科普法治建设难以满足当前科普工作的发展需求

1. 科普相关制度与政策可操作性低，科普主体的权利、责任与义务界定不明确

现行《中华人民共和国科学技术普及法》（以下简称《科普法》）自 2002 年颁布以来，20 年未做修订，也没有形成全国统一的实施细则和相关政策规章，而这 20 年来无论科学技术本身还是其传播和普及方式都发生了很大变化，修法工作亟待开展。即便按照现行《科普法》的要求，当前我国科普行政执法也严重不足，依法行政还面临很多障碍，在某些领域长期"形同虚设""疲软无力"。科普执法既不依法追究制作、传播虚假科普内容的违法者责任，也没有法律渠道追究地方政府不履行《科普法》确定的"逐步提高"和"增长"的科普经费责任。《科普法》对科普工作者、科普组织等相关科普主体的权利和义务规定得不清晰，致使各类科普主体的激励与约束

---

① 咨询课题组组长为中国科学院院士、中国科学院国家天文台研究员武向平和中国科学院院士、中国科学院地球化学研究所研究员欧阳自远。

机制没有得到应有的发挥。例如，科普基金的资助制度至今没有很好地建立起来；科普税收优惠制度没有得到很好地落实；科普产业发展的制度设计更是由于缺乏可操作性而没有得到很好地执行；等等。

2. 面对新兴科普领域与突发公共事件等，现有法治理念难以适应科普新形势的发展

随着信息技术与新媒体等领域的快速发展，在诸多科普关键和新兴领域亟须对《科普法》做进一步的补充和修订。例如，网络科普的发展对网络科普作品创作、传播提出了制度化的要求；数字新媒体的发展对大众传媒科普的法律义务和责任也有更高的要求；国家重大科技项目和重大工程的推进对科研与科普结合提出了更加迫切的需求。这些方面急需有相应的制度加以规范。农村科普和城乡社区科普急需切实可行的制度安排和推进措施，特别是面对突发公共事件（如 2019 年开始暴发的新冠肺炎疫情）等问题，都需要《科普法》面对新形势进行创新和发展。

3. 现有科普执法能力难以适应现代化科普治理水平的要求

现代化科普对提升全民科学素质，推动政府决策民主，促进社会和谐发展，构建国家创新系统，保障可持续发展意义重大，也给科学普及与传播提出了新的挑战，推动了科学普及与传播逐渐由以政府和科学共同体等为中心的权威发布模式，转变为以社会公众为中心的教学相长模式，由倡导公众理解科学知识以支持科学事业模式，转变为公众参与科学的对话协商和民主决策模式。传统科普的法律制度，难以正面回应新型科普形式所引起的权利、义务的剧烈变化，难以积极应对新型科普现象所带来的法定职权与法定边界的模糊不清，更难以有效解决新型科普纠纷所引发的观念碰撞和权利冲突等问题。

二、科研工作与科普工作的衔接缺乏制度性保障

1. 当前我国科研任务与科普工作结合不够，专家学者开展科普工作缺乏有效的制度激励

由于我国当前在制度安排上没有把科普工作真正地纳入科研任务中，因而从科研到科普的链条不够畅通，动力不足。如果不能使科普工作有效嵌入科技发展的整体设计，那么这种制度性欠缺会使科普停留在"半路上"，从而降低科技支撑的整体效率。调查显示，我国科普投入占科技总投入的比例远远低于国际平均水平，专家和学者参

与科普的人数比例偏低、科普经费偏少，而影响专家和学者参与科普工作的主要因素是动力机制和激励机制不足。这些问题更多地还是源于现有制度安排方面的不足，即没有把科普工作真正地纳入科研任务中。

2. 高层次科普人才队伍储备不足，难以满足我国当前科普事业快速发展的需求

我国严重缺乏科普专职人才，高层次科普人才占科普人才总数的比例低。科学技术部发布的《中国科普统计 2017》显示，我国共有科普专职人员 22.35 万人，还不及中国科学技术协会《中国科协科普人才发展规划纲要（2010—2020）》（简称《科普人才规划》）提出的"到 2020 年实现全国科普专职人员 50 万人"目标的一半。远不能满足我国快速增长的公众科普需求。一方面，近年来，我国科普场馆事业快速发展，数量增长迅速，科普场馆的物质建设已经成熟，但是软环境建设明显没有跟上。大多数场馆的科普展览设计、科普教育活动等仍旧停留在简单的模仿复制阶段，方式观念陈旧，特别是具备专业能力的高层次科普人才严重缺乏；另一方面，高技术企业、科技传媒、科研机构、大专院校等也急需懂科普、能够发挥科普效益的专门人才，向公众传播规范的科学知识，弘扬科学精神与科学文化。当前，我国试点的科普专门人才培养规模偏小，试点高校每年的招生总数仅百余。与我国科普场馆的发展速度及公众科普需求增长速度相比，高质量科普人才的供给量远远赶不上我国科普事业发展的需求。

三、应急科普机制不健全不利于应对重大突发事件

1. 应急科普力量分散，难以快速形成合力，无法有效应对突发事件的科普需求

从应对重大突发公共事件层面来看，我国应急科普存在明显的机制短板。科普工作在国家应急管理体系中的重视程度不高，应急科普主体之间缺乏协同机制。应急科普力量的分散主要体现在两个方面。

（1）管理权限分散。按照国内当前的应急管理体系，应急管理部是总体负责各类突发事件处置和救灾指导的核心部门，但涉及自然灾害、公共卫生和社会安全的专业应急科普资源却分散在自然资源部、国家卫生健康委员会、公安部、中国科学院、中国工程院、教育部、科学技术部、中国科学技术协会等部门，应急科普资源的统筹与共享存在严重不足。

（2）科普主体分散。各级政府及专家组、科学共同体、媒体各自掌握一定的知识、渠道和发布权限，但是没有很好的协同机制可以将其整体效能最大化，面对突发事件时易出现错位或缺位、因口径不一造成公众误解等问题。

2. 应急科普人才队伍建设滞后，科普资源缺乏，即时可用性较低

我国当前在科普知识储备、科普人才培养、科普场馆建设等方面进行了大量投入，但这些科普资源更多地服务于常态化科普活动，如"科技活动周""宣传月"等定期活动。真正的突发事件或社会热点形成后，由于缺乏有效的"战时"科普资源，应急科普更多的是依靠少数专家学者和专业机构甚至是自媒体自发开展。

3. 应急科普反应迟缓，及时性和时效性不强

一旦暴发突发事件或社会热点问题，经常会出现常态化科普没有遇到过的复杂科学问题，部分科普主题甚至对专家学者来说都具有一定的挑战。专家学者秉承严谨的科学态度，在没有充分的信息和十足的把握时往往选择不发声。这样给各类谣言提供了可乘之机，错过时效的科普往往会增加舆情引导的难度。

四、对策建议

习近平总书记在中央全面深化改革领导小组第二次会议上指出："凡属重大改革都要于法有据……确保在法治轨道上推进改革"[2]，阐述了改革与法治的辩证关系，为全面推进我国科普事业改革发展指出了方向。

1. 在法治建设方面，科学规划、统筹安排、完善国家科普法治顶层设计

建议国家尽快出台关于加强科普工作法治建设若干意见的文件。同时，建议适时启动《科普法》修法程序，科学规划，统筹安排，增加立法资源，加快对现行《科普法》的修订进程。建议科学技术部根据修订后的《科普法》内容制定实施细则，进一步明确政府、社会组织、企业及公民个人在科普中的责任、权利和义务。建议各级政府在科普规划、政策制定和监督检查方面进一步履职尽责。同时，适应实际，兼顾效率，扩大地方科普立法权限，着力解决科普领域发展不平衡、不充分的问题。

2. 在制度建设方面，使科普工作推进从重要性认识阶段转向落实绩效阶段

从我国的科技投入和科学力量的增速与体量来看，现在应该进入科研与科普结合发展的新阶段，要把科普工作纳入科研任务中，并有相应的预算、岗位和评价。建议

进一步明确国家科研机构、高校在科普中的社会职责。对事关人民健康、公益性强的国家重大科技计划、重大科技基础设施建设，要明确科普任务和相应预算，同步规划、实施和验收。建议在奖励和评价中彰显科普工作的价值，建议设立科普奖励基金，对在科普方面做出突出贡献的专家学者予以表彰与奖励。建立提升专家学者与公众有效沟通能力的长效机制，扩大高层次科普人才培养规模，着力解决人才供给与科普事业发展不协调的问题。

3. 在机制建设方面，加强应急科普的应对能力建设

建议各级政府在舆情管理系统中嵌入科普热点的侦测与分析模块，密切跟踪可能影响公共安全的重大问题，健全集体研判、专家会商等制度机制，用好大数据等信息技术手段，加强预警性分析研究，科学识别风险、有效防控风险，努力把问题解决在萌芽阶段、成灾之前。建议加强应急科普联动协调机制建设，一旦发生突发公共事件，按照突发事件的类型，协调推进部间的应急科普工作，形成统一发声、联合行动、快速反应的动态机制。建议政府部门进一步完善应急科普政策法规和组织机构，在政府应急管理体系中明确应急科普的重要地位，在现有各级应急管理预案中补充应急科普的具体工作预案，将应急科普工作纳入政府应急管理能力考核范畴。

参考文献

[1] 习近平. 为建设世界科技强国而奋斗——在全国科技创新大会、两院院士大会、中国科协第九次全国代表大会上的讲话(2016 年 5 月 30 日). http://www. xinhuanet. com/politics/2016-05/31/c_1118965169. htm[2021-01-25].
[2] 中华人民共和国中央人民政府. 习近平主持召开中央全面深化改革领导小组第二次会议. http://www. gov. cn/ldhd/2014-02/28/content_2625924. htm[2021-01-25].

Issues and Suggestions Regarding a More Effective Institutionalization Process of Science Popularization

Consultative Group of CAS Academic Division

We address the prominent issues and challenges in China's science popularization, and put forward the following suggestions. First, regarding the development of the legal system, official opinions should be issued on the development of relevant laws on science popularization, so as to make scientific plans and coordinated arrangements, to improve the top-level design of science popularization legislation, and to initiate in due course the revision process of science popularization laws. Second, in terms of the institutionalization process, we should establish norms that incorporate science popularization work into scientific and technological projects, with corresponding rules regarding budgets, staff positions, and evaluations of science popularization. Third, in terms of protocol development, it is necessary to improve the responsiveness of emergency science popularization and to establish relevant protocols.

7.2　加强我国"一带一路"建设科技支撑工作的建议

中国科学院学部咨询课题组①

随着"一带一路"建设迈向高质量发展阶段，通过科技合作，支撑和引领协议国家走创新驱动发展之路将成为新形势下"一带一路"共建工作的重中之重。

一、"一带一路"建设科技支撑工作存在的主要问题

近年来，我国与"一带一路"协议国家和地区科技创新合作的层次和水平不断提升，有效地促进了"一带一路"建设的稳步发展。但是随着建设目标的升级和工作推进，当前的科技支撑工作显现出一些薄弱环节。

1. 当前科技支撑偏重传统产业领域，前沿性和战略性领域的科技支撑需要进一步提高

当前，"一带一路"建设科技合作支撑活跃度高的领域是资源、环境、农业和能源等传统领域，而防灾减灾、航天航空和医药健康等前沿性、战略性领域活跃度较低，迫切需要拓展合作范围，调整工作重点。

2. 科技支撑统筹协调体系较薄弱，有待进一步加强和完备

当前，我国在科技与经贸、外交、教育等方面的政策协同性不高，技术、服务出口与文化输出及产能合作等方面的统筹仍不到位，科技对外开放与国际合作新体制的兼容性有待优化，不利于吸引更多全球高端创新要素参与我国科技创新。

3. 我国科技企业对建设"一带一路"的参与度不高，需要在政府和大型企业与机构的引导支持下"抱团出海"

当前，我国很多科技企业对"一带一路"协议国家的有关法律规定不熟悉，国际

---

① 咨询课题组组长为中国科学院院士、上海交通大学教授张杰。

化经验不足，抵御风险能力有限，需要在政府部门、行业组织、商会等的引导与支持下，采取战略联盟或产业集群等模式"抱团出海"。

4. 复合型科技支撑人才缺乏，属地化人才培养体系建设有待加强

"一带一路"地域广袤，国家众多，国情差异较大，属地化、复合型科技支撑人才紧缺。需要采取措施，与东道国相关组织机构紧密合作，建立属地化人才培养体系。

5. 我国技术标准的国际化水平较低，与我国的倡导者角色不匹配

近年来，中国标准的国际影响力日益增强，但还存在标准编制思路的国际通用性不足、部分技术指标低于国际标准、缺乏完整应用支撑体系、管理体系不健全等问题，需要动员多方社会力量，加大推进力度。

二、加强"一带一路"建设科技支撑的主要思路

加强"一带一路"建设科技支撑总的思路是：以服务高质量发展需求为导向，以建设创新共同体为主线，以"深化合作研究、项目开发、园区共建、平台搭建、人才交流"等方面的科技支撑模式创新为抓手，全面统筹利用国内、国外两大科技体系与资源，培育建设充满活力和竞争力的"一带一路"建设科技创新生态体系。

1. 围绕"一带一路"协议国家的科技与产业发展需求，打造一批区域性科研合作与产业协同创新中心

筛选一批影响"一带一路"协议国家未来产业整体升级、集群突破的关键技术，引导和支持这些国家编制符合本国需求和特色的科技与产业发展战略规划。以科学技术创新目标为导向，组织与这些国家联合开展重大国际科学研究。引导和鼓励科技企业与科研机构与这些国家合作建立"产学研用"协同创新基地，逐步形成若干区域性科技与产业协同创新中心。

2. 建设一批对"一带一路"协议国家科研、产业与社会发展具有重大意义的标志性工程和高新技术产业基地

以重大科研基础设施联通、优质产能合作等项目为抓手，大力推进一批对"一带一路"协议国家科研、产业与社会发展具有重大意义的标志性工程。加强重大项目的前期规划定位、中期过程管理和后期运维服务一体化的管理体系能力建设。以重大项

目建设为着力点，发挥我国企业的带动作用和极核功能，打造科技产业发展合作平台。

3. 支持"一带一路"协议国家打造一批具有良好科技创新生态的园区平台

支持我国境内开发区、高新区与"一带一路"协议国家合作，采取"一园两区"的模式，在这些国家建设主体功能突出、行业和来源地域相对集中的国际科技园和孵化器。组织我国海外园区建设主体建立行业联盟，建立全球性的科技园区开发运营平台，为入园企业提供优质的综合运营服务。

4. 打造若干一站式的"一带一路"科技成果交易市场

发挥我国在电子商务、大数据、互联网金融等方面的优势，打造融科技、贸易、金融等交易为一体的一站式科技成果交易市场。鼓励民间资本设立人民币创业投资基金、并购基金、产业基金等金融产品，支持我国出海科技企业和"一带一路"协议国家科技企业发展。

5. 建立长效科技人才交流培养机制，构建属地化科技人才培养网络

以省（自治区、直辖市）为对接窗口，和与国内城市规模相近或发展特征相似的"一带一路"协议国家的城市进行友好城市配对，拓展"一带一路"友好城市群。采取职业高等教育、网络教育等方式，强化国内教育机构、科技培训机构与境外经贸合作区、国际产业园内的企业合作。通过创业投资、科研基金资助等多种形式，鼓励和支持国外来华留学科技人才回国创新创业。

三、几点建议

1. 加强科技外交的模式与政策创新，共同打造创新驱动的"一带一路"经济体系

深入分析世界转型过渡期国际形势的演变规律和我国内外部环境特征，围绕国家的科技创新战略和总体外交大局，创新建设科技外交工作的模式和政策，协调好新老机制、发达国家与发展中国家间的科技合作关系，共同打造创新驱动的国际经济体系。

2. 全面加强"一带一路"建设科技支撑工作的统筹协调

大力加强科技支撑工作的战略规划和综合统筹，全面加快创新共同体的建设。在

组织运行方面，可考虑将科学技术部作为推进"一带一路"建设领导小组办公室副主任单位，加强我国在建设过程中科技工作与经济、教育、文化、外交等各方面工作的统筹协调，集中多方力量，精准、高效地开展科技支撑。

3. 发起设立"一带一路"协议国家科技创新合作组织

建议由科学技术部牵头，联合"一带一路"协议国家科技管理部门，发起设立科技创新合作组织。主要职能可以考虑：①加强"一带一路"协议国家科技管理部门之间的沟通，协调这些国家的科技创新政策方针，加强重点领域的科技创新合作与交流；②支持"一带一路"协议各国编制符合本国需求且具本国特色的科技行动计划；③推进专业性、区域性科技需求与成果转化对接平台的规划布局与建设，推动"一带一路"协议国家的科技成果的交流与转化。

4. 充分发挥社会力量，推动中国标准的推广和国际化工作

建议由国家标准化管理委员会和国家发展改革委等部门牵头，支持我国技术标准咨询机构、龙头企业在"一带一路"协议国家建设技术示范推广基地、技术转移中心、数据共享平台。支持行业协会以"主要窗口"身份参与国际技术标准的咨询、评议、制定认可工作，推进标准的国际互认。对使用中国标准的国家，采取贷款优惠、资金扶持、技术免费培训、后续长期技术配套服务等措施予以激励。

5. 支持在"一带一路"协议国家布局建设"一带一路"云计算中心与大数据平台

建议由国家发展改革委、工业和信息化部等部门牵头，支持国内科技、制造和金融等行业企业，在"一带一路"协议国家布局建设一批云计算中心与大数据平台，发展以我国为核心平台的"一带一路"科技与产业网络。深度挖掘我国在产业、贸易与金融等方面的数据资源，为"一带一路"协议国家提供更多高质量的信息与智力服务。

6. 密切跟踪后疫情时代国际形势变化，加强在生物安全、医疗卫生、防灾减灾等重点领域的科技支撑与合作

建议由国家卫生健康委和科学技术部等部门牵头，加快建立疫病防控国际科技合作网络，增强新发突发传染性疾病的监测预警和防控能力。根据部分"一带一路"协议国家和地区公共卫生服务能力不足的实际情况，规划建设一批"一带一路"协议国家大健康产业园和医药产业园等特色园区。

Suggestions on Strengthening S & T Support to China's Belt and Road Initiative

Consultative Group of CAS Academic Division

This paper reviewed some rough edges and problems of the scientific and technological (S&T) support to China's Belt and Road Initiative and proposed the following suggestions: ①We should bring forth some new ideas in the S&T diplomatic strategy for the Belt and Road Initiative and fully carry them out. ②We should strengthen overall planning and coordination in our work to make the development of science and technologies support the Belt and Road Initiative. ③We should take action to set up organizations for S&T innovation and cooperation in countries along the Belt and Road routes. ④Social forces should be utilized to push forward the promotion and internationalization of Chinese standards. ⑤Leading domestic companies should be encouraged to build cloud computing centers and big data platforms in countries along the Belt and Road routes. ⑥S&T support and cooperation in key areas such as biosafety, health care, as well as disaster prevention and mitigation should be enhanced.

7.3 我国石墨烯产业发展的关键问题及对策

中国科学院学部咨询课题组[①]

石墨烯是引领新一轮全球高科技竞争和产业变革的战略新兴材料之一，受到世界各国/组织的高度重视。2013年，欧盟启动为期10年的"石墨烯旗舰计划"。英国先后成立了国家石墨烯研究院和国家石墨烯工程创新中心，全力打造石墨烯产业领头羊地位。美国、日本、韩国和新加坡等制定专门计划，争夺下一个万亿级石墨烯新兴产业。中国拥有全球最大规模的石墨烯基础研究和产业大军并走在世界前列，但面临战略布局缺乏、龙头牵引不足、产业基础薄弱、创新体系分散等诸多挑战。石墨烯新材料产业面临一个重大历史机遇，当前正处在一个承上启下的关键阶段。为此，我们建议尽快做好顶层设计，体现国家意志，充分发挥我国谋划长远、集中力量办大事的制度优势。

一、全球石墨烯产业发展态势和我国所处的地位

石墨烯新材料产业处于迅猛发展阶段，大量基础研究成果走出实验室，迈向规模化和产业化探索阶段，示范性石墨烯产品已经走进市场。欧洲是石墨烯的诞生地，也是最早布局石墨烯产业的地区，通过"石墨烯旗舰计划"整合产学研资源，在石墨烯前沿技术研发方面处于国际领先地位。紧随其后的是美国和韩国，NASA、IBM、三星集团等机构和大企业的深度参与是其特色和优势。日本、澳大利亚、新加坡等也很早就布局石墨烯前沿技术研发工作。

中国是石墨烯新材料研究和应用开发最活跃的国家之一。在基础研究方面起步很早，基本上与世界同步。从2011年起，中国学者发表的学术论文稳居全球榜首。截至2020年3月，中国学者共发表学术论文101 913篇，全球占比33.2%，在《自然》及其子刊、《科学》等顶级学术期刊上频频发表独创性研究成果。高品质石墨烯材料制备居国际领先地位，代表性工作有单晶石墨烯晶圆规模化制备技术、无褶皱超平整

---

① 咨询课题组长为中国科学院院士、北京大学教授刘忠范和中国科学院院士、中国科学院金属研究所研究员成会明。

石墨烯薄膜生长技术、超洁净石墨烯、超级石墨烯玻璃等。应用基础研究也不断取得突破，代表性成果有烯碳光纤、石墨烯复合纳滤膜、氧化石墨烯海水淡化膜、声表面波类石墨烯微结构材料等。

中国对石墨烯产业的关注完全与世界同步，并拥有全球最大规模的石墨烯企业。截至 2020 年 2 月，我国企业总数已达 12 090 家，在石墨烯材料规模化生产方面居全球领先地位，2018 年石墨烯粉体年产能达 5100 t、化学气相沉积（chemical vapor deposition，CVD）薄膜年产能达 650 万 m^2。截至 2018 年底，中国申请的石墨烯专利总数达 47 397 件，全球占比 68.4%。经过近 10 年的快速发展，中国的石墨烯产业已遍布全国众多省份，初步形成了"一核两带多点"的空间分布格局。作为"一核"的北京集聚了石墨烯核心技术研发力量；作为"两带"的东部沿海地区和黑龙江-内蒙古地区，前者拥有先发优势及人才和市场优势，后者致力于发挥石墨矿资源优势；"多点"是指重庆、四川、陕西等呈分散状态但具有一定特色和优势的地区。

从总体上看，中国石墨烯产业发展与欧洲、美国、日本等发达国家/地区不在一个频道上。我们关注的是今天的石墨烯产品市场，而国外关注更多的是面向未来市场的高端应用研发。

二、我国石墨烯产业存在的关键问题

1. 低水平同质化竞争严重，缺乏整体布局和顶层设计

截至 2020 年 3 月，全国各地建立了 29 个石墨烯产业园、54 个石墨烯研究院、8 个石墨烯创新中心。但是，大部分项目未经过严密的科学论证，简单重复建设现象非常严重。受制于技术、市场等因素，近八成的下游产品集中在电加热、大健康、防腐涂料、导电添加剂等领域，技术门槛较低，同质化竞争严重。各地的石墨烯产业基本上处于各自为政、无序发展的状态，缺乏国家层面的统筹协调和长远规划。尤其是，我国在石墨烯基础前沿与变革性技术研究方面的投入力度与发达国家的差距明显。

2. 龙头牵引不足，小微企业居多，可持续发展能力有限

国内从事石墨烯生产和应用开发的企业多为初创期的小微企业。由于综合实力弱，只能关注一些投入小、产出快的短期应用产品，资金短缺和研发能力不足导致其核心竞争力和可持续发展能力严重匮乏。与发达国家显著不同的是，我国有实力的大

企业（尤其是国企和央企）对石墨烯产业的实质参与度非常低。另外，我国高技术产业的总体基础还比较薄弱，尤其是技术分工和产业配套方面与国外差距明显，导致小微企业"麻雀虽小，五脏俱全"，严重制约了自身特色的发挥和可持续发展能力。

3. 创新体系分散，产学研深度融合不足

我国拥有全球最庞大的石墨烯基础研究队伍，同时拥有最活跃的产业大军，但是两者之间尚未形成良好的协同效应。按照现行评价机制，高校和科研院所往往以发表论文为导向，缺乏真正的需求牵引，研究成果很难与市场结合。另外，大多数企业急功近利，不重视核心技术研发，缺少核心竞争力。作为"明星级"新材料，石墨烯越来越受到资本市场的关注，不少上市公司、投资机构跃跃欲试，涉足石墨烯产业。但是，这些创新要素基本上各自为战，尚未形成合力，产学研协同创新机制有待进一步完善。

三、对 策 建 议

1. 迅速推进国家石墨烯产业创新中心建设，整合全国创新资源

组建"国家石墨烯产业创新中心"的时机已经成熟，建设统领全国、服务全国的石墨烯产业国家队势在必行。该创新中心应集成代表中国石墨烯基础研究、高技术研发及产业化应用最高水平的骨干企业和研究团队，同时吸纳社会资本和具有产业引领能力的大型央企、国企和私企参与，共同打造中国石墨烯产业的旗舰。

2. 成立国家石墨烯产业联盟，统筹石墨烯标准体系建设

当前我国各地的石墨烯产业联盟多达 12 个，但是处于各自为战、发声混乱、鱼目混珠的状态，缺少权威性和专业性，进一步加剧了我国石墨烯产业发展混乱的现象，也影响了我国在国际石墨烯领域的声誉。建议成立经权威部门授权的"国家石墨烯产业联盟"，让其真正代表国家发声，改变当前的混乱局面；由"国家石墨烯产业联盟"统筹布局，加快石墨烯材料和产品的国家标准、行业标准和团体标准建设，特别是尽快完善在下游应用领域快速发展的电加热和大健康产品、防腐涂料、电池导电添加剂、石墨烯复合材料等的相关产品的定义、检测和使用标准；加快研究并制定石墨烯行业准入标准，从产业布局、生产工艺与装备、环境保护、质量管理等方面加以规范，使石墨烯的应用及其产品有标准可依、有规范可循；加强国际交流与合作，积极参与国际标准制定，确保国内的石墨烯标准体系与国际接轨。

3. 聚焦"卡脖子"技术，加大支持力度，培育核心竞争力

石墨烯材料是未来石墨烯产业的基石，也是制约石墨烯产业健康发展的"卡脖子"问题。就石墨烯材料的产能来说，中国已经高居全球榜首，并且已有产能过剩的风险。但是，由于技术尚未过关，工艺稳定性差，造成当前国内石墨烯产业的诸多乱象乃至信任危机。因此，必须整合资源，加大投入力度，久久为功，突破石墨烯材料生产的核心技术。

与此同时，我国应该积极布局未来，探索石墨烯材料的"撒手锏"级应用，而不能只关注立竿见影的扮演"味精"角色的石墨烯应用产品。真正意义上的战略性新兴材料有两种表现形式——创造全新的产业、给现有产业带来变革性的飞跃，石墨烯材料有望兼而有之。建议国家设立"石墨烯专项"，由"国家石墨烯产业创新中心"统筹布局，充分体现国家意志，打造未来石墨烯产业的核心竞争力。

4. 发挥制度优势，加强顶层设计

石墨烯新材料的特点决定了发展石墨烯产业的长期性和艰巨性，因此需要做好战略性、全局性的规划设计，而这正是我国的制度优势所在。在时间维度上，我国应制定石墨烯产业发展路线图，通过五年规划、十年规划、二十年规划，稳步推进石墨烯产业可持续发展。将石墨烯重点攻关方向与制造强国战略相统一，围绕新一代信息技术、航空航天装备、节能新能源汽车、生物医药等重点领域的发展需求布局。在空间维度上，我国应科学合理地规划全国石墨烯产业布局，推进差异化、特色化、集群化发展，避免低水平的重复建设和恶性竞争。

5. 释放政策红利，培育创新生态

石墨烯产业是处在迅速发展阶段的高科技产业，需要及时、有效的政策引导，极大限度地释放政策红利，打造创新性的文化环境和高科技研发生态。高科技产业发展的核心要素是具有创新能力、掌握核心技术的专业人才。如何最大限度地调动这些专业人才的主观能动性，最大限度地释放他们的创造力，决定着未来石墨烯产业的核心竞争力。由于现行人才和科技成果评价机制方面的原因，科研人员缺少推进成果转化的动力和勇气。国家已陆续出台新的人才和科技评价政策，相信"破四唯"改革会给石墨烯产业带来新的发展动力。建议加大力度支持产学研协同创新机制探索。当前，国内各省份已进行了大量的探索与实践，需要进行交流和总结经验，尽快建立适合我国国情的产学研协同创新体系非常必要。

Key Issues and Coping Strategies of China's Graphene Industry

Consultative Group of CAS Academic Division

This paper reviews the major issues faced by China's graphene industry and comes up with the following coping strategies: ① speed up the building of national innovation centers for graphene industry. ② establish a national-level graphene industry alliance. ③ focus on technological dependence issues of graphene materials production, increase support and cultivate core competitiveness. ④ give full play to the advantages of our systems, strengthen the top-level design, rational planning and design, and promote the differentiating, specializing and clustering development process. ⑤ foster an environment that stimulates innovation, and establish an industry-university-research collaborative innovation system that is appropriate for China's national condition.

附　录

Appendix

附录一　2020 年中国与世界十大科技进展

一、2020 年中国十大科技进展

1. "嫦娥五号"探测器完成我国首次地外天体采样返回之旅

2020 年 11 月 24 日 4 时 30 分，我国成功发射探月工程"嫦娥五号"探测器，并于 12 月 1 日晚间成功着陆在预选着陆区。在完成月壤取样后，"嫦娥五号"上升器于 12 月 3 日从月面起飞。"嫦娥五号"返回器于 12 月 17 日 1 时 59 分在内蒙古四子王旗预定区域成功着陆，标志着我国首次地外天体采样返回任务圆满完成。

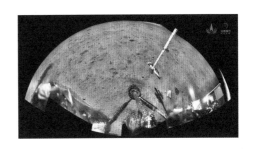

随后，重达 1731 g 的"嫦娥五号"任务月球样品移交中国科学院，将在位于中国科学院国家天文台的"月球样品实验室"中存储、处理和分析，正式开启月球样品与科学数据的应用和研究。

"嫦娥五号"任务作为我国复杂度最高、技术跨度最大的航天系统工程，对于我国提升航天技术水平、完善探月工程体系、开展月球科学研究、组织后续月球及星际探测任务具有承前启后、里程碑式的重要意义。

2. "北斗三号"最后一颗全球组网卫星发射成功

2020 年 6 月 23 日 9 时 43 分，我国在西昌卫星发射中心用"长征三号乙"运载火箭，成功发射北斗系统第五十五颗导航卫星暨"北斗三号"最后一颗全球组网卫星。

此次发射的北斗导航卫星和配套运载火箭分别由中国航天科技集团有限公司所属的中国空间技术研究院和中国运载火箭技术研究院抓总研制，中国科学院微小卫星创新研究院等多家科研院所全方位参与了研制建设。

这是长征系列运载火箭的第336次飞行。在测控、地面运控、星间链路运管、应用验证等系统的强有力支撑下，此前发射的所有在轨卫星都已入网。至此，"北斗三号"全球卫星导航系统星座部署比原计划提前半年全面完成。

3. 我国无人潜水器和载人潜水器均取得新突破

2020年6月8日，由中国科学院沈阳自动化研究所主持研制的"海斗一号"全海深自主遥控潜水器搭乘"探索一号"科考船海试归来。在此航次中，"海斗一号"全海深自主遥控潜水器在马里亚纳海沟实现近海底自主航行探测和坐底作业，最大下潜深度10 907 m，填补了我国万米级作业型无人潜水器的空白。

11月28日，由中国船舶重工集团有限公司第七〇二研究所牵头总体设计和集成建造、中国科学院深海科学与工程研究所等多家科研机构联合研发的"奋斗者号"全海深载人潜水器随"探索一号"科考船返航。

此次"奋斗者号"全海深载人潜水器在马里亚纳海沟成功坐底，创造了10 909 m的中国载人深潜新纪录，标志着我国在大深度载人深潜领域达到世界领先水平，并且有助于科学家了解深渊海底生物、矿藏、海山火山岩的物质组成和成因，以及深海海沟在调节气候方面的作用。

4. 我国率先实现水平井钻采深海可燃冰

2020年3月26日，自然资源部召开了我国海域天然气水合物第二轮试采成果汇报视频会。会议透露，此轮试采日前取得成功，并超额完成目标任务。天然气水合物通常称为"可燃冰"。在水深1225 m的南海神狐海域的试采创造了产气总量86.14万 m³、日均产气量2.87万 m³ 两项新世界纪录。

此次试采中，研究人员还自主研发了

一套实现天然气水合物勘查开采产业化的关键技术装备体系，创建了独具特色的环境保护和监测体系，自主创新形成了环境风险防控技术体系。

此次试采攻克了深海浅软地层水平井钻采核心技术，实现了从探索性试采向试验性试采的重大跨越，在产业化进程中取得标志性成果。我国也成为全球首个采用水平井钻采技术试采海域天然气水合物的国家。

5. 科学家找到小麦"癌症"克星

小麦赤霉病是世界范围内的极具毁灭性且防治困难的真菌病害，有小麦"癌症"之称。山东农业大学农学院教授、山东省现代农业产业技术体系小麦创新团队首席专家孔令让及其团队从小麦近缘植物长穗偃麦草中首次克隆出抗赤霉病主效基因 *Fhb7*，且成功将其转移至小麦品种中，首次明确并验证了其在小麦抗病育种中不仅具有稳定的赤霉病抗性，而且具有广谱的解毒功能。相关研究成果于 2020 年 4 月 10 日在线发表于《科学》上。

目前，已有 30 多家单位利用抗赤霉病的种质材料进行小麦抗赤霉病遗传改良，并在山东、河南、江苏、安徽等地进行广泛试验，效果良好。上述成果为解锁赤霉病这一世界性难题找到了"金钥匙"。

6. 科学家达到"量子计算优越性"里程碑

中国科学技术大学潘建伟、陆朝阳等与中国科学院上海微系统与信息技术研究所、国家并行计算机工程技术研究中心的研究人员合作，构建了 76 个光子的量子计

算原型机"九章"，实现了具有实用前景的"高斯玻色取样"任务的快速求解，使得我国成功达到量子计算研究的首个里程碑——量子计算优越性，为实现可解决的具有重大实用价值问题的规模化量子模拟机奠定技术基础。相关成果于 2020 年 12 月 4 日在线发表于《科学》上。

7. 科学家重现地球3亿多年生物多样性变化历史

生命起源与演化是世界十大科学之谜之一。地球上曾经生活过的生物99%以上已经灭绝，通过化石记录重建地球生物多样性变化历史是认识当今人类居住地球生物多样性现状与发展趋势的重要途径。

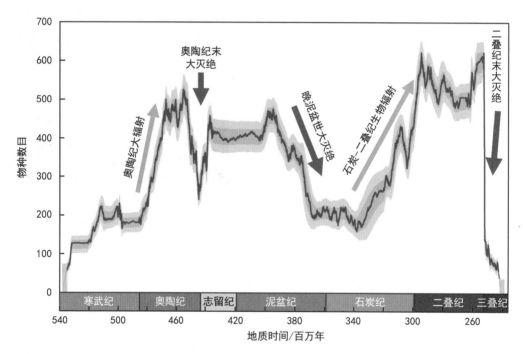

南京大学樊隽轩教授、沈树忠院士等自建大型数据库，自主研发人工智能算法，利用"天河二号"超级计算机取得突破，获得了全球第一条高精度的古生代3亿多年的海洋生物多样性变化曲线，分辨率较国际同类研究提高400倍。

新曲线精确刻画出地质历史中多次重大生物灭绝和辐射事件及其与环境变化的关系。成果于2020年1月17日以研究长文形式发表于《科学》上。

8. 我国最高参数"人造太阳"建成

我国新一代可控核聚变研究装置"中国环流器二号M"（HL-2M）于2020年12月4日在成都正式建成放电，标志着我国正式跨入全球可控核聚变研究前列，HL-2M将进一步加快人类探索未来能源的步伐。

该项目由中国核工业集团公司核工业西南物理研究院自主设计建造。据悉，该装置是我国目前规模最大、参数最高的先进托卡马克装置，是我国新一代先进磁约束核聚变实验研究装置，采用更先进的结构与控制方式，等离子体体积达到国内现有装置的 2 倍以上，等离子体电流能力提高到 2.5 MA 以上，等离子体离子温度可达到 1.5 亿℃，能实现高密

度、高比压、高自举电流运行，是实现我国核聚变能开发事业跨越式发展的重要依托装置，也是我国消化吸收国际热核聚变实验堆（International Thermonuclear Experimental Reactor，ITER）技术不可或缺的重要平台。

9. 科学家攻克 20 余年悬而未决的几何难题

中国科学技术大学教授陈秀雄、王兵发表的关于高维凯勒里奇流收敛性的论文，率先攻克了哈密尔顿-田猜想和偏零阶估计猜想——这些均为几何分析领域 20 余年来悬而未决的核心猜想。相关成果于 2020 年 11 月初发表在《微分几何学杂志》上。据了解，该篇论文的篇幅超过 120 页，从投稿到正式发表耗时 6 年。该论文引进了众多新思想和新方法，对几何分析尤其是里奇流的研究产生了深远的影响。据悉，该论文是几何分析领域内的重大进展，业已推进诸多相关工作。

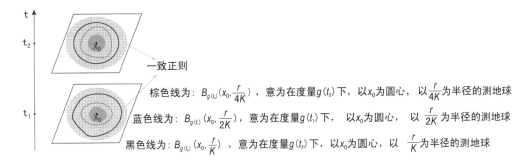

10. 机器学习模拟上亿原子

2020 年 11 月 19 日下午，由中国科学院计算技术研究所贾伟乐副研究员、中国科

本工作：分子模拟+机器学习+高性能计算

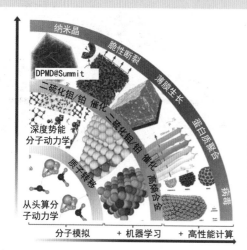

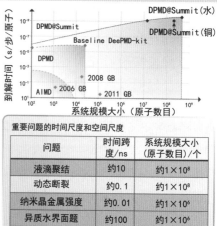

重要问题的时间尺度和空间尺度		
问题	时间跨度/ns	系统规模大小（原子数目）/个
液滴聚结	约10	约$1×10^8$
动态断裂	约0.1	约$1×10^8$
纳米晶金属强度	约0.01	约$1×10^6$
异质水界面题	约100	约$1×10^6$

学院院士鄂维南、北京大数据研究院张林峰研究员及其合作者共同完成的应用成果获得国际高性能计算应用领域最高奖——戈登贝尔奖。

该项工作在国际上首次采用智能超算与物理模型的结合，引领了科学计算从传统的计算模式朝着智能超算的方向前进。

第一性原理分子动力学以其高精度和算法复杂著称。长期以来，其计算的空间尺度和时间尺度受算法和算力限制，即使利用世界上最快的超级计算机，也只能计算数千原子体系规模。

该成果通过高性能计算和机器学习将分子动力学极限提升了数个量级，达到上亿原子的体系规模，同时仍保证了"从头计算"（ab initio calculation）的高精度，且模拟时间尺度较传统方法至少提高 1000 倍。

据了解，基于深度学习的分子动力学模拟通过高性能计算和机器学习的有机结合，将精确的物理建模带入更大尺度的材料模拟中，有望在将来为力学、化学、材料科学、生物学乃至工程领域解决实际问题发挥更大作用。

二、2020 年世界十大科技进展

1. 科学界完成迄今最全面癌症基因组分析

2020 年 2 月 5 日，英国韦尔科姆基金会桑格研究所宣布一个国际团队完成了迄今覆盖面最广的癌症全基因组分析。这将有助于加深研究人员对癌症的认识，为开发出更高效的治疗方案铺平道路。

这个被称为"泛癌症计划"的项目由来自 37 个国家的 1300 多名科学家合作开展，旨在研究可导致癌症的变异基因，绘制出这些基因的全图谱。研究团队分析了 38 种不同类型肿瘤的 2658 个全基因组，为癌症研究获取了丰富的基因数据。

相关成果在当天以 20 多篇系列报告的形式发表在《自然》及其子刊上。

2. 人造叶绿体研制成功

德国马克斯·普朗克陆地微生物研究所和法国波尔多大学的研究人员于 2020 年 5 月 8 日在《科学》上发表科研成果。他们通过将菠菜的"捕光器"与 9 种不同生物体的酶结合起来，制造了人造叶绿体。这种叶绿体可在细胞外工作、收集阳光，并利

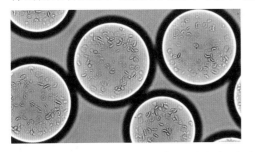

用由此产生的能量将二氧化碳转化成富含能量的分子。

研究人员希望他们制造的加强版光合作用系统最终能将二氧化碳直接转化成有用的化学物质，或者使转基因植物吸收大气中二氧化碳的量达到普通植物的 10 倍。

这种新的光合作用将为转基因作物打开新的大门，创造出比现有品种生长速度更快的新品种。在世界人口激增的背景下，这对农业发展是一个福音。

3. 人工智能首次成功解析蛋白质结构

生物学界最大的挑战之一——蛋白质三维结构解析如今有望被破解。谷歌公司旗下人工智能公司 DeepMind 开发的深度学习程序 AlphaFold 能够精确预测其三维形状。

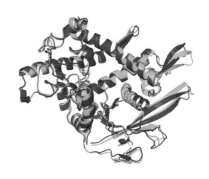

长久以来，人们需要借助实验确定完整的蛋白质结构，而这些方法往往需要耗费数月甚至数年的时间。现在，人工智能也有能力给出精确预测的计算方法，却可能只需要几天甚至半个小时。

2020 年 11 月 30 日，在蛋白质预测结构挑战赛（CASP）上，AlphaFold 程序在

上百支队伍中脱颖而出。该深度学习网络将深度学习与张力控制算法结合，并应用于结构和遗传数据，利用目前已知的 17 万种解析完毕的蛋白质进行了训练。

DeepMind 有关研发团队表示，他们还将继续对 AlphaFold 展开训练，以便更好地解析更复杂的蛋白质结构。

4. 新型催化剂将二氧化碳变为甲烷

研究人员一直尝试模仿光合作用，利用太阳的能量制造化学燃料。现在，美国的科学家开发出一种新型铜-铁基催化剂，可借助光将二氧化碳转化为天然气的主要成

分甲烷，这一方法是迄今最接近人造光合作用的方法。

研究人员称，新催化剂如果能够得到进一步改良，将降低人类对化石燃料的依赖。2020 年 1 月出版的《美国国家科学院院刊》报道了这种新型催化剂，作为将二氧化碳转化为甲烷的光驱动催化剂，其效率和产量是有史以来最高的。

5. 脑-机接口技术助瘫痪男子重获触觉

2020 年 4 月 23 日，美国巴泰尔科研中心和俄亥俄州立大学韦克斯纳医学中心的研究团队在《细胞》上发文，他们成功利用脑-机接口（BCI）系统帮助一位瘫痪患者恢复了手部触觉。

这项技术能捕捉到人类所无法感知的微弱神经信号，并通过发回受试者大脑的人工感觉反馈来增强这些信号，从而极大地优化了受试者的运动功能。

BCI 系统在改进后成为首个同时恢复运动与触觉功能的系统，不仅能让受试者仅靠触觉就能感知物体，而且能够感知握持或捡拾物体时所需的压力。

6. 科研人员绘出迄今最大三维宇宙结构图

物理学家组织网 2020 年 7 月 20 日的报道显示，在对 400 多万个星系和蕴含巨大能量的超亮类星体进行分析后，国际"斯隆数字巡天"（Sloan Digital Sky Survey,

SDSS）项目发布了迄今最大的宇宙三维结构图。

　　绘出该图的是多国科研人员组成的"扩展重子振荡光谱巡天"（eBOSS）项目。它是世界最大星系巡天项目"斯隆数字巡天"的一部分。最新成果建立在世界各地数十家机构的数百名科研人员超过 20 年合作的基础上，由 eBOSS 项目耗费数年完成。

　　目前的理论认为宇宙产生于约 138 亿年前的大爆炸。通过理论分析和天文观测，科研人员此前对宇宙的远古历史和最近的膨胀史都有相当的了解，但中间却存在一个约 110 亿年的认知缺口。有关研究人员表示，新成果终于填补了这一空白，是宇宙学领域的重大进展。

7. 美国的研究人员在超高压下实现室温超导

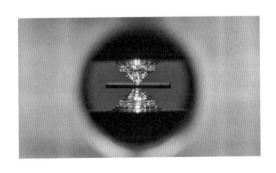

　　2020 年 10 月 16 日，美国的一个科研团队在《自然》发表研究成果。该团队在超高压下的一种氢化物材料中观察到室温超导现象，这个新突破让研究人员朝着创造出具有极优效率的电力系统迈进了一步。近年来超导研究的进展已表明，富氢材料在高压下可将超导温度提高至 $-23\,℃$ 左右。

　　美国罗切斯特大学的科研人员在实验室中将可实现零电阻的温度提高到 $15\,℃$，这个效果在 $2670\times10^8\,Pa$ 压力下的一个光化学合成三元含碳硫化氢系统中被观察到。这个压力约是典型胎压的 100 万倍，并且达到实验中实现的最高压力值。

8. "基因魔剪"首次直接用于人体试验

　　一名遗传失明症患者成为接受 CRISPR-Cas9 基因疗法直接人体试验的第一人。据 2020 年 3 月初英国《自然》网站报道，科学家首次开展临床试验，将 CRISPR-Cas9 基因疗法直接用于人体，治疗遗传性眼病——莱伯氏先天性黑蒙症（LCA10）。

　　LCA10 是导致儿童失明的主要原因，目前尚无治疗方法。CRISPR-Cas9 有"基因魔剪"之称，在最新的试验中，这种基因编辑系统的组件将被编码于病毒基因组

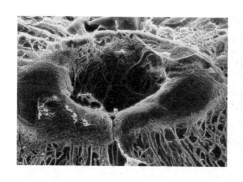

中，然后直接注入患者眼睛的近光感受器细胞内。

这项最新实验名为"光明"（BRILLIANCE），由美国俄勒冈健康与科学大学遗传性视网膜疾病专家马克·彭勒斯与美国 Editas Medicine 公司等携手开展。他们表示，这个试验旨在测试该基因编辑技术移除导致 LCA10 的基因突变的能力，具有里程碑意义。

9. 引力波探测器发现迄今最强黑洞合并事件

引力波探测器探测到天文学家未曾想到的惊人发现——迄今我们所知的最大规模的黑洞合并事件。2020 年 9 月 2 日，《物理评论快报》和《天体物理学期刊快报》分别上线文章介绍了这项发现。

此次黑洞合并事件最早被发现于 2019 年 5 月 21 日，合并产生的引力波被美国激光干涉引力波天文台（LIGO）和意大利室女座干涉仪（Virgo）探测到，合并事件被命名为 GW190521。

这是上述引力波探测器今年第二次探测到非常规的黑洞合并事件。在这次合并事件中，两个黑洞的质量分别是太阳的 85 倍和 66 倍，合并后形成的新黑洞质量接近 150 个太阳质量。

10. 冷冻电镜技术突破原子分辨率障碍

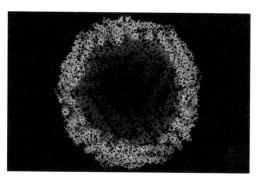

如果想绘制出蛋白质最微小的部分，科学家通常需要使数百万个单个蛋白质分子排列成晶体，然后用 X 射线晶体学分析它们，或者快速冷冻蛋白质的副本，然后用电子轰击它们。这是一种低分辨率的方法，叫作"冷冻电镜技术"。

在电子束技术、探测器和软件进一步的帮助下，来自英国和德国的两组研究人员将分辨率缩小到 1.25 埃或更小，这已经足以计算出单个原子的位置。

　　增强的分辨率或使更多的结构生物学家选择使用冷冻电镜技术。目前，这项技术只适用于异常坚硬的蛋白质。下一步，研究人员将努力在刚性较小、较大的蛋白质复合物（如剪接体）中达到类似清晰程度的分辨率。相关论文于 2020 年 10 月 21 日发表在《自然》上。

附录二　2020年香山科学会议学术讨论会一览表

序号	会次	会议主题	执行主席			会议日期
1	673	病原组国家大数据与生物安全	高　福　钱　韦　徐建国 朱宝利			1月17～18日
2	674	人工智能与中医药学	程　京　蒋田仔　李校堃 骆清铭　商洪才			9月18日
3	675	深地过程与地球宜居性	李曙光　沈树忠　徐义刚 朱日祥　侯增谦			9月21～22日
4	676	核酸生物化学与技术	谭蔚泓　王恩多　徐国良 张礼和　赵宇亮			9月24～25日
5	677	全民营养健康关键科学问题与发展战略	路福平　孟宪军　孙宝国 孙君社　杨月欣			9月28～29日
6	678	化学生物医药工程与皮肤健康	陈　坚　金　涌　钱旭红 任其龙　夏照帆　郑裕国			10月9～10日
7	S58	变革性技术关键科学问题前沿和热点研讨	方　忠　顾逸东　金之钧 吴一戎			10月9～10日
8	679	单原子催化	张　涛　何鸣元　李亚栋 李　隽　刘景月　王爱琴			10月13～14日
9	680	织物电子、传感和计算的学术前沿、核心技术与应用展望	陈东义　胡　斌　李长明 王宗敏　俞建勇　陶肖明			10月15～16日
10	681	太空治理能力现代化前沿问题与政策建议	刘继忠　王　赤　王大轶 吴伟仁　于登云			10月19～20日
11	682	制造流程物理系统与智能化	曹湘洪　干　勇　毛新平 王国栋　殷瑞钰　袁晴棠			10月21～22日
12	683	空间交通管理的前沿科学问题及关键技术	龚自正　李　明　王冀莲 王礼恒　周志鑫			10月29～30日
13	684	旱区盐碱地多学科综合治理的核心科学与技术问题	姜克隽　康绍忠　王　飞 张佳宝			11月5～6日
14	685	细胞可塑性调控与细胞工程应用	李　林　刘小龙　张先恩 张学敏　赵国屏			11月9～10日
15	686	艾滋病免疫重建和免疫恢复	董　晨　王福生　张林琦			11月12～13日

续表

序号	会次	会议主题	执行主席			会议日期
16	687	"健康中国"与智慧健康医疗体系构建	陈香美　董家鸿 吴建平　于金明	王小云		11月17日
17	688	免疫学理论前沿与技术应用：挑战与机遇	曹雪涛　董　晨 邵　峰　田志刚	高　福 沈倍奋		11月19～20日
18	689	激光驱动多束流科学及应用	龚旗煌　李儒新 张维岩	马余刚		11月21～22日
19	690	我国伴生放射性煤矿开采利用中的职业健康挑战与环境风险防控	柴之芳　丁库克 郑楚光　周平坤	肖文交 张丰收		11月24～25日
20	691	时空相干电子源关键科学问题与前沿技术	成会明　戴　庆 许宁生	李儒新		12月8～9日
21	692	环境氡污染监测、评价与防治技术	潘自强　欧阳晓平 程建平　刘森林	高　福 肖德涛		12月15～16日
22	693	离子液体功能调控及交叉融合前沿技术	韩布兴　何鸣元 王键吉　徐春明	任其龙 张锁江		12月17～18日

附录三 2020 年中国科学院学部 "科学与技术前沿论坛"一览表

序号	会次	会议主题	执行主席	举办时间
1	105	集成电路与光电芯片发展战略研究	郝 跃 龚旗煌	9 月 15 日
2	106	中-欧海洋科学与技术进展	张 经	10 月 20～21 日
3	107	中枢神经再生与临床转化研究	苏国辉	10 月 24～25 日
4	108	印度洋与青藏高原的相互作用	张 经 姚檀栋	10 月 26～27 日
5	109	新时代科技融合发展	包信和 陈和生 葛均波	11 月 22～24 日
6	111	有机光伏材料和器件	李永舫	11 月 27～29 日
7	112	离子液体科学与工程	张锁江	12 月 3～4 日
8	113	高端电子制造电子电镀	孙世刚	12 月 18～20 日
9	114	新工业·大融合——工业互联网时代的创新与发展	梅 宏 黄 维 丁 汉 陈维江	12 月 18～19 日